This book is designed as an introductory text in neuroendocrinology – the study of the interaction between the brain and endocrine system and the influence of this on be

hypothalamus and th

action of steroid and

target cell response to

peptides is discussed

behavior. The neuroin

interaction between

discussed. Finally, me

ior are outlined. Each c

tory students, and ess

ate students.

This book is writte

neuroscience, psychol

in related disciplines.

AN INTRODUCTION TO NEUROENDOCRINOLOGY

An introduction to neuroendocrinology

RICHARD E. BROWN

Professor of Psychology, Dalhousie University
Halifax, Nova Scotia, Canada

Published by the Press Syndicate of the University of Cambridge
The Pitt Building, Trumpington Street, Cambridge CB2 1RP
40 West 20th Street, New York, NY 10011–4211, USA
10 Stamford Road, Oakleigh, Melbourne 3166, Australia

First published 1994

Printed in Great Britain at the University Press, Cambridge

A catalogue record for this book is available from the British Library

Library of Congress cataloguing in publication data available

ISBN 0521 41645 0 hardback
ISBN 0521 42665 0 paperback

SE

This book is dedicated to the more than 2000 Dalhousie University students who have taken Hormones and Behaviour over the last 14 years and who have encouraged me to write this book, especially those who gave critical comments on the first three drafts of the book.

Contents

Preface

This book is an introduction to neuroendocrinology from the point of view of the behavioral neurosciences. It is intended for students in Psychology, Biology, Nursing, Health Education, and other fields of Arts and Science and for more advanced students in physiology, anatomy and medicine who have not had a course on neuroendocrinology. It is based on the first half of my Hormones and Behavior lectures at Dalhousie University. While my lectures, and thus the book, focus primarily on mammalian research, the principles outlined apply to all vertebrates.

This book provides an outline of the neuroendocrine system and will give you the vocabulary necessary to understand the interaction between hormones and the brain. This information is essential to the understanding of the effects of hormones on behavior, but contains little reference to behavior until Chapter 14. In fact, it contains mainly endocrinology, physiology and a bit of cell biology, immunology and biochemistry. Do not despair. Once you master this material, the study of how hormones influence developmental processes and behavior will be easier to understand. This book focuses primarily on the neural actions of hormones, so many of the peripheral physiological actions of hormones, such as regulation of metabolism, water balance, growth, and the regulation of calcium, sodium and potassium levels, which are the focus of traditional endocrinology texts, are referred to only in reference to their importance in the neuroendocrine system.

This book is designed for students in two levels of classes: introductory classes, in which all of the material will be new to the student, and more advanced classes, in which the students will be familiar with many of the terms and concepts through courses in biology, physiology, psychology or neuroscience, but have not studied neuroendocrinology as an integrated discipline.

The introductory student is expected to learn the material in this book at the level presented. To help in this, review questions are given at the end of each chapter. These should be treated as practice exam questions and answered after each chapter is read. The answers to the review questions are to be found within the chapter itself. For further information on the topics covered in each chapter, introductory students should use the books listed under Further Reading at the end of each chapter.

Advanced students should use this book as an introductory overview of the topics covered in each chapter. They should then use the references cited in each chapter and computer-based literature searches to find the most up-to-date information on each topic. The essay questions at the end of each chapter provide topics for discussion, analysis and directed research papers for the advanced student.

R. E. B.

Dalhousie University, Halifax, Canada

Acknowledgements

Many people have helped me with this book since I started it in September 1986. I would, therefore, like to thank the following: Heather Schellinck for keeping everything organized on the computer disks during the writing of the first draft, ensuring that back-up copies always existed and for proofreading and advice. Mary MacConnachie for typing, retyping, retyping, retyping, and retyping. Thad Murdoch for drawing the figures. The librarians of the Killam Library Science Services and the Kellogg Health Sciences Library for their help with the library research, for purchasing many of the books I requested, and for not fining me when I was overdue in returning these books. Will Moger, Ron Carr, Kazue Semba, Barry Keverne, Ed Roy and Charles Malsbury for their editorial comments. Mike Wilkinson for his thoughtful criticisms, careful proofreading, corrections, updating of references, and help in selecting the most appropriate figures. Bruce McEwen for Figure 9.1. Alan Crowden of Cambridge University Press for contracting this book and for never giving up hope that it would be finished. Dorothy for phoning me at midnight every night to remind me to come home.

Illustration credits

The author is grateful to the following publishers and individuals for permission to reproduce illustrations. The full forms of the references are given at the ends of relevant chapters.

Academic Press, Inc.: 11.4A,B (Hökfelt *et al.*, 1986), 14.1 (Brown and McFarland, 1979), 14.2 (Hyde and Sawyer, 1977), 14.3 (Balthazart and Hendrick, 1976), 14.4 (Terkel and Urbach, 1974), 14.7 (Yahr and Thiessen, 1972), 14.10 (Smith *et al.*, 1977), 14.19A (Lauder and Krebs, 1986).

Acta Physiologica Scandinavica: 5.8B,C (Ungerstedt, 1971).

American Association for the Advancement of Science: 9.6A (MacLusky and Naftolin, 1981), 9.12 (Pfaff *et al.*, 1971), 10.1D (McFarland *et al.*, 1989), 12.5 (Dekin *et al.*, 1985), 13.7 (Metcalf, 1991), 13.10 (Healy *et al.*, 1983). Copyright, according to year of publication, by the AAAS.

American Association of Immunologists: 11.5 (Roth *et al.*, 1985).

American College of Physicians: 9.18 (Oppenheimer, 1985). Reproduced with permission.

American Journal of Physiology: 10.1A (Popot and Changeux, 1984), 10.10 (Nicoll *et al.*, 1990), 13.9 (Katsuura *et al.*, 1990).

American Physiological Society: 10.5 (Nathanson, 1977), 10.10 (Nicoll *et al.*, 1990), 13.9 (Katsuura *et al.*, 1990).

American Psychosomatic Society: 14.12 (Rose *et al.*, 1975), 14.23 (Brown and Martin, 1974). © American Psychosomatic Society, according to year of publication.

American Zoologist: 9.7 (Callard *et al.*, 1978), 14.16 (Arnold, 1981).

Annual Reviews, Inc.: 11.3 (Lynch and Snyder, 1976), 14.19C (Dussault and Ruel, 1987). ©, according to year of publication, by Annual Reviews, Inc.

Appleton & Lange: 12.12 (Kandel *et al.*, 1991).

Birkhäuser Verlag: 10.7 (King and Baertschi, 1990).

Brooks/Cole Publishing Co.: 14.9 (Singh and Avery, 1975).

Cambridge University Press: 5.9 (Iversen, 1979), 8.1 (Baird, 1972).

Ediciones Doyma: 9.8 (McEwen, 1988).

Elsevier Science Publishers BV: 5.8A (Semba and Fibiger, 1989), 9.13 (Bueno and Pfaff, 1976), 9.16 (Rainbow *et al.*, 1980), 12.12 (Jessell and Kelley, 1991).

Elsevier Science Publishing Co., Inc. (NY): 10.2 and 10.3 (Spiegel, 1989). Reprinted by permission of the publisher. Copyright, according to year of publication, by Elsevier Science Publishing Co., Inc.

Elsevier Trends Journals: 11.2 and 11.11 (Khachaturian *et al.*, 1985), 12.1 and 12.3 (Lundberg and Hökfelt, 1983), 13.9 (Rothwell, 1991b).

Federation Proceedings: 12.8 (Cicero, 1980).

Garland Publishing, Inc.: 7.1, 10.8 and 13.1 (Alberts *et al.*, 1989).

John Wiley and Sons, Inc.: 9.3 (Pfaff and Keiner, 1973), 11.8 (Anderson, 1989, 12.11 (Kelley, 1989).

Journal of Endocrinology: 6.13 (Waverly and Lincoln, 1973), 9.14 (Lincoln, 1969). Reproduced by permission of the Journal of Endocrinology Ltd.

Journal of Immunology: 11.5 (Roth *et al.*, 1985) Copyright, according to year of publication, the Journal of Immunology.

S Karger AG, Basel: 6.12 (Wilson *et al.*, 1984), 8.7 (Kendall *et al.*, 1975), 8.10 (Plant, 1983), 9.4 (Stumpf *et al.*, 1975), 9.5 (Sar and Stumpf, 1975), 9.9 (Stumpf and Sar, 1975a), 9.15A,B (Grant and Stumpf, 1975), 12.12 (Kandel *et al.*, 1991).

Kluwer Academic Publishers: 13.12 (Berczi and Nagy, 1987).

Macmillan Magazines, Inc.: 10.1B (Anderson, 1989), 11.1 and 11.4A (Höckfelt *et al.*, 1980), 12.6C (Magistretti and Schorderet, 1984). Reprinted with permission from *Nature*. Copyright, according to year of publication, Macmillan Magazines, Inc.

Marcel Dekker, Inc. NY: 13.3 (Low and Goldstein, 1984).

A. S. Mason: 8.9 (Mason, 1976).

Neuropsychology Press: 11.14 (Reitan and Wolfson, 1985).

New York Academy of Sciences: 7.2 (Loh, 1987), 9.17 (Pfaff, 1989), 12.4C (Spigelman and Puel, 1991), 12.7 (Labrie *et al.*, 1987), 13.8 (Weigent *et al.*, 1990), 14.14B (Cheng, 1986), 14.17 (Crowley, 1986), 14.18 (Pfaff, 1989).

Oxford University Press: 4.1 (Martin, 1985), 4.2, 4.3 and 7.3 (Bennett and Whitehead, 1983), 8.4 (Martin, 1985), 12.4A (Kaczmarek and Levitan, 1987), 13.5 (Hamblin, 1988), 14.20 (Leshner, 1978). Reprinted by permission of Oxford University Press.

Pergamon Press Ltd: 10.1C (Venter *et al.*, 1989), 11.10 (Argiolas and Gessa, 1990), 12.14 (Jalowiec *et al.*, 1981), 14.8 (Edwards, 1969), 14.11 (Armario and Jolin, 1989), 14.19B (Toran-Allerand, 1991). Reprinted with kind permission from Pergamon Press Ltd, Headington Hill Hall, Oxford OX3 OBW, UK.

Plenum Publishing Corporation: 1.2 (Smith and Blalock, 1985), 5.6 (McGeer *et al.*, 1987), 14.15 (Johnsson *et al.*, 1983).

Prentice Hall Press: 2.3 (Hadley, 1984).

Raven Press: 5.8D (Fuxe and Jonsson, 1974), 10.1B (Goren *et al.*, 1988), 10.4 (Hemmings *et al.*, 1986), 11.9 and 11.13 (Hökfelt *et al.*, 1987).

***Scientific American*:** 3.1 (Guillemin and Burgus, 1972), 5.5A,B (Stevens, 1979), 6.1 (Segal, 1974), 8.3 (Zuckerman, 1957), 9.3 (McEwen, 1976), 9.9

(Stumpf and Sar, 1975a), 10.9 (Nathanson and Greengard, 1977). Copyright, according to year of publication, by Scientific American, Inc. All rights reserved.

G. R. Siggins: 12.4B,D (Mancillas *et al.*, 1986).

Sinauer Associates, Inc.: 12.6A,B (Hollister *et al.*, 1980).

Society for Experimental Biology and Medicine: 12.13 (Gambert *et al.*, 1981). © Society for Experimental Biology and Medicine, according to year of publication.

Springer-Verlag, Heidelberg: 14.21 and 14.22 (vom Saal, 1983).

The Endocrine Society: 8.10 (Plant, 1986), 9.2 (Walters, 1985), 9.11 (Hua and Chen, 1989), 12.10 (Bondy *et al.*, 1988), 14.13 (Samuels and Bridges, 1983). ©, according to year of publication, The Endocrine Society.

The Lancet Ltd: 8.2 (Kerin *et al.*, 1980).

Thieme Medical Publishers, Inc.: 12.9 (Grossman, 1987).

W. B. Saunders Co.: 2.2, 5.14, 8.5, 10.11, 11.6 and 14.5 (Turner and Bagnara, 1976).

R. F. Weick: 7.4 (Weick, 1981).

Abbreviations

5-HT	5-Hydroxytryptamine (serotonin)
5-HTP	5-Hydroxytryptophan
α-MSH	Alpha-melanocyte stimulating hormone
A	Adrenaline
Ach	Acetylcholine
AchE	Acetylcholinesterase
ACTH	Adrenocorticotropic hormone
ADH	Antidiuretic hormone (vasopressin)
AH	Anterior hypothalamus
ANF	Atrial natriuretic factor
ANS	Autonomic nervous system
APUD	Amine precursor uptake and decarboxylation
ARC	Arcuate nucleus
ATP	Adenosine triphosphate
AVP	Argenine vasopressin
β-END	Beta-endorphin
β-LPH	Beta-lipotropin
BBB	Blood–brain barrier
BnST	Bed nucleus of the stria terminalis
Ca^{2+}	Calcium
CBG	Corticosteroid binding globulin (transcortin)
CCK	Cholecystokinin
CG	Central gray
Cl^{-}	Chloride
CLIP	Corticotropin-like intermediate lobe peptide
CNS	Central nervous system
COMT	Catechol-*O*-methyl transferase
CRF	Corticotropin releasing factor (also called CRH)
CRH	Corticotropin releasing hormone (also called CRF)
CSF	Cerebrospinal fluid
CSFs	Colony stimulating factors
DA	Dopamine
DAG	Diacylglycerol
DBH	Dopamine beta hydroxylase
DHT	Dihydrotestosterone
DMN	Dorsomedial nucleus
DMT	Dimethyltryptamine
DNA	Deoxyribonucleic acid

dopa	Dihydroxyphenylalanine
DYN	Dynorphin
E	Estrogen
EGF	Epidermal growth factor
ENK	Enkephalin
ENS	Enteric nervous system
ER	Endoplasmic reticulum
FGF	Fibroblast growth factor
FSH	Follicle stimulating hormone
FSH-RH	Follice stimulating hormone releasing hormone
G-CSF	Granulocyte colony stimulating factor
GABA	Gamma-aminobutyric acid
GAD	Glutamic acid decarboxylase
GH	Growth hormone
GH-RH	Growth hormone releasing hormone
GH-RIH	Growth hormone release inhibiting hormone (somatostatin)
GI	Gastrointestinal
GM-CSF	Granulocyte-macrophage colony stimulating factor
GnRH	Gonadotropin releasing hormone
GR	Glucocorticoid receptor
GTP	Guanosine triphosphate
H-P-A	Hypothalamic–pituitary–adrenal
H-P-G	Hypothalamic–pituitary–gonadal
H-P-T	Hypothalamic–pituitary–thyroid
HCG	Human chorionic gonadotropin
HCS	Human chorionic somatomammotropin
HPL	Human placental lactogen
HPX	Hypophysectomy
IFN_{γ}	Interferon gamma
IGF	Insulin-like growth factor
IL	Interleukin
IP	Inositol phospholipid
IP3	Inositol triphosphate
K^+	Potassium
LH	Luteinizing hormone
lh	Lateral hypothalamus
LH-RH	Luteinizing hormone releasing hormone
LSD	Lysergic acid diethylamide
LT	Lymphotoxin
M-CSF	Macrophage colony stimulating factor
MAO	Monoamine oxydase
MBH	Mediobasal hypothalamus
MFB	Medial forebrain bundle
MHC	Major histocompatibility complex
MPOA	Medial preoptic area
MR	Mineralocorticoid receptor
mRNA	Messenger ribonucleic acid
MSH	Melanocyte stimulating hormone

MSH-RF Melanocyte stimulating hormone releasing factor
MSH-RIF Melanocyte stimulating hormone release inhibiting factor
NA Noradrenaline (also norepinephrine, NE)
Na^+ Sodium
NE Norepinephrine (also noradrenaline, NA)
NGF Nerve growth factor
NK Neurokinin
NK Natural killer cells
NMDA *N*-methyl-D-aspartate
NP Neurophysin
NPY Neuropeptide Y
OC Optic chiasm
OVLT Vascular organ of the lamina terminalis
OVX Ovariectomized
OXY Oxytocin
P Progesterone
PH Posterior hypothalamus
PIF Prolactin inhibiting factor
PIP2 Phosphatidylinositol phosphate
PLC Phospholipase C
PNMT Phenylethanolamine *N*-methyl transferase
PNS Parasympathetic nervous system
POA Preoptic area
POMC Proopiomelanocortin
PRF Prolactin releasing factor
PRL Prolactin
PTH Parathyroid hormone
PV Periventricular nucleus
PVa Anterior periventricular nucleus
PVN Paraventricular nucleus
RIA Radioimmunoassay
RNA Ribonucleic acid
SCN Suprachiasmatic nucleus
SHBG Sex hormone binding globulin (also called TeBG)
SNS Sympathetic nervous system
SOM Somatostatin
SON Supraoptic nucleus
T Testosterone
T3 Triiodothyronine
T4 Thyroxine
TBG Thyronine binding globulin
T_c Cytotoxic T cell
TeBG Testosterone-estrogen binding globulin
T_H Helper T cell
TI Tuberoinfundibular
TNF Tumor necrosis factor
TRH Thyrotropin releasing hormone (also TSH-RH)
T_s Suppressor T cell

TSH	Thyroid stimulating hormone
TSH-RH	Thyroid stimulating hormone releasing hormone (TRH)
VIP	Vasoactive intestinal peptide
VMN	Ventromedial nucleus
VP	Vasopressin

1

Classification of the chemical messengers

1.1 HORMONES, THE BRAIN AND BEHAVIOR

Research on hormones and the brain covers many fields: from cell biology and genetics to anatomy, physiology, pharmacology, medicine and psychology. This book will examine the interactions between hormones, the brain and behavior. The main focus will be on how the endocrine and nervous systems form an integrated functional neuroendocrine system which influences physiological and behavioral responses.

When you hear the term 'hormone', you think of the endocrine glands and how their secretions influence physiological responses in the body. That is, however, merely the beginning of the picture. Many of the endocrine glands (although not all of them) are influenced by the pituitary gland, the so-called 'master gland', and the pituitary is itself controlled by various hormones from the hypothalamus, a part of the brain lying above the pituitary gland. The release of hypothalamic hormones is, in turn, regulated by neurotransmitters released from nerve cells in the brain. Neurotransmitters also control behavior and the release of neurotransmitters from certain nerve cells is modulated by the level of specific hormones in the circulation. Thus, neurotransmitter release influences both hormones and behavior and hormones influence the release of neurotransmitters. This interaction between hormones, the brain and behavior involves a wide variety of chemical messengers, which are described in this chapter.

This chapter provides an introduction to the chemical messengers found in the neuroendocrine system. Later chapters describe the endocrine glands and their hormones (Chapter 2), the pituitary gland and its hormones (Chapter 3) and the regulation of the pituitary gland by the hypothalamic hormones (Chapter 4). Chapter 5 describes the role of neurotransmitters in communicating between nerve cells and Chapter 6 discusses neurotransmitter control of the hypothalamic, pituitary and other hormones. The regulation of hormone synthesis, transport, storage, release and deactivation is described in Chapter 7. Hormones from the endocrine glands, pituitary gland and hypothalamus influence each other through feedback mechanisms, which are described in Chapter 8. Hormones act on target cells in the body and the brain which have specific hormone recognition sites or receptors. The nature of the steroid and thyroid hormone receptors is discussed in Chapter 9 and the receptors for peptide hormones and neurotransmitters, which function by activating second messengers in their target cells, are described in Chapter 10. In the brain, hormones influence the release of both neurotransmitters and hypothalamic hormones by their action on neural target cells. The brain is also influenced by a number of newly discovered substances called neuropeptides, which are described in Chapter 11. Neuropeptides are important because they can act as neurotransmitters to stimulate neural activity or as neuromodulators to influence the synthesis, storage, release and action of other neurotransmitters (Chapter 12). The cells of the immune system also produce chemical messengers called cytokines or lymphokines, which interact with the neural and endocrine systems as described in Chapter 13. When hormones, neuropeptides or cytokines alter the synthesis and release of neurotransmitters in the brain, one result is a change in behavior. Methods for the study of hormones and behavior are discussed in Chapter 14, and current developments in behavioral neuroendocrinology, as well as a historical overview, are given in Chapter 15.

The neuroendocrine system, therefore, involves a complex network of hormone–brain–behavior interactions as depicted in Figure 1.1. The perception of an environmental stimulus such as a light, odor, sound, or touch occurs through the sense organs and their neural connections to the brain. These stimuli are interpreted as physical stressors, sexual stimuli, etc. by the cerebral cortex and other brain areas which influence the neuroendocrine system. Two different responses then occur. There is a rapid neuromuscular response, resulting in an immediate behavioral change: you see a truck coming and you jump out of the way. There is also a complex neuroendocrine response. The hypothalamic–pituitary–adrenal response to a stressor, for example, involves the release of many different hormones which circulate through the bloodstream to stimulate their target cells in the heart, adrenal glands, liver, skeletal muscles,

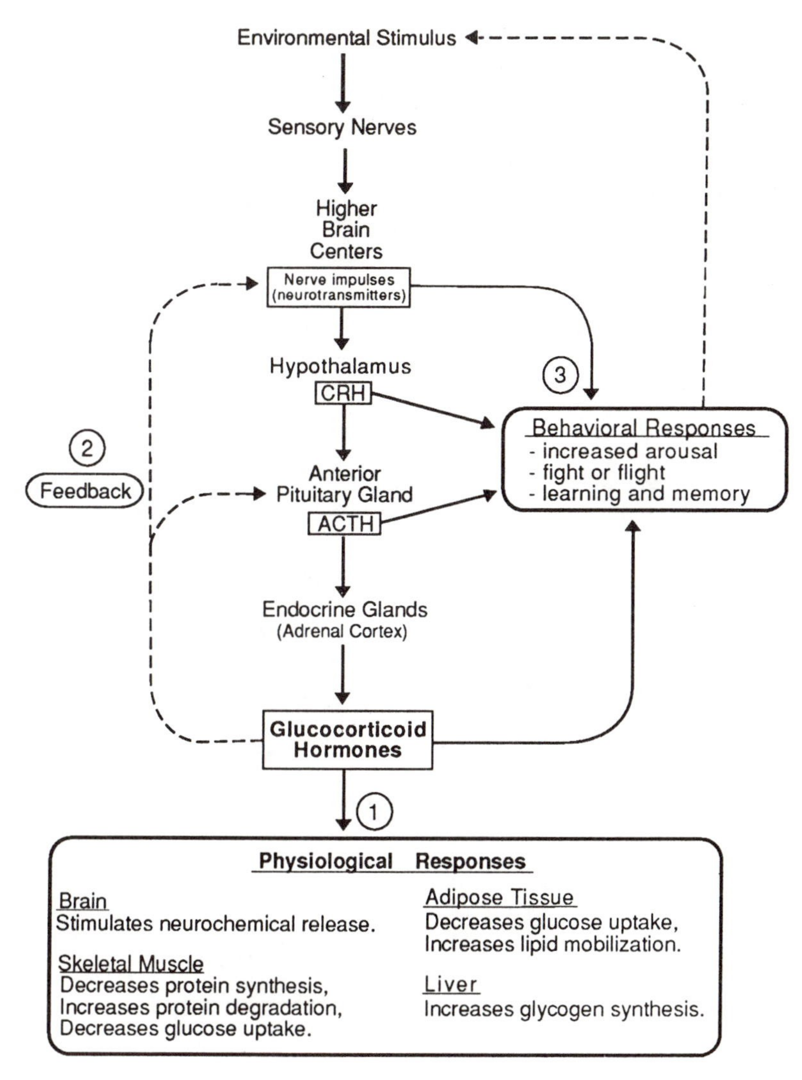

Figure 1.1. The interaction between hormones, the brain and behavior. Environmental stimuli influence the brain through sensory nerves and the brain regulates behavior and hormone secretion through the release of neurotransmitters which stimulate nerve impulses. The hormones released from the hypothalamus, pituitary gland and other endocrine glands when the neuroendocrine system is activated stimulate: (1) physiological responses in target cells in the brain and body; (2) feedback regulation of hypothalamic and pituitary hormone release; and (3) behavioral responses through their action on neurotransmitter release from neural target cells in the brain. The example used here is the hypothalamic-pituitary-adrenal response to a stressor. ACTH = adrenocorticotropic hormone; CRH = corticotropin releasing hormone.

adipose tissue and, of course, the brain. When the target cell is stimulated, it undergoes a physiological change, caused by the hormonal action. The hormones in the circulation also feed back to the hypothalamus and pituitary gland, to alter further hormone release. Finally, when the brain is a target for hormonal action, the result may be a behavioral as well as a physiological change.

1.2 THE BODY'S THREE COMMUNICATION SYSTEMS

The body has three different communication systems: the nervous system, the endocrine system and the immune system, each of which uses its own type of chemical messenger. Nerve cells communicate through the release of neurotransmitters; endocrine glands by hormones and the immune system by cytokines. These three systems are not

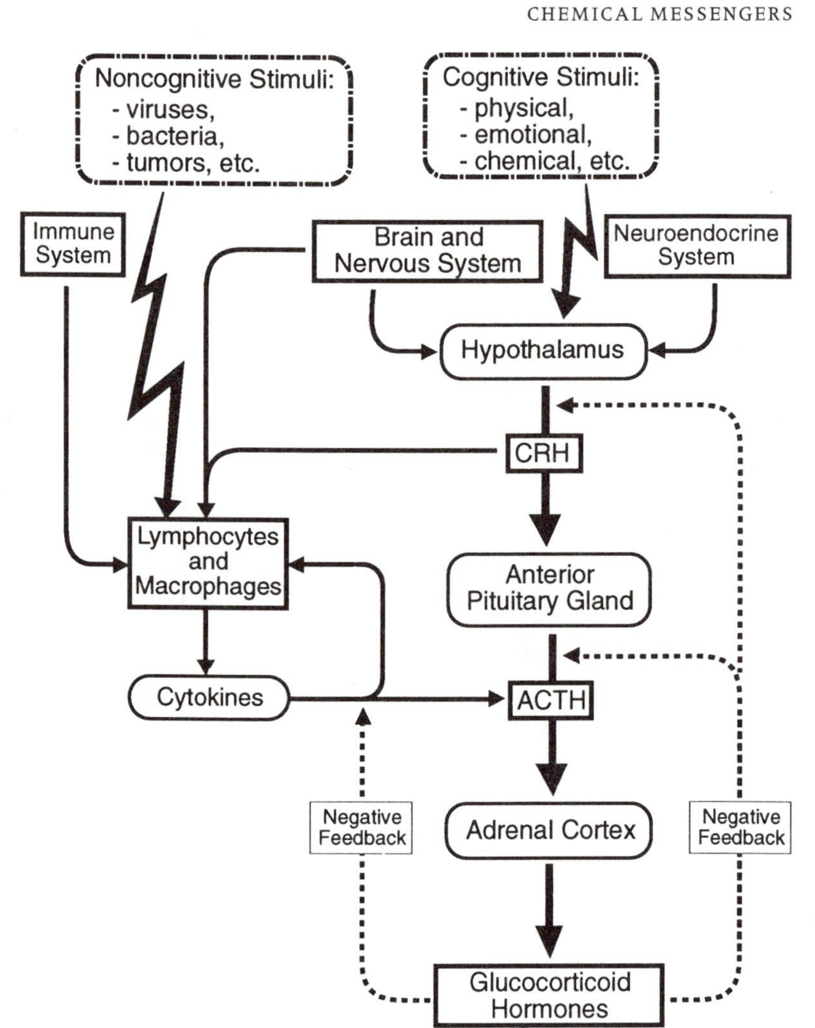

Figure 1.2. The body's three communication systems do not act independently. The brain and nervous system influence the neuroendocrine and immune systems, which also influence each other and the brain. This example shows that cognitive stimuli activate the neuroendocrine system through the brain and nervous system and the resulting neural and endocrine activation influences the release of cytokines from cells of the immune system. Non-cognitive stimuli such as viruses and bacteria first activate the immune system and the resulting release of cytokines activates the neuroendocrine system. Using the example of the hypothalamic-pituitary-adrenal system, the hypothalamic hormone CRH influences cytokine release from the cells of the immune system, which in turn influences ACTH release. Glucocorticoids have negative feedback on the cytokines as well as the hypothalamic and pituitary hormones. Abbreviations as in Figure 1.1. (Modified from Smith and Blalock, 1985.)

independent; each one interacts with the other two, as outlined in Figure 1.2 (Hall and O'Grady, 1989; Plata-Salamon, 1989).

Because these systems interact, they are often referred to as the neuroendocrine, neuroimmune or neuroimmunoendocrine systems (Blalock, 1989). To designate the influence of these systems on behavior, the terms psychoneuroendocrinology (Smythies, 1976) and psychoneuroimmunology (Ader, 1981) have been coined. All of these new fields of science are growing rapidly and have journals devoted to them (see Appendix).

The nervous system controls the release of hormones (Müller and Nistico, 1989) and can influence the release of the cytokines from the immune system (Dunn, 1989). Hormones and other chemical messengers modulate the activity of both the nervous system and the immune system (MacLean and Reichlin, 1981; Blalock, 1989). Likewise, the lymphoid system can modulate neural activity and the release of hormones by the release of cytokines. While cognitive-sensory stimuli

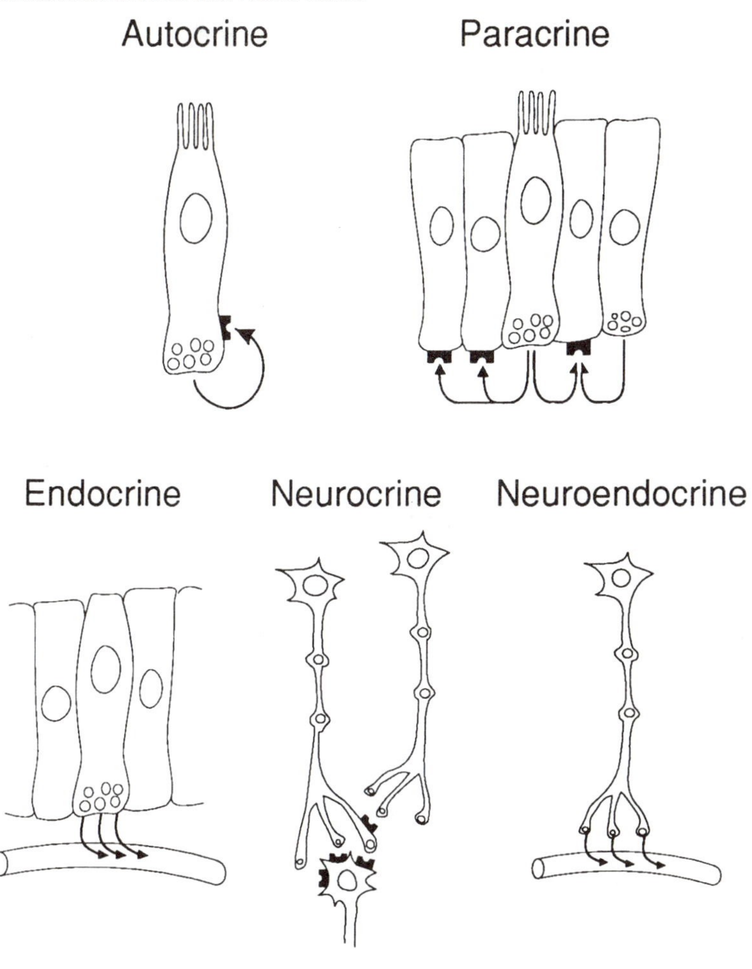

Figure 1.3. Methods of communication between cells of the neuroendocrine system. In autocrine communication, hormones act on the cells that release them. In paracrine communication, hormones act on adjacent cells as occurs in the testes, gastrointestinal tract and pituitary gland. Endocrine communication occurs when hormones are released into the bloodstream and act on cells at distant sites throughout the body. Neurocrine communication occurs when neurotransmitters are released from presynaptic cells into the synapse and act on receptors of postsynaptic cells in the central and peripheral nervous systems. Neuroendocrine secretion involves the release of neurohormones from neurosecretory cells in the hypothalamus, adrenal medulla and pineal gland. (Redrawn from Kreiger, 1983.)

influence neural, immune and endocrine activity through the brain and nervous system, non-cognitive stimuli, such as bacteria and viruses influence these systems through their action on the immune system. These three systems may be integrated through the actions of common receptor mechanisms (Kordon and Bihoreau, 1989).

1.3 METHODS OF COMMUNICATION BETWEEN CELLS

As shown in Figure 1.3, chemical messengers may communicate with their target cells through endocrine, paracrine, autocrine, neuroendocrine or neurocrine mechanisms (Krieger, 1983; Snyder, 1985; Hadley, 1992).

1.3.1 ENDOCRINE COMMUNICATION

Endocrine cells release their hormones into the bloodstream and these hormones travel through the circulation to distant target cells. For example, thyroid stimulating hormone (TSH) is released from the pituitary gland and travels through the bloodstream to stimulate its target cells in the thyroid gland.

1.3.2 PARACRINE COMMUNICATION

Endocrine cells also release hormones which act on adjacent cells. These hormones may diffuse from one cell to the next, or go into the bloodstream, but travel only a very short distance. Paracrine secretion is, therefore, a localized hormone action. This happens, for example, in the testes. In order to produce sperm, the Sertoli cells must be stimulated by testosterone from the adjacent interstitial cells. The target cell is thus next to the hormone secreting cell, resulting in a localized chemical communication within a particular tissue or organ. Paracrine secretion is also important in the immune system.

1.3.3 AUTOCRINE COMMUNICATION

If a cell releases a hormone or neurotransmitter which has a direct feedback effect on the secretory cell, this is referred to as autocrine action. A specific example of autocrine communication would be a neurotransmitter acting presynaptically to modify its own release.

1.3.4 NEUROENDOCRINE COMMUNICATION

Neuroendocrine (neurosecretory) cells are modified neurons which release neurohormones either into the peripheral circulatory system, so that they can stimulate distant target cells (e.g. the release of oxytocin by the posterior pituitary to stimulate targets cells in the uterus), or into the hypophyseal portal vessels to induce the release of pituitary hormones (e.g. luteinizing hormone releasing hormone released from the hypothalamus to stimulate the release of luteinizing hormone from the anterior pituitary).

1.3.5 NEUROCRINE COMMUNICATION

Neurons (nerve cells) release discrete chemical messengers which are active over very short distances. Neurons which release neurotransmitters into a synapse to stimulate or inhibit a postsynaptic cell are one example of neurocrine secretion (e.g. the release of acetylcholine). A second example is the case of a neuropeptide, acting as a neuromodulator, released from a neural cell, which stimulates a receptor on a nearby cell and thus regulates its activity (e.g. somatostatin or substance P).

1.4 TYPES OF CHEMICAL MESSENGER

The classification of the chemical messengers is a constantly changing endeavor as new substances and new functions for known substances are continually being discovered. The classification scheme in Table 1.1 has been derived from those of Bennett and Whitehead (1983), Krieger (1983), Hadley (1992) and Martin (1985). It is, however, unlikely to stay current for very long because the terminology keeps evolving.

Table 1.1. *Types of chemical messenger*

Phytohormones
Plant hormones: kinins, auxins, gibberellins, etc.

'True' Hormones
These are (a) chemical messengers which are (b) synthesized in ductless (endocrine) glands and (c) secreted into the blood stream. They (d) act on specific target cell receptors and (e) exert specific physiological (biochemical) regulatory actions in the target cells.

Neurohormones
Hormones which are released by neurosecretory cells (modified nerve cells) via the posterior pituitary into the circulation (e.g. oxytocin and vasopressin) or via the portal system, into the anterior pituitary (the hypothalamic releasing hormones).

Neurotransmitters
These are released by presynaptic nerve cells into a synapse (e.g. acetylcholine, dopamine, adrenaline, etc.) where they stimulate receptors on postsynaptic nerve cells.

Pheromones
These are (a) volatile chemical messengers which are (b) synthesized in exocrine (duct) glands and (c) secreted into the environment. They (d) act on other individuals, usually of the same species, through olfactory (smell) or gustatory (taste) receptors and (e) alter behavior (releaser effects) or the neuroendocrine system (primer effects).

Parahormones
Hormone-like substances which are not necessarily produced in endocrine glands (e.g. histamine, prostaglandins, vitamin D).

Prohormones
These can be (a) large peptide molecules which may be processed into single or multiple hormones (e.g. β-lipotropin to β-endorphin), or (b) steroid hormones converted to other bioactive steroids (e.g. testosterone to estrogen).

Growth factors
Hormone-like substances which promote growth of body or brain tissue e.g. nerve growth factor (NGF) or epidermal growth factor (EGF).

Cytokines
Hormone-like factors released from lymphocytes, macrophages and other cells of the immune system which regulate the activity of cells of the immune and neuroendocrine systems (e.g. interferon γ and the interleukins).

Vitamins
Chemicals which regulate metabolism, growth and development in the body. Vitamin D for example, can be synthesized in the body and has many hormone-like properties.

1.4.1 PHYTOHORMONES

Phytohormones are chemical messengers such as auxins, kinins, giberellins and other growth regulators produced by the higher plants (Chailakhyan and Khirianin, 1987). While this book is not directly concerned with the actions of phytohormones, the latter will interest us for two reasons. First, many phytohormones are similar to known mammalian hormones and neurotransmitters and may thus be important for understanding the evolution of the neuroendocrine sytem. Second, some phytohormones are used as drugs which can influence the human neuroendocrine system (Krieger, 1983; Roth *et al.*, 1986).

Thus, plant substances such as muscarine, nicotine and morphine stimulate highly specific receptors on mammalian target cells, whereas atropine, ergocornine and strychnine block target cell receptors, preventing their response to hormones and neurotransmitters (Roth *et al.*, 1986). In some cases, the same chemical may be found in both plants and animals. Such is the case for abscisic acid, a phytohormone which causes leaves and fruit to fall from trees and is also found in the brains of mammals, where its function is unknown (Le Page-Degivry *et al.*, 1986).

1.4.2 HORMONES

A hormone is defined as: (a) a chemical messenger which is effective in minute quantities; (b) synthesized in a ductless or endocrine gland; (c) secreted into the circulatory system, and transported through the body in the blood; (d) acts on receptors on specific target cells located at a distance from the site of synthesis; (e) exerts a specific physiological or biochemical regulatory action on the target cell. During the last decade, this 'classical' definition of a hormone has been challenged and there is a controversy about what constitutes a 'true' hormone. Martin (1976, 1985) has commented on the problems of the 'classical' definition of a hormone, indicating difficulties with each of the five points in the definition above.

(a) Chemical messenger

Hormones are chemical messengers which regulate the physiological actions of their target cells, but not all physiological regulators are hormones. For example, there are a number of non-hormonal chemicals, such as carbon dioxide, glucose (blood sugar), and the parahormones, histamine and the prostaglandins, which also regulate the physiological actions of their target cells.

Although hormones are effective in minute quantities, the physiological concentrations vary, depending on the hormone. Normally, the hypothalamic releasing hormones are secreted in very small quantities (femtograms, 10^{-15} g). Pituitary hormones are released in greater quantities (picograms, 10^{-12} g) and gonadal, adrenal and thyroid hormones are released in much larger quantities (nanograms, 10^{-9} g). This means that it is relatively easy to detect and identify the gonadal hormones. It is more difficult to quantify the pituitary hormones because they are secreted in smaller quantities, and it is correspondingly more difficult to isolate and assay hypothalamic releasing hormones because they are secreted in such minute quantities.

(b) Biosynthesis in an endocrine gland

Although hormones are synthesized in endocrine glands, some hormone-like chemicals are produced in other locations. The production of angiotensin I, for example, occurs in the bloodstream, not in an endocrine gland. Neurohormones (e.g. hypothalamic releasing hormones, oxytocin and vasopressin) are synthesized in neurosecretory cells which are modified nerve cells. Growth factors such as somatomedin and nerve growth factor act like growth hormones to promote tissue growth, but are not synthesized in endocrine glands. Likewise, the cytokines, which have hormone-like activity are synthesized by lymphocytes.

One hormone may be synthesized in a number of locations. Insulin, for example, is synthesized in both the pancreatic islets and the brain and somatostatin is produced in both the gastrointestinal tract and the brain. Estrogen is synthesized in the ovaries, testes, adrenal cortex, and placenta and by tumor cells. Peptides, such as somatostatin, are called hormones when they are secreted from endocrine glands, but if they are produced by nerve cells in the brain, they are called neuropeptides. Finally, some hormones, such as ACTH, are secreted from both endocrine glands and from lymphocytes and other cells of the immune system.

(c) Secreted into the bloodstream

The traditional definition of a hormone is that it is secreted into the bloodstream and transported to its target cells through the circulatory system. But many chemical messengers are not secreted into the bloodstream. Hormones can activate the cell next to the one that releases them (paracrine action) or even the same cell that releases them (autocrine action). Neurotransmitters and neuropeptides are secreted from nerve cells into a synapse (i.e. the junction between two nerve cells). Neurohormones, neuropeptides and neurotransmitters may be transported by the cerebral spinal fluid (CSF) as well as by the circulatory system, and, the small quantities which enter the blood are quickly degraded. Pheromones are released into the air by one individual to act on another individual.

(d) Act on specific target cells located at a distance from the site of synthesis

Although hormones are defined as acting on specific target cells, there are a number of hormones which act on several different cell types in the body, rather than on a specific cell type. For example, growth hormone promotes growth in a variety of cells, so it is hard to say that there is a specific target cell for this hormone. Likewise, glucocorticoids act on virtually every type of cell in the body.

Hormones are defined as acting at target cells located at a distance from the site of synthesis, but how large is this distance? Hypothalamic hormones travel only a short distance down the portal venous system to the anterior pituitary gland. Paracrine hormones influence the very next cell. Autocrine hormones influence the same cell that secretes them. On the other hand, pituitary hormones influence the gonads, so they travel a long distance through the bloodstream. Gonadal and other hormones which influence the brain travel through the whole circulatory system before reaching their target cells.

(e) Specific physiological actions

Although hormones are defined as having a specific action at their target cell, this action may vary according to the type of target cell stimulated. For example, prolactin stimulates the production of milk in the breast;

this is an example of a specific hormonal function. Estrogen, on the other hand, can have general effects such as priming cells to respond to other hormones or increasing cell growth. Moreover, some hormones can interact with multiple receptor types such that they have different functions in different target cells. For example, noradrenaline released from the adrenal medulla can bind to either alpha-adrenergic or beta-adrenergic receptors on target cells in the muscles and the resulting physiological action may differ depending on which receptor was stimulated. On the other hand, many different hormones, no matter what their target cell, induce the same cellular response, i.e. an increase in intracellular cyclic AMP production. Others may modulate nerve activity by altering the amount of a neurotransmitter synthesized, stored and released from that nerve cell.

1.4.3 NEUROHORMONES

A neurohormone differs from a 'true' hormone because it is produced in and released from a neurosecretory cell, which is a modified nerve cell. Oxytocin and vasopressin are neurohormones manufactured by hypothalamic neurosecretory cells, but are stored and released from nerve terminals in the posterior pituitary gland. The releasing hormones of the hypothalamus are also neurohormones. Chapters 3 and 4 discuss the production of neurohormones in neurosecretory cells and the complex interactions between the hypothalamus and the pituitary gland (see also Brownstein, Russell and Gainer, 1980).

Now is a convenient time to address the problem of many hormones having two or more different names. Vasopressin, for example, is also known as antidiuretic hormone (ADH) and adrenaline is also known as epinephrine. Many hormones were discovered independently by different researchers and, since these researchers each thought that they had found something new, they gave them different names. Later on they found out that the same hormone had two functions, and there were not two different hormones, but both names remain in use to describe the same chemical. We will see many examples of this in the next few chapters. Tausk (1975) gives a brief history of the discovery of each hormone.

1.4.4 NEUROTRANSMITTERS

Neurotransmitters differ from 'true' hormones because they are synthesized by nerve cells rather than by endocrine glands. They are also released into the synapse between nerve cells rather than into the bloodstream. Some examples of neurotransmitters are acetylcholine, dopamine and noradrenaline. Neurotransmitters function primarily to communicate between nerve cells in the brain, central nervous system and autonomic nervous system (Chapter 5). The definition of a neurotransmitter, like that of a hormone, is currently undergoing revisions as new discoveries force a reformulation of many concepts in chemical communication (Bloom, 1988). This is discussed in Chapter 11.

1.4.5 PHEROMONES

Pheromones differ from 'true' hormones in many respects (Martin, 1985, p. 21). Pheromones are: (a) chemical messengers; (b) produced in exocrine or ducted glands; (c) secreted to the outside environment rather than into the bloodstream; (d) act on other individuals, usually of the same species, rather than on other cells within the secreting individual; (e) stimulate the olfactory (smell) or gustatory (taste) receptors (Karlson and Luscher, 1959). When pheromones were first discovered they were called 'external hormones' (ectohormones) because they act like hormones, but on an animal different from the one that produces them. The secretion of pheromones in mammals is regulated by a wide variety of hormones.

There is a complex terminology to describe the various actions of pheromones in both vertebrates and invertebrates but this need not concern us here (see Macdonald and Brown (1985) for more details). Pheromones have two general effects on the receiving individual: 'releaser effects' which result in rapid behavioral changes, (i.e. they 'release' behavior), or 'primer effects', which alter the neuroendocrine system, resulting in a later behavioral change (i.e. they 'prime' the behavioral change). It normally takes about 48 hours for the 'primer effect' of a pheromone to cause a change of behavior.

The pheromone is detected by the main olfactory receptors or the special 'vomeronasal organ', both of which stimulate the olfactory nerves, causing neurotransmitter release in the olfactory bulb and other brain areas which influence the neuroendocrine system. This results in the release of hypothalamic, pituitary and gonadal hormones, which stimulate physiological and behavioral changes in the receiver animal. Thus, there is a complex series of neuroendocrine events which must occur before a pheromone affects behavior through its primer action on the neuroendocrine system. Little more will be said about pheromones in this book.

1.4.6 PARAHORMONES

A parahormone is a hormone-like chemical which is generally not produced in an endocrine gland, but has all of the other characteristics of a 'true' hormone (Martin, 1976, p. 1). Two examples of parahormones are histamine and the prostaglandins. Prostaglandins are often formed in response to hormones (see Section 1.4.2(e)). Histamine and the prostaglandins also have many autocrine functions, affecting the cells that produce them (Martin, 1985).

1.4.7 PROHORMONES

Prohormones are hormone precursors. They can be large molecules that are split to form hormones, or molecules which are in other ways modified to form hormones. Large molecules can be split into one or more hormones, or a hormone and a residual chemical. One important prohormone is proopiomelanocortin (POMC), a giant molecule which, when divided into different segments, produces a number of different hormones. One segment of this prohormone becomes beta-lipotropin which in turn is the prohormone for the endogenous opiate, beta-endorphin. Beta-lipotropin itself may act as a hormone, while beta-endorphin functions both as a hormone and a neurotransmitter (see Chapter 4). Other segments of POMC become the pituitary hormones melanocyte stimulating hormone (MSH) and adrenocorticotropic hormone (ACTH). Testosterone, the primary male gonadal hormone is also a prohormone, as it is the precursor of two other hormones: estrogen and dihydrotestosterone (DHT). The significance of the conversion of testosterone to estrogen and DHT is discussed in Chapter 9.

Other prohormones are molecules which are modified to form hormones in some structure other than an endocrine gland. One example is the renin–angiotensin system which is important in thirst and water regulation. The production of the 'hormone' occurs in the bloodstream not in an endocrine gland. Angiotensinogen, the prohormone, is released by the liver into the blood stream and converted by the enzyme renin, which is produced in the kidneys, into angiotensin I and then to angiotensin II, which acts like a hormone. Because the synthesis of angiotensin I occurs in the bloodstream rather than an endocrine gland, angiotensin is not a 'true' hormone, even though it has all the characteristic hormonal actions.

1.4.8 GROWTH FACTORS

Growth factors are chemical messengers which are synthesized in various types of cell and act to stimulate tissue growth (Hadley, 1992; Sporn and Roberts, 1988). The three best known growth factors are nerve growth factor (NGF), which is synthesized in the peripheral nerves, epidermal growth factor (EGF), which is synthesized in the mouse submaxillary gland, and fibroblast growth factor (FGF), which is synthesized throughout the neural and endocrine systems.

1.4.9 CYTOKINES

The circulating cells of the immune system consist of a wide variety of lymphocytes and phagocytes (monocytes and macrophages).These cells produce a number of peptides which have been termed 'immunomodulators' (Plata-Salaman, 1989) or 'immunotransmitters' (Hall and O'Grady, 1989). These immunomodulators regulate the activities of the immune system to defend the body from bacteria and viruses and to produce inflammatory responses. Collectively, these immunomodulators are called cytokines or lymphokines (Dinarello and Mier, 1987; Hamblin, 1988) and they regulate the growth and function of a number of different target cells. Cytokines are synthesized and released from lymphocytes, whereas monokines are released by macrophages (monocytes). The cytokines include interferon γ, and the interleukins, which encompass tumor necrosis factors, T and B cell growth factors and platelet activating factors (see Chapter 13).

Because neurotransmitters, hormones and neuropeptides can influence the immune system, they are also considered to be immunomodulators (Plata-Salamon, 1989). The brain can regulate the immune system through direct nerve pathways to the thymus gland, through the release of neurotransmitters, or through its control over the release of hormones (Maclean and Reichlin, 1981; Dunn, 1989). Figure 1.2 summarizes the mechanisms through which the neuroendocrine system can influence the immune system. The details of the interaction between the neuroendocrine and immune systems is described in Chapter 13.

1.4.10 VITAMINS

Most vitamins, such as the B vitamins, do not qualify as hormones because they are not produced in the body, but are consumed as nutrients in food. Vitamins A, D and K, however, are formed in the body and have hormone-like actions. Vitamin A is formed in the body from carotene; vitamin D in the skin through the action of sunlight; and vitamin K by bacteria in the intestines. After synthesis in the body, vitamin D travels in the blood and acts at target cells in the intestines, kidney and bone to help to regulate calcium levels in the body (Martin, 1985).

1.5 NEUROREGULATORS: NEUROMODULATORS AND NEUROPEPTIDES

The categorization of the chemical messengers as shown in Table 1.1 is gradually becoming antiquated. As mentioned above, there are problems in determining what is a 'true' hormone and what is not. Certain chemicals that do not quite fit the definition of a hormone still appear to act like hormones. Some hormones synthesized in endocrine glands are also produced in the brain and act as neurotransmitters (e.g. cholecystokinin); many chemical messengers have receptors in the brain and thus influence brain function. These chemicals may be called neuroregulators, neuropeptides, or neuromodulators. The same chemicals may fall under two or more of these classifications, as the differences between neuroregulators, neuromodulators and neuropeptides are fairly vague. In fact, the terminology used to describe communication among neurons is a source of some controversy (see Dismukes (1979) and the ensuing discussion). To reduce some of this confusion, the following definitions will be used in this book.

1.5.1 NEUROREGULATORS

A 'neuroregulator' is a general term for any chemical messenger which regulates the activity of a nerve cell. A neuroregulator can be either a neurotransmitter or a neuromodulator. As defined in Table 1.2, a neurotransmitter is a chemical messenger released by a presynaptic nerve

Table 1.2. *Neuroregulators: neurotransmitters, neuromodulators and neuropeptides*

Neuroregulator A neurotransmitter or other chemical (neuromodulator) which alters nerve cell activity

Neurotransmitter	Neuromodulator
A chemical messenger	A chemical messenger
Alters neural activity	Alters neural activity
Released from a presynaptic cell	Released from neural and non-neural cells (supporting cells, neuroendocrine cells, endocrine glands).
Acts via the synapse on the receptors of the postsynaptic cell	Acts non-synaptically on both the presynaptic and postsynaptic cell to alter synthesis, storage, release and reuptake of the neurotransmitter
Can be a monamine, indoleamine, or catecholamine	Can be a steroid or neuropeptide hormone or non-hormonal peptide

Neuropeptide A hormonal or non-hormonal peptide which acts as a neuromodulator

cell that acts via the synapse to stimulate receptors on a postsynaptic nerve cell (see Chapter 5).

1.5.2 NEUROMODULATORS

A neuromodulator is a chemical released by a neural, endocrine, or other type of cell and acts on a neuron to modulate its response to a neurotransmitter. One mechanism through which neuromodulators influence neural activity is by altering the permeability of the nerve cell membrane to ions such as sodium or chloride. A neuromodulator can be a true hormone, neurohormone, parahormone or other type of chemical messenger. Steroid hormones such as estrogen or testosterone act as neuromodulators, as do peptide hormones such as cholecystokinin. Finally, non-hormone peptides such as substance P and bombesin also act as neuromodulators. The concept of neuromodulation has become central to our understanding of the action of the neuroendocrine system (Kaczamarek and Levitan, 1987).

1.5.3 NEUROPEPTIDES

Neuropeptides are hormone or non-hormone peptides that act as neuromodulators. Thus, adrenocorticotropic hormone (ACTH) and insulin are both peptide hormones in the body and neuropeptides in the brain. Luteinizing hormone releasing hormone (LH-RH) is a peptide hormone and also a neuropeptide which acts as a neuromodulator, and is thus a neuroregulator. Bombesin and neurotensin are non-hormonal peptides which modulate neural activity. Neuropeptides are discussed in detail in Chapters 11 and 12.

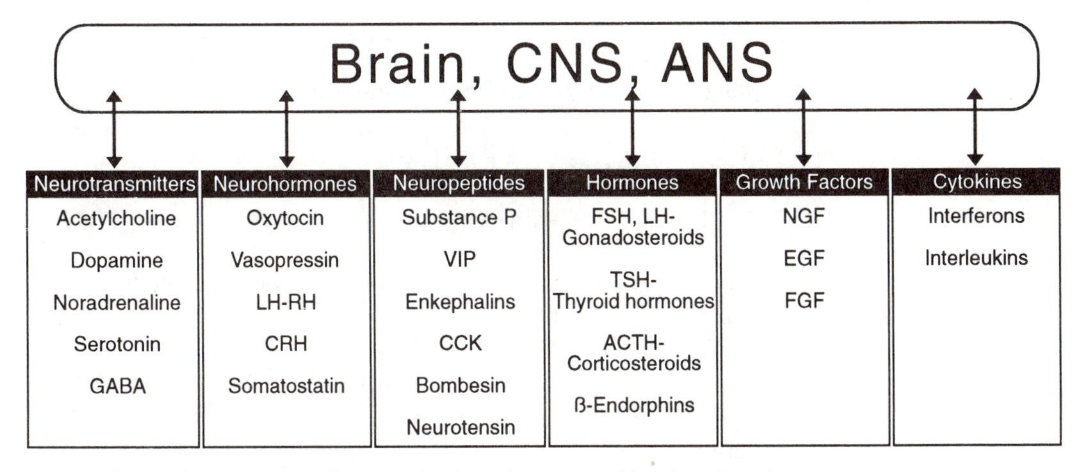

Figure 1.4. Some of the mechanisms through which the brain interacts with and regulates the secretion of the chemical messengers of the neuroendocrine and neuroimmune systems. The brain, central nervous system (CNS) and autonomic nervous system (ANS) stimulate the release of neurotransmitters, neurohormones and neuropeptides, and the release of other secretions such as hormones, growth factors and cytokines. Each of the different types of chemical messenger can also influence the functioning of the brain and nervous systems.

1.6 SUMMARY

This chapter introduces the concept of hormone, brain and behavioral interactions through the body's three communication systems: the nervous, endocrine and immune systems. Some of the mechanisms through which the brain interacts with and regulates the secretions of the neuroendocrine and neuroimmune systems are shown in Figure 1.4. The types of chemical messenger in these three systems are described, with reference to the definition of 'true' hormones and the distinctions between hormones, neurohormones, neurotransmitters, pheromones, parahormones, and prohormones are discussed. Neuroregulators are defined as chemical messengers which regulate the activity of a nerve cell and the distinction is made between neuromodulators, neuropeptides, and neurotransmitters. The differences between endocrine, paracrine, autocrine, neuroendocrine, and neurocrine communication are discussed and the mechanisms for the neuroendocrine influence on the immune system are outlined. The concept of an immunomodulator is discussed and the cytokines, peptides produced by the cells of the immune system, are defined. Because neurotransmitters, hormones and neuropeptides can regulate the immune system, they are also considered immunomodulators.

FURTHER READING

Ader, R. (1981). *Psychoneuroimmunology*. New York: Academic Press.

Ader, R., Felton, D. L. and Cohen, N. (1991). *Psychoneuroimmunology*, 2nd edn. San Diego: Academic Press.

Bennett, G. W. and Whitehead, S. A. (1983). *Mammalian Neuroendocrinology*. New York: Oxford University Press.

Hadley, M. E. (1992). *Endocrinology*, 3rd edn. Englewood Cliffs, NJ: Prentice-Hall.
Martin, C. R. (1985). *Endocrine Physiology*. Oxford: Oxford University Press.

REVIEW QUESTIONS

1.1 What is the difference between a prohormone and a phytohormone?
1.2 What is the difference between a neurohormone and a neurotransmitter?
1.3 What is a prohormone?
1.4 Which hormone could be considered a prohormone for estrogen?
1.5 What is the difference between a 'true' hormone and a 'pheromone'?
1.6 Why is a neurohormone different from a 'true' hormone?
1.7 What is the difference between paracrine and endocrine communication?
1.8 What is the difference between neurocrine and neuroendocrine communication?
1.9 What is a parahormone?
1.10 What is the difference between a neurotransmitter and a neuromodulator?
1.11 What are cytokines?

ESSAY QUESTIONS

1.1 Define the term 'neuroendocrine' and describe the components of the neuroendocrine system. How does the term 'psychoneuroendocrine' relate to this system?
1.2 Describe the 'common receptor mechanisms' which might act to integrate the neural, endocrine and immune systems.
1.3 Discuss the differences between endocrine, autocrine and paracrine communication.
1.4 Discuss the problems in the classical definition of a 'hormone'.
1.5 Discuss the problems in differentiating a true hormone from neurohormones, parahormones and prohormones.
1.6 Discuss the problems in differentiating between neurotransmitters, neuropeptides and neuromodulators.
1.7 Discuss the similarities and differences between neurohormones and neurotransmitters.
1.8 Why might cytokines be called 'immunotransmitters'?

REFERENCES

Ader, R. (1981). *Psychoneuroimmunology*. New York: Academic Press.
Bennett, G. W. and Whitehead, S. A. (1983). *Mammalian Neuroendocrinology*. New York: Oxford University Press.
Blalock, J. E. (1989). A molecular basis for bidirectional communication between the immune and neuroendocrine systems. *Physiological*

Reviews, **69**, 1–32.
Bloom, F. E. (1988). Neurotransmitters: past, present, and future directions. *Federation Proceedings*, **2**, 32–41.
Brownstein, M. J., Russell, J. T. and Gainer, H. (1980). Synthesis, transport and release of posterior pituitary hormones. *Science*, **207**, 373–378.
Chailakhyan, M. K. H. and Khrianin, V. N. (1987). *Sexuality in Plants and its Hormonal Regulation*. Berlin: Springer-Verlag.
Dinarello, C. A. and Mier, J. W. (1987). Lymphokines. *New England Journal of Medicine*, **317**, 940–945.
Dismukes, R. K. (1979). New concepts of molecular communication among neurons. *Behavioral and Brain Sciences*, **2**, 409–448. (This includes the 'target' article, a number of critical commentaries, and the author's reply.)
Dunn, A. J. (1989). Psychoneuroimmunology for the psychoneuro-endocrinologist: A review of animal studies of nervous system-immune system interactions. *Psychoneuroendocrinology*, **14**, 251–274.
Hadley, M. E. (1992). *Endocrinology*, 3rd edn. Englewood Cliffs, NJ: Prentice-Hall.
Hall, N. R. S. and O'Grady, M. P. (1989). Regulation of pituitary peptides by the immune system: Historical and current perspectives. *Progress in Neuroendocrinimmunology*, **2**, 4–10.
Hamblin, A. S. (1988). *Lymphokines*. Oxford: IRL Press.
Kaczamerek, L. K. and Levitan, I. B. (1987). *Neuromodulation: The Biochemical Control of Neuronal Excitability*. New York: Oxford University Press.
Karlson, P. and Luscher, M. (1959). 'Pheromones': A new term for a class of biologically active substances. *Nature*, **183**, 545–546.
Kordon, C. and Bihoreau, C. (1989). Integrated communication between the neurons, endocrine and immune systems. *Hormone Research*, **31**, 100–104.
Kreiger, D. T. (1983). Brain peptides: what, where, and why? *Science*, **222**, 975–985.
Le Page-Degivry, M.-Th., Bidard, J.-N., Rouvier, E., Bulard, C. and Lazdunski, M. (1986). Presence of abscisic acid, a phytohormone, in the mammalian brain. *Proceedings of the National Academy of Sciences, USA*, **83**, 1155–1158.
Macdonald, D. W. and Brown, R. E. (1985). Introduction: The pheromone concept in mammalian chemical communication. In R. E. Brown and D. W. Macdonald (eds.) *Social Odours in Mammals*, vol. 1, pp. 1–18. Oxford: Clarendon Press.
MacLean, D. and Reichlin, S. (1981). Neuroendocrinology and the immune process. In R. Ader (ed.) *Psychoneuroimmunology*, pp. 475–520. Orlando, FL: Academic Press.
Martin, C. R. (1976). *Textbook of Endocrine Physiology*. Baltimore: Williams and Wilkins.
Martin, C. R. (1985). *Endocrine Physiology*. Oxford: Oxford University Press.
Müller, E. E. and Nistico, G. (1989). *Brain Messengers and the Pituitary*. San Diego: Academic Press.
Plata-Salaman, C. R. (1989). Immunomodulators and feeding regulation: a humoral link between the immune and nervous systems. *Brain, Behavior and Immunity*, **3**, 193–213.
Roth, J., LeRoith, D., Lesniak, M. A., de Pablo, F., Bassas, L. and Collier,

E. (1986). Molecules of intercellular communication in vertebrates, invertebrates and microbes: do they share common origins? *Progress in Brain Research*, **68**, 71–79.

Smith, E. M. and Blalock, J. E. (1985). A complete regulatory loop between the immune and neuroendocrine systems operates through common signal molecules (hormones) and receptors. In: N. P. Plotnikoff, R. E. Faith, A. J. Murgo and R. A. Good (eds.) *Enkephalins and Endorphins: Stress and the Immune System*, pp. 119–127. New York: Plenum Press.

Smythies, J. R. (1976). Perspectives in psychoneuroendocrinology. *Psychoneuroendocrinology*, **1**, 317–319.

Snyder, S. (1985). The molecular basis of communication between cells. *Scientific American*, **253** (4), 114–123.

Sporn, M. B. and Roberts, A. B. (1988). Peptide growth factors are multifunctional. *Nature*, **332**, 217–219.

Tausk, M. (1975). *Pharmacology of Hormones*. Chicago: Year Book Medical Publishers.

2

The endocrine glands and their hormones

2.1 THE ENDOCRINE GLANDS

The endocrine glands of the human are shown in Figure 2.1. The pineal gland is a small gland lying between the cerebral cortex and the cerebellum at the posterior end of the third ventricle in the middle of the brain. The pituitary gland hangs from the bottom of the hypothalamus at the base of the brain and sits in a small cavity of bone above the roof of the mouth. The thyroid gland is located in the neck and the small parathyroid glands are embedded in the surface of the thyroid. In the chest is the thymus gland, which is very important in the development of the immunological system and in immunity. The heart, stomach and small intestine (duodenum) are also endocrine glands which secrete hormones. The adrenal gland is a complex endocrine gland situated on top of the kidney. The pancreas secretes hormones involved in regulating blood sugar levels and the kidney also produces hormone-like chemicals. The testes and ovaries produce the gonadal hormones or sex hormones which, in addition to the maintenance of fertility and sex characteristics, have important effects on behavior. Finally, during pregnancy the

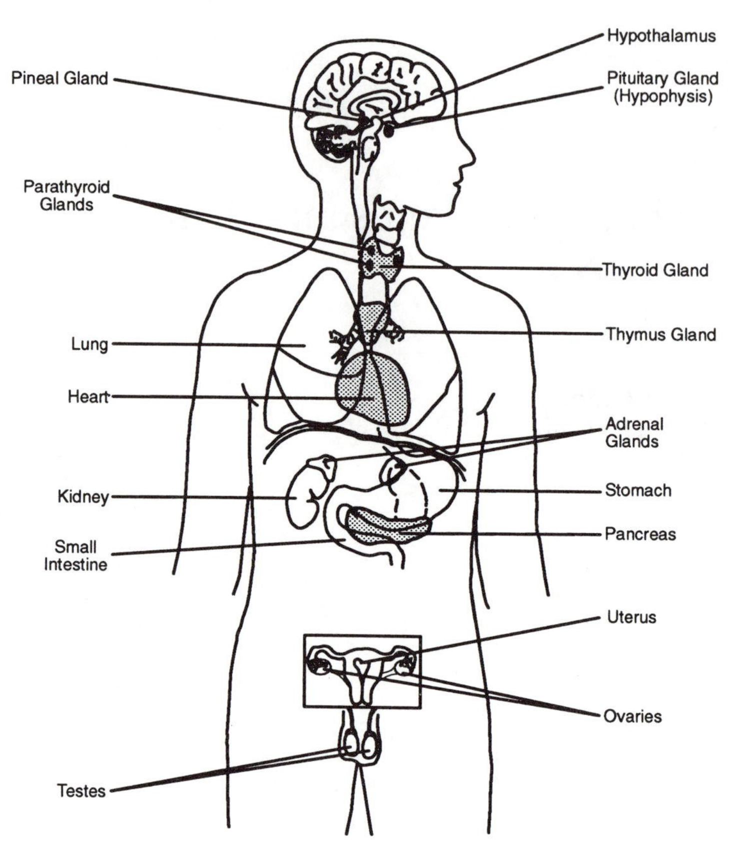

Figure 2.1. The endocrine glands of the human body and their approximate location.

placenta acts as an endocrine gland. The endocrine glands occur in similar locations in all vertebrates.

2.2 THE HORMONES OF THE ENDOCRINE GLANDS

Each endocrine gland secretes one or more hormones. Table 2.1 summarizes the main hormones produced by each gland and some of the functions of these hormones.

2.2.1 THE PINEAL GLAND

The main hormone secreted by the pineal gland is **melatonin**, but the pineal also produces a number of other hormones, including arginine vasotocin. In many mammals, melatonin mediates reproductive activity in response to changes in environmental light cycles (Reiter, 1983). The pineal gland may also help to regulate the timing of puberty in mammals such as sheep. In non-mammalian vertebrates, such as frogs, melatonin causes lightening of pigment coloration.

2.2.2 THE PITUITARY GLAND

The pituitary is a complex gland which produces at least ten hormones. These are discussed in Chapter 3.

Table 2.1. *Endocrine glands and their hormones*

Gland	Hormone	Action
Pineal	Melatonin	Modulates gonadal activity Mediates responses to light Alters pigment coloration
Pituitary	Many hormones which are discussed in Chapter 3	
Thyroid	Thyroxine (T4) and triiodothyronine (T3)	Regulates cell growth and differentiation; especially bone growth and neural development Regulates metabolic rate
	Calcitonin	Lowers blood calcium levels
Parathyroid	Parathyroid hormone (PTH)	Raises blood calcium levels
Thymus	Thymosin	Stimulates lymphocyte production and differentiation (T cells)
	Thymostatin	Inhibits lymphocyte production
Heart	Atrial natriuretic factor (ANF)	Regulates blood pressure, blood volume and electrolyte balance
Stomach	Gastrin	Stimulates secretion of hydrochloric acid and increases intestinal motility
Duodenum (small intestine)	Secretin	Stimulates secretion of pancreatic juice
	Cholecystokinin (CCK)	Stimulates gall bladder contraction and pancreatic enzyme secretions
Pancreas (islets of Langerhans)	Insulin (beta cells)	Lowers blood glucose and promotes synthesis of fat and protein
	Glucagon (alpha cells)	Increases blood glucose and promotes breakdown of fat and protein
Adrenal cortex	Glucocorticoids, (cortisol, corticosterone)	Converts stored fats and proteins to carbohydrates; anti-inflammatory, antiallergy, anti-immune function
	Mineralocorticoids (aldosterone)	Increases sodium retention and potassium loss in kidney
	Sex steroids	Estrogen, androgens and progesterone
Adrenal medulla	Adrenaline (Epinephrine)	Increases heart rate, oxygen consumption and glycogen mobilization
	Noradrenaline (Norepinephrine)	Increases blood pressure, constricts blood vessels
Testes		
Leydig cells	Androgens (testosterone)	Male sexual characteristics
Sertoli cells	Inhibin	Inhibits FSH secretion from pituitary
Ovaries		
Follicles	Estrogens	Female sexual characteristics
Corpus luteum	Progesterone	Maintains pregnancy; facilitates uterine and breast development
	Relaxin	Relaxes birth canal and dilates uterine cervix at birth
Placenta	Human chorionic gonadotropin (HCG)	Maintains progesterone synthesis from ovaries; acts like pituitary LH
	Human Placental lactogen (HPL)	Mammary gland growth and development; acts like pituitary LH and prolactin
	Progesterone	Maintains pregnancy

2.2.3 THE THYROID GLAND

The thyroid gland produces three main hormones. Two of these, **triiodothyronine** (T3) and **thyroxine** (T4), are quite similar and depend on iodine for their synthesis. T4 is much more prevalent in the blood (98%) than is T3 (2%). These hormones regulate body metabolism and are also important in bone growth and in the development and maturation of the brain and nervous system (Dussault and Ruel, 1987). Congenital lack of thyroid hormones leads to reduced brain development, a syndrome called cretinism. The third thyroid hormone, **calcitonin**, reduces blood calcium levels.

2.2.4 THE PARATHYROID GLANDS

The parathyroids are small glands embedded in the posterior surface of the thyroid gland. **Parathyroid hormone** (PTH) raises blood calcium levels through its action on bone, kidney and, indirectly, the gut (Habener, Roseblatt and Potts, 1984).

This is a good time to point out that hormones often occur in pairs that have antagonistic actions. One hormone will stimulate a response and the other will inhibit it. Thus, calcitonin lowers blood calcium levels and parathyroid hormone raises blood calcium levels. Many pairs of hormones have such opposing effects.

2.2.5 THE THYMUS GLAND

The thymus gland, located in the upper chest cavity, produces a number of thymus hormones including thymosin and thymostatin. **Thymosin** stimulates the production and differentiation of lymphocytes, whereas **thymostatin** inhibits the production of lymphocytes (Martin, 1976). The T cells, which are involved in cellular immunity, depend on the thymus gland for their development. Thymosin is essential for the conversion of prethymic cells into mature thymus cells and the development of immunocompetence, the ability to respond to antigenic stimulation. The thymus is also important for the neonatal production of antibodies. The thymus gland is one of the points of interaction between the endocrine, neural and immune systems (see Chapter 13).

2.2.6 THE HEART

As well as its function as a pump for the blood, the heart is an endocrine gland. Granular cells in the heart muscle secrete a hormone called **atrial natriuretic factor** (ANF) which regulates blood pressure, blood volume and the excretion of water, sodium and potassium through its actions on the kidneys and adrenal glands (Cantin and Genest, 1986). ANF also acts as a neuropeptide in the brain, where it regulates salt and water intake, heart rate and vasopressin secretion (Quirion, 1988).

2.2.7 GASTROINTESTINAL HORMONES

The mucosa of the gastrointestinal (GI) tract secretes over a dozen peptide hormones (Johnson, 1977; Martin, 1985). The three best known GI hormones are **secretin, gastrin** and **cholecystokinin** (CCK). Other peptides secreted by endocrine cells in the GI tract include gastric inhibitory peptide (GIP), vasoactive intestinal peptide (VIP), substance P, somatostatin, gastrin releasing peptide, bombesin, neurotensin, motilin, and others (Martin, 1985; Hadley, 1992). These hormones are of great interest, since they are also synthesized in the brain and function as neuropeptides (Krieger, 1983, 1986; see discussion in Chapter 11). Other GI hormones are postulated to exist only on the basis of physiological evidence (see Johnson, 1977).

Many hormones of the GI tract do not meet the criterion of 'true' hormones because they are not secreted from specific endocrine glands and their action at target cells has not been specified (Turner and Bagnara, 1976; Johnson, 1977). The release of many GI hormones is controlled by nerves of the autonomic nervous system, so such hormones form an integral part of the neuroendocrine system. The three primary gastrointestinal hormones, secretin, gastrin and cholecystokinin, are produced in endocrine cells scattered throughout the walls of the alimentary tract, rather than in specialized endocrine glands. Figure 2.2 shows the source and action of these three hormones.

Gastrin. The walls of the stomach produce gastrin in response to distention caused by the presence of food (Figure 2.2). Gastrin has many actions, including stimulation of hydrochloric acid secretion by the stomach, stimulation of pancreatic enzyme secretion and increasing intestinal motility. Gastrin is thus an important hormone for digesting food (Turner and Bagnara, 1976; Hadley, 1992).

Secretin and cholecystokinin. Secretin and cholecystokinin are produced in the duodenum or small intestine (Figure 2.2). The passage of partially digested food from the stomach to the duodenum stimulates the release of secretin from the mucosa of the duodenum. Secretin stimulates the secretion of pancreatic bicarbonate and has many other functions in digestion, including the stimulation of hepatic bile flow and the potentiation of CCK-stimulated pancreatic enzyme secretion (Turner and Bagnara, 1976; Figure 2.2). Secretin was the first hormone to be discovered (Baylis and Starling, 1902). Cholecystokinin is released from cells in the walls of the duodenum by the presence of food, particularly fats and fatty acids. It has many functions in common with gastrin, including the stimulation of gall bladder contraction and pancreatic enzyme secretion and the inhibition of gastric emptying (Figure 2.2).

2.2.8 THE PANCREAS

The endocrine cells of the pancreas consist of islets of tissue (islets of Langerhans) surrounded by cells which secrete the pancreatic juices

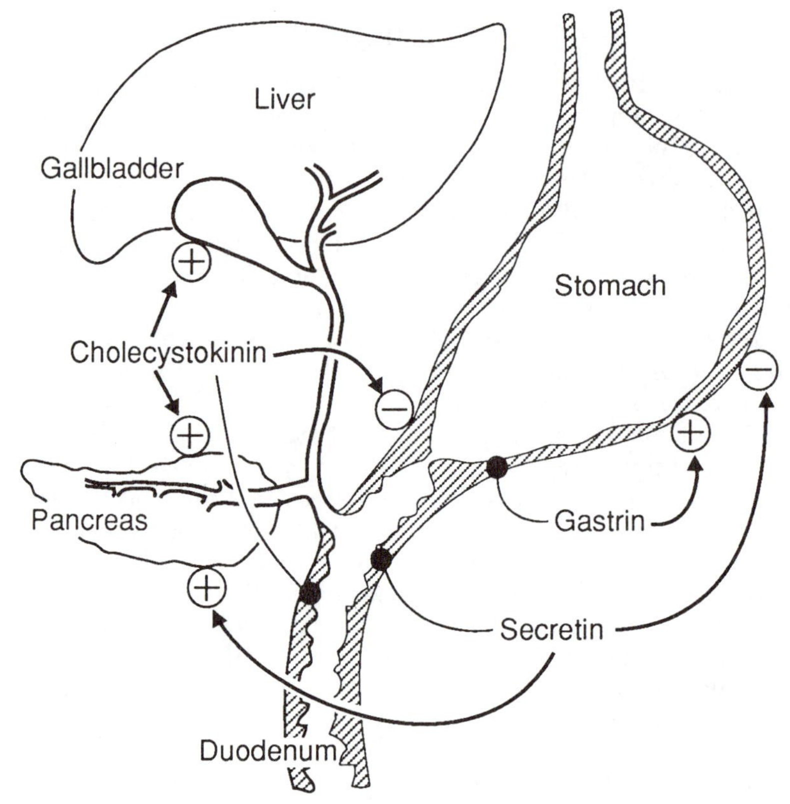

Figure 2.2. The source and action of the gastrointestinal hormones: secretin, gastrin and cholecystokinin. The arrows extend from the source of the hormones (indicated by a dot) and point toward the target organs. + indicates stimulation; − indicates inhibition. (Redrawn from Turner and Bagnara, 1976.)

essential for the digestion of food. The β cells of the islets of Langerhans secrete **insulin** in response to increased blood glucose levels which result from carbohydrate intake. Insulin lowers glucose levels in the blood by increasing glucose uptake in adipose, hepatic or muscle cells, where it is stored as glycogen or utilized as an energy source.

The α-cells of the islets of Langerhans secrete the hormone **glucagon** which increases blood glucose levels, thus having the opposite action to that of insulin. Glucagon increases blood glucose levels by stimulating the conversion of glycogen to glucose in the liver. Other endocrine cells of the pancreas produce the hormones somatostatin and pancreatic polypeptide (Hadley, 1992).

2.2.9 THE ADRENAL GLANDS

The adrenal glands sit on top of the kidneys (Figure 2.1) and consist of two distinct types of tissue: a medulla surrounded by a cortex (Figure 2.3).

The adrenal cortex

The adrenal cortex is a true endocrine gland which secretes three categories of steroid hormones: mineralocorticoids, glucocorticoids and sex steroids. **Aldosterone** is the primary mineralocorticoid produced by

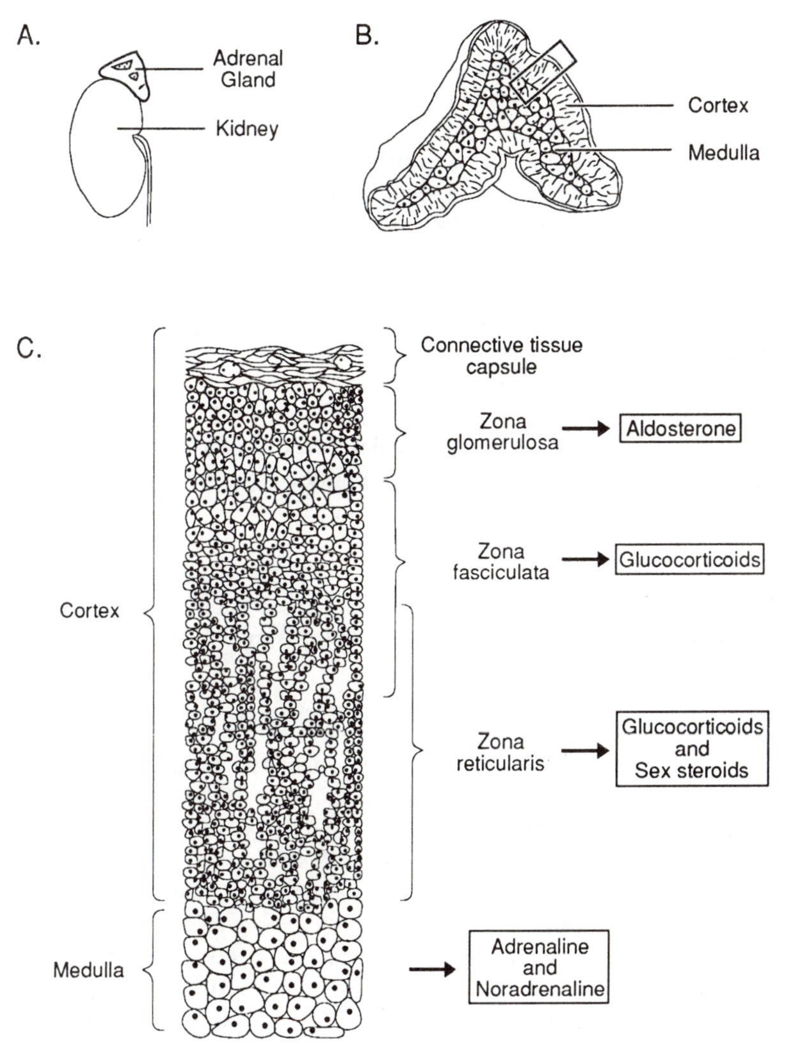

Figure 2.3. The adrenal gland. A. The location of the adrenal glands above the kidney. B. A cross-section of the adrenal gland showing the cortex and medulla. C. A cross-section of the anatomy of the adrenal gland, showing the three layers of the adrenal cortex surrounding the adrenal medulla and the hormones secreted by each layer. (Modified from Hadley, 1992.)

the adrenal cortex. Aldosterone secretion is stimulated by sodium deprivation and, when released, it acts to increase the reabsorption of sodium ions (Na^+) in the kidneys, salivary glands and sweat glands.

The synthesis and release of the glucocorticoids (for example, **cortisol** and **corticosterone**) is stimulated by adrenocorticotropic hormone (ACTH) from the anterior pituitary gland (Chapter 3). Glucocorticoids modulate carbohydrate metabolism, converting stored proteins to carbohydrates. Glucocorticoids are released by stressful stimuli and have antiinflammatory and immunosuppressive functions; that is, they inhibit inflammatory and allergic reactions and inhibit the production of lymphocytes by the immune system.

The adrenal cortex also produces small amounts of the gonadal steroids: **androgen**, **estrogen** and **progesterone**, and these adrenal sex steroids may influence sexual differentiation and the bodily changes which occur at puberty.

The adrenal medulla

The adrenal medulla is surrounded by the adrenal cortex (Figure 2.3) and resembles brain tissue more than an endocrine gland, i.e. the cells behave like neurons. Secretion of hormones from the adrenal medulla is controlled by the sympathetic branch of the autonomic nervous system. Two hormones are released: **adrenaline** (epinephrine) and **noradrenaline** (norepinephrine). These two chemicals are also produced in the brain where they act as neurotransmitters. Adrenaline is released following stress due to environmental extremes (cold), physical exertion, or fear and it acts to increase heart rate and blood glucose levels, thus increasing the amount of work the muscles can do. Noradrenaline acts to increase blood pressure and to constrict blood vessels. Under chronic high stress there is hyperactivity of both the adrenal cortex and the adrenal medulla and, as a result, high levels of adrenal hormones are secreted.

2.2.10 THE GONADS

The gonads (testes and ovaries) secrete three categories of steroid hormones: androgens, estrogens and progestins, which are referred to as the 'sex steroids'.

The testes

The male gonads (the testes) produce **androgens** from the Leydig or interstitial cells. The primary androgen is testosterone but there are many other androgens, such as dihydrotestosterone and androstenedione which are less potent than testosterone. Testosterone is important for masculinization during sexual differentiation, for the control of sperm production, the development of male secondary sexual characteristics at puberty and for the activation of sexual, aggressive and other behaviors in adulthood. The Sertoli cells of the testes produce **inhibin**, a peptide hormone which inhibits the secretion of follicle stimulating hormone (FSH) from the pituitary gland.

The ovaries

The ovaries are the female gonads which produce two major classes of hormones, **estrogens** and **progestins**. The primary estrogen is estradiol but there are a variety of other estrogenic hormones including estrone and estriol. There are also a large number of synthetic estrogens which are used in birth control pills. Estrogens are produced in the granulosa cells of the ovarian follicle and are important at puberty for the development of female secondary sex characteristics. Estrogens also function to influence metabolic rate, body temperature, skin texture, fat distribution and many enzyme, circulatory and immune functions. Estrogens also influence sexual, parental, and other behaviors in the female.

The progestins (e.g. progesterone) are produced in the corpus luteum of the ovary and are important for uterine, vaginal and mammary gland growth. Progesterone helps to stimulate the breast and uterine enlargement at puberty and during the menstrual cycle. In maintaining pregnancy, progesterone inhibits the menstrual or reproductive cycle and inhibits the sexual behavior associated with it in rats, mice, and other mammals.

Estrogen and progesterone usually act synergistically. Most actions of progesterone require estrogen priming of the target cells. For example, breast development at puberty requires estrogen to prime the cells and then progesterone causes cell differentiation and growth. Estrogen also stimulates progesterone secretion in the ovary. Progesterone acts on the target cells to stimulate growth and it also feeds back and inhibits the secretion of estrogen. Progesterone also has a variety of influences on behavior.

Near the end of pregnancy, the ovary secretes **relaxin**, a peptide hormone which acts to prepare the birth canal for parturition. Relaxin increases the flexibility of the ligaments of the cervix and connective tissue of the pubic area (Martin, 1985). Small quantities of relaxin are also secreted by the placenta and uterus. The ovary also produces **inhibin**, which acts to inhibit FSH secretion in the female.

2.2.11 THE PLACENTA

When pregnancy occurs, specific hormones are secreted by the fertilized egg and are thus useful in pregnancy tests. The first of these is **human chorionic gonadotropin** (HCG). HCG is released coincident with the formation of the implantation site. HCG stimulates the corpus luteum of the mother's ovary to keep progesterone secretion at levels high enough to maintain the uterine lining so that placental development can proceed. If progesterone levels drop, the pregnancy is aborted. HCG stimulates progesterone release for only a certain period of time, after which the placenta begins to produce its own progesterone to maintain the pregnancy. Once the placenta develops, the fetus, placenta and mother form an integrated materno-feto-placental unit which produces a number of hormones critical for the maintainance of pregnancy (Goebelsmann, 1979; Jaffe, 1986).

Another hormone unique to human pregnancy is **human placental lactogen** (HPL) which is also called human chorionic somatomammotropin (HCS). HPL has functions similar to those of growth hormone and prolactin and stimulates the mammary glands to differentiate and to begin to secrete milk. HPL is not secreted until the pregnancy is well established. The placenta also produces **estrogens**, **androgens** and **relaxin**.

2.3 SUMMARY

This chapter provides a brief overview of the endocrine glands, their location in the body, and the hormones they produce. You should become familiar with each of these glands and be able to identify the hormones produced by each and the primary functions of each hormone as outlined in Table 2.1. In Chapter 3 you will become familiar with the stimuli which regulate the release of hormones. Thus, understanding of material in future chapters depends on knowledge of the information in this chapter.

FURTHER READING

Greenspan, F. S. (1991). *Basic and Clinical Endocrinology*, 3rd edn. Norwalk, CN: Appleton and Lange.

Hadley, M. E. (1992). *Endocrinology*, 3rd edn. Englewood Cliffs, NJ: Prentice-Hall.

Martin, C. R. (1985). *Endocrine Physiology*. Oxford: Oxford University Press.

Turner, C. D. and Bagnara, J. T. (1976). *General Endocrinology*, 6th edn. Philadelphia: Saunders.

REVIEW QUESTIONS

2.1 Which endocrine glands secrete the following hormones: (a) calcitonin, (b) melatonin, (c) glucagon, (d) gastrin?

2.2 Each of the following glands secretes two primary hormones. Name the hormones from (a) duodenum, (b) ovary.

2.3 Name the two parts of the adrenal gland and the two hormones secreted from each part.

2.4 Name the five steroid hormones.

2.5 Name the hormones produced in the following glands: (a) testis, (b) stomach, (c) thyroid, (d) pancreas (β cells).

2.6 Which endocrine gland is important for the development of the immune system?

2.7 Which placental hormone stimulates the ovaries to keep producing progesterone in the early stages of pregnancy?

2.8 Which endocrine glands secrete the following hormones: (a) human placental lactogen, (b) thymosin, (c) aldosterone, (d) adrenaline?

2.9 Name two hormones which are also neurotransmitters.

2.10 As well as the gonads, the sex steroids are produced in which other endocrine gland?

ESSAY QUESTIONS

2.1 Discuss the concept of opposing actions between pairs of hormones such as calcitonin and parathyroid hormone or insulin and glucagon.

2.2 Discuss the relationship between the thymus gland and the immune system.

2.3 Discuss the gastrointestinal hormones with respect to the definition of a 'true' hormone given in Chapter 1. Do they meet the criteria or not?

2.4 How is it that the sex steroids are produced in both the gonads and the adrenal cortex?

2.5 Why are adrenaline and noradrenaline classed both as hormones and as neurotransmitters?

2.6 Discuss the different roles of the pineal gland in amphibians and mammals.
2.7 Describe the changes in the ovarian follicle that regulate the timing of estrogen and progesterone secretion.
2.8 Discuss the placental hormones, the timing of their secretion and their functions.

REFERENCES

Bayliss, W. M. and Starling, E. M. (1902). The mechanism of pancreatic secretion. *Journal of Physiology, London*, **280**, 325–353.

Cantin, M. and Genest, J. (1986). The heart as an endocrine gland. *Scientific American*, **254** (2), 62–67.

Dussault, J. H. and Ruel, J. (1987). Thyroid hormones and brain development. *Annual Review of Physiology*, **49**, 321–334.

Goebelsmann, U. (1979). Protein and steroid hormones in pregnancy. *Journal of Reproductive Medicine*, **23**, 166–177.

Habener, J. F., Rosenblatt, M. and Potts, J. T., Jr (1984). Parathyroid hormone: Biochemical aspects of biosynthesis, secretion, action, and metabolism. *Physiological Reviews*, **64**, 985–1053.

Hadley, M. E. (1992). *Endocrinology*, 3rd edn. Englewood Cliffs, NJ: Prentice-Hall.

Jaffe, R. B. (1986). Endocrine physiology of the fetus and fetoplacental unit. In S. S. C. Yen and R. B. Jaffe (eds.) *Reproductive Endocrinology*, 2nd edn, pp. 737–757. Philadelphia: W. B. Saunders.

Johnson, L. R. (1977). Gastrointestinal hormones and their functions. *Annual Review of Physiology*, **39**, 135–158.

Kreiger, D. T. (1983). Brain peptides: What, where, and why? *Science*, **222**, 975–985.

Kreiger, D. T. (1986). An overview of neuropeptides. In J. B. Martin and J. D. Barchas (eds.) *Neuropeptides in Neurologic and Psychiatric Disease*, pp. 1–32. New York: Raven Press.

Martin, C. R. (1976). *Textbook of Endocrine Physiology*. Baltimore: Williams and Wilkins.

Martin, C. R. (1985). *Endocrine Physiology*. Oxford: Oxford University Press.

Quirion, R. (1988). Atrial natriuretic factors and the brain: an update. *Trends in Neurosciences*, **11**, 58–62.

Reiter, R. J. (1983). The pineal gland: an intermediary between the environment and the endocrine system. *Psychoneuroendocrinology*, **8**, 31–40.

Starling, E. H. (1905). The chemical correlation of the functions of the body. Lecture 1. *Lancet*, **2**, 339–341.

Turner, C. D. and Bagnara, J. T. (1976). *General Endocrinology*, 6th edn. Philadelphia: Saunders.

3

The pituitary gland and its hormones

3.1 THE PITUITARY GLAND

The pituitary gland, which is also called the hypophysis, is attached to the hypothalamus at the base of the brain (see Figure 3.1). Secretion of the hormones of the pituitary gland is regulated by the hypothalamus and it is through the hypothalamic-pituitary connection that external and internal stimuli can influence the release of the pituitary hormones, thus producing the neural-endocrine interaction. The pituitary has been called the body's 'master gland' because its secretions stimulate other endocrine glands to synthesize and secrete their hormones, but it is really the hypothalamus that is the master gland, because it controls the pituitary.

The pituitary gland is a complex organ (Figure 3.2) which is divided into three parts: the anterior lobe (pars distalis), the intermediate lobe (pars intermedia) and the posterior lobe (Pars nervosa). Together, the anterior and intermediate lobes form a true endocrine gland, the **adenohypophysis**. The posterior lobe, also called the **neurohypophysis**, is really neural tissue and is an extension of the hypothalamus. The pituitary is attached to the hypothalamus by the **hypophyseal stalk**.

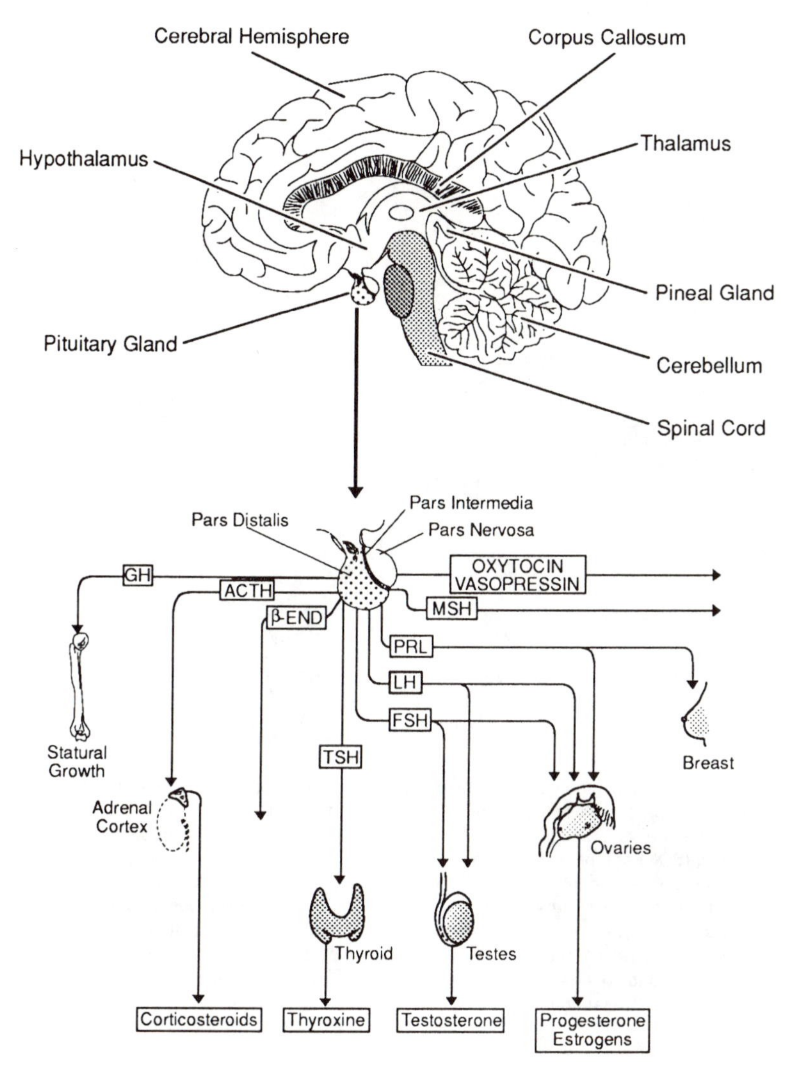

Figure 3.1. The pituitary gland is connected to the median eminence of the hypothalamus by the hypophyseal stalk. The pituitary gland secretes ten hormones from its three lobes, as described in Table 3.1. β-END = β-endorphin. For other abbreviations see Table 3.1. (Redrawn from Guillemin and Burgus, 1972.)

Detailed anatomy of the pituitary gland is provided by Turner and Bagnara (1976) and Hadley (1992).

3.1.1 THE NEUROPHYPOPHYSIS

The neurohypophysis consists of neural tissue and is essentially a projection of the brain. The neurohypophysis consists of the posterior lobe of the pituitary, called the pars nervosa, because of its neural tissue, and the portion of the hypophyseal stalk, termed the **infundibulum**, which contains the axons of the hypothalamic neurosecretory cells (Figure 3.2). These neurosecretory cells are located in the paraventricular nucleus (PVN) and supraoptic nucleus (SON) of the hypothalamus. These two nuclei manufacture the hormones oxytocin and vasopressin (antidiuretic hormone).

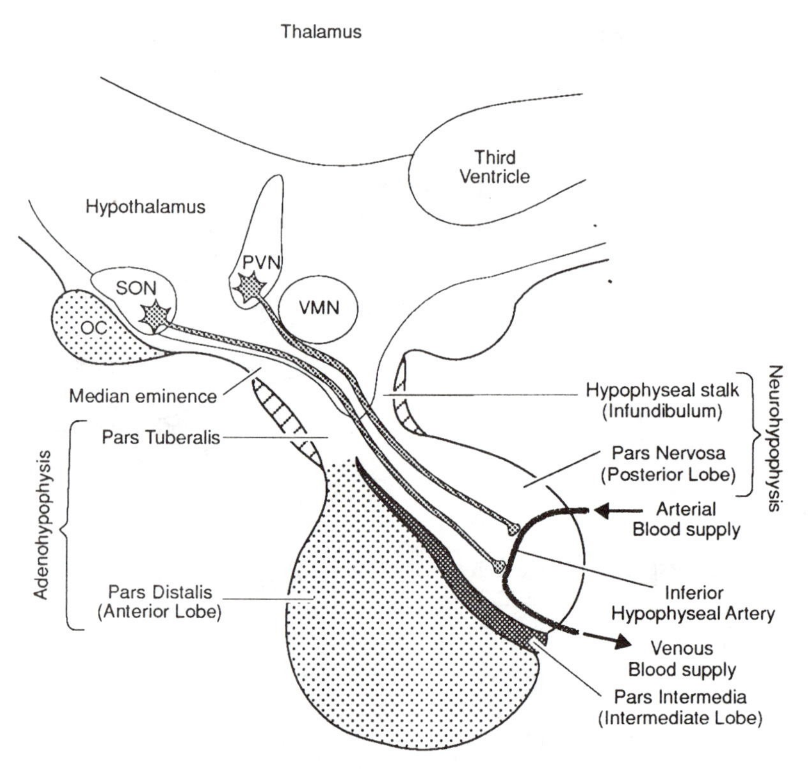

Figure 3.2. The major subdivisions of the pituitary gland or hypophysis. The neurohypophysis consists of the pars nervosa (posterior lobe), and the infundibulum of the hypophyseal stalk. It contains the nerve endings of the neurosecretory cells whose cell bodies are in the supraoptic (SON) and paraventricular nuclei (PVN) of the hypothalamus. The axons of the neurosecretory cells in the PVN and SON project through the infundibulum of the hypophyseal stalk to the pars nervosa where they release their hormones into the inferior hypophyseal artery. The adenohypophysis has two elements: the pars distalis (anterior lobe) and the pars tuberalis of the hypophyseal stalk. The pars intermedia (intermediate lobe) is also considered to be a part of the adenohypophysis in many species. VMN = ventromedial nucleus of the hypothalamus; OC = optic chiasm.

3.1.2 THE ADENOHYPOPHYSIS

The adenohypophysis is a true endocrine gland involving the anterior lobe (pars distalis) and the intermediate lobe (pars intermedia) of the pituitary. The adenohypophysis is attached to the hypothalamus by that part of the hypophyseal stalk called the pars tuberalis which contains the hypophyseal portal system of blood vessels (Figure 3.2). The nerve endings of the neurosecretory cells of the hypothalamus (described in Chapter 4) terminate at the median eminence, where their hormones are released into the hypophyseal portal system, through which they are carried to the adenohypophysis. The hypophyseal stalk thus contains both nerve axons and blood vessels which connect the hypothalamus and pituitary gland (Figure 3.3).

This vascular connection between the hypothalamus and the adenohypophysis (Figure 3.3) is relatively complex. The superior hypophyseal artery delivers blood to the median eminence of the hypothalamus where it forms a series of tiny blood vessels (capillaries), called the 'primary

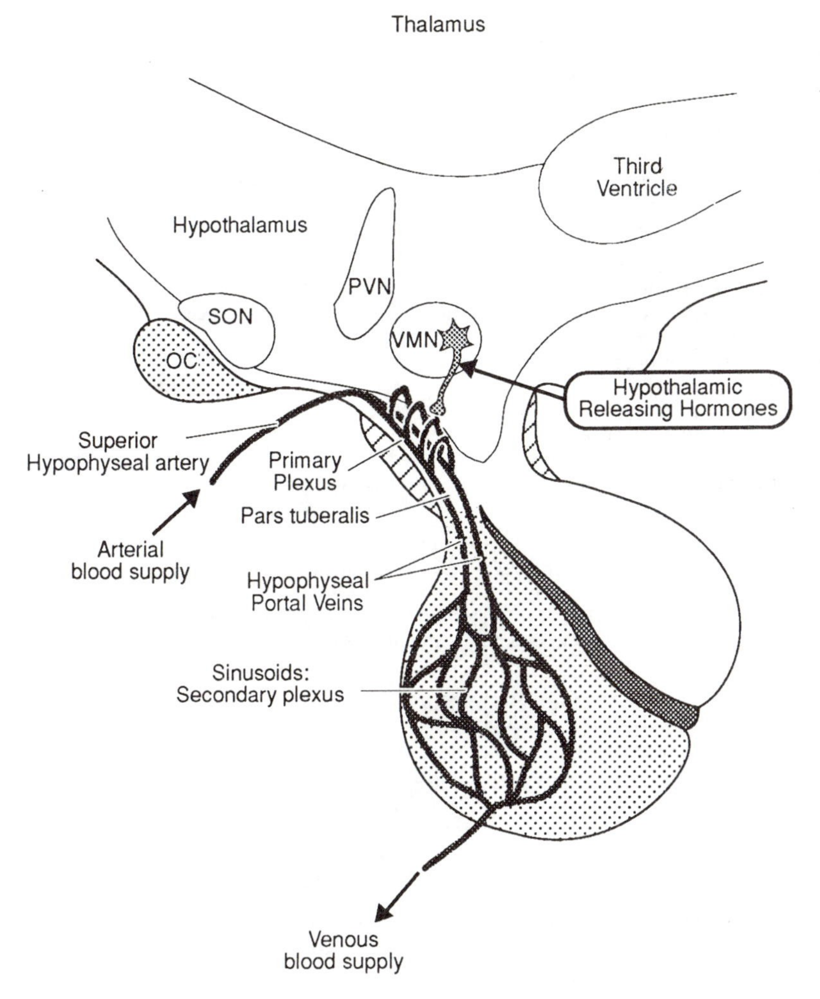

Figure 3.3. The connections between the hypothalamus and the adenohypophysis. The hypothalamic releasing hormones are secreted by the neurosecretory cells in the ventromedial nucleus (VMN) and other areas of the hypothalamus. The axons of these neurosecretory cells project to the primary plexus in the median eminence of the hypothalamus. The hypophyseal portal veins carry the hypothalamic hormones from the primary plexus through the pars tuberalis of the hypophyseal stalk to the secondary plexus in the adenohypophysis. PVN = paraventricular nuclei; SON = supraoptic nuclei; OC = optic chiasm.

plexus', into which the hypothalamic hormones are released. These hormones then travel through the hypophyseal portal veins of the pars tuberalis to the secondary plexus, another series of capillaries in the adenohypophysis. Here, the hypothalamic hormones stimulate pituitary cells to release their hormones into the secondary plexus from which they enter the general circulation (Figure 3.3). As well as being released into the blood in the secondary plexus, some adenohypophyseal hormones such as prolactin are thought to be secreted into the cerebrospinal fluid (Lenhard and Deftos, 1982).

3.2 THE HORMONES OF THE PITUITARY GLAND (Table 3.1)

3.2.1 PARS NERVOSA

The two hormones of the pars nervosa, oxytocin and vasopressin, are really hypothalamic hormones. They are manufactured in the neurosecretory cells of the paraventricular (PVN) and supraoptic nuclei (SON) and

Table 3.1. *The hormones of the pituitary gland*

Pars Nervosa
Oxytocin: Stimulates uterine contractions and milk ejection from the mammary glands
Vasopressin = antidiuretic hormone (ADH): Elevates blood pressure and promotes reabsorption of water by the kidney
Neurophysins: Carrier proteins for oxytocin and vasopressin.

Pars distalis
Growth hormone (GH) = somatotropin = somatotropic hormone: Promotes protein synthesis and carbohydrate metabolism and growth of bone and muscle by stimulating somatomedins
Adrenocorticotropic hormone (ACTH): Stimulates glucocorticoid secretion from the adrenal cortex
Thyroid stimulating hormone (TSH) = thyrotropin: Stimulates thyroxine (T4) and triiodothyronine (T3) secretion from the thyroid gland
Prolactin (PRL): Initiates milk production and secretion in the mammary glands and has many other functions, including stimulation of the gonads

Gonadotropic hormones
Follicle stimulating hormone (FSH): Stimulates growth of the primary follicle and estrogen secretion from the ovary in females; sperm production and inhibin secretion in the testis of males
Luteinizing hormone (LH): Stimulates ovulation, formation of the corpora lutea and progesterone secretion in females; stimulates Leydig (interstitial) cells to secrete androgens in males

Pars intermedia (not a distinct gland in adult humans, but is present in the fetus)
Melanocyte stimulating hormone (MSH): Stimulates melanophores to darken skin color in amphibia. Some evidence for a similar effect in humans
β-Endorphin: Acts as a neuromodulator in the brain to regulate neurotransmitter release, and possibly as a circulating analgesic

transported through the infundibulum in the axons of these neurosecretory cells to the pars nervosa where they are stored in nerve terminals and then released into the inferior hypophyseal artery through which they enter the bloodstream (see Figure 3.3).

Oxytocin has two primary functions: it promotes uterine contractions at the time of birth and it stimulates milk ejection from the mammary glands during lactation. Oxytocin also has a number of neuropeptide functions in the brain (see Chapter 12). **Vasopressin** or antidiuretic hormone (ADH) acts to raise blood pressure and promote water reabsorption in the kidneys, i.e. it acts as an anti-diuretic. As a central neuropeptide, vasopressin may enhance memory (see Chapter 12). As well as these hormones, the pars nervosa releases two large proteins called **neurophysins** which function as carrier proteins for oxytocin and vasopressin (see Chapter 7). A detailed description of the neurohypophysis and its hormones is provided by Turner and Bagnara (1976), Hadley (1992) and Bennett and Whitehead (1983).

3.2.2 PARS DISTALIS

There are six hormones produced and released from the pars distalis (Table 3.1). This section provides a brief outline of the functions of these hormones. More detailed descriptions of the hormones of the pars distalis are given by Turner and Bagnara (1976), Hadley (1992), and Martin (1985).

Growth hormone (GH). Growth hormone is also known as somatotropin or somatotropic hormone. The suffix *-tropin* refers to a substance which has a stimulating effect on its target organ, thus somatotropin is body (soma) stimulating hormone. Growth hormone is produced in somatotroph cells of the adenohypophysis and promotes growth in almost all body cells: bone, muscle, brain, heart, etc. Growth hormone does not stimulate cell growth directly, but does so by stimulating **somatomedins**, peptide growth factors which mediate the growth-promoting actions of GH (see Spencer, 1991).

Adrenocorticotropic hormone (ACTH). ACTH is produced in the corticotroph cells of the adenohypophysis and acts to stimulate the synthesis and release of glucocorticoid hormones (cortisol, corticosterone, etc.) in the adrenal cortex.

Thyroid stimulating hormone (TSH). TSH, also known as thyrotropin or thyrotropic hormone, is produced in thyrotroph cells of the adenohypophysis. TSH stimulates the synthesis and release of thyroxine (T4) and triiodothyronine (T3) from the thyroid gland.

The gonadotropic hormones. The gonad stimulating or gonadotropic hormones, follicle stimulating hormone (FSH) and luteinizing hormone (LH) are produced in the gonadotroph cells of the adenohypophysis.

Follicle stimulating hormone (FSH). Follicle stimulating hormone has a similar function in both sexes: it promotes the development of the gametes and the secretion of gonadal hormones. In the female, FSH stimulates the growth of the primary follicle in the ovary, promoting development of the ova and the secretion of estrogen. In the male, FSH stimulates sperm production (spermatogenesis) and the secretion of the hormone inhibin by acting on the Sertoli cells of the testis.

Luteinizing hormone (LH). In the female, luteinizing hormone stimulates ovulation and the formation of the progesterone-secreting luteal cells (corpora lutea) in the ovary. In the male, luteinizing hormone stimulates the Leydig cells (also called interstitial cells) to secrete androgens such as testosterone.

Prolactin (PRL). Prolactin is produced in lactotroph (or mammotroph) cells in the adenohypophysis. Prolactin has also been called luteotropin, luteotropic hormone and lactogenic hormone, but these terms are rarely used now. Prolactin is essential for initiating milk synthesis in the mammary glands and also has many functions related to growth, osmoregulation, fat and carbohydrate metabolism, reproduction and parental behavior (Turner and Bagnara, 1976, pp. 104–10). In many of

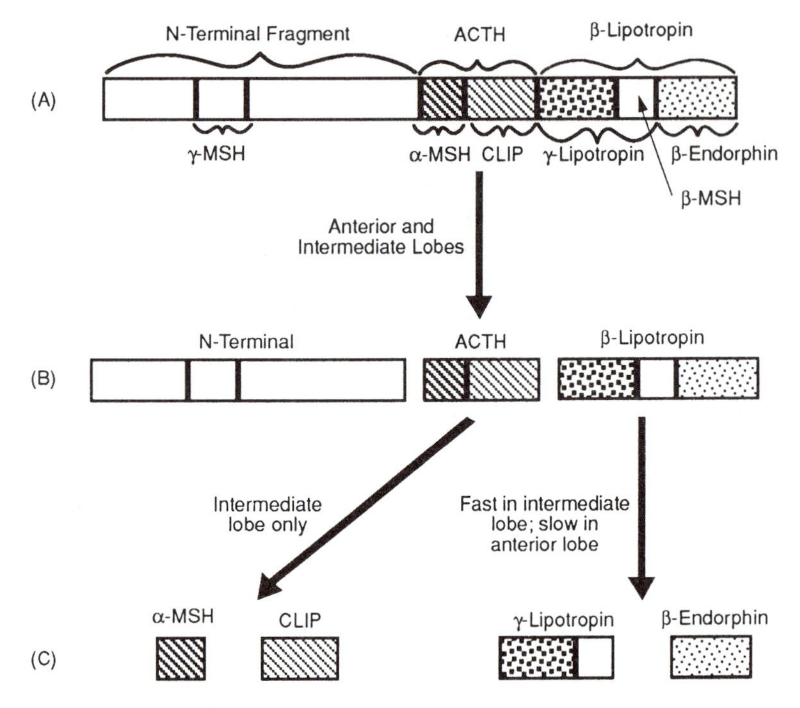

Figure 3.4. The proopiomelanocortin (POMC) molecule and its conversion to active hormones in the anterior and intermediate lobes of the pituitary gland. The dark vertical bars indicate places where enzymes split the prohormone into active peptides. The first cleavage (from A to B) occurs in both lobes of the adenohypophysis, resulting in the synthesis of ACTH and β-lipotropin, which are released from the anterior lobe. A second cleavage (from B to C) is necessary in the intermediate lobe to produce its final secretory products, α-MSH, β-endorphin and γ-lipotropin. MSH = melanocyte stimulating hormone; CLIP = corticotropin-like intermediate lobe peptide; ACTH = adrenocorticotropic hormone.

these actions, prolactin interacts with other hormones including estrogen, progesterone and oxytocin. Because prolactin can act on the gonads, it has also been classed as a 'gonadotropic hormone'.

3.2.3 PARS INTERMEDIA

Adult humans do not have a distinct pars intermedia, but it is well developed in fetal humans and in other mammals. The pars intermedia synthesizes the hormone melanotropin or **melanocyte stimulating hormone** (MSH) in the melanotroph cells. MSH acts on the melanophores of amphibia to change their skin color to match their background. MSH may have several forms (see Section 3.3) which are similar in structure to adrenocorticotropic hormone (ACTH). α-MSH functions as a neuropeptide to influence learning and memory (see Chapter 12). Details on the hormones of the pars intermedia are given by Turner and Bagnara (1976), Bennett and Whitehead (1983) and Hadley (1992).

3.3 THE ENDORPHINS

As well as the 'traditional' hormones discussed above, the adenohypophysis also manufactures an opioid peptide, β-endorphin. β-Endorphin, ACTH and MSH, are all derived from the same prohormone, proopiomelanocortin (POMC), as outlined in Figure 3.4. β-Endorphin has pronounced morphine-like activity.

POMC is a large polypeptide which is synthesized in the pars distalis and pars intermedia (as well as in the brain). It is broken down into active hormones by enzymes in the two lobes of the adenohypophysis. The POMC molecule contains the sequences for seven pituitary peptides (ACTH, α-MSH, β-MSH, γ-MSH, CLIP, β-lipotropin, and β-endorphin). The conversion of POMC into these active peptides occurs in two stages as shown in Figure 3.4. First, ACTH and β-lipotropin are cleaved off and this occurs in both the anterior and intermediate pituitary (Figure 3.4B). These are both secreted by the anterior lobe, but the intermediate lobe secretes no ACTH or β-lipotropin. All of the ACTH in the intermediate lobe is converted to α-MSH and CLIP (corticotropin-like intermediate lobe peptide), and all of the β-lipotropin is converted to β-endorphin and γ-lipotropin (Smith and Funder, 1988).

Thus, the anterior lobe of the pituitary secretes both ACTH and β-lipotropin from the corticotroph cells. The melanotroph cells of the intermediate lobe secrete α-MSH, β-MSH, CLIP, β-endorphin and γ-lipotropin (Figure 3.4C). β-Endorphin has a wide range of neuropeptide functions in analgesia, learning and memory, psychiatric diseases, feeding, thermoregulation, blood pressure regulation and reproductive behavior (Krieger 1986), which are discussed in detail in Chapter 12.

3.4 PITUITARY HORMONES IN THE BRAIN

Many of the peptide hormones synthesized in the pituitary gland are also produced as neuropeptides in nerves of the central nervous system, where they function as neuromodulators. The functions of these neuropeptides are discussed in Chapter 12.

3.5 SUMMARY

The pituitary gland consists of the anterior, intermediate and posterior lobes which are connected to the hypothalamus by the hypophyseal stalk. Axons from the PVN and SON of the hypothalamus project through the infundibulum to terminate in the posterior lobe, which consists of neural tissue. The anterior and intermediate lobes of the pituitary consist of endocrine tissue, which receive hormonal stimulation from the hypothalamus through the blood vessels of the hypophyseal portal system in the pars tuberalis of the infundibular stalk. The pituitary gland produces ten hormones: six from the pars distalis, two from the pars nervosa and one from the pars intermedia, while β-endorphin is secreted from the pars intermedia and the pars distalis. The hormones of the pars nervosa (oxytocin and vasopressin) are hypothalamic hormones which are stored and released from nerve terminals in the posterior pituitary. Some of the hormones of the adenohypophysis, including TSH, ACTH, LH and FSH stimulate the release of hormones from other endocrine glands, such as the thyroid, adrenal cortex and gonads. Other pituitary hormones (PRL, GH, MSH) act directly on non-endocrine target cells in the brain and body. The pituitary gland produces a number of other peptides, including the neurophysins from the pars nervosa. All of the pituitary hormones, including β-endorphin, function as central neuropeptides (neuromodulators) as well as hormones.

FURTHER READING

Bennett, G. W. and Whitehead, S. A. (1983). *Mammalian Neuroendocrinology*. New York: Oxford University Press.
Greenspan, F. S. (1991). *Basic and Clinical Endocrinology*. 3d edn. Norwalk, CN: Appleton and Lange.
Hadley, M. E. (1992). *Endocrinology*, 3rd edn. Englewood Cliffs, N J: Prentice-Hall.
Martin, C. R. (1985). *Endocrine Physiology*. Oxford: Oxford University Press.
Turner, C. D. and Bagnara, J. T. (1976). *General Endocrinology*, 6th edn. Philadelphia: Saunders.

REVIEW QUESTIONS

3.1 What is the hypophysis?
3.2 Which two pituitary hormones are neurohormones?
3.3 Describe the connections between the hypothalamus and the neurohypophysis.
3.4 Which pituitary hormones serve the following functions: (a) stimulating ovulation, (b) stimulating corticosteroid secretion, (c) stimulating milk secretion in the breast, (d) stimulating uterine contractions at childbirth.
3.5 Name the six hormones of the pars distalis.
3.6 What does 'tropic' mean?
3.7 Which pituitary hormones have the following functions: (a) cause skin color changes in amphibia, (b) stimulate T4 secretion?
3.8 Which two pituitary hormones are gonadotropins?
3.9 Give the Latin names of the three lobes of the pituitary gland.
3.10 Name the hormone released by the intermediate pituitary gland.
3.11 Which three adenohypophyseal hormones are synthesized from the prohormone POMC?

ESSAY QUESTIONS

3.1 Discuss the anatomy of the pituitary stalk and its importance for hypothalamic-pituitary connections.
3.2 How does it come about that the pituitary gland is made up of both endocrine and neural tissue? Discuss the embryological development of the pituitary gland.
3.3 Discuss the functions of oxytocin in male and female mammals.
3.4 Compare and contrast the functions of prolactin in humans, fish and birds.
3.5 Discuss the relationship between ACTH, MSH and β-endorphin.
3.6 Discuss the role of the somatomedins in mediating the growth-promoting functions of growth hormone.

REFERENCES

Bennett, G. W. and Whitehead, S. A. (1983). *Mammalian Neuroendocrinology*. New York: Oxford University Press.

Guillemin, R. and Burgus, R. (1972). The hormones of the hypothalamus, *Scientific American*, **227** (Nov), 24–33.

Hadley, M. E. (1992). *Endocrinology*, 2nd edn. Englewood Cliffs, NJ: Prentice-Hall.

Krieger, D. T. (1986). An overview of neuropeptides. In J. B. Martin and J. D. Barchas (eds.) *Neuropeptides in Neurologic and Psychiatric Disease*, pp. 1–32. New York: Raven Press.

Lenhard, L. and Deftos, L. J. (1982). Adenohypophyseal hormones in the CSF. *Neuroendocrinology*, **34**, 303–308.

Martin, C. R. (1985). *Endocrine Physiology*. Oxford: Oxford University Press.

Smith, A. I. and Funder, J. W. (1988) Proopiomelanocortin processing in the pituitary, central nervous system, and peripheral tissues. *Endocrine Reviews*, **9**, 159–179.

Spencer, E. M. (1991). Somatomedins. In F. S. Greenspan (ed.), *Basic and Clinical Endocrinology*, 3rd edn, pp. 133–146. Norwalk, CN: Appleton and Lange.

Turner, C. D. and Bagnara, J. T. (1976). *General Endocrinology*, 6th edn. Philadelphia: Saunders.

4

The hypothalamic hormones

Chapters 2 and 3 have surveyed the hormones of the endocrine glands and the pituitary gland. This chapter outlines the functions of the hypothalamus and the hypothalamic neurosecretory cells, and examines the role of the hypothalamus in controlling the release of pituitary hormones.

4.1 THE FUNCTIONS OF THE HYPOTHALAMUS

The hypothalamus is located at the base of the forebrain, below the thalamus (Figure 3.1), and is divided in two by the third ventricle, which is filled with cerebrospinal fluid (CSF). As shown in sagittal (sideways) section in Figure 4.1, the hypothalamus contains many groups of nerve cell bodies (nuclei). These nuclei are paired, one on either side of the third ventricle, as shown in coronal (frontal) sections in Figure 9.4 (p. 156). The medial basal hypothalamus, comprising the ventromedial nuclei (VMN), arcuate nuclei and the median eminence, is often referred to as 'the endocrine hypothalamus' because of its neuroendocrine functions. Details of the anatomy of the hypothalamus are given by Everett (1978); Zaborszky (1982); Bennet and Whitehead (1983).

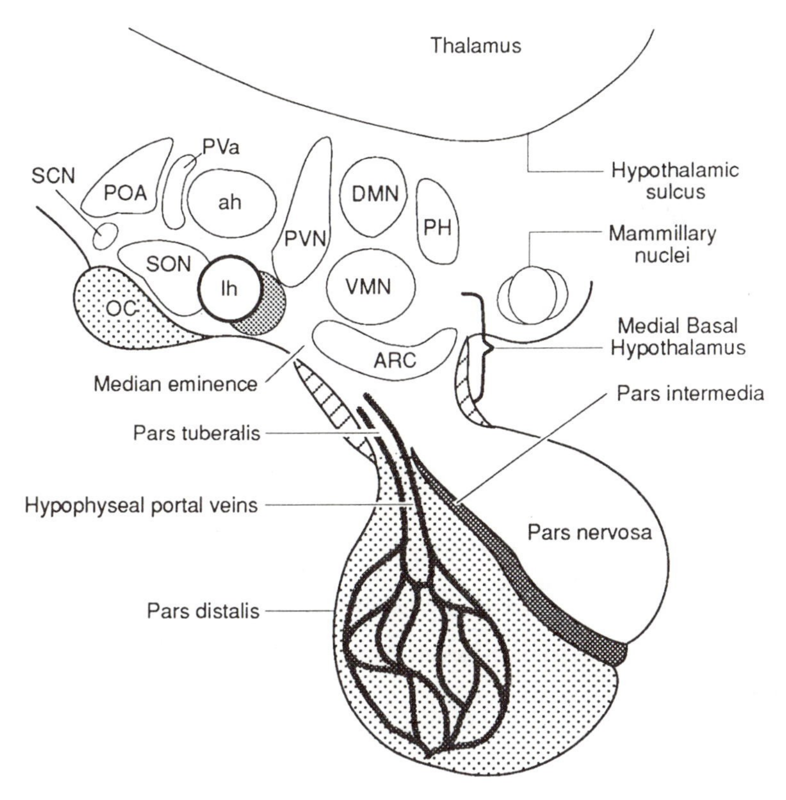

Figure 4.1. The hypothalamic nuclei. The hypothalamus is bounded anteriorly by the optic chiasm (OC), dorsally by the hypothalamic sulcus, which separates it from the thalamus, and caudally by the mammillary nuclei. The median eminence contains the primary plexus and the hypophyseal portal veins which go through the pars tuberalis to the adenohypophysis. ah = anterior hypothalamus; ARC = arcuate nucleus = DMN, dorsomedial nucleus; lh = lateral hypothalamus; PH = posterior hypothalamus; POA = preoptic area; PVa = anterior periventricular nucleus; PVN = paraventricular nucleus; SCN = suprachiasmatic nucleus; SON = supraoptic nucleus; VMN = ventromedial nucleus. (Redrawn from Martin, 1985.)

Afferent and efferent nerve fibers connect the hypothalamus to the cerebral cortex, thalamus, other parts of the limbic system (the hippocampus, amygdala and septum) and the spinal cord (Martin, 1985; Martin and Reichlin, 1987) and neurotransmitters released by neurons in these areas can regulate the cells of the hypothalamus. The hypothalamus is also well supplied with blood vessels and the hypothalamic nuclei are influenced by a wide variety of chemical messengers from both the blood and CSF as well as neurotransmitters from other neurons.

The hypothalamus has a multitude of functions which can be 'localized' to particular nuclei, although any boundaries, such as those drawn in Figure 4.1, are arbitrary. As well as synthesizing hormones, the hypothalamus: (a) regulates the sympathetic and parasympathetic branches of the autonomic nervous system which control visceral functions; (b) controls the temperature regulation mechanisms of the body; (c) contains a 'biological clock' which determines many biological rhythms; (d) regulates electrolyte balance; (e) controls emotional behavior (anger, fear, euphoria); and (f) mediates motivational arousal (hunger, thirst, aggression and sexual arousal). The nuclei associated with these functions are described in Table 4.1.

4.2 HYPOTHALAMIC NEUROSECRETORY CELLS

Neurosecretory cells are modified nerve cells which, rather than secreting a neurotransmitter, release a hormone into the circulation for neuroen-

Table 4.1. *Functions of the hypothalamic nuclei*

Preoptic area (POA) and anterior hypothalamus (AH)
Synthesis of LH-RH and TRH
Stimulates LH-RH and PRL surges
Coordinates parasympathetic nervous system functions
Temperature regulation; vasodilation responses to heat
Regulates male sexual behavior and female parental behavior

Suprachiasmatic nuclei (SCN)
Biological clock, regulates rhythmic release of glucocorticoids, melatonin and other hormones
Regulates sleep–wake and other body rhythms

Periventricular nuclei (PV)
Synthesis of CRH, TRH and SOM

Supraoptic nuclei (SON)
Synthesis of vasopressin (ADH), oxytocin and neurophysins from the magnocellular division
Regulation of thirst and drinking
Synthesis of CRH from the parvicellular division

Paraventricular nuclei (PVN)
Synthesis of oxytocin, vasopressin and neurophysins from the magnocellular division
Synthesis of TRH and CRH from the parvicellular division

Lateral hypothalamus (lh)
Control of hunger
Regulation of sodium balance

Dorsomedial nuclei (DMN)
Synthesis of TRH, CRH and somatostatin
Regulates autonomic nervous system activity
Control of aggression

Ventromedial nuclei (VMN)
Synthesis of GH-RH, somatostatin, CRH, PRF, and TRH
Regulates insulin and glucagon secretion
Controls digestive system functions
Detects blood glucose levels (glucoreceptors) and regulates food intake
Regulates female sexual behavior

Posterior hypothalamus (PH)
Temperature regulation – responses to cold
Regulates sympathetic nervous system and visceral functions
Regulates 'fight or flight' response
Influences sleep and arousal

Arcuate Nucleus (ARC) and median eminence
Synthesis of TRH, CRH and GH-RH
LH-RH nerve terminals (tonic release)
Dopamine released into portal veins from the tuberoinfundibular dopaminergic neurons

Other functions of the hypothalamus
Controls heart rate and blood pressure
Controls respiration
Controls 'emotions' – anger, fear, euphoria
Regulates calcium balance
Influences the immune system through the thymus gland
Influences the release of pancreatic and other gut hormones

Source: From Martin, 1985; Zaborszky, 1982.

docrine communication (Figure 1.4). There are two groups of hypothalamic neurosecretory cells: the magnocellular and parvicellular systems.

- *The magnocellular system.* The large magnocellular neurosecretory cells are located in the paraventricular (PVN) and supraoptic nuclei (SON). The paraventricular nucleus consists of two cell types, one produces the hormone oxytocin and the other vasopressin (antidiuretic hormone). Similarly, the supraoptic nucleus (SON) produces both oxytocin and vasopressin. These hormones are released from the nerve terminals of the axons of the magnocellular neurosecretory cells in the pars nervosa (neurohypophysis), as described in Chapter 3. Details of the organization of the magnocellular neurosecretory system are given by Silverman and Zimmerman (1983) and Swanson and Sawchenko (1983).
- *The parvicellular system.* The smaller parvicellular neurosecretory cells are found in the preoptic area, VMN, and arcuate nucleus, as well as various other hypothalamic areas and project to the median eminence. As well as having magnocellular neurosecretory cells, the PVN and SON also have a number of the smaller parvicellular neurosecretory cells. The parvicellular neurosecretory cells terminating at the median eminence release their hypothalamic hormones into the primary plexus, from which they enter the hypophyseal portal veins. These hormones modulate the release of adenohypophyseal hormones and are thus referred to as the 'hypophysiotropic hormones'.

Following considerable controversy about the concept of neurosecretion between 1934 and 1955, Geoffry Harris postulated that the hypothalamus controlled the release of adenohypophyseal hormones by the release of neurohormones into the hypophyseal portal veins. A historical review of this research is given by Harris (1972). After Harris's death, Guillemin and Schally were awarded the 1977 Nobel Prize for Physiology and Medicine for isolating several of the hypothalamic releasing hormones and proving that Harris's theory was correct. Harris (1972) set three criteria for the definition of hypothalamic hypophysiotropic hormones: (a) the hormone is present in the median eminence of the hypothalamus; (b) it is present in higher levels in the hypophyseal portal blood than in the rest of the circulatory system; and, (c) the level of the hormone in the hypophyseal portal blood is correlated with the secretory rate of particular adenohypophyseal hormones (see Sarkar, 1983).

In a neurosecretory cell (Figure 4.2) the neurohormones are synthesized as prohormones and packaged into neurosecretory granules in the cell body. The granules are then transported down the nerve axon and stored at the nerve terminal. When the cell is stimulated, the action potential depolarizes the nerve terminal (see Chapter 5) and releases the neurohormone into the circulation, cerebrospinal fluid, or into a synapse (Brownstein, Russell and Gainer, 1980; Bennett and Whitehead, 1983). The release of hypothalamic neurohormones is stimulated by neurotransmitters from other nerve cells (Chapter 6) and regulated by feedback from hormones, neuropeptides, and other chemical messengers (see Chapters 8 and 12).

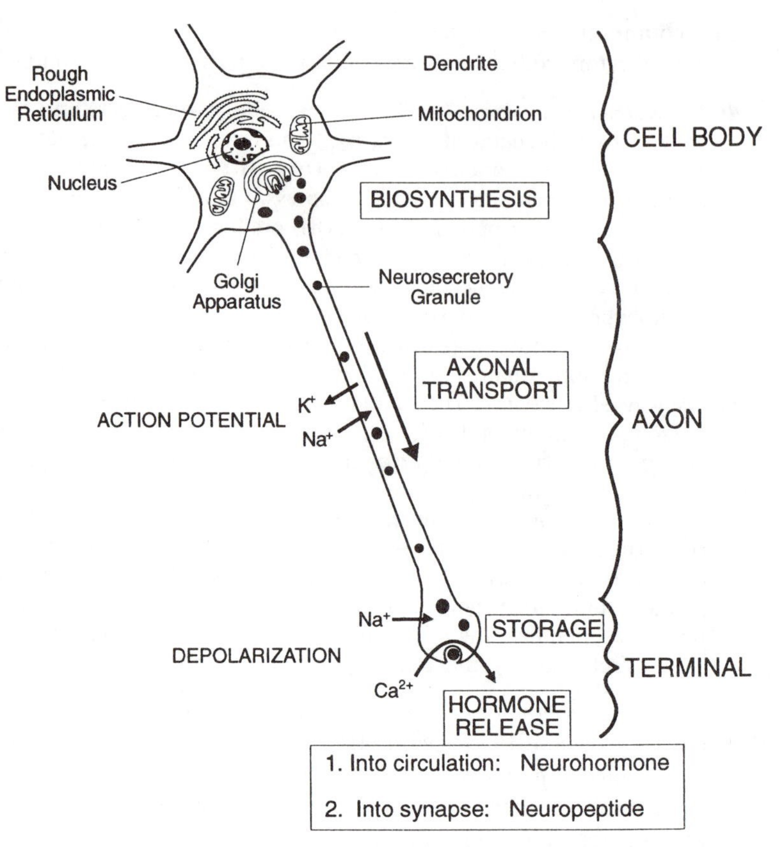

Figure 4.2. The components of a neurosecretory cell indicating sites of neurohormone biosynthesis (usually as a propeptide), axonal transport, storage and release. Neurosecretory cells can synthesize hormones or neuropeptides and release them into the circulation, CSF or into a synapse. (Redrawn from Bennett and Whitehead, 1983.)

4.3 THE NEUROENDOCRINE TRANSDUCER CONCEPT

Because hypothalamic neurosecretory cells are stimulated by neurotransmitters (e.g. dopamine, noradrenaline and serotonin) to release their hormones, they are able to convert neural information to hormonal output and have thus been termed 'neuroendocrine transducers' (Wurtman and Anton-Tay, 1969). The neurosecretory cells thus provide a mechanism for bringing the endocrine system under the influence of the nervous system and, therefore, under the influence of external and internal stimuli. Light, taste, sound or touch stimuli, which reach the brain via sensory nerves, stimulate neurotransmitter release, which can alter the secretion of hypothalamic neurohormones, resulting in hormonal responses to environmental changes (as outlined in Figure 1.1, p. 3). Similarly, psychological states such as fear, anger, sexual arousal, happiness and depression, which alter neurotransmitter levels, can also influence hormone secretion. Finally, external chemicals such as drugs, hormones or even nutrients in food which alter neurotransmitter levels will also alter hormone levels (Chapter 5).

A neuroendocrine transducer is, therefore, a modified nerve cell, with neurotransmitter input and neurohormone output. The body has four different neuroendocrine transducers: (a) the magnocellular neurosecretory cells of the SON and PVN which synthesize and release oxytocin and

Table 4.2. *Hypothalamic hypophysiotropic hormones*

Releasing hormones
Thyrotropin releasing hormone (TRH) = TSH-RH
Corticotropin releasing hormone (CRH)
Gonadotropin releasing hormone (GnRH) = LH-RH (possibly also FSH-RH)
Paired releasing and inhibiting hormones
Growth hormone releasing hormone (GH-RH)
Growth hormone release inhibiting hormone (GH-RIH) = somatostatin
Prolactin releasing factor (PRF)
Prolactin release inhibiting factor (PIF) (probably dopamine)
Melanocyte stimulating hormone releasing factor (MSH-RF)
Melanocyte stimulating Hormone release inhibiting factor (MSH-RIF) (probably dopamine)

vasopressin; (b) the parvicellular hypothalamic neurosecretory cells, which secrete the hypophysiotropic hormones into the primary plexus in the median eminence; (c) the adrenal medulla, which is stimulated by sympathetic nerves to secrete adrenaline and noradrenaline into the bloodstream; and (d) the pineal gland, which is stimulated by adrenergic nerves to release melatonin into the bloodstream.

4.4 THE HYPOTHALAMIC HYPOPHYSIOTROPIC HORMONES

There are nine hypopthalamic hypophysiotropic hormones which regulate the release of the adenohypophyseal hormones (Table 4.2). Three adenohypophyseal hormones (PRL, GH, and MSH) are controlled by paired hypothalamic hormones, one stimulatory (releasing) and one inhibitory (release inhibiting). The other four (TSH, ACTH, LH and FSH) are regulated only by hypothalamic releasing hormones. The chemical structure and mode of action of the hypothalamic hypophysiotropic hormones are reviewed by Schally (1978) and Schally, Coy and Meyers (1978).

Hypothalamic hormones are often called factors rather than hormones. There is no hard and fast rule for this, but the term 'factor' usually applies to a hypothalamic substance whose chemical structure is unknown. Hypothalamic substances of known chemical structure are called hormones (Schally, 1978; Schally *et al.*, 1978). As of 1987, five hypothalamic hormones (TRH, CRH, GnRH, GH-RH and somatostatin) were chemically identified (Martin and Reichlin, 1987). As yet, the hypothalamic hormones regulating melanocyte stimulating hormone (MSH-RF and MSH-RIF) and prolactin (PRF and PIF) have not been identified, although there is good evidence that PIF is dopamine.

The hypothalamic hormones are secreted from neuroendocrine cells in a number of different hypothalamic nuclei (Table 4.3); thus the secretion of a single hypothalamic hormone may be regulated through a number of different neural pathways (see Chapter 6). Details of the hypothalamic

Table 4.3. *Location of the hypothalamic neurosecretory cells which synthesize the hypothalamic hormones*

The magnocellular neurosecretory cells which synthesize the hormones released from the pars nervosa	
Oxytocin	Paraventricular (PVN) and supraoptic (SON) nuclei
Vasopressin (ADH)	Paraventricular (PVN) and supraoptic (SON) nuclei
The parvicellular neurosecretory cells which synthesize the hypothalamic hypophyseal hormones	
TRH (TSH-RH)	Primarily in the PVN and PV (also in the POA-AH, DMN, VMN and ARC)
CRH	Primarily in the PVa and PVN (also in the DMN, VMN, SON and ARC)
GnRH (LH-RH)	POA-AH, SCN and MBH (in rodents) MBH-ARC (in primates)
FSH-RH (?)	Dorsal AH (?)
GH-RH	Primarily in the MBH-ARC (also in the VMN)
SOM	Primarily in the PVa (also from the DMN, VMN, POA-AH, PVN and ARC)
PRF	PVN and POA-AH (= oxytocin, TRH or VIP?)
PIF (dopamine)	ARC (tuberoinfundibular dopaminergic neurons)
MSH-RF	PVN ?
MSH-RIF (dopamine)	ARC (tuberoinfundibular dopaminergic neurons)

Note:
Abbreviations are as in Tables 4.1 and 4.2.

hormones and their sites of secretion are given by Müller and Nistico (1989).

4.4.1 THYROTROPIN RELEASING HORMONE (TRH)

TRH is also known as thyroid stimulating hormone releasing hormone (TSH-RH). TRH stimulates the thyrotroph cells of the anterior pituitary to produce and release TSH. TRH also acts as a neuromodulator in the brain. TRH is synthesized primarily in the paraventricular nucleus (PVN) and the anterior periventricular nuclei (PVa). Several other nuclei of the hypothalamus, including the preoptic-anterior hypothalamic area, the dorsomedial, ventromedial, and suprachiasmatic nuclei and arcuate nuclei produce smaller amounts of TRH (Zaborszky, 1982; Müller and Nistico, 1989). TRH secretion is regulated by catecholaminergic neurotransmitters as well as neuropeptides such as somatostatin and the opioids (Bennett and Whitehead, 1983; Martin, 1985). Environmental factors which stimulate the release of TRH include acute cold exposure and stress (Martin and Reichlin, 1987).

4.4.2 CORTICOTROPIN RELEASING HORMONE (CRH)

CRH stimulates the release of ACTH from the corticotroph cells of the anterior pituitary and acts as a neuromodulator in the brain. CRH secretion is regulated by a number of neurotransmitters and neuropeptides, including acetylcholine, serotonin, histamine and the opioids (Martin, 1985; Rivier and Plotsky, 1986). CRH is synthesized primarily in the paraventricular nucleus (PVN) and the anterior periventricular nuclei (PVa) of the hypothalamus. Several other nuclei of the hypothalamus, including the supraoptic (SON), dorsomedial (DMN), and ventromedial nuclei (VMN) also synthesize CRH, which is released from the axon terminals in the median eminence in a distinct day–night rhythm and in response to pain or stress (Bennett and Whitehead, 1983; Martin and Reichlin, 1987).

4.4.3 GONADOTROPIN RELEASING HORMONE (GnRH)

The release of the gonadotropins FSH and LH from the gonadotroph cells of the adenohypophysis is regulated by GnRH. GnRH also functions as a neuromodulator in the brain (Krieger, 1983, 1986). There has been considerable controversy over whether there is one GnRH that stimulates both LH and FSH or whether there is a separate, as yet unidentified, FSH releasing factor. Some researchers believe that there are two different gonadotropin releasing hormones, LH-RH and FSH-RH, while others argue that there is only one GnRH (LH-RH).

- *The one GnRH theory.* The one GnRH theory suggests that the pulsatile release of GnRH can stimulate the secretion of both LH and FSH. There are a number of different types of evidence for this theory. First, LH-RH stimulates the release of both LH and FSH in rats, chimpanzees, humans and a number of other animals. Second, LH-RH stimulates the simultaneous release of LH and FSH from *in vitro* preparations of rat pituitary glands. Third, inactivation of LH-RH using antagonists or antiserum is accompanied by an inhibition of FSH release. Fourth, the amino acid sequence of LH-RH has been determined, allowing for the production of synthetic LH-RH and the same fraction of this synthetic LH-RH releases LH and FSH in both *in vivo* and *in vitro* preparations (Schally *et al.*, 1971; Schally, 1978). If there is only one GnRH, how is the differential stimulation of LH and FSH release effected? Different patterns of LH and FSH release in response to a single GnRH may arise in a number of ways, four of which are mentioned here. First, different groups of pituitary gonadotroph cells may secrete varying ratios of LH and FSH in response to GnRH stimulation. Second, differences in LH and FSH release patterns may be influenced by changes in the frequency or amplitude of GnRH secretion. Third, positive and negative feedback from gonadal steroids and inhibin may exert different effects on the release of LH and FSH from pituitary gonadotrophs (Ory, 1983; see further discussion in Martin, 1985, pp. 593–5). Fourth, there may be a specific FSH releasing protein secreted from the gonads rather than the hypothalamus, which stimulates FSH release from the gonadotrophs (Vale *et al.*, 1986).
- *The two GnRH theory.* Whereas FSH release is often correlated with LH release, particularly in humans, FSH release does occur in the absence of LH release, and vice versa, suggesting a separate FSH releasing hormone (Levine and Duffy, 1988). There are four lines of evidence to suggest that this is a hypothalamic hormone distinct from LH-RH. (a) Electrical stimulation of the dorsal anterior hypothalamus evokes FSH

release while stimulation of the preoptic area evokes LH release. (b) Lesioning the dorsal anterior hypothalamus interferes with FSH secretion but not with LH secretion in response to gonadal steroids. Lesions of the preoptic area, on the other hand, suppress LH responses to gonadal steroids but not FSH responses. (c) Neutralization of LH-RH by injecting GnRH antiserum or GnRH antagonists abolishes LH pulses but not FSH pulses and inhibition of LH-RH pulses using catecholaminergic drugs inhibits LH pulses but not FSH pulses, although the remaining FSH pulses are of lower amplitude. (d) FSH release is stimulated by injections of hypothalamic extracts having no detectable LH-RH activity (McCann *et al.*, 1983; McCann and Rettori, 1987).

Thus, even though there is evidence that one GnRH controls the release of both LH and FSH, particularly in humans, there is considerable evidence for the existence of a separate FSH releasing hormone in many laboratory animals (Chappel, 1985; Clarke, 1987). This FSH-RH may have a chemical structure very similar to that of LH-RH and be synthesized within the neurons of the dorsal anterior hypothalamus, while LH-RH is synthesized in the preoptic area (McCann and Rettori, 1987). Final proof awaits the isolation and identification of the putative FSH-RH.

Tonic (basal) versus cyclic (pulsatile) GnRH secretion

In rodents, LH-RH (GnRH) is synthesized in neurons of the preoptic area-anterior hypothalamus (POA-AH) (Merchantaler *et al.*, 1989), whereas in primates the cell bodies are located in the medial basal hypothalamus (MBH) (Knobil, 1980). These neurosecretory cells regulate the tonic or basal secretion of LH and FSH and are in turn modulated by negative feedback from gonadal steroids (Chapter 7). In the female rat, the LH-RH secreting cells of the preoptic anterior-hypothalamic area also stimulate the pre-ovulatory surge of LH in response to positive feedback from ovarian estrogen (Sharp and Fraser, 1978; Zaborszky, 1982). In primates, positive feedback from estrogen acts on the LH-RH neurosecretory cells of the MBH to stimulate the pre-ovulatory LH surge (Müller and Nistico, 1989). In male rodents and primates, LH-RH is released in tonic pulses (see Figure 8.10), but there are no LH surges in males due to the lack of positive feedback from estrogen.

The pulsatile nature of GnRH release is essential for the maintenance of LH secretion. Non-pulsatile, continuous high doses of GnRH have, paradoxically, the opposite effects of pulsatile GnRH administration: they inhibit LH release and inhibit gonadal functions (Knobil, 1980; Crowley *et al.*, 1985). Thus, synthetic GnRH agonists can be used as antifertility drugs (Brodie and Crowley, 1984).

Regulation of GnRH release

GnRH release is regulated by a plethora of neurotransmitters and neuropeptides, many of which interact with each other and the gonadal steroids to modulate GnRH release (Weiner, Findell and Kordon, 1988). This multiple control mechanism allows a wide variety of external and internal stimuli to influence the neural control of GnRH release. For example, GnRH release is modulated by: neurons projecting from the suprachiasmatic nucleus of the hypothalamus, which regulates circadian rhythms; neurons from the paraventricular nucleus, which process visceral afferent input and may activate stress-induced changes in GnRH release; neural input from the olfactory and vomeronasal pathways, through which pheromones can influence GnRH release (Sharp and Fraser, 1978; Zaborszky, 1982).

Extra-hypothalamic GnRH release

Some of the LH-RH neurosecretory cells in the anterior hypothalamus do not release LH-RH into the hypophyseal portal system, but send their axons to other brain areas, particularly the limbic system (Merchenthaler *et al.*, 1989). As well as the neural projections of these hypothalamic GnRH secreting cells, there are a number of GnRH releasing neurons in other regions of the brain. These extra-hypothalamic LH-RH neurons occur in the accessory olfactory bulb, medial olfactory tract, septum, bed nucleus of the stria terminalis and the corpus callosum (Witkin, Paden and Silverman, 1982). The high concentration of LH-RH neurons associated with the olfactory pathways explains why pheromones have such potent 'primer effects' on the neuroendocrine system. The functions of GnRH as a neuromodulator in the brain and CNS are discussed in Chapter 12.

4.4.4 GROWTH HORMONE RELEASING AND INHIBITING HORMONES

Growth hormone secretion from the somatotroph cells of the adenohypophysis is regulated by growth hormone releasing hormone (GH-RH) or somatocrinin (Guillemin *et al.*, 1984) and growth hormone release inhibiting hormone (GH-RIH) which is also called somatostatin (SOM). Because of the similarity of their names, somatostatin (GH-RIH) may be confused with somatotropin (GH). GH-RH is released in bursts from neurosecretory cells in the medial basal (ventromedial nucleus) region of the hypothalamus and the arcuate nucleus. These neurosecretory cells are regulated by catecholaminergic and serotonergic neurotransmitters as well as by a number of neuropeptides, including the opioids and TRH (Bennett and Whitehead, 1983; Martin, 1985).

Somatostatin is synthesized primarily from neurosecretory cells of the periventricular nuclei and the preoptic-anterior hypothalamus, which send axons to the hypophyseal portal system in the median eminence (Zaborszky, 1982). Somatostatin is also released from neurosecretory cells of the ventromedial and dorsomedial hypothalamus, which send their axons to other areas of the brain where SOM is released as a neuromodulator (Krieger, 1983, 1986; Merchenthaler *et al.*, 1989). Somatostatin release is regulated by a number of neurotransmitters and neuropeptides (see Chapter 6).

4.4.5 PROLACTIN RELEASING AND INHIBITING FACTORS

Prolactin secretion from the lactotroph cells of the adenohypophysis is stimulated by prolactin releasing factor (PRF) and inhibited by prolactin inhibiting factor (PIF) (Ben-Jonathan, Arbogast and Hyde, 1989). A specific prolactin releasing factor has not yet been identified, but many neuropeptides have been shown to elevate prolactin release, including TRH, vasoactive intestinal peptide (VIP), oxytocin and β-endorphin (Leong, Frawley and Neill, 1983; Ben-Jonathan *et al.*, 1989).

Likewise, a specific prolactin inhibiting factor has not yet been identified, although there are a number of candidates. The neurotransmitter dopamine, which is released from the tuberoinfundibular dopaminergic neurons of the arcuate nucleus-median eminence region into the hypophyseal portal veins (Chapter 5) is a major prolactin inhibiting factor (Ben-Jonathan *et al.*, 1989). The inhibitory neurotransmitter GABA also acts as a prolactin inhibitory factor (Leong *et al.*, 1983; Ben-Jonathan *et al.*, 1989).

Sex differences in the control of prolactin secretion

The hypothalamic control of prolactin, like that of LH, differs in males and females. In the male, PRL is released in a tonic, acyclic pattern while in the female, PRL release is cyclic, with periodic surges (Neill, 1972). These sex differences in prolactin and LH secretion appear to be the result of sex differences in the organization of the medial preoptic, dorsomedial and ventromedial areas of the hypothalamus due to 'maculinization' of the brain by androgens during prenatal development (Gunnett and Freeman, 1982).

4.4.6 MELANOCYTE STIMULATING HORMONE RELEASING AND INHIBITING FACTORS

Melanocyte stimulating hormone is released from the melanotroph cells of the pars intermedia by melanocyte stimulating hormone releasing factor (MSH-RF) and inhibited by melanocyte stimulating hormone release inhibiting factor (MSH-RIF). Neither MSH-RF nor MSH-RIF have been identified, but there are a number of hormones which may influence the release of MSH. Part of the oxytocin molecule (the C-terminal fragment) may act as an MSH releasing factor, while dopamine acts as an MSH inhibiting factor, similar to its action in inhibiting PRL (Schally *et al.*, 1978; Taleisnik, 1978).

4.5 COMPLEXITIES OF HYPOTHALAMIC–PITUITARY INTERACTIONS

The relationship between the endocrine hypothalamus and the pituitary gland involves many complexities, five of which are mentioned here.

1. Hypothalamic hormones do not always have a one-to-one relationship with the pituitary hormones. While many hypothalamic hormones influence only one pituitary hormone, TRH stimulates both prolactin and TSH release and somatostatin inhibits the release of TSH as well as GH (Figure 4.3). Dopamine inhibits both PRL and MSH secretion and GnRH stimulates the secretion of both LH and FSH.
2. Pituitary hormones may be transported back to the hypothalamus to modify neural activity by acting as neuromodulators. Figure 3.3

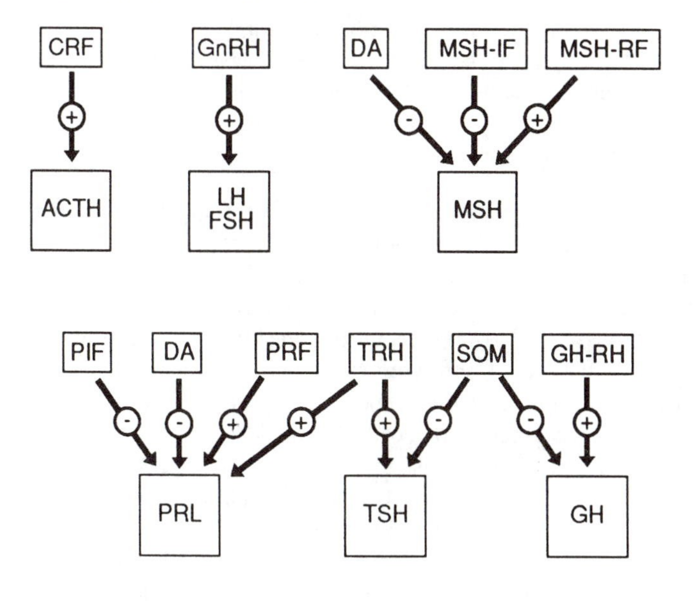

Figure 4.3. Hypothalamic control of the adenohypophyseal hormones. CRF = corticotrophin releasing hormone; DA = dopamine; FSH = follicle stimulating hormone; GnRH = gonadotropin releasing hormone; GH-RH = growth hormone releasing hormone; MSH = melanocyte stimulating hormone (−IF = inhibiting factor and −RF = releasing factor); PIF = prolactin release inhibiting factor; PRF = prolactin releasing factor; PRL = prolactin; SOM = somatostatin; TRH = thyrotropin releasing hormone; TSH = thyroid stimulating hormone. + indicates a stimulating (releasing) effect; − indicates an inhibiting effect. (From Bennett and Whitehead, 1983.)

described the hypophyseal stalk as carrying hormones in only one direction, from the hypothalamus to the pituitary, but there are three mechanisms by which pituitary hormones may be transported back to the hypothalamus. Neurohypophyseal hormones may be transported from the posterior pituitary to the hypothalamic nuclei by retrograde axonal transport (see Figure 5.3) and adenohypophyseal hormones may be carried to the hypothalamus by efferent portal vessels (Berglund and Page, 1979). Pituitary hormones may also be released into the CSF and stimulate hypothalamic nuclei around the third ventricle. β-Endorphin and ACTH, for example, are secreted directly into the third ventricle while other adenohypophyseal hormones enter the CSF from the blood through capillaries of the choroid plexus in the ventricles (Lenhard and Deftos, 1982).

3. Not all hypothalamic hormones are secreted into the portal system. Several of the 'hypothalamic hormones' are also secreted from brain cells in regions other than the hypothalamus. This is particularly true for GnRH (see p. 49), TRH and somatostatin, whose neurosecretory cells send axons to various regions of the limbic system (Bennett and Whitehead, 1983; Merchenthaler *et al.*, 1989). When secreted to other brain regions, the hypothalamic hormones act as neuropeptides to modulate neural excitability, and thus influence neurotransmitter release and behavior (Moss, 1979).
4. The pituitary hormones are regulated by a number of neuropeptides and neurotransmitters as well as by the hypothalamic hypophysiotropic hormones. Dopamine and GABA may be released into the hypophyseal portal veins to act directly on pituitary endocrine cells. Likewise, oxytocin, VIP, enkephalins, substance P and other neuropeptides also regulate the release of adenohypophyseal hormones (Bennett and Whitehead, 1983; Weiner *et al.*, 1988; Ben-Jonathan *et al.*, 1989).
5. The hypothalamic hormones interact with other hormones, particularly the gonadal steroids, in controlling the release of pituitary hormones. Estrogen and progesterone regulate the release of prolactin, LH and FSH by their direct effects on the lactotroph and gonadotroph

cells of the anterior pituitary. Likewise, inhibin regulates the release of FSH through its action on pituitary gonadotrophs (Knobil, 1980; McCann and Rettori, 1987; Ben-Jonathan *et al.*, 1989).

4.6 SUMMARY

This chapter examines the functions of the hypothalamus with particular reference to the hypothalamic control of pituitary hormones. The hypothalamus contains nuclei which control the autonomic nervous system, temperature regulation, biological rhythms, emotional responses, motivational arousal and hormone secretion. The hypothalamic hormones are synthesized in neurosecretory (neuroendocrine) cells which act as neuroendocrine transducers, converting neural input to hormonal output. The magnocellular neurosecretory cells of the SON and PVN produce the neurohypophyseal hormones oxytocin and vasopressin which are released from the pars nervosa. The parvicellular neurosecretory cells of the hypothalamus release hypophysiotropic hormones into the hypophyseal portal veins in the median eminence. Three hypothalamic hormones act individually to stimulate the release of pituitary hormones (TRH, CRH and GnRH(LH-RH)), while six act in pairs to release or inhibit pituitary hormones (GH-RH and GH-RIH(SOM), PRF and PIF, and MSH-RF and MSH-RIF). Considerable controversy exists over whether there is a FSH-RH separate from LH-RH and there are a number of complications in the hypothalamic control of pituitary hormones, including sex differences, multiple effects of hypothalamic hormones on the pituitary, retrograde transport of pituitary hormones to the hypothalamus, and the stimulation of pituitary hormones by neuropeptides and neurotransmitters acting directly on the endocrine cells of the pituitary.

FURTHER READING

Bennett, G. W. and Whitehead, S. A. (1983). *Mammalian Neuroendocrinology*. New York: Oxford University Press.

Martin, C. R. (1985). *Endocrine Physiology*. Oxford: Oxford University Press.

Müller, E. E. and Nistico, G. (1989). *Brain messengers and the pituitary*. San Diego: Academic Press.

Schally, A. V., Coy, D. H. and Meyers, C. A. (1978). Hypothalamic regulatory hormones. *Annual Review of Biochemistry*, **47**, 89–128.

REVIEW QUESTIONS

4.1 Name the three areas of the medial basal hypothalamus which are referred to as the 'endocrine hypothalamus'.

4.2 What are the two types of neurosecretory cell in the hypothalamus and what are their functions?

4.3 What is a neuroendocrine transducer?

4.4 Name the nine hypothalamic hypophysiotropic hormones and describe how they influence the release of the adenohypophyseal hormones.

4.5 How do the hormones of the hypothalamus reach the anterior pituitary?

4.6 What is the difference between tonic and cyclic GnRH secretion?
4.7 What two neurotransmitters act as prolactin inhibiting factors?
4.8 Which two pituitary hormones do each of the following hypothalamic hormones affect, and are these effects stimulatory or inhibitory (a) TRH, (b) somatostatin, (c) GnRH?

ESSAY QUESTIONS

4.1 Discuss the functional anatomy of the nuclei of the hypothalamus, with particular emphasis on describing 'the endocrine hypothalamus'.
4.2 Discuss the neuroanatomy and neural connections of the magnocellular neurosecretory cells.
4.3 Outline the history of the discovery of the hypothalamic hormones from 1935 to the present.
4.4 Discuss the mechanism of the neuroendocrine transducer using one of the four different neuroendocrine transducers as an example.
4.5 Discuss the evidence for the existence of one versus two GnRHs. Which theory do you believe?
4.6 Discuss the sex differences in the secretion of prolactin and the gonadtropins.
4.7 Discuss the extra-hypothalamic sources of GnRH and their possible functions.
4.8 Discuss some of the problems in trying to understand the hypothalamic control of the anterior pituitary gland.
4.9 Imagine that you have discovered a new hypothalamic hormone which was the mysterious prolactin releasing factor, and you named it prolactinotropin. How would you prove that this hormone was what you said it was? Describe the experiments necessary to demonstrate that you had really discovered a new hypothalamic hypophysiotropic hormone.

REFERENCES

Ben-Jonathan, N., Arbogast, L. A. and Hyde, J. F. (1989). Neuroendocrine regulation of prolactin release. *Progress in Neurobiology*, **33**, 399–447.

Bennett, G. W. and Whitehead, S. A. (1983). *Mammalian Neuroendocrinology*. New York: Oxford University Press.

Bergland, R. M. and Page, R. B. (1979). Pituitary–brain vascular relations: a new paradigm. *Science*, **204**, 18–24.

Brodie, T. D. and Crowley, W. F., Jr (1984). Neuroendocrine control of reproduction and its manipulation with LHRH and its analogs. *Trends in Neuroscience*, **7**, 340–342.

Brownstein, M. J., Russell, J. T. and Gainer, H. (1980). Synthesis, transport and release of posterior pituitary hormones. *Science*, **207**, 373–378.

Chappel, S. C. (1985). Neuroendocrine regulation of luteinizing hormone and follicle stimulating hormone: a review. *Life Sciences*, **36**, 97–103.

Clarke, I. J. (1987). New concepts in gonadotropin-releasing hormone action on the pituitary gland. *Seminars in Reproductive Endocrinology*, **5**, 345–352.

Crowley, W. F., Jr, Filicori, M., Spratt, D. I. and Santoro, N. F. (1985). The physiology of gonadotropin-releasing hormone (GnRH) secretion in men and women. *Recent Progress in Hormone Research*, **41**, 473–531.

Everett, J. W. (1978). The mammalian hypothalamo-hypophysial system. In S. L. Jeffcoate and J. S. M. Hutchinson (eds.), *The Endocrine Hypothalamus*, pp. 1–34. London: Academic Press.

Guilleman, R., Brazeau, P., Bohlen, P, *et al.* (1984). Somatocrinin, the growth hormone releasing factor. *Recent Progress in Hormone Research*, **40**, 233–299.

Gunnet, J. W. and Freeman, M. E. (1982). Sexual differences in regulation of prolactin secretion by two hypothalamic areas. *Endocrinology*, **110**, 697–702.

Harris, G. W. (1972). Humours and hormones. *Journal of Endocrinology*, **53**, ii–xxiii.

Knobil, E. (1980). The neuroendocrine control of the menstrual cycle. *Recent Progress in Hormone Research*, **36**, 53–88.

Kreiger, D. T. (1983). Brain peptides: what, where, and why? *Science*, **222**, 975–985.

Kreiger, D. T. (1986). An overview of neuropeptides. In J. B. Martin and J. D. Barchas (eds.), *Neuropeptides in Neurologic and Psychiatric Disease*, pp. 1–32. New York: Raven Press.

Lenhard, L. and Deftos, L. J. (1982). Adenohypophyseal hormones in the CSF. *Neuroendocrinology*, **34**, 303–308.

Leong, D. A., Frawley, L. S. and Neill, J. D. (1983). Neuroendocrine control of prolactin secretion. *Annual Review of Physiology*, **45**, 109–127.

Levine, J. E. and Duffy, M. T. (1988). Simultaneous measurement of luteinizing hormone (LH)-releasing hormone, LH, and follicle-stimulating hormone release in intact and short-term castrate rats. *Endocrinology*, **122**, 2211–2221.

Martin, C. R. (1985). *Endocrine Physiology*. Oxford: Oxford University Press.

Martin, J. B. and Reichlin, S. (1987). *Clinical Neuroendocrinology*, 2nd edn. Philadelphia: F. A. Davis.

McCann, S. M., Mizunuma, H., Samson, W. K. and Lumpkin, M. D. (1983). Differential hypothalamic control of FSH secretion: a review. *Psychoneuroendocrinology*, **8**, 299–308.

McCann, S. M. and Rettori, V. (1987). Physiology of luteinizing hormone-releasing hormone. *Seminars in Reproductive Endocrinology*, **5**, 333–343.

Merchenthaler, I., Setalo, G., Csontos, C., Petrusz, P., Flerko, B. and Negro-Vilar, A. (1989). Combined retrograde tracing and immunocytochemical identification of luteinizing hormone-releasing hormone- and somatostatin-containing neurons projecting to the median eminence of the rat. *Endocrinology*, **125**, 2812–2821.

Moss, R. L. (1979). Actions of hypothalamic-hypophysiotropic hormones on the brain. *Annual Review of Physiology*, **41**, 617–631.

Müller, E. E. and Nistico, G. (1989). *Brain messengers and the pituitary*. San Diego: Academic Press.

Neill, J. D. (1972). Sexual differences in the hypothalamic regulation of prolactin. *Endocrinology*, **90**, 1154–1159.

Ory, S. J. (1983). Clinical uses of luteinizing hormone-releasing hormone. *Fertility and Sterility*, **39**, 577–591.

Rivier, C. L. and Plotsky, P. M. (1986). Mediation by corticotropin releasing factor (CRF) of adenohypophysial hormone secretion. *Annual Review of Physiology*, **48**, 475–494.

Sarkar, D. K. (1983). Does LHRH meet the criteria for a hypothalamic releasing factor? *Psychoneuroendocrinology*, **8**, 259–275.

Schally, A. V. (1978). Aspects of hypothalamic regulation of the pituitary gland. *Science*, **202**, 18–28.

Schally, A. V., Arimura, A., Kastin, A. J. *et al.* (1971). Gonadotropin-releasing hormone: one polypeptide regulates secretion of luteinizing and follicle-stimulating hormones. *Science*, **173**, 1036–1038.

Schally, A. V., Coy, D. H. and Meyers, C. A. (1978). Hypothalamic regulatory hormones. *Annual Review of Biochemistry*, **47**, 89–128.

Sharp, P. J. and Fraser, H. M. (1978). Control of reproduction. In S. L. Jeffcoate and J. S. M. Hutchinson (eds.), *The Endocrine Hypothalamus*, pp. 271–332. London: Academic Press

Silverman, A. J. and Zimmerman, E. A. (1983). Magnocellular neurosecretory system. *Annual Review of Neuroscience*, **6**, 357–380.

Swanson, L. W. and Sawchenko, P. E. (1983). Hypothalamic integration: Organization of the paraventricular and supraoptic nucleii. *Annual Review of Neuroscience*, **6**, 269–324.

Taleisnik, S. (1978). Control of melanocyte stimulating hormone (MSH) secretion. In S. L. Jeffcoate and J. S. M. Hutchinson (eds.) *The Endocrine Hypothalamus*. pp. 421–439. London: Academic Press.

Vale, W., Rivier, J., Vaughn, J., McClintock, R., Corrigan, A., Woo, W., Karr, D. and Spiess, J. (1986). Purification and characterization of an FSH releasing protein from porcine ovarian follicular fluid. *Nature*, **321**, 776–779.

Weiner, R. I., Findell, P. R. and Kordon, C. (1988). Role of classic and peptide neuromediators in the neuroendocrine regulation of LH and prolactin. In E. Knobil, J. D. Neill *et al.* (eds.) *The Physiology of Reproduction*, vol 1, pp. 1235–1281. New York: Raven Press.

Witkin, J. W., Paden, C. M. and Silverman, A.-J. (1982). The luteinizing hormone-releasing hormone (LHRH) systems in the rat brain. *Neuroendocrinology*, **35**, 429–438.

Wurtman, R. J. and Anton-Tay, F. (1969). The mammalian pineal as a neuroendocrine transducer. *Recent Progress in Hormone Research*, **25**, 493–522.

Zaborszky, L. (1982). Afferent connections of the medial basal hypothalamus. *Advances in Anatomy, Embryology and Cell Biology*, **69**, 1–107.

5

Neurotransmitters

Neurotransmitters are synthesized in nerve cells, released into the synapse and bind to receptors on the postsynaptic cell (Figure 5.1). This chapter will examine the different categories of neurotransmitters, the structure of the nerve cell, the synthesis, storage, transport and release of neurotransmitters, their action at receptors and their deactivation. The influence of drugs on neurotransmitter activity will also be discussed. Chapter 6 will examine the effects of neurotransmitters on the neuroendocrine system and Chapter 10 covers the actions of neurotransmitters at their receptors on postsynaptic cells.

To be considered as a neurotransmitter, a chemical messenger should meet the eight criteria listed in Table 5.1 (an expanded discussion of these criteria is provided by McGeer, Eccles and McGeer, 1987, pp. 152–4). These criteria apply equally well to 'classical' neurotransmitters such as acetylcholine and to peptide transmitters such as substance P, and they are used to distinguish 'true' neurotransmitters from neuromodulators and other neuroregulators which are discussed in Chapter 11 (Barchas *et al.*, 1978; Elliott and Barchas, 1979; Osborne, 1981). A detailed introduction to the neurotransmitter systems of the brain is given by Cooper,

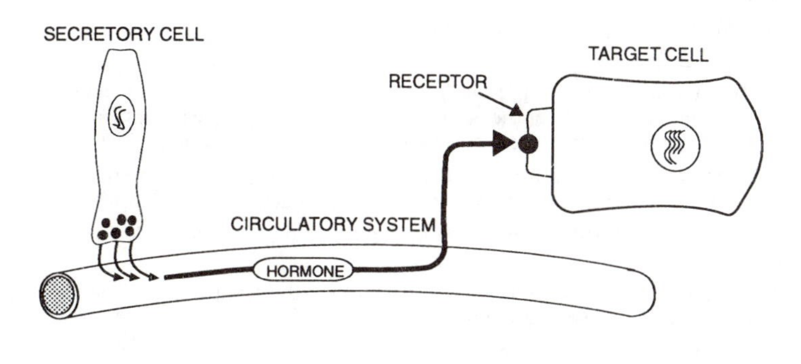

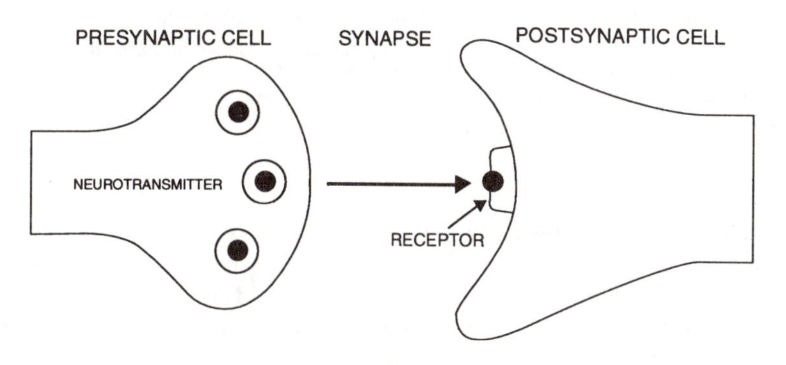

Figure 5.1. A comparison of the mechanisms of communication by hormones and neurotransmitters. (A) Hormones are released from endocrine cells into the circulatory system to stimulate receptors on target cells at a distance. (B) Neurotransmitters are released from presynaptic cells into the synapse to stimulate receptors on postsynaptic cells at very close range. (Redrawn from Nathanson and Greengard, 1977.)

Bloom and Roth (1991) and by Bradford (1986). The definition of a neurotransmitter, like that of a hormone, is constantly changing as new discoveries are made on the nature of neurotransmitter action (Bloom, 1988).

5.1 CATEGORIES OF NEUROTRANSMITTERS

On the basis of the criteria presented in Table 5.1, a number of neurotransmitters have been identified. As summarized in Table 5.2, these belong to five general categories: amino acids, acetylcholine, monoamines, peptides and putative transmitters (McGeer *et al.*, 1987).

5.1.1 THE AMINO ACID TRANSMITTERS

The amino acid transmitters are the major neurotransmitters in the mammalian central nervous system (CNS) and occur in neurons throughout the brain and spinal cord. They stimulate their receptors to open ion channels in the membrane of their postsynaptic target cells and thus induce rapid excitatory or inhibitory actions on these postsynaptic cells (see Figure 5.6). Gamma-aminobutyric acid (GABA) acts as an inhibitory neurotransmitter, inhibiting electrophysiological activity in postsynaptic

Table 5.1. *Eight criteria for determining whether or not a neuroregulatory chemical is a 'true' neurotransmitter*

1. The substance must be present in presynaptic neurons, usually in an uneven distribution throughout the brain, i.e. certain brain areas will make one neurotransmitter, whereas others will make an alternate one.
2. Neurotransmitter precursors and synthetic enzymes must be present in the neuron, usually in close proximity to the site of action.
3. Stimulation of nerve afferents (dendrites) should cause release of the substance in physiologically significant amounts.
4. Effects of direct application of the substance to the synapse should be identical to those produced by stimulating nerve afferents.
5. Specific receptors that interact with the substance should be present in close proximity to the presynaptic neurons.
6. Interaction of the substance with its receptors should induce changes in postsynaptic membrane permeability leading to excitatory or inhibitory postsynaptic potentials in the postsynaptic cell.
7. Specific inactivating mechanisms should exist which stop interactions of the substance with its receptor in a physiologically reasonable time frame.
8. Interventions at postsynaptic sites using agonist drugs should mimic the action of the transmitter and antagonists should block its effects.

Source: Barchas *et al.*, 1978; Elliot and Barchas, 1979.

cells and suppressing many behavioral responses (Panksepp, 1986). Two other amino acids which act as inhibitory transmitters are glycine and taurine. Other amino acids, such as glutamic acid and aspartic acid act as excitatory neurotransmitters in the brain and spinal cord (Cooper *et al.*, 1991; McGeer *et al.*, 1987).

5.1.2 ACETYLCHOLINE (ACH)

Acetylcholine is released by nerves of the 'cholinergic' pathways. As well as acting in the brain and central nervous system (CNS), acetylcholine is the neurotransmitter used at the neuromuscular junction, in the parasympathetic branch of the autonomic nervous system (ANS) and in autonomic ganglia (Cooper *et al.*, 1991; McGeer *et al.*, 1987). Acetylcholine is important in attentional and memory processes and may be involved in diseases of aging such as Alzheimer's disease. Acetylcholine also functions in motivated behaviors including aggression, sexual behavior and the regulation of thirst and drinking (Panksepp, 1986).

5.1.3 THE MONOAMINE NEUROTRANSMITTERS

The monoamine neurotransmitters occur at a much lower concentration in the brain than do the amino acid transmitters and exist along specific neural pathways. When these transmitters bind to their receptors, they activate a series of chemical changes involving second messenger systems in the cytoplasm of the cell. These second messengers, such as cyclic AMP, cause short-term changes in membrane potential or long-

Table 5.2. *Categories of neurotransmitters and 'putative' neurotransmitters*

Category	Neurotransmitter
A. Amino acid transmitters	
Excitatory	Aspartic acid Glutamic acid
Inhibitory	Gamma-aminobutyric acid (GABA) Glycine (spinal cord)
B. 'Cholinergic' neurotransmitter	Acetylcholine (Ach)
C. Monoamine neurotransmitters	
'Adrenergic' (catecholamines)	Dopamine (DA) Noradrenaline (NA) or norepinephrine (NE) Adrenaline or epinephrine
Indoleamine	Serotonin (5-HT)
Other	Histamine
D. Peptide transmitters	Substance P Somatostatin Neurotensin Cholecystokinin Enkephalins, endorphins
E. Putative neurotransmitters	
	Endogenous benzodiazepines Prostaglandin

term changes in the structure of the cell, such as those involved in protein synthesis. Considerable evidence now indicates that histamine should be considered as a true monoamine neurotransmitter along with dopamine, noradrenaline and serotonin (McGeer *et al.*, 1987; Cooper *et al.*, 1991).

The catecholamines. The catecholamines: dopamine (DA), noradrenaline (NA) (or norepinephrine) and adrenaline (A) (or epinephrine) are the neurotransmitters synthesized and released by nerves of the 'adrenergic' pathways. Adrenergic pathways occur in a number of brain areas, particularly the limbic system, and activate the sympathetic branch of the autonomic nervous system. Catecholamines are important in the arousal of emotional and motivated behavior (Panksepp, 1986) and in the regulation of the endocrine hypothalamus (see Chapter 6).

The indoleamines. There are two indoleamines, the neurotransmitter serotonin, which is also called 5-hydroxytryptamine (5-HT) and its close relative, melatonin, the hormone secreted from the pineal gland. The cell bodies of serotonin-secreting neurons are found primarily in the midline raphe region and the reticular system of the medulla, pons and upper brain stem (Cooper *et al.*, 1991; McGeer *et al.*, 1987). Serotonin promotes sleep and reduces or inhibits emotional and motivated behavior (Panksepp, 1986).

Histamine. Histamine neurons occur primarily in the hypothalamus and other areas of the limbic system and send fibers to the cortex and the brain stem. Histamine acts on smooth muscle, gastric secretion and in the allergic reaction (McGeer *et al.*, 1987).

5.1.4 PEPTIDE TRANSMITTERS

Certain peptides occur in very low concentrations in neural pathways where they have neurotransmitter-like activity (Nieuwenhuys, 1985; Bradford, 1986). Neuropeptides, such as substance P, neurotensin, somatostatin and the enkephalins are produced in neural cells and meet many of the criteria of neurotransmitter action given in Table 5.1. The enkephalins, for example, are synthesized in the brain and spinal cord and have neurotransmitter-like activity, particularly in the control of pain (Akil *et al.*, 1984; Panksepp, 1986). Many other neuropeptides act as neuromodulators, but whether or not their functions and modes of action meet the criteria for a 'true' neurotransmitter is a source of debate. This is discussed in Chapter 11 (Barchas *et al.*, 1978; Dismukes, 1979; Osborne, 1981; McGeer *et al.*, 1987).

5.1.5 PUTATIVE NEUROTRANSMITTERS

Substances such as the prostaglandins, endogenous benzodiazepines and numerous neuropeptides are classed as putative transmitters because they do not meet the majority of the criteria for neurotransmitters listed in Table 5.1. The existence of endogenous benzodiazepine-like substances, for example, has been postulated, but none has yet been isolated. Synthetic benzodiazepines, such as librium and valium, have anti-anxiety, muscle relaxant and sedative-hypnotic properties and are used clinically to treat anxiety and insomnia (Tallman *et al.*, 1980).

5.2 THE NERVE CELL AND THE SYNAPSE

Nerve cells consist of the dendrites, the cell body, which contains the nucleus, and the axon, as shown in Figure 5.2. The dendrites receive information from other cells while the axon transmits information to other cells. Although each nerve cell has only one axon, this axon has a number of branches and the nerve terminals at the end of each branch in detail by Cooper *et al.*, (1991) and by McGeer *et al.*, (1987).

Nerve cells communicate with each other by the release of neurotransmitters from the nerve terminals of the axon into the synapse, which separates the presynaptic and postsynaptic cells. The neurotransmitters released into the synapse are picked up by receptors on the postsynaptic cell. As shown in Figure 5.2, synapses can form between the axon of the presynaptic cell and a number of different sites on the postsynaptic cell, including the dendrites (axo-dendritic synapses), dendritic 'spines' (spine synapses), the cell body (axo-somatic synapses), and the axons (axo-axonal synapses).

Synapses can be either excitatory or inhibitory. At an excitatory synapse, the neurotransmitter released from the presynaptic cell excites the postsynaptic cell, changing the electrical potential of the cell mem-

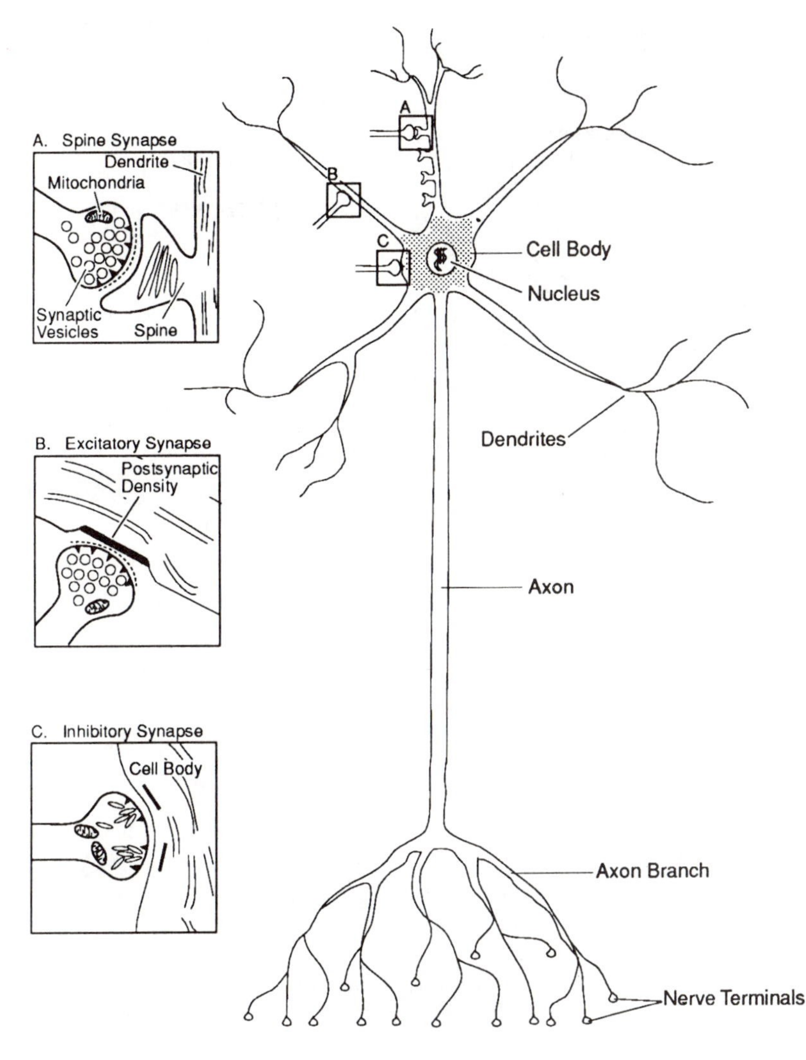

Figure 5.2. The structure of the nerve cell. The three components of the nerve cell are the dendrites, the cell body or soma, and the axon. Synapses can be classified according to their position on the surface of the receiving neuron. They may be on 'spines' projecting from the dendrites, as shown in inset A, on the trunk of the dendrites (axo-dendritic), as shown in inset B, on the cell body (axo-somatic), as shown in inset C, or on the axon (axo-axonal). Synapses impinging on the neuron are either excitatory or inhibitory. These synapses can be distinguished using an electron microscope to identify their structure. Excitatory synapses tend to have round vesicles and a dense thickening (postsynaptic density) of the postsynaptic membrane, as shown in inset B, while inhibitory synapses tend to have flattened vesicles and a discontinuous postsynaptic density, as shown in inset C. (Redrawn from Iversen, 1979.)

brane and causing it to become depolarized or 'fire'. At inhibitory synapses, the neurotransmitter released from the presynaptic cell inhibits the postsynaptic cell from firing (Figure 5.2). Excitatory synapses are usually found on the dendrites, whereas inhibitory synapses occur on the cell body. Despite the morphological differences between excitatory and inhibitory synapses shown in Figure 5.2, the postsynaptic receptors at excitatory and inhibitory synapses function in the same way (see McGeer *et al.*, 1987).

5.3 NEUROTRANSMITTER BIOSYNTHESIS AND STORAGE

The precursor molecules and the enzymes required for amino acid and monoamine neurotransmitter biosynthesis are produced on the ribosomes of the endoplasmic reticulum in the nerve cell body and transported in synaptic vesicles down the axon to the nerve terminals (Figure

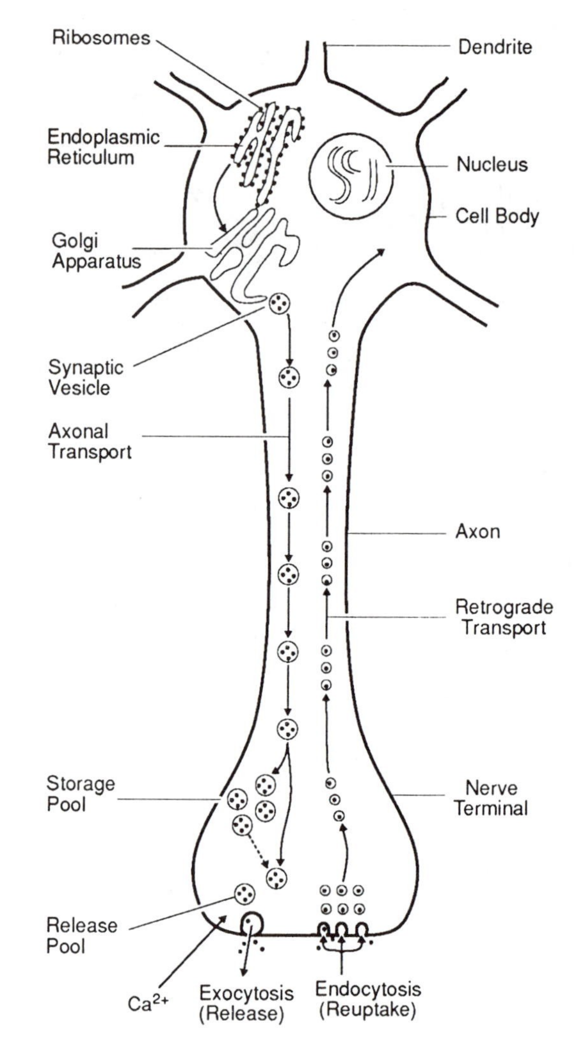

Figure 5.3. The synthesis, transport and storage of neurotransmitters in the nerve cell. Biosynthetic enzymes are manufactured in the endoplasmic reticulum of the cell body and transported to the nerve terminal, via axonal transport, in synaptic vesicles, where they manufacture transmitters which are then stored until their release (exocytosis) under the influence of calcium (Ca^{2+}). Within the nerve terminal, synaptic vesicles maintain neurotransmitters in storage and release pools. Vesicles are also involved in the reuptake of transmitters from the synapse (endocytosis) and the retrograde transport of these transmitters to the cell body, where they are either reused or destroyed. (Redrawn from Iversen, 1979.)

5.3). The synthesis of these neurotransmitters is completed in the synaptic vesicles. The precursors (propeptides) of the peptide transmitters are also synthesized on the ribosomes in the cell body and transported, along with the enzymes required for the completion of peptide synthesis, to the nerve terminals in the synaptic vesicles, where peptide synthesis is completed. Axonal transport is important for moving the synaptic vesicles to the nerve terminals (Schwartz, 1980). Inside the nerve terminal, the neurotransmitters remain stored in synaptic vesicles (the storage pool) before they are released into the synapse (Figure 5.3).

The synaptic vesicles perform a number of essential functions in the nerve cell. (a) They transport the neurotransmitter precursors and biosynthetic enzymes from the cell body along the axon to the nerve terminal. (b) Transmitter synthesis is often completed within the vesicles. (c) The vesicles store the transmitters until they are released. (d) The vesicles protect the transmitters from deactivation. (e) Transmitters are released from the nerve terminal by exocytosis when the vesicle contacts the depolarized cell membrane in the presence of calcium ions (Figure 5.3). (f) The vesicles also help to regulate the rate of neurotransmitter synthesis through a negative feedback mechanism (Iversen, 1979; Stevens, 1979).

5.3.1 SYNTHESIS AND STORAGE OF PARTICULAR NEUROTRANSMITTERS

Amino acid transmitters

Glutamate and aspartate are synthesized from glucose and other precursors in neurons and glial cells. GABA is formed from its precursor, glutamic acid, by the enzyme glutamic acid decarboxylase (GAD) in the nerve terminals. The biosynthesis of the amino acid transmitters is given in detail by Cooper *et al.*, (1986) and by McGeer *et al.*, (1987).

Acetylcholine

Acetylcholine is synthesized from choline and acetyl coenzyme A through the action of the enzyme choline acetyltransferase (Blusztajn and Wurtman, 1983). Choline acetyltransferase is synthesized in the cell body and transported through the axon to the nerve terminals, where acetylcholine is synthesized (details are given in Blusztajn and Wurtman, 1983; Cooper *et al.*, 1991 and McGeer *et al.*, 1987).

Monoamine neurotransmitters

The synthesis of the catecholamines (Figure 5.4) starts when the amino acid tyrosine enters the nerve cell from the blood and is converted to dopa through the action of the enzyme tyrosine hydroxylase. Dopa is then converted to dopamine in the cell body through the action of the enzyme dopa decarboxylase and dopamine is transported to the nerve terminal in the vesicles. In a dopaminergic neuron, biosynthesis does not proceed further. At a noradrenergic nerve terminal, however, dopamine is converted to noradrenaline in the synaptic vesicles (by the action of dopamine beta-hydroxylase) and, at an adrenergic nerve terminal, noradrenaline is converted to adrenaline, as shown in Figure 5.4. Specific enzymes are required for each conversion and, if any of these enzymes are missing, the next transmitter in the series will not be synthesized. The drug L-dopa readily enters the brain to act as a precursor for dopamine and will facilitate dopamine synthesis. Details of catecholamine synthesis are given by Cooper *et al.*, (1991) and in McGeer *et al.*, (1987). Serotonin or 5-hydroxytryptamine (5-HT) is synthesized from the amino acid tryptophan, which is converted to 5-hydroxytryptophan (5-HTP) by the enzyme tryptophan hydroxylase. 5-HTP is then converted to serotonin by the enzyme 5-HTP decarboxylase. This is described in more detail by Cooper *et al.*, (1991) and McGeer *et al.*, (1987).

Peptide transmitters

The precursors for the neuropeptides are prepropeptides (prohormones) which are synthesized in the endoplasmic reticulum in the cell body and packaged into secretory vesicles in the Golgi bodies. The endogenous

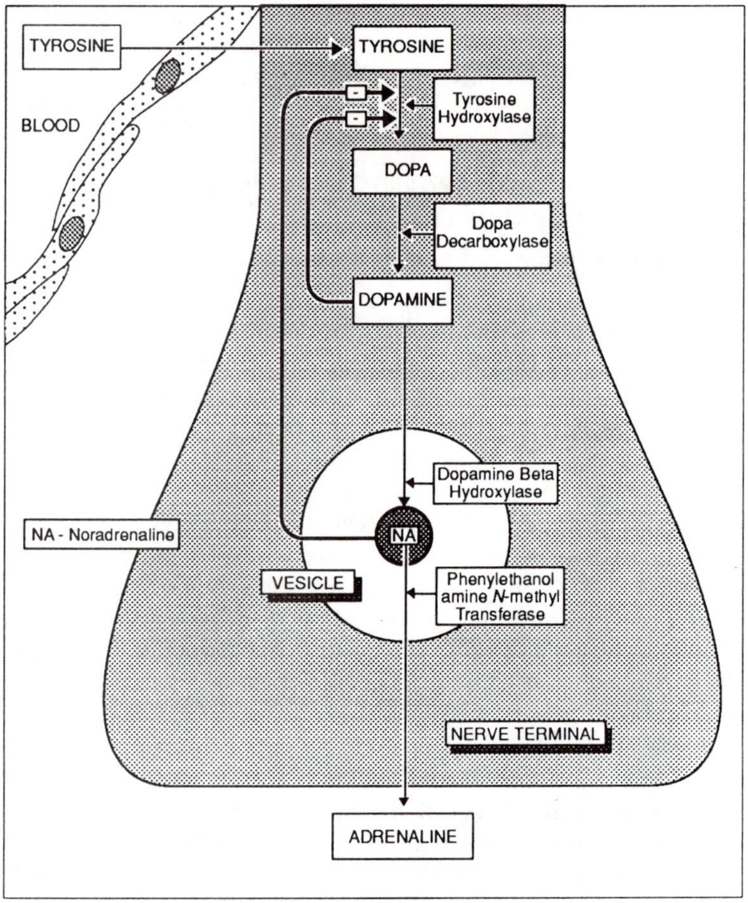

Figure 5.4. Synthesis of catecholomines. Catecholamine synthesis begins when tyrosine enters the neuron from the blood and is converted to dopa by the enzyme tyrosine hydroxylase. The enzyme dopa decarboxylase converts dopa to dopamine, which is stored in the synaptic vesicle. Within the synaptic vesicle, the enzyme dopamine beta hydroxylase converts dopamine to noradrenaline (NA). The enzyme phenylethanolamine *N*-methyl transferase (PNMT) converts noradrenaline to adrenaline. Regulation of catecholamine synthesis is accomplished by a feedback mechanism: a build-up of dopamine or noradrenaline in storage pools inhibits the activity of tyrosine hydroxylase, which is essential for the first step in catecholamine synthesis. An increase in nerve activity stimulates the release of catecholamines, reducing the amount stored in the nerve terminal. This removes the inhibition of tyrosine hydroxylase activity and more catecholamines are synthesized. (Redrawn from Axelrod, 1974.)

opioids, for example, are synthesized from three different precursors: proenkephalin, prodynorphin and proopiomelanocortin (Akil *et al.*, 1984). Peptide transmitter synthesis is completed in the synaptic vesicles, which are transported down the axon to the nerve terminal for the storage and release of the neuropeptide. Synthesis of the neuropeptides is discussed in more detail in Chapter 11.

5.3.2 REGULATION OF NEUROTRANSMITTER SYNTHESIS AND STORAGE

Neurotransmitters are continually being synthesized, stored, released and deactivated. Neurotransmitter synthesis and storage is regulated in a number of ways. Figure 5.4 shows that the amount of neurotransmitter stored in the vesicles can regulate its own rate of synthesis by a negative feedback loop. The amount of neurotransmitter recaptured by presynaptic reuptake (Figure 5.3) also regulates the synthesis of that neurotransmitter, particularly in adrenergic neurons (Westfall, 1984). Regulation of neurotransmitter synthesis is accomplished by altering the levels of the enzymes necessary for their biosynthesis. For example, stimulation of sympathetic nerves speeds up the conversion of tyrosine to noradrenaline by increasing the activity of the enzyme tyrosine hydroxylase which converts tyrosine to dopa. The activity of this enzyme is inhibited by negative feedback from noradrenaline and dopamine stored in the vesicles. The synthesis of neurotransmitters can also be regulated by neuromodulators, neural input from other cells, hormones and the availability of amino acids in the bloodstream, which may be related to the diet.

5.4 THE RELEASE OF NEUROTRANSMITTERS AND THEIR ACTION AT RECEPTORS

Each neuron is both a post- and a presynaptic cell. When stimulated postsynaptically by neurotransmitters, the neuron 'fires', releasing its transmitters presynaptically into the next synapse. When these neurotransmitters stimulate their postsynaptic receptors at an excitatory synapse, they cause ion channels to open, and the nerve membrane is depolarized, i.e. sodium ions (Na^+) flow in and potassium ions (K^+) move out of the cell. This depolarization causes a measurable change in the electrical activity of the cell, called an action potential, which travels along the nerve cell membrane with a wave action until it reaches the nerve terminal of the axon (Figure 5.5A). When depolarization of the nerve terminal occurs, the synaptic vesicles in which the neurotransmitters are stored fuse with the cell membrane, and, under the influence of calcium ions, there is a change in the permeability of the cell membrane, releasing the neurotransmitters from the vesicle into the synapse (Stevens, 1979). The release of the neurotransmitter into the synapse involves a number of complex biochemical actions (Kelley *et al.*, 1979).

After the cell fires, it returns to its resting potential until stimulated by another neurotransmitter to fire again. The cell cannot, however, fire again instantaneously, as the resting potential is reinstated through the action of the ion pump, which pumps Na^+ ions out of the cell and K^+ ions back into the cell (Figure 5.5B). During the time that this ion pump takes to reinstate the resting potential, the cell cannot fire and is in a refractory period (Figure 5.5A).

5.5 RECEPTORS FOR NEUROTRANSMITTERS

After the neurotransmitter is released into the synapse, it binds to a receptor protein on the surface membrane of the postsynaptic cell. Neurotransmitters are, in general, too polar to enter their target cells (i.e. they will not cross the cell membrane) and, therefore, they stimulate membrane receptors. As shown in Figure 5.6, these receptors are of two general types. Ionotropic receptors, which open ion channels, are stimulated primarily by the amino acid transmitters. Metabotropic receptors, which activate second messenger systems to stimulate metabolic changes within the postsynaptic cell, are stimulated by amine and peptide transmitters (McGeer *et al.*, 1987).

Each neurotransmitter may have more than one type of receptor and

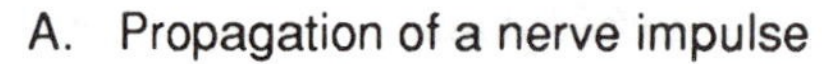

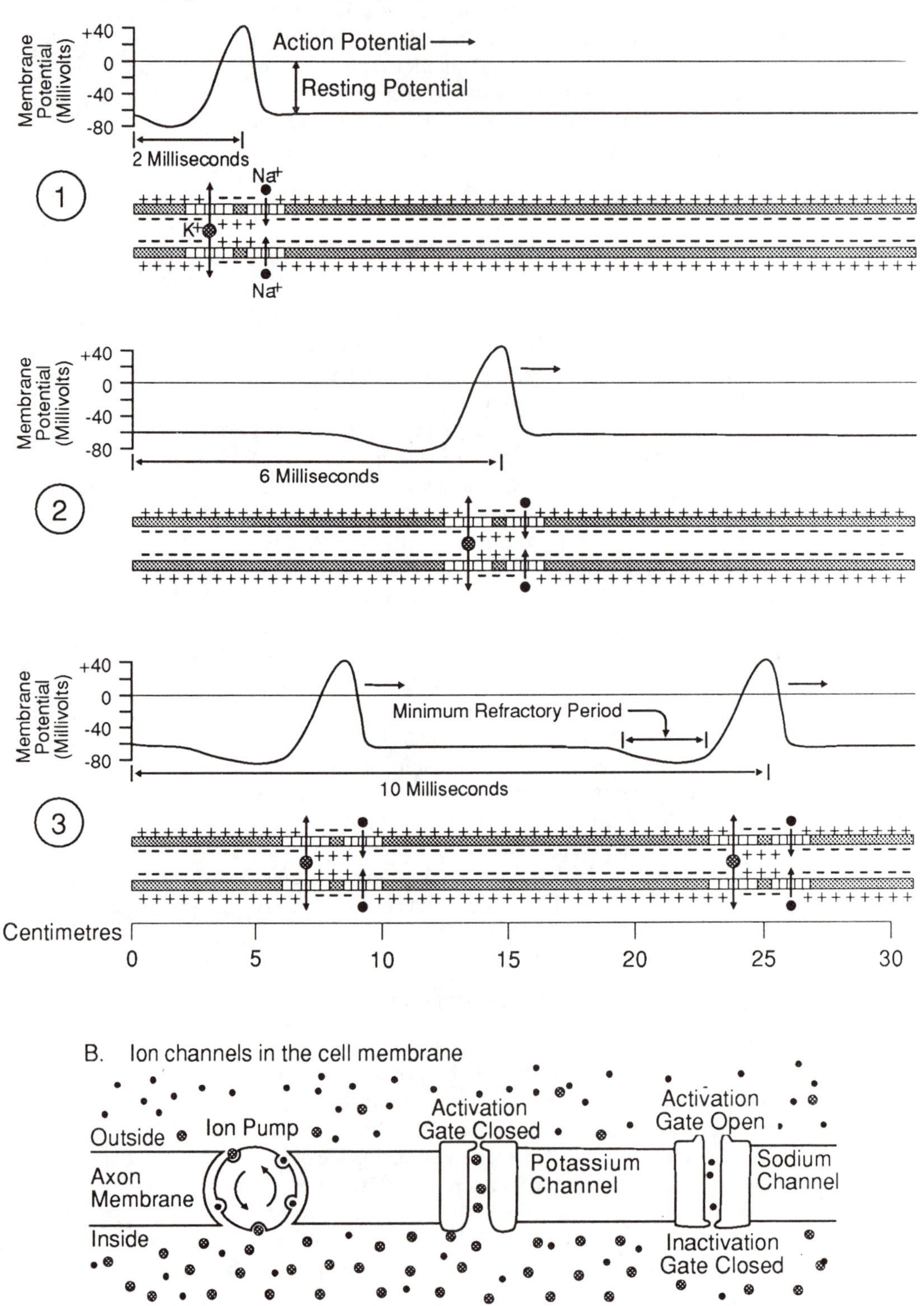
A. Propagation of a nerve impulse
Membrane Potential (Millivolts)
+40
0
-40
-80
Action Potential
Resting Potential
2 Milliseconds
1
Na+
K+
6 Milliseconds
2
Minimum Refractory Period
10 Milliseconds
3
Centimetres
0
5
10
15
20
25
30
B. Ion channels in the cell membrane
Outside
Ion Pump
Axon Membrane
Inside
Activation Gate Closed
Potassium Channel
Activation Gate Open
Sodium Channel
Inactivation Gate Closed

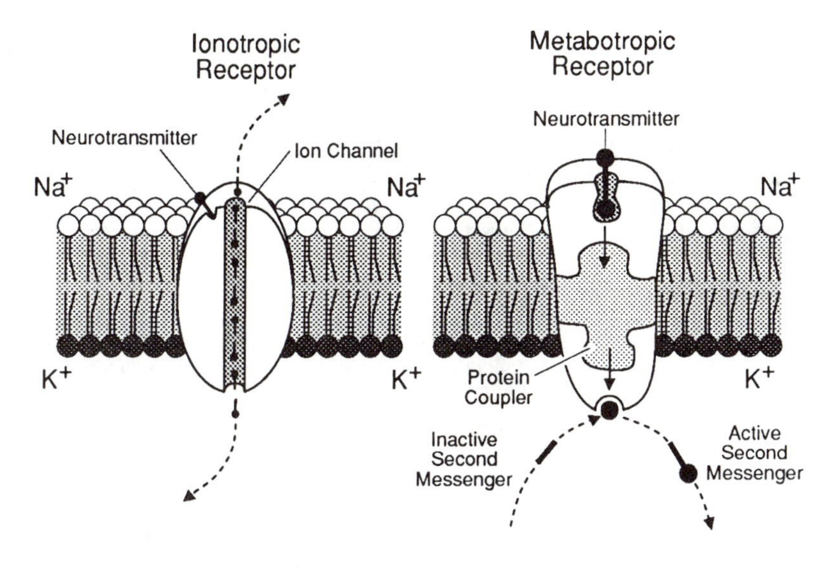

Figure 5.6. Two general types of receptor for neurotransmitters. Ionotropic receptors open ion channels when activated by a neurotransmitter and allow rapid passage of sodium, potassium, and other ions through the cell membrane. Metabotropic receptors are more complex. When the transmitter binds to its receptor, a protein coupler (G-protein) is activated which causes enzymes within the membrane to activate second messengers, which alter the metabolic activity of the cell. (Redrawn from McGeer *et al.*, 1987.)

these receptors may be on different types of cells, including the presynaptic nerve terminal. As a result, cells may respond in different ways to the same transmitter (Table 5.3). For example, there are both α-(alpha) and β-(beta) adrenergic receptors for noradrenaline and these receptors sometimes act antagonistically, for example α-adrenergic receptors cause contraction and β-adrenergic receptors cause relaxation of smooth muscle (Hadley, 1992). There are also different subtypes of α- and β-adrenergic receptors. There are a number of criteria for determining the existence of different receptor types and their specificity for particular neurotransmitters (see McGeer *et al.*, 1987). Receptor localization or 'receptor mapping' in the brain has become a specialized field of

Figure 5.5. A. Propagation of a nerve impulse (action potential) along the axon. This coincides with a localized inflow of sodium ions (Na^+) followed by an outflow of potassium ions (K^+) through channels that are controlled by voltage changes across the axon membrane. The cell membrane separates fluids that differ in their content of Na^+ and K^+. The exterior fluid has about ten times more Na^+ ions than K^+; but, in the interior fluid the ratio is the reverse. The axon membrane has a number of selective ion channels which allow Na^+ and K^+ to pass through the cell membrane. (1) The electrical event that sends a nerve impulse traveling down the axon normally originates in the cell body. The action potential begins with a depolarization, or reduction in the negative potential, across the axon membrane. This voltage shift opens some of the Na^+ channels, increasing the voltage still further. The inflow of Na^+ accelerates until the inner surface of the membrane shifts from a negative to a positive charge. This voltage reversal closes the Na^+ channel and opens the K^+ channel. (2) The outflow of K^+ ions quickly restores the negative potential inside the axon and the action potential propagates itself down the axon. (3) After a brief refractory period, during which Na^+ is pumped out of the cell and K^+ into the cell by the ion pump, a second impulse can flow. (Redrawn from Stevens, 1979.)

Figure 5.5. B. In the resting state, when no action potential is being transmitted, the sodium and potassium channels are closed and an ion pump maintains the ionic disequilibrium by pumping out Na^+ in exchange for K^+. The interior of the axon is normally about 70 mV negative with respect to the exterior. If this voltage difference is reduced by the arrival of a nerve impulse, the sodium channel opens, allowing Na^+ to flow into the axon. An instant later the sodium channel closes and the potassium channel opens, allowing an outflow of K^+. The sequential opening and closing the these two channels effects the propagation of the nerve impulse, which is illustrated in A. (Redrawn from Stevens, 1979.)

Table 5.3. *Neurotransmitter receptor types and their properties*

Receptor	Properties
1. Glutamate and aspartate receptors	
A_1 (NMDA), A_2, A_3, A_4	Located in the brain and bind to glutamate and aspartate
2. GABA receptors	
(Types A and B)	Located on postsynaptic nerves (A type) and may regulate the release of neurotransmitters (B type)
3. Cholinergic receptors	
Muscarinic (M_1 to M_5)	Located on both pre- and postsynaptic neurons in the sympathetic nervous system, corpus striatum, hindbrain, hypothalamus, cerebellum, heart and stomach
Nicotinic (N_1 to N_4)	Located on skeletal muscles at neuromuscular junctions and in autonomic ganglia
4. Adrenergic receptors	
(a) Dopamine receptors	
D_1 and D_5	Located in parathyroid gland and brain (cortex and limbic system); stimulates cyclic AMP synthesis
D_2	Located in brain (limbic system) and anterior pituitary; inhibits cyclic AMP synthesis. Responsible for actions of antipsychotic drugs and inhibitory effects of dopamine on the release of adenohypophyseal hormones
D_3 and D_4	Located in the limbic system and cortex
(b) Noradrenaline and adrenaline receptors	
Alpha-adrenergic (α_1 and α_2)	Located on postsynaptic cells in heart and brain (α_1) and presynapic cells in pancreas, duodenum and brain (α_2). Cause contraction of vascular smooth muscle (α_1).
Beta-adrenergic (β_1 and β_2)	Located in heart, lung and brain. Stimulate the contraction of heart muscle and relax smooth muscle of intestine and lungs. Stimulates cyclic AMP synthesis
5. Serotonin receptors	
(5-HT_1, 5-HT_2, and 5-HT_3)	Located in brain and circulatory system to regulate contractions of blood vessels. 5-HT_2 receptors in brain regulate behavioral effects of serotonin. 5-HT_3 seems to be important in the vomiting reflex
6. Histamine receptors	
(H_1, H_2 and H_3)	Mediate bronchial constriction in lungs (H_1) and gastric secretion in stomach (H_2). Also occur in muscle, cardiovascular system, brain and immune system
7. Opiate receptors	
Mu (μ)	In brain and spinal cord to regulate pain; implicated in analgesia; morphine selective
Delta (δ)	Located in limbic system of brain; may mediate epilepsy and rewarded behavior; selective for enkephalins
Kappa (κ)	Located in cerebral cortex; mediate sedative analgesia; selective for dynorphin
Sigma (σ)	Located in hippocampus; mediates psychomimetic opiate effects
Epsilon (ϵ)	β-Endorphin selective
8. Benzodiazepine receptors	
Type 1	Located on postsynaptic cells

Table 5.3. (*cont.*)

Type 2	Located on presynaptic cells
GABA receptors	Benzodiazepine receptors are also part of the GABA receptor complex

Sources: Iversen and Iversen, 1981; Minneman, Pittman and Molinoff, 1981; Akil *et al.*, 1984; Snyder, 1984; Cooper *et al.*, 1991; McGeer *et al.*, 1987; Hartig, 1989; Sunahara, *et al.*, 1991.

neuroscience (Kuhar, de Souza and Unnerstall, 1986). The details of known receptor types for each class of neurotransmitters are shown in Table 5.3.

When monoamine and peptide transmitters stimulate their receptors at an excitatory synapse, two events occur in the postsynaptic cell. First, there is a rapid depolarization of the cell, i.e. the cell 'fires', and, second, there is the synthesis of intracellular 'second messengers', such as cyclic AMP, which alter the long-term permeability of the cell membrane and stimulate protein synthesis in the cell. At an inhibitory synapse, cell firing and second messenger synthesis are inhibited. The actions of second messenger systems in postsynaptic cells are discussed in Chapter 10.

5.6 DEACTIVATION OF NEUROTRANSMITTERS

After the neurotransmitter has stimulated its postsynaptic receptor, it is released into the synapse and deactivated. There are two general ways of deactivating a neurotransmitter (Figure 5.7). The first is reuptake (endocytosis) by receptors on the presynaptic cell which take the transmitter into vesicles so that it can be reused or deactivated (Figures 5.3 and 5.7). Up to 80% of the catecholamines released into the synapse are deactivated by reuptake into the nerve terminal of the presynaptic cell. This reuptake is facilitated by special presynaptic receptors which transport the transmitter from the synapse back into the nerve terminal. The amino acid transmitters are also deactivated by presynaptic reuptake. This neural recycling saves energy required for the synthesis of new transmitters.

The second method of deactivation of neurotransmitters is to have them broken down (catabolized) by enzymes. Each transmitter is deactivated by specific enzymes. Acetylcholine, for example, is rapidly deactivated by the enzyme acetylcholinesterase. The two products of acetylcholine deactivation, acetic acid and choline may then be picked up by the presynaptic cell and used to synthesize new transmitter molecules.

As shown in Figure 5.7, the catecholamines are deactivated by two different enzymes, catechol-*O*-methyltransferase (COMT) and monoamine oxidase (MAO). Peptide transmitters are deactivated by peptidase enzymes (McKelvey, 1986). Other transmitters also have specific deactivating enzymes (Table 5.4). Because these enzymes exist in the synapse (Figure 5.7), they can deactivate neurotransmitters before they bind to their receptors as well as after. Thus, enough transmitter must be released to stimulate the postsynaptic cell, even though some will be deactivated before it has a chance to reach the receptor.

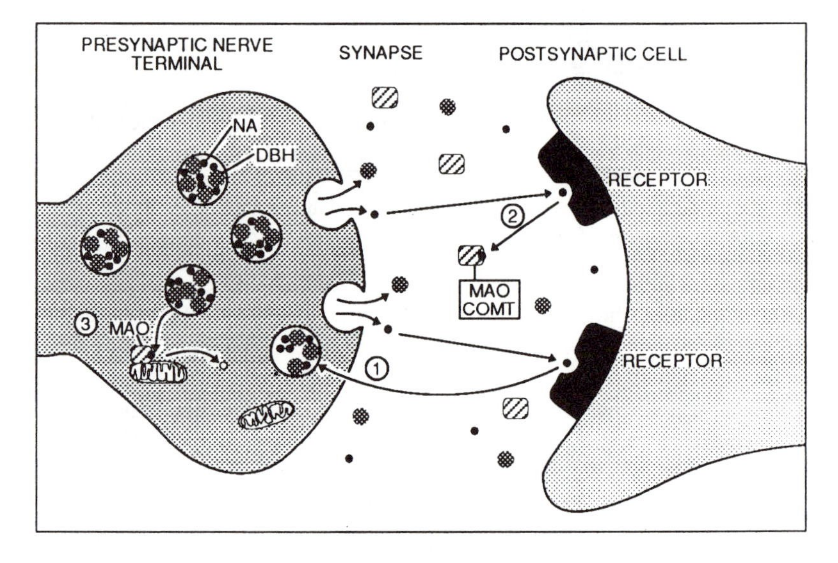

Figure 5.7. The release and inactivation of the neurotransmitter, noradrenaline. As shown in Figure 5.4, the enzyme dopamine beta-hydroxylase (DBH) is stored in the synaptic vesicles of the presynaptic nerve terminal and converts dopamine to noradrenaline (NA), which is released into the synapse when the presynaptic nerve is stimulated. The noradrenaline binds to its receptors on the postsynaptic cell, stimulating an action potential. The action of noradrenaline is then terminated as follows. (1) By reuptake and storage in the presynaptic nerve terminal. (2) Through metabolism by the enzymes monoamine oxidase (MAO) and/or catechol-*O*-methyltransferase (COMT). The former is the more important process. (3) By MAO, stored in the presynaptic cell, also inactivating noradrenaline that leaks out of synaptic vesicles. (Redrawn from Axelrod, 1974.)

Any neurotransmitter which leaks out of the synaptic vesicles in the presynaptic nerve terminal is deactivated by MAO in the nerve terminal (Figure 5.7). This prevents the uncontrolled release of neurotransmitter from the nerve terminal. Finally, unused transmitters in the storage pool of synaptic vesicles are also deactivated by enzymes in the nerve terminal (Figure 5.7).

5.7 NEUROTRANSMITTER PATHWAYS

Neurotransmitters occur in specific nerve pathways in the brain. Figure 5.8 shows the neural pathways for (a) acetylcholine, (b) dopamine, (c) noradrenaline and (d) serotonin in the rat brain. Some of the transmitter pathways innervating the endocrine hypothalamus are described below.

The excitatory amino acids, glutamate and aspartate have pathways in the cerebellum and the cerebral cortex. These pathways connect the cortex to the amygdala, thalamus, midbrain and spinal cord. Other excitatory amino acid pathways connect the hippocampus to the hypothalamus and other limbic system structures (see McGeer *et al.*, 1987).

There are five major cholinergic pathways in the brain, with nuclei in the medial septal nucleus, the nucleus of the diagonal band of Broca, the magnocellular preoptic area and the medulla (McGeer *et al.*, 1987; Semba and Fibiger, 1989). Some of these cholinergic neurons project to the hypothalamus and other areas of the limbic system (Figure 5.8A).

Table 5.4. *Some enzymes which are known to deactivate neurotransmitters*

Neurotransmitter	Deactivating enzyme
GABA	GABA-transaminase (GABA-T)
Acetylcholine	Acetylcholinesterase (cholinesterase)
Catecholamines	Monoamine oxidase (MAO) Catechol-*O*-methyl transferase (COMT)
Serotonin	Monoamine oxidase
Histamine	Histaminase
Neuropeptides	Peptidases

Sources: Cooper *et al.*, 1991; McGeer *et al.*, 1987.

Dopamine pathways have their cell bodies concentrated in the substantia nigra and project to a number of sites in the midbrain and limbic system (Cooper *et al.*, 1991; McGeer *et al.*, 1987). The short dopaminergic fibers of the tuberoinfundibular dopaminergic system in the median eminence regulate the secretion of pituitary hormones (Figure 5.8B). Figure 5.9 shows the dopamine pathways in the human brain (since you are probably tired of looking at rat brains by now.) Dopamine pathways are important for the integration of movement and a lack of dopamine in the corpus striatum causes the rigidity and tremor of patients who suffer from Parkinson's disease. Injection of L-dopa, a dopamine precursor which is able to cross the blood–brain barrier, relieves many of these symptoms. An excess of dopamine in the limbic forebrain is one of the neurochemical mechanisms postulated to underlie schizophrenia. Information on the neurotransmitter pathways in the human brain is given by Nieuwenhuys (1985).

The noradrenaline pathways have their cell bodies located primarily in the locus ceruleus and the medulla and form five major tracts. These include the central tegmental tract (labelled the dorsal bundle in Figure 5.8C) and the ventral tegmental-medial forebrain bundle (labelled the ventral bundle in Figure 5.8C) which innervate the cortex, olfactory bulb, thalamus and hypothalamus. Other noradrenergic fibers innervate the cerebellum, the brain stem and spinal cord (Cooper *et al.*, 1991; Moore and Bloom, 1979). Some of these hypothalamic projections regulate emotional and motivated behavior, while others regulate the secretion of hypothalamic hormones.

Serotonin pathways extend from the midline raphe region through the midbrain in the medial forebrain bundle (MFB) to the cortex (Figure 5.8D). Other serotonin pathways extend from the pons and upper brain stem down the spinal cord. The ascending serotonin pathways innervate the limbic system, especially the amygdala, septum and hypothalamus (McGeer *et al.*, 1987). The serotonin pathways are involved in the stimulation of sleep (catecholamine pathways stimulate wakefulness), and the control of cyclic GnRH release and disruption of these pathways may be involved in mental illness.

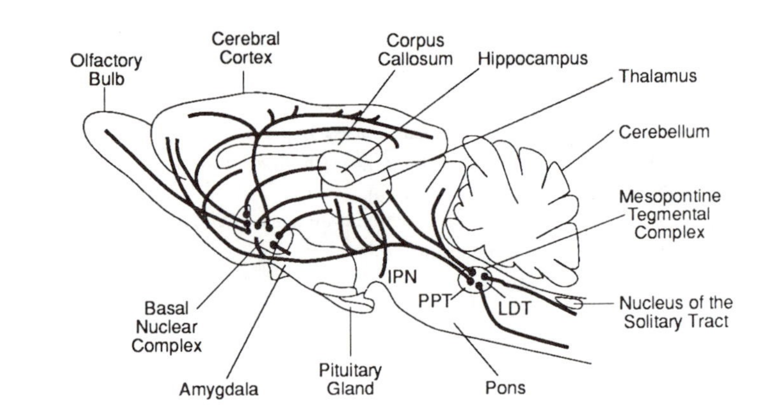

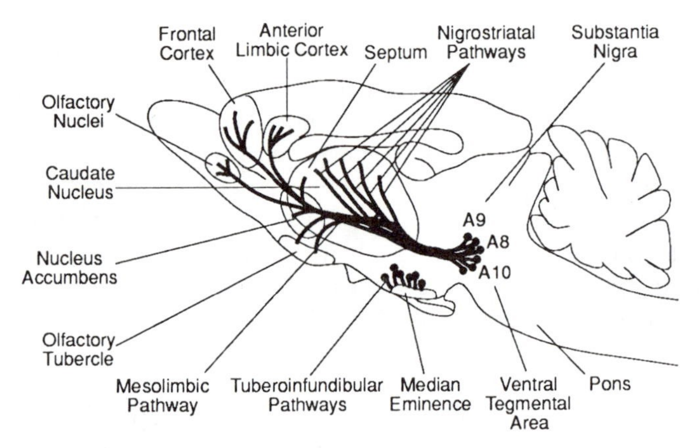

Figure 5.8. Neurotransmitter pathways in the rat brain, shown in sagittal (longitudinal) sections along the midline of the brain. (A) Cholinergic pathways. Cholinergic neurons occur in two principal brain regions: the basal nuclear complex, which sends axons to the olfactory bulb, cerebral cortex, hippocampus, thalamus and interpeduncular nucleus (IPN); and the pedunculopontine tegmental nucleus (PPT) and laterodorsal tegmental nucleus (LDT) of the mesopontine tegmental region, which send axons to the thalamus, olfactory bulb and cerebral cortex. (Redrawn from Semba and Fibiger, 1989.)

(B) Dopamine pathways. The dopamine cell bodies in the substantia nigra (A9), send axons along the nigrostriatal pathways to the caudate nucleus and other midbrain regions. Cell bodies in the ventral tegmentum (A10) send axons along the mesolimbic pathways to the nucleus accumbens, olfactory tubercule and cerebral cortex. The tuberoinfundibular dopamine pathways of the median eminence influence the release of hypothalamic and pituitary hormones. (Redrawn from Ungerstedt, 1971.)

The locations of the neuropeptide secreting cells and their neural pathways are discussed in Chapter 11.

Because neurotransmitter pathways make many diffuse connections, they are able to integrate information from many brain areas. Thus, stimulation from nerves of the olfactory bulb, cerebral cortex and spinal cord may all alter noradrenaline secretion in the hypothalamus and thus influence the release of neurohormones (Chapter 6). Noradrenaline and serotonin pathways in the hypothalamus also cause changes in arousal, mood and motivational states. Drugs have many and varied effects on neurotransmitters and can thus alter moods and emotional states as well as hypothalamic hormone secretion.

There are also steroid and peptide hormone-sensitive neurons along neurotransmitter pathways where these hormones can act as modulators of neurotransmitter release. For example, estrogen stimulation of its receptors in the brain can modify noradrenaline and dopamine release in a number of specific locations, causing changes in emotionality, sexual arousal, hunger, and parental behavior and altering the release of the

C. Noradrenergic pathways

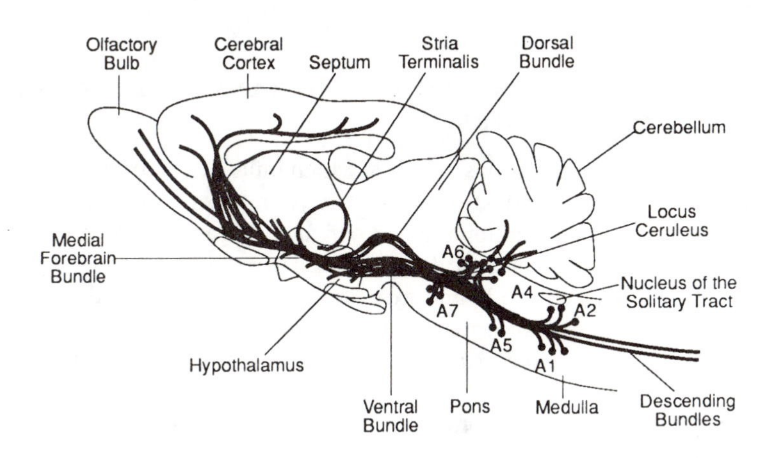

(C) Noradrenergic pathways. The noradrenergic cells in the pons and medulla (A1–A7) give rise to two major ascending pathways: the dorsal bundle, which projects from the locus ceruleus to the midbrain reticular formation, medial forebrain bundle, olfactory bulb and cerebral cortex; and the ventral bundle, which projects to the hypothalamic nuclei and median eminence and to the stria terminalis. (Redrawn from Ungerstedt, 1971.)

D. Serotonin pathways

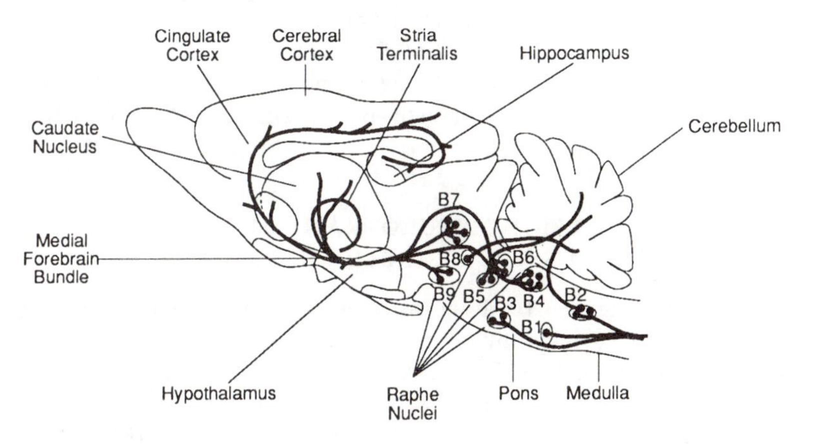

(D) Serotonin pathways. The primary serotonin (5-HT) nuclei occur in the dorsal and caudal raphe nuclei of the pons and medulla (B1–B9). The axons from the caudal nuclei (B1–B3) project down the spinal cord, while the axons from the dorsal nuclei (B4–B9) project to the cerebral cortex, caudate nucleus, hippocampus, and hypothalamus. (Redrawn from Fuxe and Jonsson, 1974.)

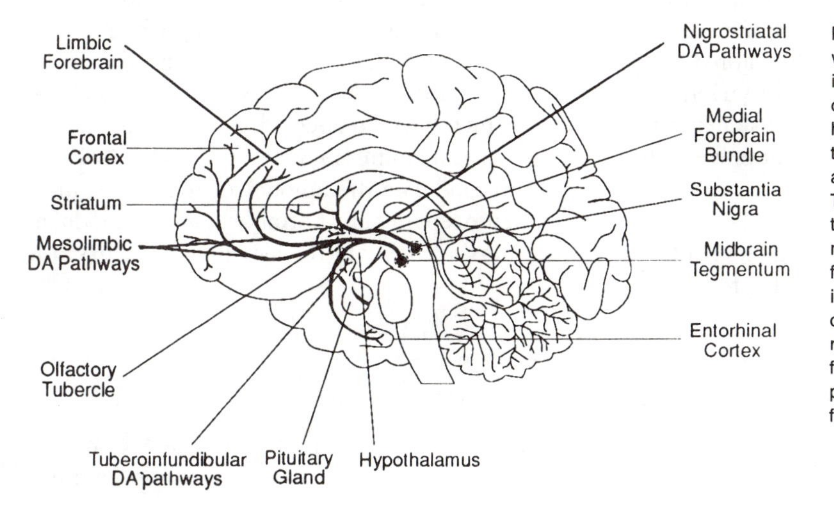

Figure 5.9. Dopamine pathways in the human brain. As in the rat (Fig. 5.8B), the neurons that contain dopamine have their cell bodies clustered in the substantia nigra and the midbrain tegmentum. These neurons send axons to the striatum, which regulates motor activity, and the limbic forebrain, which is involved in emotion. The tuberoinfundibular dopamine pathways regulate hormone secretion from the hypothalamus and pituitary gland. (Redrawn from Iversen, 1979.)

gonadotropic hormones and prolactin. The ability of hormones to modulate neurotransmitter release is discussed in Chapters 9 and 10.

5.8 DRUGS INFLUENCING NEUROTRANSMITTERS AND THEIR RECEPTORS

There are innumerable drugs available which influence neurotransmitter release and receptor activation. These drugs are used clinically to treat disorders of mood, emotionality, motor function and neuroendocrine secretion. Since many of these drugs are used to regulate the neuroendocrine system, both clinically and experimentally, they will be discussed here. More complete information on the role of drugs in the nervous system is given by Cooper *et al.*, (1991); McGeer *et al.*, (1987) and Gilman *et al.*, (1985).

Drugs can influence neurotransmitter action in three ways: (a) they can act as neuromodulators to influence the synthesis, storage, release and reuptake of the neurotransmitter by the presynaptic cell. (b) They can alter the activity of the deactivating enzymes, such as MAO and COMT in the synapse, thus influencing synaptic concentrations of the transmitter. (c) They can mimic (or block) the neurotransmitter at the synapse by attaching directly to the receptor on the postsynaptic cell (Axelrod, 1974; Iversen and Iversen, 1981; Cooper *et al.*, 1991; McGeer, *et al.*, 1987).

5.8.1 DRUGS ALTERING NEUROTRANSMITTER SYNTHESIS, STORAGE, RELEASE AND REUPTAKE

As depicted in Figure 5.10, drugs can increase the amount of transmitter released into the synapse by (a) promoting neurotransmitter synthesis; (b) increasing the amount of transmitter stored, (c) increasing the amount of transmitter released when the nerve is depolarized, (d) blocking reuptake into the presynaptic nerve terminal, or (e) inhibition of MAO metabolism of the transmitter. Examples of drugs having these effects are given in Table 5.5.

Other drugs reduce the amount of neurotransmitter in the synapse by (a) inhibiting transmitter synthesis, (b) reducing the amount of transmitter stored, or (c) reducing the amount of transmitter released when the presynaptic cell is depolarized. Drugs may also (d) be stored in vesicles as 'false neurotransmitters', which have no effect on the postsynaptic cell but prevent storage of the real transmitter in the vesicle (Table 5.5).

All of these drugs act to modulate the amount of neurotransmitter available in the synapse. Neuropeptides and hormones can also modulate the level of neurotransmitters available in the synapse as discussed in Chapter 12.

5.8.2 DRUGS INFLUENCING THE ENZYMES WHICH DEACTIVATE NEUROTRANSMITTERS

The enzymes which deactivate neurotransmitters (Table 5.4) can be influenced by a number of drugs (Table 5.5). One of the most common

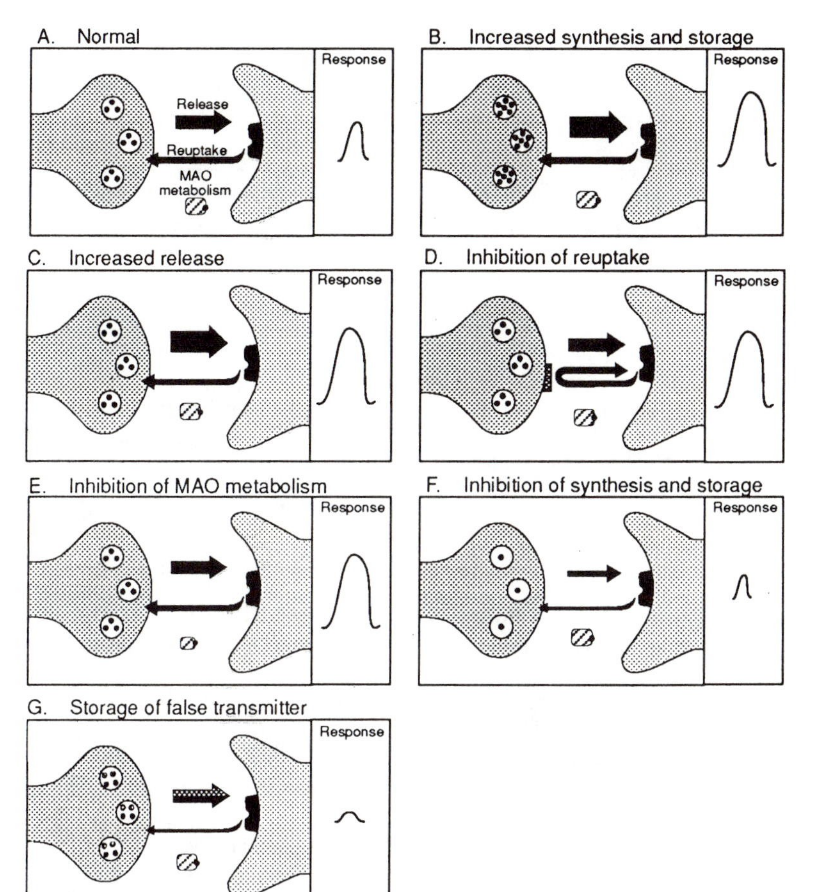

Figure 5.10. The mechanisms through which drugs alter the amount of adrenergic transmitters available in the synapse to postsynaptic receptors and the resultant response of the postsynaptic cell. (A) The normal storage, release, reuptake and MAO metabolism of noradrenaline with a curve representing the normal response of the postsynaptic cell. (B) Drugs such as tyrosine and L-dopa increase the amount of catecholamines synthesized and stored in the nerve terminal. (C) Drugs such as amphetamines increase the release of catecholamines and (D) block reuptake. (D) Antidepressant drugs, such as imipramine, a tricyclic antidepressant, increase the amount of catecholamines available in the synapse by blocking reuptake in to the presynaptic cell, while (E) MAO inhibitors, which are also used as antidepressants, increase the amount of catecholamines in the synapse by inhibition of MAO metabolism of the transmitters. (F) Conversely, reserpine, which reduces blood pressure and may induce depression, reduces the availability of the transmitter in the synapse by decreasing the amount stored and drugs such as alpha-methotyrosine reduce the amount of catecholamines in the synapse by inhibiting catecholamine synthesis, thus reducing the amount of stimulation of the postsynaptic cell. (G) Alpha-methyldopa is a 'false transmitter' which is stored in the synaptic vesicles with noradrenaline and released with it, diluting its effect. (Redrawn from Axelrod, 1974.)

types of drug influencing these enzymes is the monoamine oxidase (MAO) inhibitors, such as pargyline and clorgyline. The result of blocking MAO with these drugs is an increase in neurotransmitter levels in the synapse (Figure 5.10). Likewise, drugs which inhibit catechol-*O*-methyl transferase (COMT), such as tropolone, will result in an elevated level of noradrenaline in the synapse. Acetylcholinesterase-inhibiting

Table 5.5. *Drugs altering the amount of neurotransmitter available in the synapse*

Transmitter	Drug	Action
GABA	Valproate	(+) Inhibits GABA-T deactivation
	Tetanus toxin	(−) Inhibits release
	Isoniazid	(−) Inhibits synthesis
Acetylcholine	Choline	(+) Stimulates release
	Black-widow spider venom	(+) Facilitates synthesis
	Acetylcholinesterase inhibitors	(+) Reduce deactivation by AchE
	Pyridines	(−) Blocks synthesis
	Botulinus toxin	(−) Inhibits release
Dopamine	Tyrosine and L-dopa	(+) Increase dopamine synthesis
	Amphetamines	(+) Increase release and block reuptake
	MAO inhibitors (e.g. pargyline)	(+) Reduce deactivation by MAO
	Alpha-methyltyrosine	(−) Inhibits dopamine synthesis
	Reserpine	(−) Reduces amount stored
	Alpha-methyldopa	(−) Stored as a false transmitter
Noradrenaline	Amphetamines	(+) Increase release and block reuptake
	Tricyclic antidepressants (e.g. imipramine)	(+) Block reuptake
	MAO inhibitors	(+) Reduce deactivation by MAO
	Cocaine	(+) Blocks reuptake
	COMT inhibitors	(+) Inhibit deactivation by COMT
	Reserpine	(−) Reduces amount stored
	Alpha-methyltyrosine	(−) Inhibits synthesis
	Alpha-methyldopa	(−) Stored as a false transmitter
Serotonin	Tryptophan	(+) Increases synthesis
	Tricyclic antidepressants (e.g. imipramine)	(+) Block reuptake
	MAO inhibitors	(+) Reduce deactivation by MAO
	Para-chlorophenylalanine (pcpa)	(−) Inhibits synthesis
	Reserpine	(−) Reduces amount stored

Sources: Iversen and Iversen, 1981; Cooper *et al.*, 1991; McGeer *et al.*, 1987.
+ = increased, − = decreased amounts.

drugs such as physostigmine and neostigmine elevate acetylcholine levels in the synapse.

5.8.3 DRUGS WHICH ACT AT POSTSYNAPTIC RECEPTORS

Drugs can also bind to the receptor on the postsynaptic cell to act as neurotransmitter agonists or antagonists (Figure 5.11). The chemical structure of these drugs is very similar to that of the neurotransmitter so the receptor accepts them as if they were the transmitter. **Agonists** are drugs that can mimic or even magnify the effect of the transmitter. At an excitatory synapse an agonist will increase the excitation of the postsynaptic cell. If the synapse is inhibitory, an agonist will cause increased inhibition. A neurotransmitter **antagonist**, on the other hand, blocks

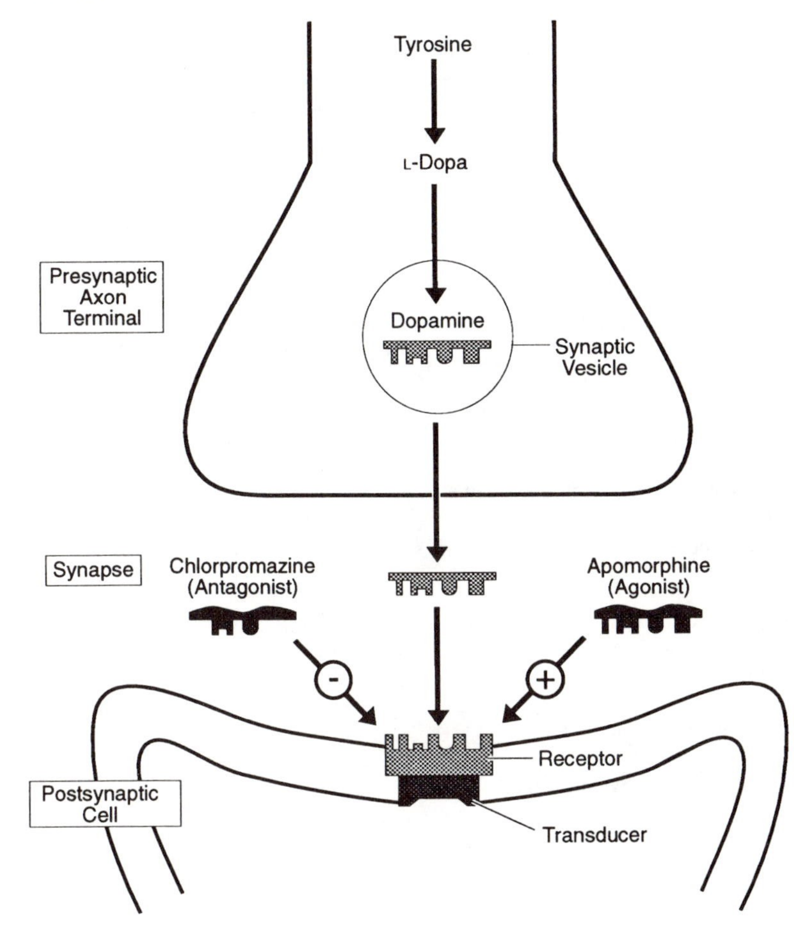

Figure 5.11. Dopamine receptor agonists and antagonists. Agonist drugs such as apomorphine mimic the action of dopamine by activating the receptor. Antagonist drugs such as chlorpromazine (Thorazine) block the receptor and prevent its activation by dopamine. (Redrawn from Nathanson and Greengard, 1977.)

the receptors for that transmitter. The antagonist has a chemical structure very much like that of the neurotransmitter but, when it binds to the receptor, it does not stimulate the cell. The real transmitter will be present in the synapse but will not be able to stimulate the receptor because it is blocked by the antagonist.

Treatment with a transmitter agonist is like that with the transmitter: the agonist binds to the postsynaptic cell and either excites or inhibits it. An antagonist also binds to the receptor, but blocks it, preventing the transmitter in the synapse from attaching to the receptor, and thus preventing any stimulation of the postsynaptic cell. Examples of drugs which act as specific agonists or antagonists for particular neurotransmitters are given in Table 5.6.

5.8.4 DRUGS, THE NEUROENDOCRINE SYSTEM AND BEHAVIOR

Changes in the release of neurotransmitters caused by drugs influence hormone release (Chapter 6), and alter arousal levels, emotional states, and motivated behavior. For example, depression is associated with low

Table 5.6. *Drugs acting as neurotransmitter agonists and antagonists*

Transmitter	Drug	Action
Excitatory amino acids: glutamate and aspartate	Kainic acid	(+) Agonist
	Quinolinic acid	(+) Agonist
GABA	Baclofen	(+) Agonist
	Muscimol	(+) Agonist
	Bicuculline	(−) Antagonist
	Picrotoxin	(−) Antagonist
Acetylcholine	Nicotine and carbachol	(+) Nicotinic receptor agonists
	Muscarine and pilocarpine	(+) Muscarinic receptor agonists
	Atropine and scopolamine	(−) Muscarinic receptor antagonists
	Curare	(−) Nicotinic receptor antagonist
Dopamine	Apomorphine	(+) Agonist
	Bromocriptime	(+) Agonist (D_2)
	Antipsychotic drugs (chlorpromazine)	(−) Antagonist
	Pimozide and haloperidol	(−) Antagonist
Noradrenaline	Adrenaline	(+) α and β agonist
	Clonidine	(+) α_2 Agonist
	Isoproterenol	(+) β Agonist
	Propranolol	(−) β Antagonist
	Phentolamine	(−) α Antagonist
	Yohimbine	(−) α_2 Antagonist
Serotonin	Quipazine	(+) Agonist
	Dimethyltryptamine (DMT)	(+) Agonist
	Methysergide	(−) Antagonist
	Lysergic acid diethylamide (LSD)	(−) Antagonist
Histamine	Impromadine	(+) Agonist
	Diamaprit	(+) Agonist
	Chlorpheniramine (antihistamines)	(−) Antagonist
	Metiamide	(−) Antagonist
Opioids	Morphine	(+) Agonist
	Heroin	(+) Agonist
	Methadone	(+) Weak agonist
	Naloxone, naltrexone	(−) Antagonist
	Nalorphine	(−) Antagonist
Benzodiazepines	Diazepam (Valium)	(+) Agonist
	Chlordiazepoxide (Librium)	(+) Agonist

Sources: Tallman, *et al.*, 1980; Iversen and Iversen, 1981; Cooper *et al.*, 1991; McGeer *et al.*, 1987.

catecholamine levels, schizophrenia with high levels, so drugs which alter catecholaminergic activity also influence these clinical symptoms. Altering catecholamine levels also alters sexual and aggressive motivation, and the activity of the neuroendocrine system.

5.9 NUTRIENTS MODIFYING NEUROTRANSMITTER LEVELS

Many nutrients in foods are believed to facilitate the synthesis of neurotransmitters (Wurtman, 1982; Milner and Wurtman, 1986). Thus, foods such as carbohydrates, which elevate the level of tryptophan, the amino acid precursor for serotonin, may facilitate serotonin synthesis. Increased serotonin facilitates changes in sleep, mood and appetite; thus, a daily rhythm of serotonin synthesis may be based on the timing of meals and the type of food eaten. High carbohydrate meals increase serotonin levels by increasing insulin levels, which decreases the levels of other amino acids. Thus, a high carbohydrate intake may be followed by tiredness and lethargy.

A high protein meal elevates levels of all amino acids and, therefore, lowers uptake of tryptophan due to competition of amino acids for uptake into the brain. High protein meals elevate levels of choline, the precursor for acetylcholine, and may influence attentional and memory mechanisms regulated by acetylcholine. According to Wurtman (1982), increasing lecithin consumption can also elevate choline levels and the resultant increase in acetylcholine can reduce the symptoms of neuromuscular diseases such as tardive dyskinesia and reduce the memory deficits associated with Alzheimer's disease. Wurtman (1982) also suggests that foods high in tyrosine may stimulate the synthesis of dopamine and other catecholamines and, therefore, reduce the uncontrolled movements symptomatic of Parkinson's disease. Such diet therapy may also be of use in treating affective disorders such as depression, hyperactivity and hypertension.

To what degree changes in diet can modulate neurotransmitter synthesis in humans is not yet known. This type of 'nutrition therapy' is controversial and critical analyis indicates that it has little or no neurological effects (DeFeudis, 1987). This controversy is bound to continue for some time before it is resolved.

5.10 THE DIVISIONS OF THE NERVOUS SYSTEM

The nervous system can be divided into a number of divisions and subdivisions, as shown in Figure 5.12. The central nervous system (CNS) consists of the nerve cells and neural pathways of the brain and spinal cord described in Section 5.7 and the peripheral nervous system consists of the peripheral sensory and motor nerves which extend throughout the body from the central nervous system. The peripheral nervous system has a somatic (skeletal) division and an autonomic division.

5.10.1 THE SOMATIC NERVOUS SYSTEM

The somatic division of the peripheral nervous system receives sensory (afferent) input from the sense organs and controls motor (efferent) output to the skeletal muscles through the 12 pairs of cranial nerves and the 31 pairs of spinal nerves shown in Figure 5.13. The somatic sensory nerves receive tactile input from the skin and muscles, while the somatic motor nerves form neuromuscular synapses which use acetylcholine as a neurotransmitter. The somatic nervous system is concerned with conscious (voluntary) actions and reflexes and sends motor nerves to those muscles which are under voluntary control. The cranial sensory and motor nerves include the **olfactory** nerve (smell), the **optic** nerve

Figure 5.12. The divisions of the nervous system.

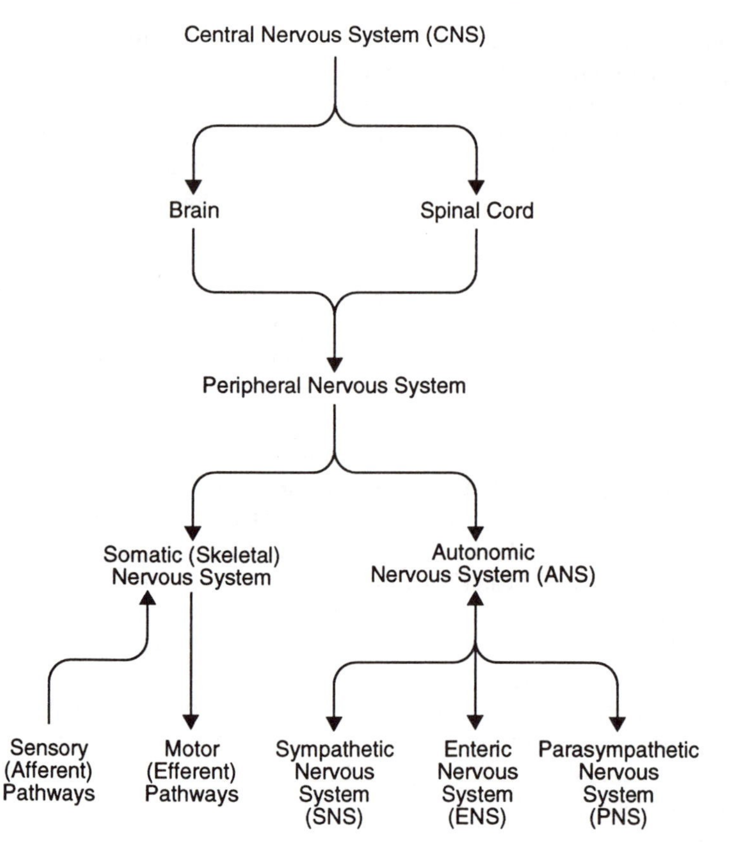

(vision), the **acoustic** nerve (hearing and balance), the **oculomotor**, **trochlear** and **abducens** nerves (eye movement), the **trigeminal** nerve (receives facial sensations and controls jaw muscles), the **facial** nerve (receives taste stimuli and controls the facial muscles), the **glosso-pharyngeal** nerve (receives taste stimuli and controls the muscles of the larynx used in speech), the **vagus** nerve (supplies heart, lungs, stomach, kidneys and intestines), the **accessory** nerve (controls head and neck movement), and the **hypoglossal** nerve (controls tongue movements).

The 31 pairs of somatic spinal sensory and motor nerves extend from the spinal cord in four groups as shown in Figure 5.13. These are the **cervical** (8 pairs of nerves which innervate the the throat, chest, arms and hands), **thoracic** (12 pairs of nerves which innervate the the trunk of the body), **lumbar** (5 pairs of nerves which innervate the front of the legs and the feet), and **sacral** (5 pairs of nerves which innervate the soles of the feet and back of the legs and the coccygeal nerve). The nerves of the somatic nervous system are myelinated and monosynaptic, extending directly from synapses in the spinal cord to their target tissues.

5.10.2 THE AUTONOMIC NERVOUS SYSTEM

The autonomic nervous system (ANS), as shown in Figure 5.14, receives sensory input from the viscera and sends motor nerves to the visceral

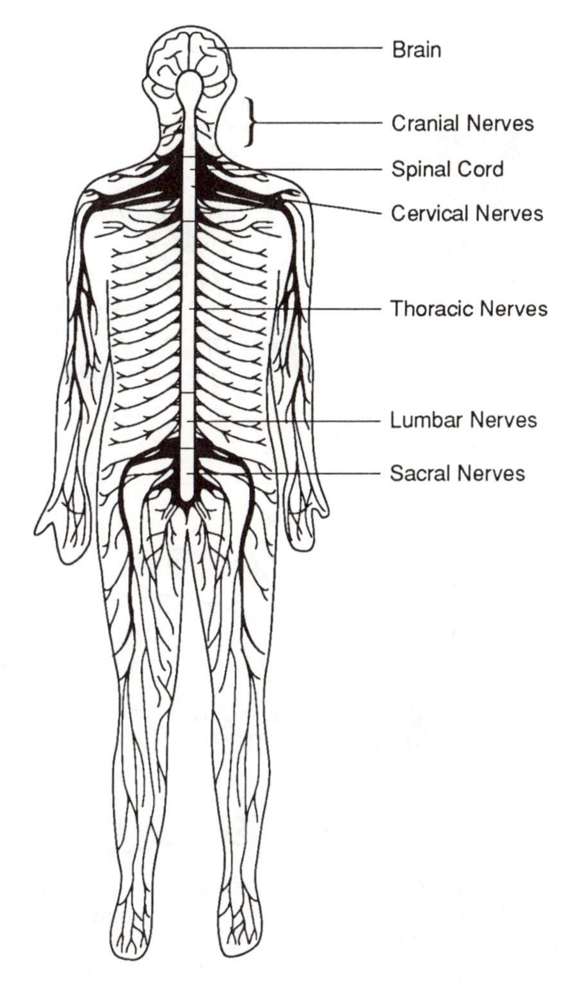

Figure 5.13. The central nervous system (CNS) and the peripheral somatic nervous system. The CNS consists of the brain and spinal cord. The peripheral somatic nervous system includes the cranial and spinal sensory and motor nerves. The cranial nerves connect directly to the brain while the spinal nerves extend from the cervical, thoracic, lumbar and sacral regions of the spinal cord.

(smooth) muscles of the heart, arteries and gastrointestinal system, etc. to the glandular tissues of the body, such as the salivary glands, lachrymal (tear) glands, sweat glands, and to the peripheral endocrine glands (thymus, adrenal, thyroid and gonads). The sensory nerves of the ANS are non-myelinated nerves which project from the heart, lungs, gastrointestinal tract, etc. to the autonomic sensory ganglia. Sensory afferents are carried to the spinal cord and brain by the vagus, pelvic, splanchnic and other autonomic nerves. Neural transmission in the visceral afferent nerves involves a wide range of neuropeptides including substance P, the endogenous opioid peptides, cholecystokinin, neuropeptide Y and others (see Chapters 11 and 12).The efferent branch of the ANS has three divisions: the **sympathetic** (SNS), **parasympathetic** (PNS) and **enteric** (ENS) nervous systems, which regulate those activities over which there is no conscious control, the involuntary muscles and glands. The visceral muscles and glands receive paired inputs, one from the sympathetic and one from the parasympathetic branches of the ANS. The enteric nervous system is that branch of the ANS which innervates the gastrointestinal tract.

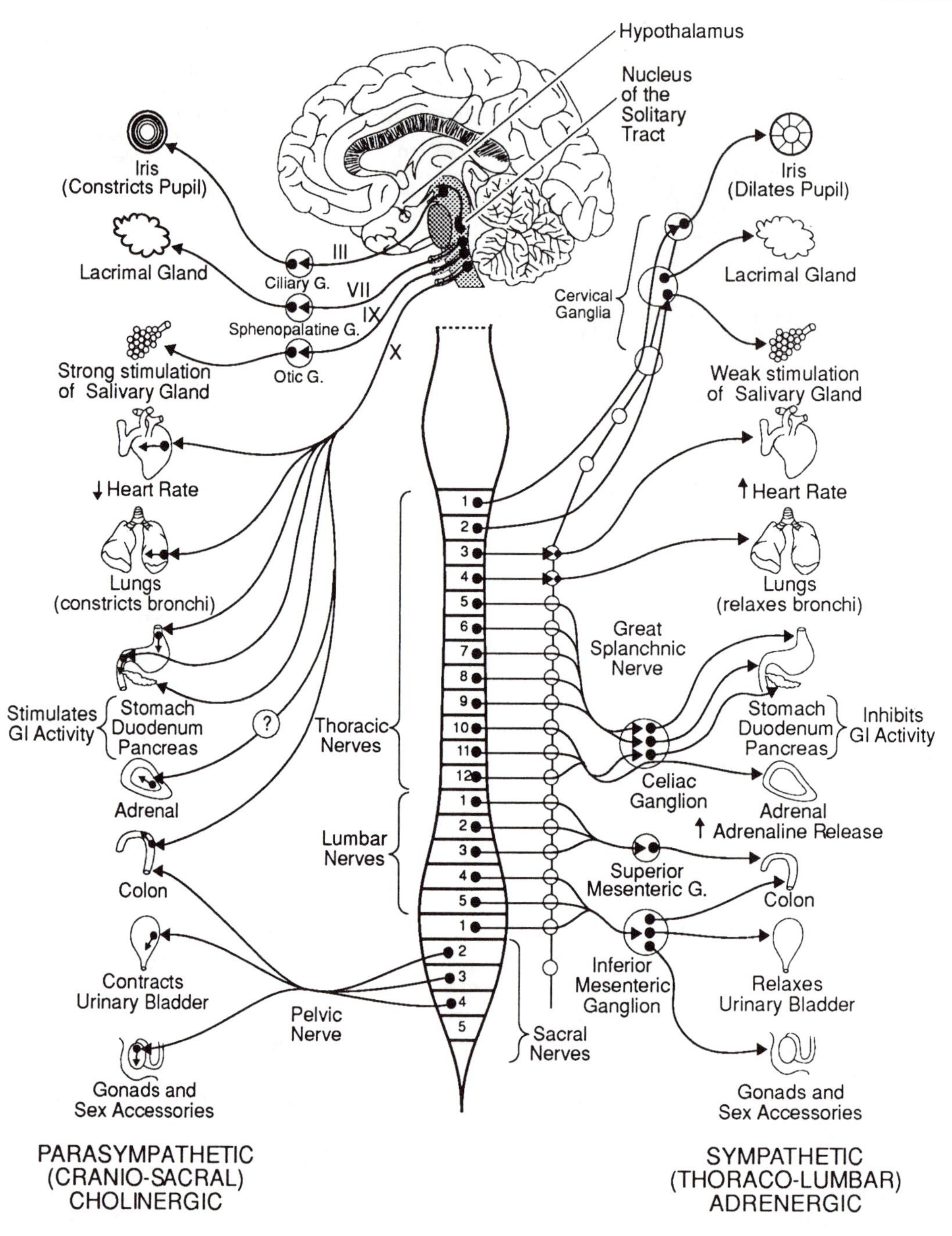

Figure 5.14. The two branches of the autonomic nervous system (ANS). The sympathetic nerves extend from the thoracic and lumbar regions of the spinal cord and act via adrenergic synapses. The parasympathetic nerves extend from the cranial and sacral regions of the spinal cord and act via cholinergic synapses. The ANS is regulated by the CNS via neurons in the hypothalamus and limbic system which project to the nucleus of the solitary tract in the brain stem, and then to the spinal cord and autonomic ganglia (G) which lie outside the spinal cord. Visceral afferents send projections to the autonomic sensory ganglia and then to the spinal cord, where sensory information is carried to the nucleus of the solitary tract, hypothalamus and limbic system. (Redrawn from Turner and Bagnara, 1976.)

The nerves of the ANS synapse at ganglia outside of the spinal cord and projections are sent from these autonomic ganglia to the target organs. The autonomic ganglia include the cervical ganglia, splanchnic ganglia, celiac ganglia, and the mesenteric ganglia (see Figure 5.14). The sympathetic nerves extend from the thoracic and lumbar regions of the spinal cord to the sympathetic ganglia and then to their target organs. The parasympathetic nerves extend from the cranial and sacral regions of the spinal cord to the autonomic ganglia and then to their target organs. The SNS activates the visceral muscles and glands during stress and thus the SNS deals with emergency responses by increasing blood pressure, causing sweating and stimulating adrenaline release. The PNS acts to maintain the body's systems at a steady state and opposes the excitatory actions of the SNS. The PNS is involved in growth promoting and energy conserving functions, while the SNS increases energy expenditure. The SNS causes pupil dilation, relaxes the bronchi in the lungs, increases heart rate, stimulates adrenaline secretion, inhibits gastrointestinal activity and relaxes the bladder. The PNS constricts the pupils, stimulates salivation, constricts the bronchi in the lungs, slows heart rate, stimulates gastrointestinal activity and constricts the bladder. The SNS nerves are adrenergic, while the PNS nerves are cholinergic (see Weiner and Taylor, 1985; Dodd and Role, 1991).

5.10.3 THE ENTERIC NERVOUS SYSTEM

The third branch of the ANS is the enteric nervous system, which innervates the gastrointestinal tract, pancreas and gall bladder through autonomic nerves of the superior and inferior mesenteric ganglia (see Figure 5.14). The sensory nerves of the enteric nervous system send input from the intestinal walls and monitor the chemical environment of the gastrointestinal tract, while the motor nerves control the muscles, vasculature and glandular secretions of the gastrointestinal tract. The enteric nervous system can act independently, but is also regulated by CNS reflexes and by the sympathetic and parasympathetic branches of the ANS (Dodd and Role, 1991).

5.10.4 CENTRAL REGULATION OF THE AUTONOMIC NERVOUS SYSTEM

All three branches of the ANS are regulated by the central nervous system as shown in Figure 5.14. Nuclei of the hypothalamus and limbic system regulate the efferent pathways of the SNS, PNS and ENS through the nucleus of the solitary tract which sends fibers to the brain stem nuclei and spinal autonomic ganglia (see Figure 5.8). Visceral afferents from the periphery project to the autonomic ganglia, which send fibers to the spinal cord and brain stem nuclei, and on to the hypothalamic and limbic nuclei through the pathways shown in Figure 5.8. The hypothalamus thus integrates the activity of the ANS as well as the neuroendocrine system as discussed in Section 4.1 (see Dodd and Role, 1991).

5.11 SUMMARY

This chapter has examined the criteria necessary to define a neurotransmitter and the function of neurotransmitters in communication between nerve cells through their synaptic activity. Details of the synthesis, storage, release and deactivation of amino acid, monoamine and peptide neurotransmitters are described as well as the mechanism for neurotransmitter release and the actions of transmitters at their postsynaptic receptors. The different types of receptor for neurotransmitters are described and the enzymes which deactivate the transmitters are summarized. Neurotransmitters are found in specific nerve pathways in the brain and the location of these pathways is outlined. Finally, the effects of drugs on the synthesis, storage, release, reuptake and deactivation of neurotransmitters are discussed. Other drugs act on neurotransmitter receptors as agonists or antagonists and the mechanism of action of these drugs is examined. Neurotransmitters and the drugs that influence them regulate the release of hypothalamic hormones and thus affect hormone-dependent behaviors. Because drugs are used both clinically and experimentally to alter neuroendocrine activity, they are an important tool in the study of neuroendocrinology. Since dietary amino acids may influence neurotransmitter synthesis, diet may also influence neuroendocrine secretion.

FURTHER READING

Bradford, H. F. (1986). *Chemical neurobiology: An introduction to neurochemistry*. New York: W. H. Freeman.

Cooper, J. R., Bloom, F. E. and Roth, R. H. (1991). *The Biochemical Basis of Neuropharmacology*, 6th edn. New York: Oxford University Press.

Iversen, S. D. and Iversen, L. I. (1981). *Behavioral Pharmacology*, 2nd edn. New York: Oxford University Press

McGeer, P. L., Eccles, J. C. and McGeer, E. G. (1987). *Molecular neurobiology of the mammalian brain*, 2nd edn. New York: Plenum Press.

Niewenhuys, R. (1985). *Chemoarchitecture of the brain*. Berlin: Springer-Verlag.

REVIEW QUESTIONS

5.1 Name the three catecholamine neurotransmitters.

5.2 Name three ways that drugs can increase the amount of neurotransmitter available at a synapse.

5.3 Is GABA an excitatory or inhibitory neurotransmitter?

5.4 What is the cholinergic neurotransmitter?

5.5 Which neurotransmitter is synthesized from dopamine by the enzyme dopamine beta-hydroxylase (DBH)?

5.6 What is the difference between a neurotransmitter agonist and an antagonist?

5.7 Which two enzymes deactivate noradrenaline?

5.8 Which neurotransmitter is synthesized from the amino acid tryptophan?
5.9 Would an antagonist for dopamine stimulate production of the second messenger cyclic AMP in a postsynaptic cell? (yes or no).
5.10 Where are neurotransmitters stored before release and what triggers their release?
5.11 In what two general ways can neurotransmitters be deactivated?
5.12 How does a transmitter cause a postsynaptic cell to 'fire'?
5.13 Which neurotransmitter acts at nicotinic receptors?
5.14 Which neurotransmitter has its primary cell bodies in the locus ceruleus?

ESSAY QUESTIONS

5.1 Are the eight criteria defining a neurotransmitter necessary and sufficient? Could they be reduced or expanded?
5.2 Discuss the mechanisms involved in the release of a neurotransmitter into the synapse.
5.3 Discuss the functions of synaptic vesicles.
5.4 How do the α- and β-adrenergic receptors differ in their functions?
5.5 Discuss the five major acetylcholine pathways in the brain with reference to the neuroendocrine system.
5.6 MAO inhibitors and tricyclic antidepressants are both used to treat depression, yet act through different mechanisms. Discuss these mechanisms and explain how these two types of drug both reduce depression.
5.7 Discuss the problems in determining how a particular type of drug such as β blockers actually affect the person treated with them. That is, what receptors are influenced and what are the physiological results? Why do two drugs with the same action, such as two β blockers, have different physiological effects?
5.8 Does 'nutrition therapy' alter neurotransmitter levels enough to influence memory and other neural processes in humans?

REFERENCES

Akil, H., Watson, S. J., Young, E., Lewis, M. E., Khachaturian, H. and Walker, J. M. (1984). Endogenous opioids: Biology and function. *Annual Review of Neuroscience*, **7**, 223–255.

Axelrod, J. (1974). Neurotransmitters. *Scientific American*, **230** (6), 58–71.

Barchas, J. D., Akil, H., Elliott, G. R., Holman, R. B. and Watson, S. J. (1978). Behavioral neurochemistry: neuroregulators and behavioral states. *Science*, **200**, 964–973.

Bloom, F. E. (1988). Neurotransmitters: past, present, and future directions. *Federation Proceedings*, **2**, 32–41.

Blusztajn, J. K. and Wurtman, R. J. (1983). Choline and cholinergic neurons. *Science*, **221**, 614–620.

Bradford, H. F. (1986). *Chemical neurobiology: An introduction to neurochemistry*. New York: W. H. Freeman.

Cooper, J. R., Bloom, F. E. and Roth, R. H. (1991). *The Biochemical Basis of Neuropharmacology*, 6th edn. New York: Oxford University Press.
DeFeudis, F. V. (1987). The brain is protected from nutrient access. *Life Sciences*, **40**, 1–9.
Dismukes, R. K. (1979). New concepts of molecular communication among neurons. *Behavioral and Brain Sciences*, **2**, 409–448. (This includes the 'target' article, a number of critical commentaries, and the author's reply.)
Dodd, J. and Role, L. W. (1991). The autonomic nervous system. In E. R. Kandel. J. H. Schwartz, and T. M. Jessell (eds.) *Principles of Neural Science*, pp. 761–775. New York: Elsevier.
Elliott, G. R. and Barchas, J. D. (1979). Neuroregulators: neurotransmitters and neuromodulators. *Behavioural and Brain Sciences*, **2**, 423–424.
Fuxe, K. and Jonsson, G. (1974). Further mapping of central 5-hydroxy-tryptamine neurons: Studies with the neurotoxic dihydroxytryptamines. *Advances in Biochemical Psychopharmacology*, **10**, 1–12.
Gilman, A. G., Goodman, L. S., Rall, T. W. and Murad, F. (1985). *Goodman and Gilman's The Pharmacological Basis of Therapeutics*, 7th edn. New York: Macmillan.
Hadley, M. E. (1992). *Endocrinology*. 3rd edn. Englewood Cliffs, NJ: Prentice-Hall.
Hartig, P. R. (1989). Molecular biology of 5-HT receptors. *Trends in Pharmacological Sciences*, **10**, 64–69.
Iversen, L. L. (1979). The chemistry of the brain. *Scientific American*, **241**(3), 134–149.
Iversen, S. D. and Iversen, L. I. (1981). *Behavioral Pharmacology*, 2nd edn. New York: Oxford University Press.
Kelley, R. B., Deutsch, J. W., Carlson, S. S. and Wagner, J. A. (1979). Biochemistry of neurotransmitter release. *Annual Review of Neuroscience*, **2**, 399–446.
Kuhar, M. J., de Souza, E. B. and Unnerstall, J. R. (1986). Neurotransmitter receptor mapping by autoradiography and other methods. *Annual Review of Neuroscience*, **9**, 27–59.
McGeer, P. L., Eccles, J. C. and McGeer, E. G. (1987). *Molecular Neurobiology of the Mammalian Brain*, 2nd edn. New York: Plenum Press.
McKelvy, J. F. (1986). Inactivation and metabolism of neuropeptides. *Annual Review of Neuroscience*, **9**, 415–434.
Milner, J. D. and Wurtman, R. J. (1986). Catecholamine synthesis: Physiological coupling to precursor supply. *Biochemical Pharmacology*, **35**, 875–881.
Minneman, K. P., Pittman, R. N. and Molinoff, P. B. (1981). β-Adrenergic receptor subtypes: properties, distribution, and regulation. *Annual Review of Neuroscience*, **4**, 419–461.
Moore, R. Y. and Bloom, F. E. (1979). Central catecholamine neuron systems: anatomy and physiology of the norepinephrine and epinephrine systems. *Annual Review of Neuroscience*, **2**, 113–168.
Nathanson, J. A. and Greengard, P. (1977). 'Second messengers' in the brain. *Scientific American*, **237**(2), 108–119.
Niewenhuys, R. (1985). *Chemoarchitecture of the Brain*. Berlin: Springer-Verlag.
Osborne, N. N. (1981). Communication between neurones: current concepts. *Neurochemistry International*, **3**, 3–16.

Panksepp, J. (1986). The neurochemistry of behavior. *Annual Review of Psychology*, **37**, 77–107.
Schwartz, J. H. (1980). The transport of substances in nerve cells. *Scientific American*, **242** (4), 152–171.
Semba, K. and Fibiger, H. C. (1989). Organization of central cholinergic system. *Progress in Brain Research*, **79**, 37–63.
Snyder, S. (1984). Drug and neurotransmitter receptors in the brain. *Science*, **224**, 22–31.
Stevens, C. F. (1979). The neuron. *Scientific American*, **241** (3), 55–65.
Sunahara, R. K., Guan, H.-C., O'Dowd, B. F., Seeman, P., Laurier, L. G., Ng, G., George, S. R., Torchia, J., van Tol, H. H. M. and Niznik, H. B. (1991). Cloning the gene for a human dopamine D_5 receptor with higher affinity for dopamine than D_1. *Nature*, **350**, 614–619.
Tallman, J. F., Paul, S. M., Skolnick, P. and Gallager, D. W. (1980). Receptors for the age of anxiety: pharmacology of the benzodiazepines. *Science*, **207**, 274–281.
Turner, C. D. and Bagnara, J. T. (1976). *General Endocrinology*, 6th edn. Philadelphia: Saunders.
Ungerstedt, U. (1971). Stereotaxic mapping of the monoamine pathways in the rat brain. *Acta Physiologica Scandinavica, Supplement*. **367**, 1–48.
Weiner, N. and Taylor, P. (1985). Neurohumoral transmission: the autonomic and somatic motor nervous systems. In A. G. Gilman, L. S. Goodman, T. W. Rall, and F. Murad (eds.) *Goodman and Gilman's The Pharmacological Basis of Therapeutics*, pp. 66–99. New York: Macmillan.
Westfall, T. C. (1984). Evidence that noradrenergic transmitter release is regulated by presynaptic receptors. *Federation Proceedings*, **43**, 1352–1357.
Wurtman, R. J. (1982). Nutrients that modify brain function. *Scientific American*, **246**(4), 50–59.

6

Neurotransmitter control of hypothalamic, pituitary and other hormones

Previous chapters have discussed the endocrine glands and their hormones (Chapter 2), the hormones of the pituitary gland (Chapter 3), the hypothalamic hormones (Chapter 4) and the neurotransmitters (Chapter 5). This chapter will describe how neurotransmitters influence the release of the hypothalamic and pituitary hormones and the hormones of the adrenal medulla, pancreas, thymus and gastrointestinal tract. It will also examine the electrophysiological properties of neurosecretory cells and the effects of drugs on the release of neurohormones.

6.1 THE CASCADE OF CHEMICAL MESSENGERS

As shown in Figure 6.1, there is a cascade of chemical messengers: a nerve cell releases a neurotransmitter which regulates the release of neurohormones from hypothalamic neurosecretory cells. These hypothalamic hormones stimulate the cells of the adenohypophysis to synthesize and release their hormones. Many of these pituitary hormones, such as LH and FSH, act on endocrine target cells, such as the gonads, causing

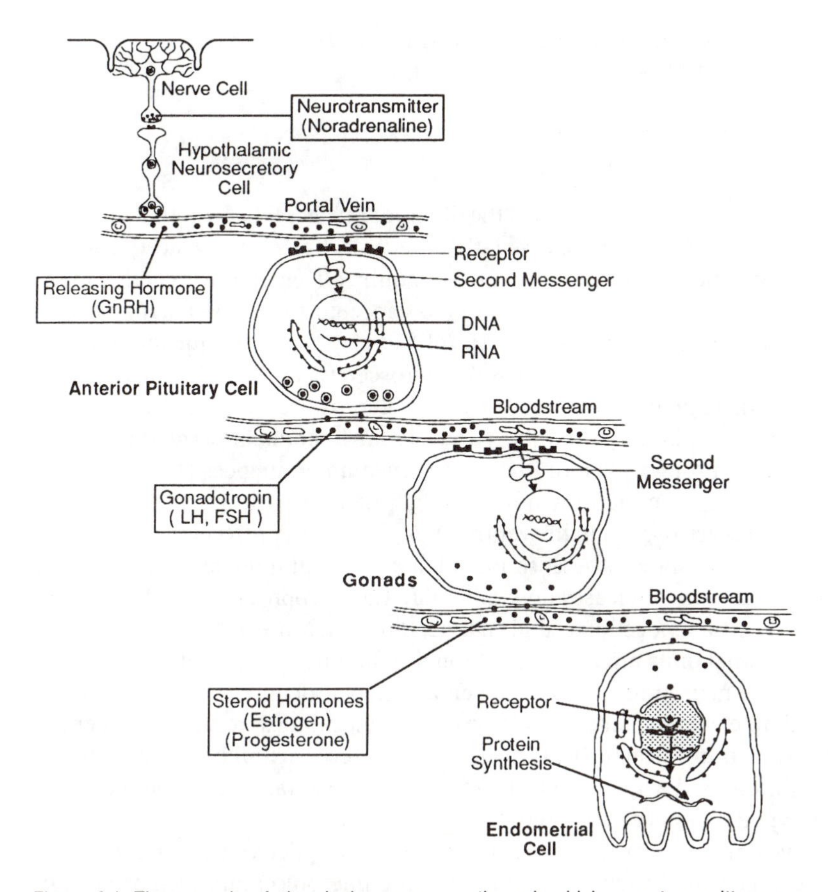

Figure 6.1. The cascade of chemical messengers through which neurotransmitters can regulate the neuroendocrine system. A neurotransmitter (perhaps noradrenaline) is released from a neuron and stimulates a hypothalamic neurosecretory cell to secrete a neurohormone, such as GnRH, into the hypophyseal portal veins, which transport it to the anterior pituitary, where it stimulates receptors on gonadotroph cells. These receptors activate a second messenger system within the cell which causes the synthesis and release of pituitary hormones, such as LH and FSH. These pituitary hormones enter the general circulation where they are picked up by receptors on their target cells in the gonads where they act, by way of the cyclic AMP second messenger system, to stimulate the synthesis of the gonadal steroid hormones, estrogen and progesterone. The gonadal steroids move through the blood-stream to their target cells, some of which are in the endometrium of the uterus, where they interact with the cell nucleus via intracellular receptors to stimulate protein synthesis. (Redrawn from Segal, 1974.)

them to synthesize and release their own hormones (estrogen and progesterone), which stimulate biochemical changes in target cells in the body or brain. In each step of this pathway the neurotransmitters and peptide hormones stimulate membrane receptors which activate a second messenger, such as cyclic AMP, within the target cell (Chapter 10). The steroid hormones act on receptors located inside the target cells (Chapter 9). This chapter describes the effects of neurotransmitters on neurosecretory cells.

6.2 NEURAL CONTROL OF HYPOTHALAMIC NEUROSECRETORY CELLS

6.2.1 NEURAL INPUT TO THE ENDOCRINE HYPOTHALAMUS

Figure 4.1 (p. 41) described the different nuclei of the hypothalamus and Figure 5.8 (p. 72) showed that nerve cells releasing acetylcholine, dopamine, noradrenaline and serotonin all enter the hypothalamus. These neurotransmitters all stimulate receptors on hypothalamic neurosecretory cells. Figure 6.2 is a simplified view of the relationship between the magnocellular hypothalamic neurosecretory cells of the paraventricular and supraoptic nuclei (PVN and SON), whose axons terminate in the posterior pituitary, and the parvicellular hypothalamic neurosecretory cells, whose axons terminate in the median eminence. This figure also gives an example of a noradrenergic neuron with a synapse on to a neurosecretory cell in the ventromedial nucleus (VMN) of the hypothalamus and shows the location of a tuberoinfundibular dopaminergic neuron which releases dopamine into the hypophyseal portal veins.

The magnocellular and parvicellular neurosecretory cells of the hypothalamus are regulated through many neurotransmitter pathways. These neurotransmitters, as well as the neuropeptides (as described in Chapter 11) regulate the release of neurohormones from the neuroendocrine 'transducer' cells; i.e. cells which receive neural input (neurotransmitters) and which secrete neurohormones. Neural input to the endocrine hypothalamus comes from: the sensory receptors (nose, taste buds, ears, eyes and touch); the exteroreceptors, which detect pain, temperature changes, and suckling stimulation; the interoreceptors such as chemoreceptors, blood volume receptors (baroreceptors), osmoreceptors and glucoreceptors; the sensory receptors in the uterus and cervix; other nuclei in the brain which control time-dependent and sleep–wake rhythms; psychological arousal as occurs during emotional states such as stress (Hutchinson, 1978).

Figure 6.3 indicates how different neurotransmitters (serotonin, dopamine and noradrenaline) from nerve cell axons might regulate the secretion of hypothalamic releasing and inhibiting hormones through their action on the synaptic receptors of the neurosecretory cells. Neurosecretory cells occur in different neurotransmitter pathways and the neurotransmitters in these pathways may stimulate or inhibit hormone release. One of the main problems in studying the neurotransmitters regulating hypothalamic and pituitary hormone release is that the same transmitters may be involved in different pathways of hormone secretion. Thus, different transmitters may regulate the baseline secretion of a hormone, circadian rhythms of hormone secretion, hormone responses to stress, pain or cold, and feedback regulation by other hormones.

Because hypothalamic neurohormones are released in small quantities (approximately 10^{-15} g) and have very short half-lives in the bloodstream, it is easier to measure circulating levels of pituitary hormones

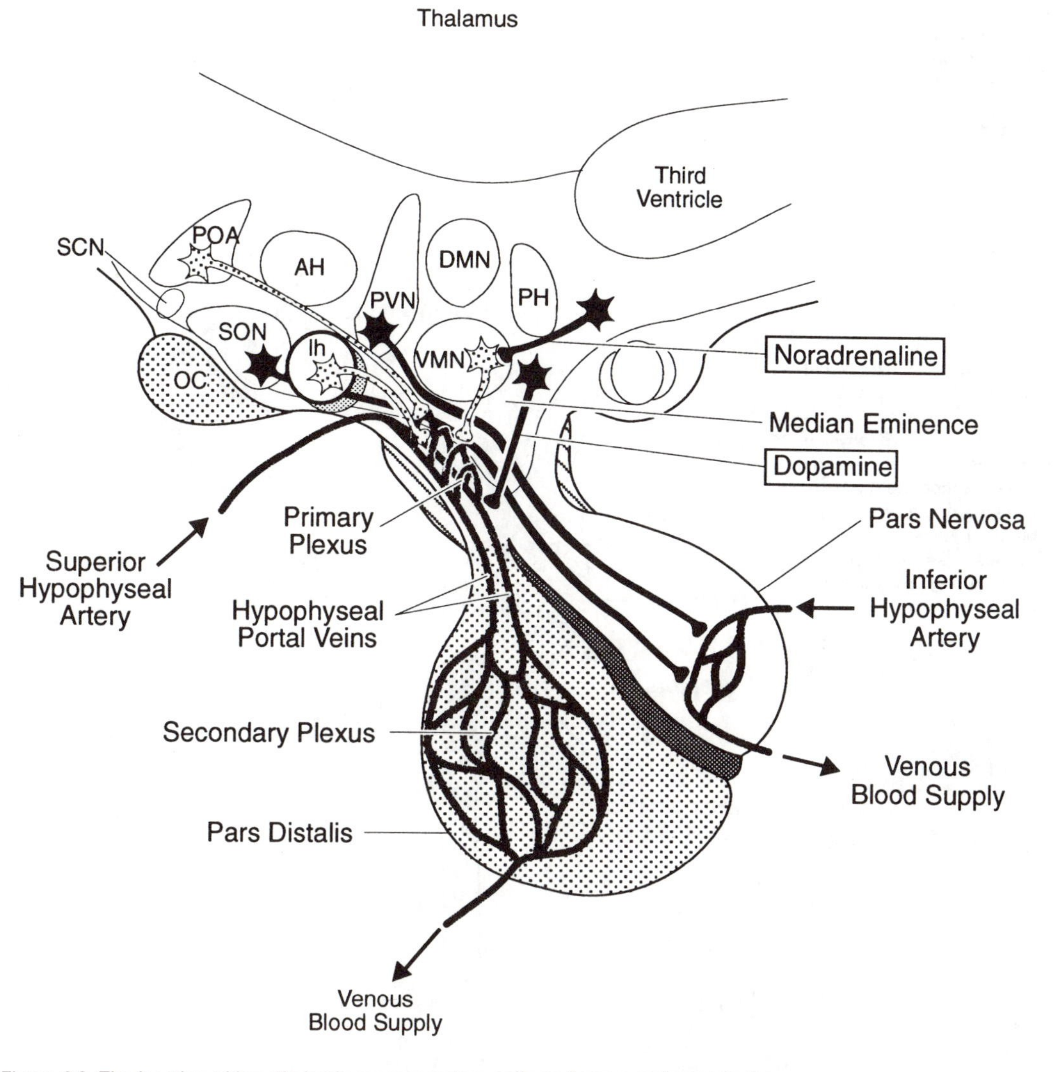

Figure 6.2. The location of hypothalamic neurosecretory cells and nerve pathways in the hypothalamus. Parvicellular neurosecretory cells which have their cell bodies in the POA, lh and VMN are shown. High concentrations of dopamine and noradrenaline are found in the ventromedial nucleus and the median eminence of the hypothalamus. These transmitters may alter neurohormone secretion from neurosecretory cells, as is shown for noradrenaline in the VMN. Transmitters may also be released directly into the primary plexus of the hypophyseal portal system, as is shown for dopamine. The magnocellular neurosecretory cells, whose axons extend to the posterior pituitary, are also shown. AH = anterior hypothalamus; DMN = dorsomedial nucleus; lh = lateral hypothalamus; oc = optic chiasm; PH = posterior hypothalamus; POA = preoptic area; PVN = paraventricular nucleus; SON = Supra optic nucleus; VMN = ventromedial nucleus.

after the manipulation of neurotransmitter levels. Thus, the following discussion will focus on pituitary hormone responses to changes in neurotransmitter levels. Keep in mind, however, that the plasma levels of pituitary hormones reflect changes in hypothalamic hormone release via the cascade of chemical messengers illustrated in Figure 6.1.

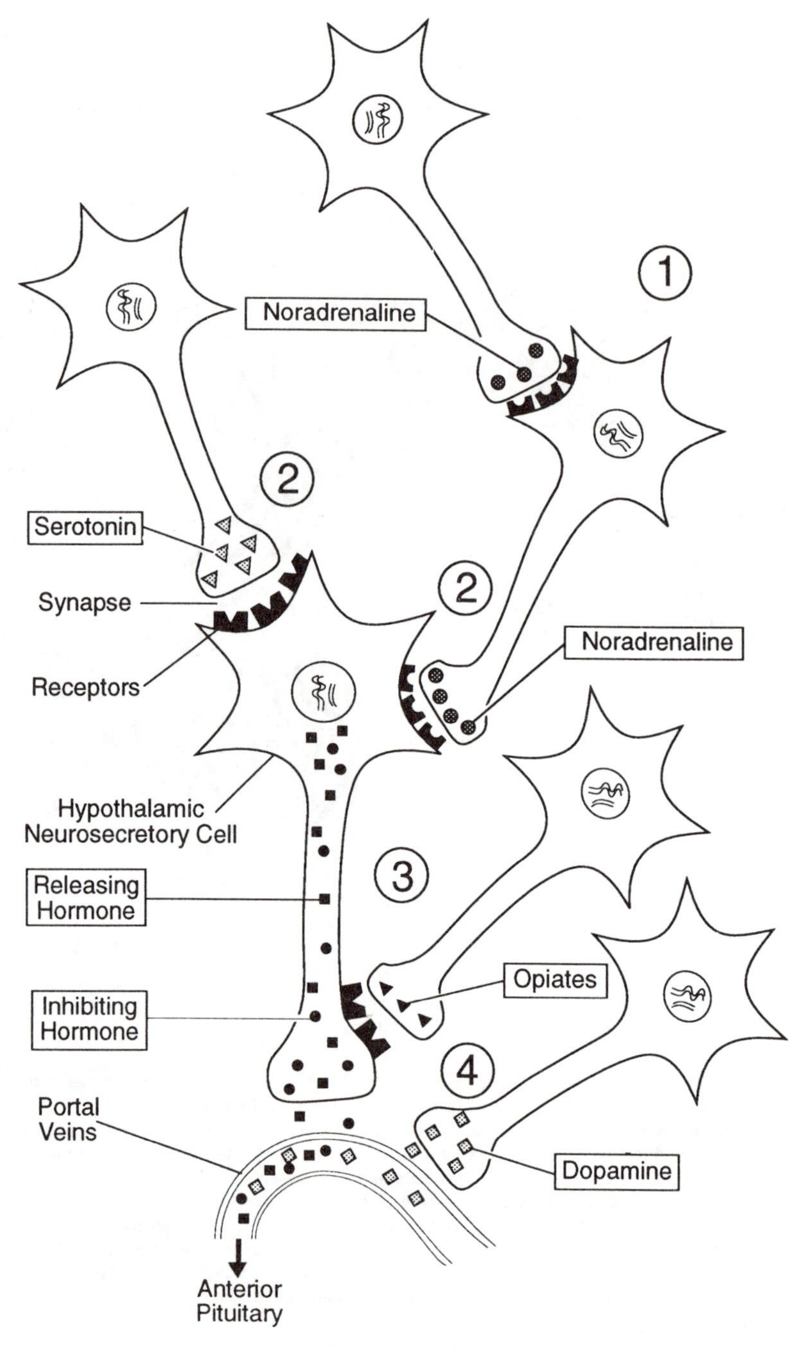

Figure 6.3. Four possible mechanisms through which neurotransmitters can regulate the release of hypothalamic and pituitary hormones. (1) Indirect stimulation of the neurosecretory cell through sensory or other nerve pathways which synapse on an interneuron which then synapses onto the neurosecretory cell. (2) Direct axo-dendritic or axo-somatic synapses on the neurosecretory cell. (3) Axo-axonal synapses on the neurosecretory cell. (4) Release of transmitters into the hypophyseal portal veins to act directly on the endocrine cells of the pituitary. Upon stimulation, the neurosecretory cell releases its releasing or inhibiting hormones into the portal veins.

6.2.2 MECHANICANISMS THROUGH WHICH NEUROTRANSMITTERS CAN REGULATE HYPOTHALAMIC AND PITUITARY HORMONES

As shown in Figure 6.3, there are four mechanisms through which neurotransmitters can act to regulate hormone release from the hypotha-

Table 6.1. *A summary of the neurotransmitter control of the hormones of the pituitary gland*

Transmitter	ACTH	TSH	LH FSH	PRL	GH	α MSH	Oxytocin	Vasopressin
Noradrenaline	↑↓	↑	↑	↑↓?	↑	↑↓	↓	↓
Dopamine	↑↓	↓	↑↓	↓	↑	↓	↓?	↑
5-HT	↑	↑↓	↑↓	↑	↑↓	↑?	0	0
Acetylcholine	↑	0↑	↓?	↓?	↑	↑?	↑	↑
Histamine	↑	↑?	↑↓?	↑	↑	?	↑	↑
GABA	↓	↓	↓	↑↓	↑↓	↓	↓	↓

Notes:
↑ = increase; ↓ = decrease; 0 = no effect; ? = contradictory results; 5-HT = serotonin.

lamus and pituitary gland. (a) The transmitter can be released at a synapse in a neural pathway carrying input from the sensory receptors, from other brain regions, such as the limbic system, or from the spinal cord, to the neurosecretory cell (indirect stimulation). (b) The transmitter can be released at a synapse between axons from extra-hypothalamic neurons and the dendrites or cell body of the neurosecretory cell (direct stimulation). (c) The transmitter can be released from nerve cells having synapses on the axons or nerve terminals of the neurosecretory cell. (d) Neurotransmitters can be released into the hypophyseal portal veins to act directly on the pituitary gland (Hutchinson, 1978; Bennett and Whitehead, 1983). The secretion of both adenohypophyseal and neurohypophyseal hormones is regulated by neurotransmitters acting through these mechanisms.

6.3 NEUROTRANSMITTER REGULATION OF ADENOHYPOPHYSEAL HORMONES

In general, the determination of the neurotransmitters regulating adenohypophyseal hormone secretion is a complex and confused field of study. The information provided here is merely an indication of some of the studies which have been attempted. Although there have been numerous reviews of these studies (Weiner and Ganong, 1978; Krulich, 1979; Frohman, 1980; McCann, 1980; Racagni *et al.*, 1982; Steger and Johns, 1985 and Müller and Nistico, 1989), final conclusions can not yet be established (see Table 6.1).

6.3.1 ACTH AND β-ENDORPHIN

Figure 6.4 shows the variety of neurotransmitters which regulate the CRH stimulation of ACTH secretion (Buckingham, 1981). The secretion of CRH and ACTH occurs in bursts and the rate of these bursts shows a circadian (daily) rhythm which is related to the light–dark and sleep–wake cycles of the subject. This cyclic release of CRH and ACTH, as well as the stress-induced release of these hormones, is stimulated by the

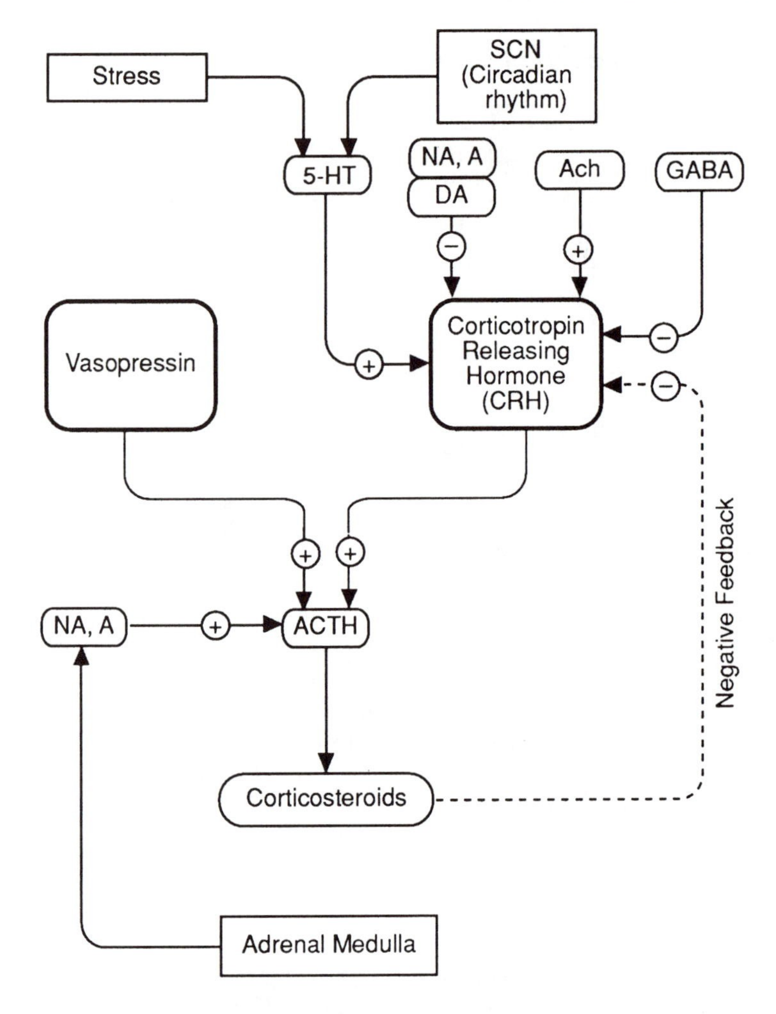

Figure 6.4. Summary of the neurotransmitter control of CRH and ACTH secretion. Details are given in the text. A = adrenaline; Ach = acetylcholine; DA = dopamine; NA = noradrenaline; SCN = suprachiasmatic nucleus; 5-HT = serotonin.

release of serotonin (5-HT) from axons in the medial forebrain bundle (see Figure 5.8D). Serotonin receptor antagonist drugs inhibit the rhythmic release of ACTH (Bennett and Whitehead, 1983; Müller and Nistico, 1989).

Noradrenaline and adrenaline inhibit baseline CRH release through their action on α-adrenergic receptors on CRH neurosecretory cells (Simpkins *et al.*, 1985; Müller and Nistico, 1989). Dopamine also inhibits CRH release (Steger and Morgan, 1985). Under stressful conditions, adrenaline and noradrenaline, released from the adrenal medulla or from hypothalamic neurons directly into the hypophyseal portal veins, may stimulate ACTH secretion by activating the α_1- or β-adrenergic receptors on the corticotroph cells in the anterior pituitary. Thus, catecholamines may inhibit CRH-induced ACTH release, but stimulate ACTH release directly from the pituitary gland. This contradiction is yet to be resolved.

Acetylcholine increases both the basal and stress-induced release of ACTH through its action on muscarinic and nicotinic receptors on the

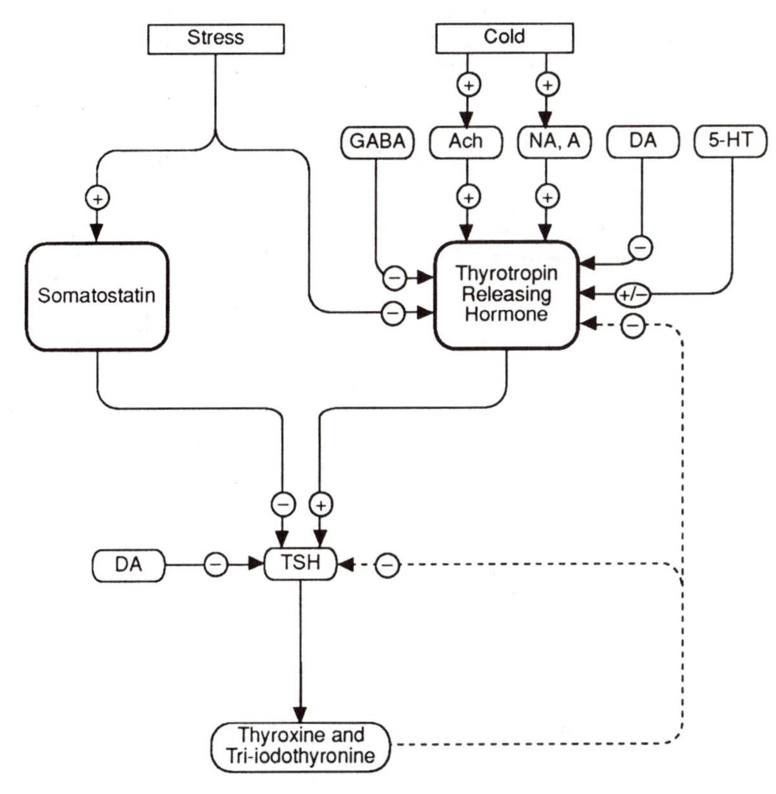

Figure 6.5. Summary of the neurotransmitter control of thyrotropin releasing hormone (TRH) and thyroid stimulating hormone (TSH) secretion. Details are given in the text. Abbreviations as in Figure 6.4.

CRH neurons. GABA inhibits stress-induced CRH and ACTH release and may be important in the negative feedback of adrenal corticosteroids on CRH and ACTH (Müller and Nistico, 1989). Other neuroregulators also influence the release of ACTH. Histamine and vasopressin elevate ACTH release, while the enkephalins have either no effect or show a slight stimulation of ACTH release (see Chapter 12).

The release of β-endorphin (and related peptides such as β-lipotropin) from the anterior pituitary is stimulated by serotonin and noradrenaline and inhibited by GABA (Müller and Nistico, 1989).

6.3.2 THYROID STIMULATING HORMONE

As shown in Figure 6.5, TSH release from the thyrotroph cells of the anterior pituitary is stimulated by TRH from the neurosecretory cells in the dorsomedial hypothalamus and inhibited by somatostatin from neurosecretory cells in the medial preoptic area (Bennett and Whitehead, 1983).

Dopamine reduces baseline TSH levels and inhibits the cold-induced release of TSH in two ways: by inhibiting the release of TRH from the hypothalamus and by inhibition of the thyrotroph cells in the pituitary. Adrenaline and noradrenaline stimulate both baseline and cold-induced TRH and TSH release. The role of serotonin in TSH release is controversial, with some studies reporting serotonin elevation of TSH, some

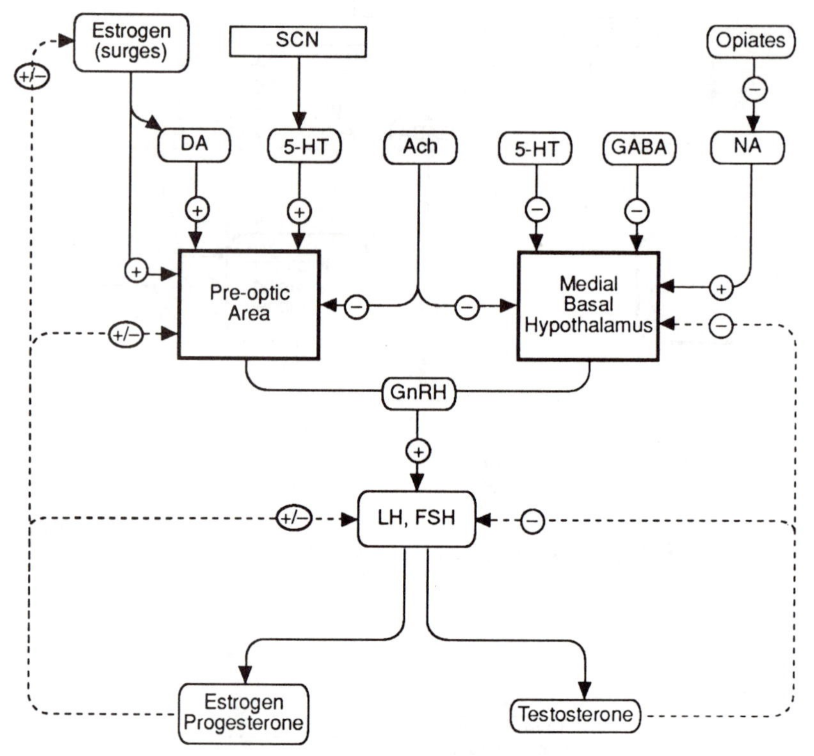

Figure 6.6. Summary of the neurotransmitter control of gonodotropin releasing hormone (GnRH) and gonadotropin secretion. Details are given in the text. Abbreviations as in Figure 6.4.

inhibition, and some no effect (Krulich, 1979; Jiminez and Walker, 1985). Part of this confusion may be due to the interaction of serotonin and thyroid hormone (T3 and T4) feedback in the regulation of TSH release. In the absence of thyroid hormones (i.e. in hypothyroid animals), serotonin stimulates TRH and TSH release, whereas in animals with normal thyroid hormone levels, serotonin inhibits TSH release. Thus, it is possible that serotonin is involved in the negative feedback effects of thyroid hormones on the secretion of TRH (Müller and Nistico, 1989).

Acetylcholine has little effect on the baseline release of TSH, but may be involved in cold-stimulated TSH release. Histamine appears to stimulate TRH release, while GABA and the enkephalins inhibit TRH and TSH release.

6.3.3 LUTEINIZING HORMONE AND FOLLICLE STIMULATING HORMONE

Several transmitters are involved in regulating the secretion of the gonadotropins (Figure 6.6). The exact influence of each transmitter is difficult to determine because of species and sex differences in the neural control of GnRH release, changes in the pulsatile pattern of GnRH secretion over the female reproductive cycle, seasonal variations in GnRH secretion, interactions between transmitters, and the effects of steroid hormone feedback (see Section 4.4.3). Basal secretion of the gonadotropins is stimulated by GnRH from neurons in the medial basal hypothalamus (MBH) and the pre-ovulatory surge of GnRH in the female is released from neurons in the preoptic area (POA). Thus, cyclic release of GnRH from the preoptic area appears to be mediated by the clock mechanism in the suprachiasmatic nucleus (SCN) and by other extra-hypothalamic pathways.

Information on the effects of serotonin on GnRH secretion is contradictory, but suggests that the 5-HT pathways in the MBH inhibit GnRH secretion whereas the 5-HT pathways in the POA facilitate GnRH release. Since the stimulatory effect of 5-HT on GnRH release is closely associated with the GnRH surge which triggers ovulation, the timing of ovulation may be determined by the 5-HT pathways from the suprachiasmatic nucleus (Müller and Nistico, 1989). Serotonin appears to have little or no effect on the release of FSH (Table 6.1).

The role of the catecholamines in stimulating GnRH release in the female depends on the presence of ovarian hormones. In estrogen–progesterone primed females, dopamine stimulates LH release, whereas in ovariectomized females, it may inhibit LH release. Dopamine stimulation of the LH surge is most effective when: (a) it is injected directly into the third ventricle, (b) the level of circulating gonadal hormones is high, and (c) it is given in pulsatile rather than tonic administration (Steger and Morgan, 1985). Likewise, the effect of noradrenaline on LH and FSH release depends on the presence of gonadal steroids. Following ovariectomy, when gonadal steroid levels are low, noradrenaline inhibits LH release (Simpkins *et al.*, 1985). Estrogen appears to facilitate the release of noradrenaline and increase the number of α-adrenergic receptors, thus increasing the sensitivity of the neurosecretory cells to the transmitter.

The effects of dopamine and noradrenaline on GnRH secretion in the male are confusing. Some studies find that these catecholamines stimulate GnRH release in males, some find no effect, and some find an inhibition of GnRH release and suggest that dopamine in the median eminence may be involved in the negative feedback of testosterone on GnRH release (Müller and Nistico, 1989).

Acetylcholine and GABA appear to inhibit LH and FSH release. Histamine stimulates LH release in estrogen-primed females, but inhibits LH release in males. The opiates inhibit LH and FSH release by inhibiting noradrenaline release, rather than by direct action on the GnRH neurosecretory cells as shown in Figure 6.3 (see further discussion in Chapter 12).

6.3.4 PROLACTIN

Prolactin release is controlled by a complex array of excitatory and inhibitory factors. The primary prolactin inhibitory factor is dopamine, which is released directly into the hypophyseal portal veins from the tuberoinfundibular dopamine pathways, as shown in Figures 6.2 and 6.3, to inhibit prolactin release from the lactotroph cells of the pituitary (Figure 6.7).

While the majority of studies indicate that adrenaline and noradrenaline inhibit prolactin release through their action on α-adrenergic recep-

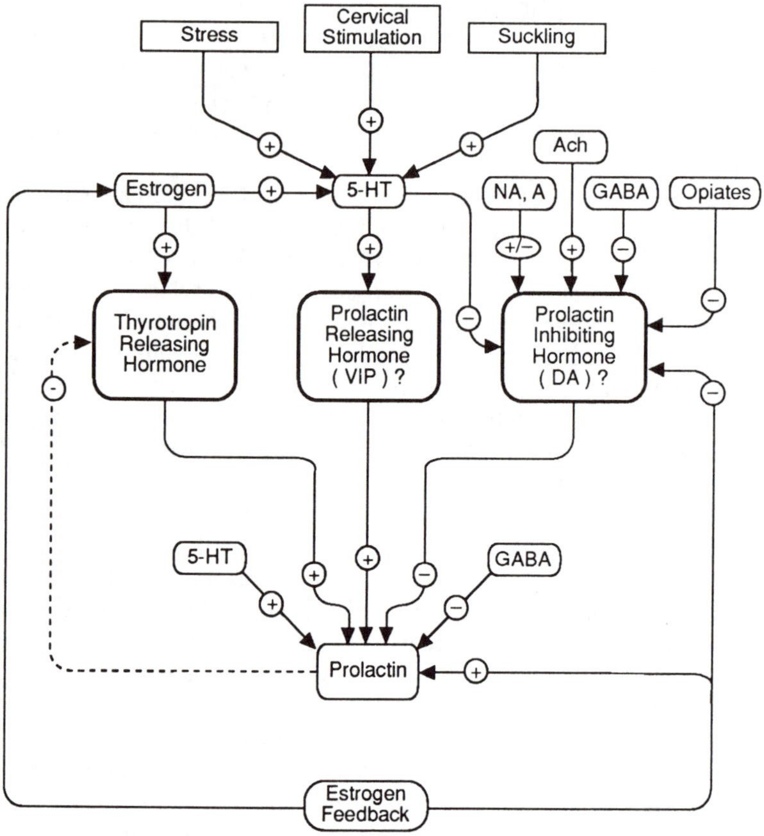

Figure 6.7. Summary of the neurotransmitter control of prolactin secretion. Details are given in the text. Abbreviations as in Figure 6.4.

tors, some evidence suggests that these catecholamines stimulate prolactin release through activation of β-adrenergic receptors (Müller and Nistico, 1989). The effect of noradrenaline on prolactin release is particularly difficult to evaluate because of the predominant inhibitory effect of dopamine. Since many adrenergic drugs influence both noradrenaline and dopamine levels, drugs given to alter noradrenaline may alter prolactin release by altering dopamine levels. The effect of noradrenergic drugs on prolactin depends on the dose given, the route of administration and the method used to sample blood (Simpkins *et al.*, 1985).

Serotonin stimulates prolactin release and serotonin pathways may be responsible for estrogen-stimulated prolactin release, suckling-induced prolactin release, and stress-induced prolactin release (Müller and Nistico, 1989). Serotonin may stimulate prolactin release by inhibiting dopamine secretion or by increasing the secretion of vasoactive intestinal peptide (VIP), a neuropeptide which stimulates prolactin release (Figure 6.7).

Most evidence suggests that acetylcholine inhibits the release of prolactin, possibly by elevating dopamine levels, through its action at both muscarinic and nicotinic receptors (Table 6.1) (Grandison, 1985). Histamine stimulates prolactin release, while GABA has contradictory effects. Large doses of GABA stimulate PRL secretion, whereas small

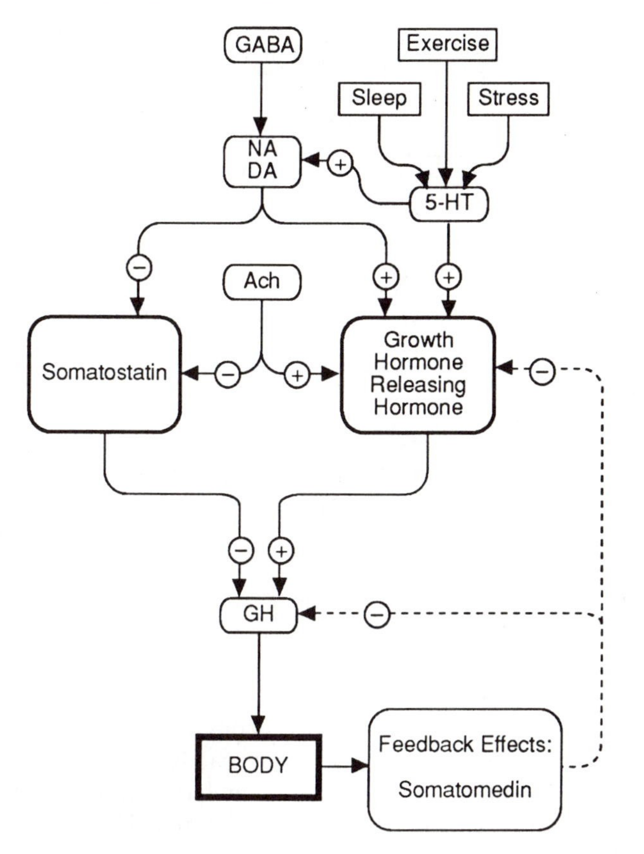

Figure 6.8. Summary of the neurotransmitter control of growth hormone secretion. Details are given in the text. Abbreviations as in Figure 6.4.

doses are inhibitory. There is some evidence that GABA stimulates prolactin secretion by inhibiting dopamine release in the hypothalamus and inhibits PRL synthesis and release via direct action on the lactotrophs in the anterior pituitary (Racagni *et al.*, 1982). Like GABA, the opiates may stimulate prolactin release by inhibition of dopamine levels. Since TRH also stimulates prolactin release, transmitters influencing TRH levels will also influence prolactin release.

6.3.5 GROWTH HORMONE

Because GH is controlled by two hypothalamic hormones, GH-RH and somatostatin, one must consider whether transmitters stimulate GH release by stimulating GH-RH or by inhibiting somatostatin, and vice versa (Figure 6.8). Dopamine and noradrenaline both appear to elevate GH release. Drugs which act as dopamine receptor agonists or α_2-adrenergic receptor agonists elevate GH secretion, whereas drugs which reduce catecholamine levels inhibit GH secetion. As well as stimulating GH-RH release through their action at α_2-adrenergic receptors, adrenergic drugs may also elevate GH levels by inhibiting somatostatin release through their action at β-adrenergic receptors (Simpkins *et al.*, 1985; Steger and Morgan, 1985).

The role of serotonin in regulating GH secretion is controversial as both

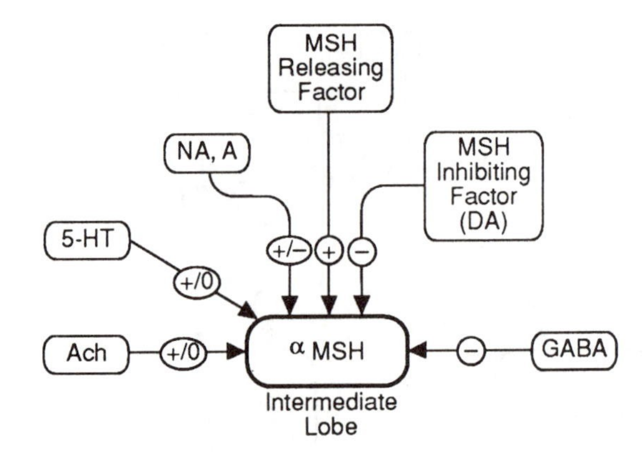

Figure 6.9. Summary of the neurotransmitter control of α-melanocyte-stimulating hormone (α-MSH) secretion. Details are given in the text. Abbreviations as in Figure 6.4.

stimulatory and inhibitory effects have been demonstrated. Serotonin stimulates the baseline release of GH and may be involved in exercise and stress-related increases in GH as well as sleep-related GH surges. The stimulatory effect of serotonin on GH may occur through the stimulation of GH-RH release or through inhibition of somatostatin release and these effects may be due to 5-HT stimulation of noradrenaline release rather than direct action on the neurosecretory cells. (Müller and Nistico, 1989).

Acetylcholine appears to stimulate the release of GH-RH and to inhibit the release of somatostatin through its actions on muscarinic receptors. GABA generally elevates GH release, but may inhibit GH under some physiological conditions. As GABA is an inhibitory neurotransmitter, it may inhibit somatostatin or GH-RH release, or both, depending on other factors (Müller and Nistico, 1989). GABA may not act directly on the neurosecretory neurons in the hypothalamus, but may regulate dopamine activity, further complicating attempts to understand the effects of either transmitter in isolation. Histamine and the opiates appear to elevate GH release.

6.3.6 α-MELANOCYTE STIMULATING HORMONE

The release of α-MSH from the intermediate lobe of the pituitary is controlled by both excitatory and inhibitory factors (Figure 6.9). Dopamine is an important α-MSH inhibiting factor, and, like prolactin, α-MSH is under the inhibitory control of dopamine. Dopamine appears to act directly on the melanotroph cells of the pituitary to inhibit α-MSH release. Low doses of adrenaline and noradrenaline stimulate α-MSH release, probably through their action at β-adrenergic receptors, but high doses of these catecholamines inhibit α-MSH release (Bennett and Whitehead, 1983; Müller and Nistico, 1989). GABA inhibits α-MSH release, while serotonin and acetylcholine generally stimulate or 'have little effect on α-MSH release (Kastin, Schalley and Kostrzewa, 1980).

6.3.7 OTHER NEUROREGULATORS

The prostaglandins and a variety of neuropeptides also regulate the release of the adenohypophyseal hormones. Some of the effects of

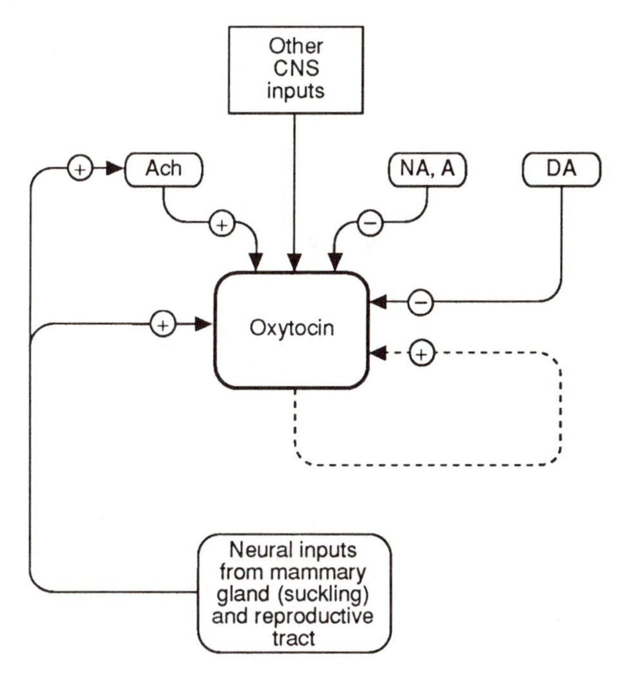

Figure 6.10. Summary of the neurotransmitter control of oxytocin secretion. Details are given in the text. Abbreviations as in Figure 6.4.

neuropeptides on pituitary hormones are discussed in Chapter 12 and the effects of the prostaglandins on pituitary hormone release are summarized by Ojeda and Aquado (1985).

6.4 NEUROTRANSMITTER REGULATION OF NEUROHYPOPHYSEAL HORMONES

The release of oxytocin and vasopressin is stimulated by nerve axons from a number of brain areas, including the amygdala, hippocampus and the spinal nerves (see Figures 6.10 and 6.11). Acetylcholine stimulates oxytocin and vasopressin release through both muscarinic and nicotinic receptors (Table 6.1). Acetylcholine appears to be the transmitter which stimulates oxytocin release in response to suckling during lactation (Bennett and Whitehead, 1983).

Noradrenaline and adrenaline appear to inhibit the release of the neurohypophyseal hormones (Bennett and Whitehead, 1983; Sklar and Schrier, 1983), whereas the effect of dopamine is controversial. L-Dopa inhibits both the milk ejection and milk secretion responses to suckling, indicating inhibition of both oxytocin and prolactin, however, other studies report no effects or a stimulating effect of dopamine on the release of oxytocin. Vasopressin release is stimulated by dopamine (Sklar and Schrier, 1983; Steger and Morgan, 1985).

Serotonin appears to have little effect on the release of the neurohypophyseal hormones, while histamine stimulates their release, particularly that of vasopressin. Angiotensin II also stimulates vasopressin release. GABA and the opiates inhibit the release of the neurohypophyseal hormones (Bennett and Whitehead, 1983; Sklar and Schrier, 1983; Bicknell, 1985).

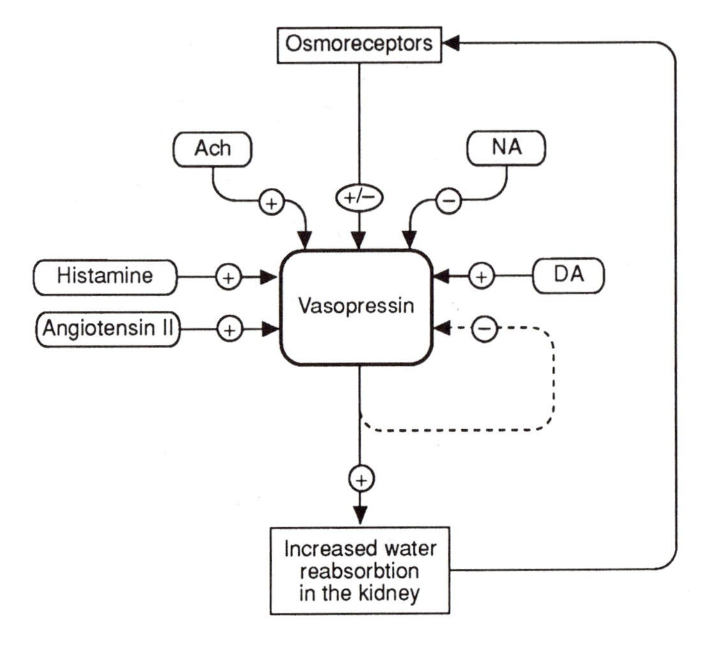

Figure 6.11. Summary of the neurotransmitter control of vasopressin secretion. Details are given in the text. Abbreviations as in Figure 6.4.

6.5 ELECTROPHYSIOLOGY OF NEUROSECRETORY CELLS

As discussed in Chapter 5 (see Figure 5.5), the response of a postsynaptic cell to neurotransmitter stimulation at excitatory synapses is a change of electrical potential, causing the cell to 'fire'. When a nerve cell fires, it releases neurotransmitters from its axon terminals into the synapse. When a neurosecretory cell fires, it releases neurohormones into the circulation. The resting potential of neurosecretory cells, like that of other neurons, is in the range of −40 to −90 mV, as shown in Figure 5.5 (Summerlee, 1986). When these cells fire, the amplitude of the 'spike' of the action potential ranges from 55 to 95 mV, and each action potential is followed by a refractory period of about 8 ms. The duration of the action potential of a neurosecretory cell is about 2·5 ms, which is much longer than a regular nerve cell, thus the firing of a neurosecretory cell can be identified by the duration of this spike (Summerlee, 1986).

Neurosecretory cells show different baseline patterns of firing, so by taking recordings from these cells and measuring changes in hormone levels, one can show how the release of neurohormones is related to the electrical activity of the neurosecretory cell. For example, bursts of electrical activity can be recorded from neurons in the arcuate nucleus of the rhesus monkey just before a rise in LH is detected (Figure 6.12). These bursts of electrical activity last for about 11 min and occur from 2 to 5 min before the surge in LH secretion is detected (Wilson *et al.*, 1984). This suggests that the firing pattern of the neurosecretory cells in the hypothalamus controls the release of 'bursts' of pituitary hormones into the blood.

Another way of investigating the functional relationship between nerve cell activity and neurohormone release is to stimulate individual

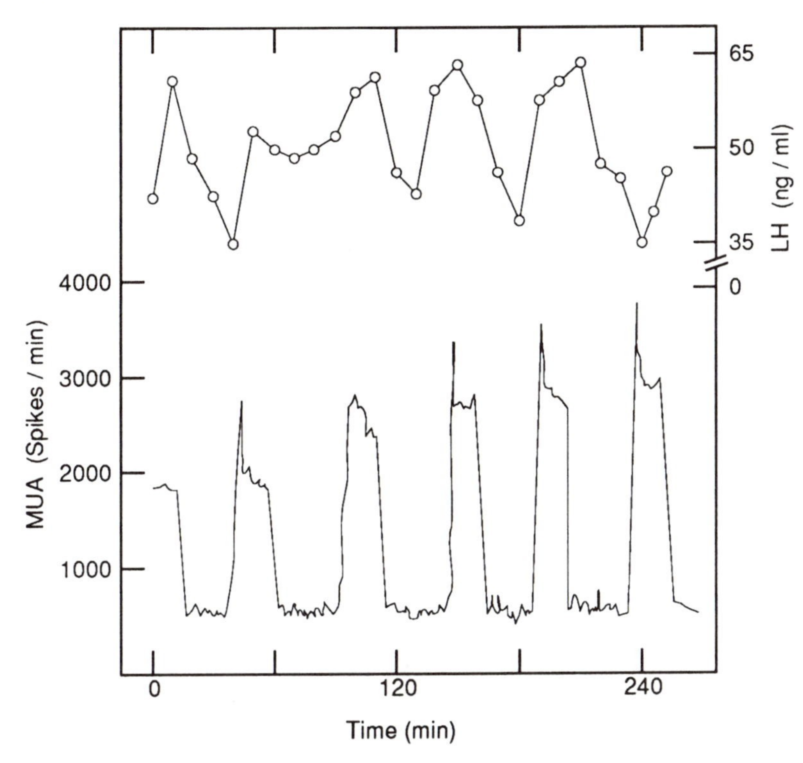

Figure 6.12. Correlation between bursts of multiunit activity (MUA) in hypothalamic neurosecretory cells and the release of LH from the anterior pituitary gland in awake, unanaesthetized rhesus monkeys. Note that the pulses occur at about hourly intervals. (Redrawn from Wilson *et al.*, 1984.)

neurons electrically and measure hormone release. If the neuron stimulated is involved in the control of the hormone being measured, increased (or decreased) hormone release should be correlated with the electrical activity of the nerve cell.

Stimulus-induced neurohormone secretion can also be examined through the use of electrophysiological recordings. The best example of this is the suckling-induced release of oxytocin in the lactating female rat (Lincoln and Paisley, 1982). In this case, the intramammary pressure induced by the pups suckling stimulates the ascending mammary nerves in the spinal cord which stimulate the release of acetylcholine in the cholinergic CNS pathways. This activation of the cholinergic system stimulates a burst of firing in the magnocellular neurosecretory cells of the SON and PVN (Waverly and Lincoln, 1973). These bursts last from 5 to 15 s and are followed about 18 s later by milk ejection (Figure 6.13). Noradrenaline and the opiates inhibit this suckling-induced firing of the neurosecretory cells, probably by inhibition of acetylcholine release (see Chapter 12).

6.6 NEUROTRANSMITTER REGULATION OF OTHER ENDOCRINE GLANDS

Neurotransmitters also regulate hormone release from the adrenal medulla,the pineal gland, thymus and thyroid glands, the gastrointestinal tract and the pancreas. The release of adrenaline from the adrenal medulla is stimulated by catecholamines released from the sympathetic

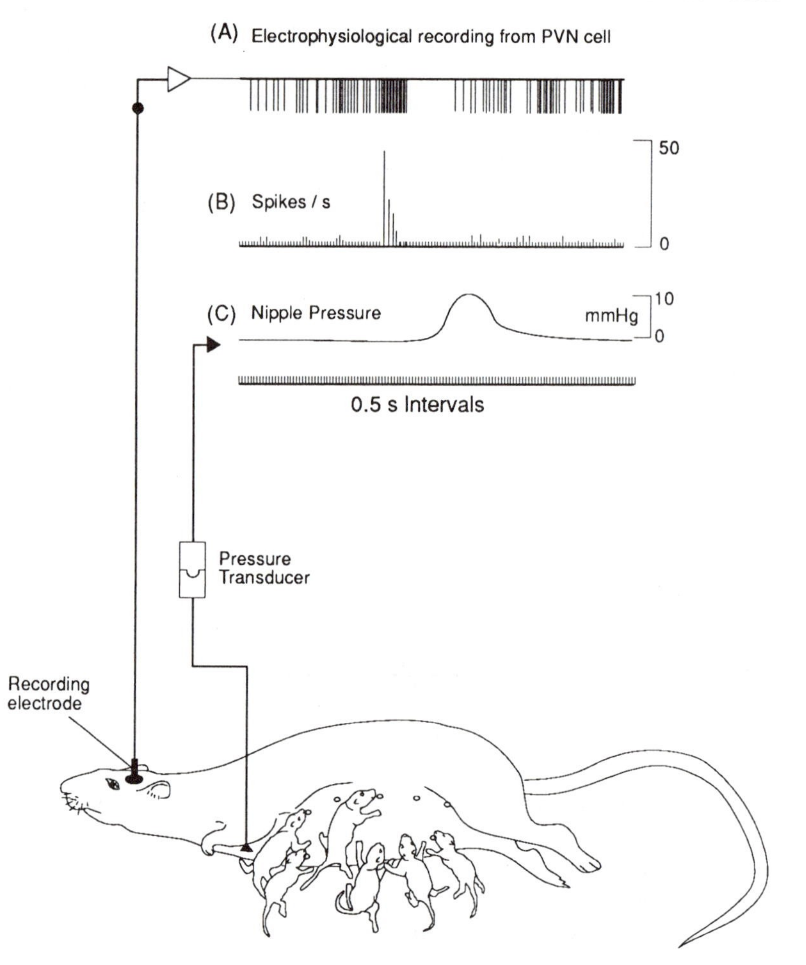

Figure 6.13. A recording of the electrical activity of an oxytocin neurosecretory cell in the PVN of a lactating female rat, showing the rapid increase in firing rate at the time of oxytocin release. The milk ejection response at the nipple is measured in mm of mercury (Hg) by a pressure transducer. (A) A pen recording of the electrical activity of the PVN cell which indicates the rapid burst of activity associated with oxytocin release and the period of inhibition of electrical activity following this burst. (B) The peak firing rate of this cell during the firing burst is about 50 pulses/s. (C) There is a latency of about 18 s between the burst of firing in the PVN cell and the onset of the milk ejection response. (Redrawn from Waverly and Lincoln, 1973.)

nerves which innervate this gland. Likewise, the synthesis and release of melatonin from the pineal gland is stimulated by noradrenaline released from the superior cervical postganglionic neurons (see Hadley, 1992). Catecholamines are also important in the neural stimulation of the thymus gland (Marchetti, Morale and Pelletier, 1990). The thyroid gland is innervated by adrenergic, cholinergic and peptide releasing neurons. Noradrenaline stimulates thyroid hormone secretion and VIP facilitates TSH-induced thyroid hormone release, while acetylcholine inhibits thyroid hormone secretion (Ahrén, 1986). The gastrointestinal tract and pancreas are also innervated by cholinergic and adrenergic nerves from the autonomic nervous system (Weekley, 1981; Hadley, 1992). Acetylcholine stimulates gastric acid secretion and insulin release, while noradrenaline inhibits insulin release and stimulates glucagon release.

Thus, the 'true' endocrine glands are stimulated by neural activity as well as hypothalamic hormones and other chemicals, so should be considered as neuroendocrine transducers. This neural innervation of endocrine glands through the autonomic nervous system, which is controlled by the hypothalamus, means that hormone release from these glands can be stimulated by psychological factors, such as stress or depression, and conditioned to external stimuli.

6.7 COMPLICATIONS IN THE STUDY OF THE NEUROTRANSMITTER CONTROL OF HYPOTHALAMIC HORMONE RELEASE

As illustrated in Table 6.1, many of the effects of neurotransmitters on the hypothalamic and pituitary hormones are contradictory or unknown. Why should this be so, when we know so much about neurotransmitters and their pathways in the brain? Nine possible reasons for these complications are given here.

1. *Multiple neurotransmitter interactions.* It is naive to assume that a single neurotransmitter controls the release of a specific hypothalamic hormone. Interactions between neurotransmitters may be necessary to stimulate the synthesis and release of hypothalamic hormones. McCann's hypothesis for the neural control of the pulsatile LH release illustrates this point:

   We would postulate that pulsatile LH release involves at least several steps: (1) discharge of cholinergic neurons in the arcuate region which would synapse with the tuberoinfundibular dopaminergic neurons in the arcuate nucleus and trigger the release of dopamine; (2) dopamine released from terminals in axo-axonal contact with LH-RH terminals in the median eminence would then depolarize these terminals resulting in the release of LH-RH which would evoke the pulsatile release of LH; (3) . . . noradrenergic tone would be necessary to maintain the central excitatory state of the LH-RH neurons' (McCann, 1980, pp. 357–8).

   With the discovery of the peptidergic neurons, this view is itself incomplete as we now know that neuropeptides such as the opiates and VIP interact with other neurotransmitters to regulate the release of the hypothalamic and pituitary hormones.
2. *Interactions with other hormones.* The response of the hypothalamic neurosecretory cells to particular transmitters may depend on the presence or absence of other hormones. Thus, the effects of noradrenaline and dopamine on the release of LH depend on the presence of gonadal steroids, i.e. intact females respond differently from ovariectomized females (Simpkins *et al.*, 1985). Likewise, the effect of serotonin on TRH release depends on the level of thyroid hormones. In the absense of T3 and T4, serotonin stimulates TRH release, whereas, in the presence of the thyroid hormones, serotonin inhibits TRH release.
3. *Non-specific drug effects.* Most studies use drugs to manipulate neurotransmitter levels, and these drugs may not be specific in their action. For example, drugs given to alter noradrenaline levels will often alter dopamine levels as well (Simpkins *et al.*, 1985). Likewise, two different drugs, both of which influence the same neurotransmitter pathways, may produce different results on the neuroendocrine system. For example, drugs which increase serotonin biosynthesis may not affect the release of serotonin and, therefore, may result in effects different from those obtained with serotonin receptor agonists. This principle would be equally true for other transmitters. Different doses of the same drug may also produce different effects. Thus, large doses of GABA stimulate PRL release, while smaller doses inhibit PRL release

(Krulich, 1979). Low doses of noradrenaline given intraventricularly stimulate PRL release, whereas higher doses inhibit PRL release (Simpkins *et al.*, 1985). The most obvious reason for this is the liklihood of various pathways being stimulated as the drug dosage is increased. The route of administration of drugs may also influence their neuroendocrine effects. For example, intravenous administration of apomorphine, surprisingly, elevates LH levels, but intraperitoneal (IP) injection reduces LH levels (Krulich, 1979). Other drugs can not enter the brain other than by direct injection, because of the blood–brain barrier.

4. *Anaesthetic use.* The neuroendocrine response to a drug may depend on whether or not the animal is anaesthetized and the type of anaesthetic used. If the anaesthetic is stressful (e.g. urethane), levels of GH are suppressed and PRL elevated (in the rat), independently of any neurotransmitter manipulation (Steger and Morgan, 1985). Many anaesthetics influence the secretion of adenohypophyseal and neurohypophyseal hormones, primarily through their actions on the nervous system, but possibly also by direct effects on the pituitary (Lawson, 1985).
5. *In vivo* versus *in vitro* studies. *In vivo* studies may give results different from *in vitro* studies. For example, the hypothalamus *in vitro* would be separated from all incoming neuronal influences, both excitatory and inhibitory. Nevertheless, an *in vitro* technique can provide data not obtainable *in vivo* (Bennett and Whitehead, 1983).
6. *Species differences.* There are species differences in the responses of the neuroendocrine system to neural input. For example, stress elevates GH levels in primates, but inhibits GH release in rodents.
7. *Sex differences.* Sex differences in neuroendocrine secretion occur in response to certain drugs, especially in the mechanism controlling the release of prolactin and the gonadotropins. Thus, males and females may show different neuroendocrine responses to changes in the same neurotransmitter.
8. *Psychological factors.* Psychological factors play an important role in the endocrine response to neurotransmitters. The emotional state of the animal (e.g. fear or stress), the time of the day–night cycle when the test is done, and the relationship with feeding patterns all influence hormone responses to changes in neurotransmitter levels. Likewise, the test environment and the stimuli presented may influence the responses.
9. *Multiple hypothalamic-pituitary interactions.* One must also remember that some 'hypothalamic hormones' such as LH-RH, TRH and somatostatin are released as neuropeptides throughout the brain, where they act as neuromodulators to regulate neurotransmitter release. A good example of this is the finding that CRH, β-endorphin and LH-RH may all be involved in the estrogen-facilitated activation of sexual behavior in the female rat through their neuromodulatory action on noradrenergic neurons (Sirinathsinghji, 1985).

Understanding the effects of neurotransmitters on specific pituitary hormones is further complicated by the fact that the hypothalamic hormones are not as specific in their pituitary stimulation as was once thought (see Figure 4.3, p. 51). For example, TRH releases PRL and GH as well as TSH; and somatostatin inhibits TSH and PRL release as well as GH. Thus, it is very difficult to determine the exact relationship between changes in neurotransmitter release and changes in hypothalamic-pituitary hormone levels for the following reasons. (a) More than one neurotransmitter affects a single neurosecretory cell type. (b) One neurotransmitter may affect several types of hypothalamic neurosecretory cells. (c) More than one hypothalamic releasing hormone may affect a single pituitary hormone. (d) Other hormones and neuropeptides may also alter neurohormone release.

6.8 NEUROENDOCRINE CORRELATES OF PSYCHIATRIC DISORDERS AND PSYCHOTROPIC DRUG TREATMENT OF THESE DISORDERS

6.8.1 PSYCHOTROPIC DRUGS ALTER NEUROHORMONE RELEASE

As shown in Tables 5.5 and 5.6, the psychotropic drugs used as antidepressants (MAO inhibitors and tricyclic antidepressants such as imipramine) and antipsychotics (e.g. reserpine and chlorpromazine) alter the levels of dopamine and noradrenaline in the central nervous system. One side effect of altering these catecholamines is a disruption of the neuroendocrine system.

Dopamine synthesis stimulators, such as L-dopa, which is used to treat Parkinson's disease, decrease prolactin secretion and increase GH secretion. The antipsychotic drugs reserpine and chlorpromazine block dopamine receptors in the hypothalamus and pituitary gland, increasing prolactin levels to the point that breast enlargement and galactorrhea (milk secretion) may occur, even in male patients. Chlorpromazine also inhibits GH and TSH release, interferes with the stress-induced release of CRH and may inhibit GnRH secretion. Some of the effects of chlorpromazine, such as reduced sex drive and disruption of the menstrual cycle in some female patients, may be related to elevated prolactin levels. The inhibition of GH and TSH by neuroleptic drugs is not as pronounced as the effects of these drugs on prolactin. (Wurtman and Fernstrom, 1976; Gilman *et al.*, 1985; Gunnett and Moore, 1988). Tricyclic antidepressants and MAO inhibitors may alter GH and ACTH secretion by increasing serotonin levels and may also alter the release of other neurohormones through their effects on cholinergic and histaminergic pathways (Gilman *et al.*, 1985).

6.8.2 NEUROENDOCRINE CORRELATES OF PSYCHIATRIC DISORDERS

The disorders of neurotransmitter systems which underlie psychiatric disorders may also cause neuroendocrine disorders since certain psychiatric disorders are associated with abnormal neuroendocrine responses. Depressed patients, for example, show elevated ACTH and corticosteroid levels, and blunted TSH responses to TRH, i.e. TRH stimulates a much lower TSH response in depressed patients than in normal control subjects. Depression may also reduce GnRH levels in some patients (Freedman and Carter, 1982; Jackson, 1984).

Although schizophrenia seems to be related to abnormally high dopamine levels, there is little endocrine abnormality in schizophrenic patients, suggesting that the disruption may be in the mesolimbic dopamine pathways rather than the tuberoinfundibular pathways (Freedman and Carter, 1982). Further discussion of endocrine abnormalities associated with psychiatric disorders is provided by Martin and Reichlin (1987).

6.9 SUMMARY

This chapter has examined the ways in which neurotransmitters can alter the release of hypothalamic and pituitary hormones, resulting in a 'cascade' of chemical messengers. Neural input to the endocrine hypothalamus arises from neurotransmitter pathways from many parts of the brain. The endocrine hypothalamus is stimulated by changes in acetylcholine, dopamine, noradrenaline, serotonin, GABA, enkephalin and other transmitters. These transmitters can influence neuroendocrine cells indirectly, through neural pathways passing through the hypothalamus, directly, through axo-dendritic or axo-axonal synapses, or by release into the hypophyseal portal veins to act directly on the pituitary gland. While the effects of some transmitters on neurohormone release are clear cut, others are contradictory and controversial. The use of electrophysiological stimulation and recording methods demonstrates some aspects of the neural control of pituitary hormone release. The effects of transmitters on hormone release are, however, complicated by multiple neurotransmitter interactions, the non-specific effects of the drugs given to alter transmitter levels, the species and sex of subject used, interactions with other hormones and psychological factors. The assay of neurotransmitter effects is further complicated by the multiple interactions between hypothalamic and pituitary hormones. Finally, neuroendocrine abnormalities may be associated with psychiatric disorders and the drugs used to treat these disorders.

FURTHER READING

Bennett, G. W. and Whitehead, S. A. (1983). *Mammalian Neuroendocrinology*. New York: Oxford University Press.

Müller, E. E. and Nistico, G. (1989). *Brain messengers and the pituitary*. New York: Academic Press.

Steger, R. W. and Johns, A. (1985). *CRC Handbook of Pharmacologic Methodologies for the Study of the Neuroendocrine System*. Boca Raton, FL: CRC Press.

Summerlee, A. J. S. (1986). The neural control of luteinizing hormone release. *Progress in Neurobiology*, **26**, 147–178.

REVIEW QUESTIONS

6.1 What is a neuroendocrine transducer?

6.2 Give two reasons why pituitary hormones rather than hypothalamic hormones are measured in response to changes in neurotransmitter levels.

6.3 What are the four ways by which neurotransmitters can regulate the release of hypothalamic and pituitary hormones?

6.4 Which neurotransmitter appears to regulate the cyclic and stress-related release of ACTH?

6.5 What effect does noradrenaline have on TSH release?

6.6 How does GABA influence GnRH release?

6.7 How does serotonin influence prolactin release?

6.8 The catecholamines and acetylcholine appear to have the same effect on the release of GH. What is it?

6.9 Which transmitter releases oxytocin in response to infant suckling?

6.10 Which transmitter is released through the sympathetic nervous system to stimulate hormone release from the adrenal medulla and the pineal gland?

6.11 What effect would an injection of morphine, an opiate agonist, have on the electrical activity of the magnocellular neurosecretory cells of a lactating female rat during a bout of suckling?

6.12 The release of prolactin and thyroid stimulating hormone is inhibited by which neurotransmitter?

6.13 Do the enkephalins stimulate or inhibit the release of (a) prolactin and (b) LH?

6.14 Neurotransmitters can be released into the hypophyseal portal veins to stimulate or inhibit the pituitary gland. True or false?

6.15 Noradrenaline inhibits the release of LH and FSH. True or false?

6.16 Why might *in vitro* studies of the effects of DA or LH release give results different from those of *in vivo* studies?

6.17 How do psychiatric drugs such as chlorpromazine affect prolactin levels?

ESSAY QUESTIONS

6.1 You want to do a study on the effects of prolactin on parental behavior in rats and decide to use three groups of subjects: a control group, a group with elevated prolactin levels, and a group with prolactin release inhibited. What procedures should you use to alter prolactin levels in these three groups, what side effects should you worry about, and what experimental and control procedures should you consider with respect to the use of drugs to alter prolactin release?

6.2 You are under a great deal of stress (from cold, exercise, or psychological pressure). Compare the effects of this stress on your ACTH and LH levels and explain the neural pathways and neurotransmitters involved in these neuroendocrine changes.

6.3 Summarize the effects of acetylcholine on the release of hormones from the anterior pituitary gland.

6.4 Discuss the neurotransmitter control of the release of TSH, explaining how cold activates transmitter pathways controlling TRH release and how the thyroid hormones modulate the transmitter control of TSH release by negative feedback.

6.5 Describe the neuroendocrine reflex involved in suckling-induced milk ejection, outlining the nerve pathways, neurons, and transmitters involved.

6.6 Describe the neural control of the pineal and thyroid glands.

6.7 Present a learning theory model for the conditioned release of hormones, such as oxytocin.

6.8 Outline the costs and benefits of using *in vivo* versus *in vitro* methods for the study of the neurotransmitter control of neurohormone release.

6.9 Discuss the advantages and disadvantages of using electrophysiological methods in the study of the neural control of neurohormone secretion.

6.10 Discuss the hormonal changes associated with depression.

REFERENCES

Ahrén, B. (1986). Thyroid neuroendocrinology: neural regulation of thyroid hormone secretion. *Endocrine Reviews*, **7**, 149–155.

Bennett, G. W. and Whitehead, S. A. (1983). *Mammalian Neuroendocrinology*. New York: Oxford University Press.

Bicknell, R. J. (1985). Endogenous opioid peptides and hypothalamic neuroendocrine neurones. *Journal of Endocrinology*, **107**, 437–446.

Buckingham, J. C. (1981). The control of ACTH secretion. *Stress*, **2**, 23–31.

Freedman, R. and Carter, D. B. (1982). Neuroendocrine strategies in psychiatric research. In A. Vernadakis and P. S. Timiras (eds.) *Hormones in Development and Aging*, pp. 619–636. New York: Spectrum.

Frohman, L. A. (1980). Neurotransmitters as regulators of endocrine function. In D. T. Krieger and J. C. Hughes (eds.) *Neuroendocrinology*, pp. 44–57. Sunderland, MA: Sinauer.

Gilman, A. G., Goodman, L. S., Rall, T. W. and Murad, F. (1985). *Goodman and Gilman's The Pharmacological Basis of Therapeutics*, 7th edn. New York: MacMillan.

Grandison, L. (1985). Effects of cholinergic agonists and antagonists on anterior pituitary hormone secretion. In R. W. Steger and A. Johns (eds.) *Handbook of Pharmacologic Methodologies for the Study of the Neuroendocrine System*, pp. 155–172. Boca Raton, FL: CRC Press.

Gunnett, J. W. and Moore, K. E. (1988). Neuroleptics and neuroendocrine function. *Annual Review of Pharmacology and Toxicology*, **28**, 347–366.

Hadley, M. E. (1992). *Endocrinology*, 3rd edn. New York: Prentice-Hall.

Hutchinson, J. S. M. (1978). Control of the endocrine hypothalamus. In S. L. Jeffcoate and J. S. M. Hutchinson (eds.) *The Endocrine Hypothalamus*, pp. 75–106. London: Academic Press.

Jackson, I. M. D. (1984). Hypothalamic releasing hormones: mechanisms underlying neuroendocrine dysfunction in affective disorders. In G. M. Brown, S. H. Koslow and S. Reichlin (eds.) *Neuroendocrinology and Psychiatric Disorder*, pp. 255–266, New York: Raven Press.

Jimanez, A. and Walker, R. F. (1985). The serotoninergic system. In R. W. Steger and A. Johns (eds.) *Handbook of Pharmacologic Methodologies for the Study of the Neuroendocrine System*, pp. 109–154. Boca Raton, FL: CRC Press.

Kastin, A. J., Schally, A. V. and Kostrzewa, R. M. (1980). Possible aminergic mediation of MSH release and of CNS effects of MSH and MIF-I. *Federation Proceedings*, **39**, 2931–2936.

Krulich, L. (1979). Central neurotransmitters and the secretion of prolactin, GH, LH and TSH. *Annual Review of Physiology*, **21**, 603–615.

Lawson, D. M. (1985). Anesthetics and neuroendocrine function. In R. W. Steger and A. Johns (eds.) *Handbook of Pharmacologic Methodologies for the Study of the Neuroendocrine System*, pp. 385–406. Boca Raton, FL: CRC Press.

Lincoln, D. W. and Paisley, A. C. (1982). Neuroendocrine control of milk ejection. *Journal of Reproduction and Fertility*, **65**, 571–586.

Marchetti, B., Morale, M. C. and Pelletier, G. (1990). Sympathetic nervous system control of rat thymus gland maturation: Autoradiographic localization of the β_2-adrenergic receptor in the thymus and presence of sexual dimporphism during ontogeny. *Progress in NeuroendocrinImmunology*, **3**, 103–115.

Martin, J. B. and Reichlin, S. (1987). *Clinical Neuroendocrinology*, 2nd edn. Philadelphia: F. A. Davis.

McCann, S. M. (1980). Control of anterior pituitary hormone release by brain peptides. *Neuroendocrinology*, **31**, 355–363.

Müller, E. E. and Nistico, G. (1989). *Brain messengers and the pituitary*. New York: Academic Press.

Ojeda, S. R. and Aguado, L. I. (1985). Prostaglandins in reproductive neuroendocrinology. In R. W. Steger and A. Johns (eds.) *Handbook of Pharmacologic Methodologies for the Study of the Neuroendocrine System*, pp. 205–243. Boca Raton, FL: CRC Press.

Racagni, G., Apud, J. A., Cocchi, D., Locatelli, V, and Muller, E. E. (1982). GABAergic control of anterior pituitary hormone secretion. *Life Sciences*, **31**, 823–838.

Segal, S. (1974). The physiology of human reproduction. *Scientific American*, **231**, 53–62.

Simpkins, J. W., Millard, W. J., Gabriel, S. M. and Soltis, E. E. (1985). Noradrenergic methods in neuroendocrinology. In R. W. Steger and A. Johns (eds.) *Handbook of Pharmacologic Methodologies for the Study of the Neuroendocrine System*, pp. 1–63. Boca Raton, FL: CRC Press.

Sirinathsinghji, D. J. S. (1985). Modulation of lordosis behaviour in the female rat by corticotropin releasing factor, β-endorphin and gonadotropin releasing hormone in the mesencephalic central gray. *Brain Research*, **336**, 45–55.

Sklar, A. H. and Schrier, R. W. (1983). Central nervous system mediators of vasopressin release. *Physiological Reviews*, **63**, 1243–1280.

Steger, R. W. and Johns, A. (1985). *CRC Handbook of Pharmacologic Methodologies for the Study of the Neuroendocrine System*. Boca Raton, FL: CRC Press.

Steger, R. W. and Morgan, W. (1985). Dopaminergic control of pituitary hormone release. In R. W. Steger and A. Johns (eds.) *Handbook of Pharmacologic Methodologies for the Study of the Neuroendocrine System*, pp. 65–108. Boca Raton, FL: CRC Press.

Summcrlcc, A. J. S. (1986). The neural control of luteinizing hormone release. *Progress in Neurobiology*, **26**, 147–178.

Waverley, J. B. and Lincoln, D. W. (1973). The milk-ejection reflex of the rat: A 20- to 40-fold acceleration in the firing of paraventricular neurones during oxytocin release. *Journal of Endocrinology*, **57**, 477–493.

Weekly, L. B. (1981). On the neurogenic-neuroendocrine regulation of the pancreas. *Journal of Theoretical Biology*, **88**, 275–278.

Weiner, R. I. and Ganong, W. F. (1978). Role of brain monoamines and histamine in regulation of anterior pituitary secretion. *Physiological Reviews*, **58**, 905–976.

Wilson, R. C., Kesner, J. S., Kaufman, J.-M., Uemura, T. Akema, T. and Knobil, E. (1984). Central electrophysiologic correlates of pulsatile luteinizing hormone secretion in the rhesus monkey. *Neuroendocrinology*, **39**, 256–260.

Wurtman, R. J. and Fernstrom, J. D. (1976). Neuroendocrine effects of psychotropic drugs. In E. J. Sachar (ed.) *Hormones, Behavior and Psychopathology*, pp. 145–151. New York: Raven Press.

7

Regulation of hormone synthesis, storage, release, transport and deactivation

Hormones are synthesized and stored in endocrine cells and, on cell stimulation, they are released into the circulatory system. A number of hormones are transported in the bloodstream by special carrier proteins and, after they have been used, they are deactivated by lysosomes contained within the secretory cell or by special deactivating enzymes outside the secretory cell. Hormone synthesis, storage, release, transport and deactivation occur through a variety of different mechanisms, depending on the chemical structure of the hormone, so the first section of this chapter examines the chemical structure of hormones.

7.1 THE CHEMICAL STRUCTURE OF HORMONES

The first distinction to make when considering chemical structure is that hormones can be either steroids or non-steroids (Table 7.1). The endocrine glands synthesizing steroid hormones develop from different embryonic tissues than do the glands secreting non-steroid hormones. The glands derived from the mesoderm (adrenal cortex and gonads) produce steroid hormones, while those derived from the ectoderm or endoderm secrete non-steroid hormones (Turner and Bagnara, 1976).

Table 7.1. *Chemical structure of hormones*

1. *Non-steroid hormones*
   - A. Modified amino acids: hormones synthesized from single amino acids
     - Catecholamines. Adrenaline and noradrenaline are synthesized from tyrosine as described in Section 5.1.2
     - Indoleamines. Melatonin (and serotonin) are synthesized from tryptophan
     - Thyroid hormones. Thyroxine and triiodothyronine are synthesized from iodinated tyrosine.
   - B. Peptides: hormones formed from chains of amino acids
     - Small peptides
       - Thyrotropin releasing hormone (3 amino acid residues)
       - Angiotensin II (8 amino acid residues)
       - Vasopressin (9 amino acid residues)
       - Oxytocin (9 amino acid residues)
       - LH-RH (10 amino acid residues)
     - Somatostatin (14 amino acid residues)
     - Large peptides
       - Gastrins (17 and 34 amino acid residues)
       - Secretin (27 amino acid residues)
       - Glucagon (29 amino acid residues)
       - Calcitonin (32 amino acid residues)
       - Adrenocorticotropic hormone (39 amino acid residues)
     - Proteins (polypeptides)
       - Insulin (51 amino acid residues)
       - Parathyroid hormone (84 amino acid residues)
       - Human growth hormone (191 amino acid residues)
   - C. Glycoproteins: hormones synthesized from proteins and carbohydrates
     - Follicle stimulating hormone
     - Luteinizing hormone
     - Thyroid stimulating hormone
     - Human chorionic gonadotropin
   - D. Prostaglandins: synthesized from essential fatty acids
2. *Steroid hormones:* synthesized from cholesterol
   - A. Gonadal steroids
     - Progestins (progesterone)
     - Androgens (testosterone)
     - Estrogens
   - B. Adrenal steroids
     - Glucocorticoids (corticosterone, cortisol)
     - Mineralocorticoids (aldosterone)

7.1.1 THE STEROID HORMONES

The steroid hormones are synthesized from cholesterol in the adrenal cortex and gonads. The adrenal steroids include corticosterone and aldosterone and the gonadal steroids are progestins, androgens and estrogens (see Table 7.1).

7.1.2 THE NON-STEROID HORMONES

The non-steroid hormones can take many forms, including modified amino acids, peptides, glycoproteins and prostaglandins (see Table 7.1).

Modified amino acids

The modified amino acids include the catecholamines, indoleamines, and thyroid hormones. The catecholamines, adrenaline and nordrenaline (epinephrine and norepinephrine) are synthesized from the amino acid tyrosine in the adrenal medulla, from which they are secreted as hormones, and in the central and peripheral nervous systems, where they act as neurotransmitters. The synthesis of the catecholamines is described in Chapter 5 (see Figure 5.4).

The indoleamines include serotonin and melatonin. The neurotransmitter serotonin (5-HT) is synthesized from the amino acid tryptophan in the central nervous system. The hormone melatonin is synthesized from tryptophan, via serotonin, in the pineal gland. The thyroid hormones, thyroxine (T4) and triiodothyronine (T3) are synthesized, in the thyroid gland, from the glycoprotein precursor thyroglobulin. The thyroid hormones are formed by adding iodine to the amino acid tyrosine, which is an integral part of the thyroglobulin molecule, via a complex metabolic pathway (Degroot and Niepomniszcze, 1977).

Peptides

Peptides are synthesized from amino acids in many different organs, including the peripheral endocrine glands, the pituitary gland, the hypothalamus and nerve cells. They function as peptide hormones, neurohormones, or neuropeptides, as described in Chapter 1. Peptides consist of amino acid chains which vary in length from three to more than 100 amino acid residues and are categorized according to their size into small peptides, large peptides, and proteins or polypeptides (Table 7.1). The structure of protein and that of peptide hormones are described by Goldman (1981).

Glycoproteins

The glycoproteins are hormones synthesized from amino acids and carbohydrates. They have two amino acid chains (α and β) with carbohydrate subunits (see Table 7.1).

Prostaglandins

Prostaglandins are short-acting (unstable) 20-carbon fatty acids which are synthesized from polyunsaturated fatty acids in almost every tissue of the body, including the reproductive system and brain. They can act as parahormones or paracrine hormones (Chapter 1), or as second messengers to regulate intracellular activity. The primary precursor of the prostaglandins is arachidonic acid, which is produced from the essential fatty acid linoleic acid (see Martin, 1985).

7.2 HORMONE SYNTHESIS

7.2.1 SYNTHESIS OF NON-STEROID HORMONES

The non-steroid hormones are synthesized from amino acids in the ribosomes of the rough endoplasmic reticulum of the endocrine cell body

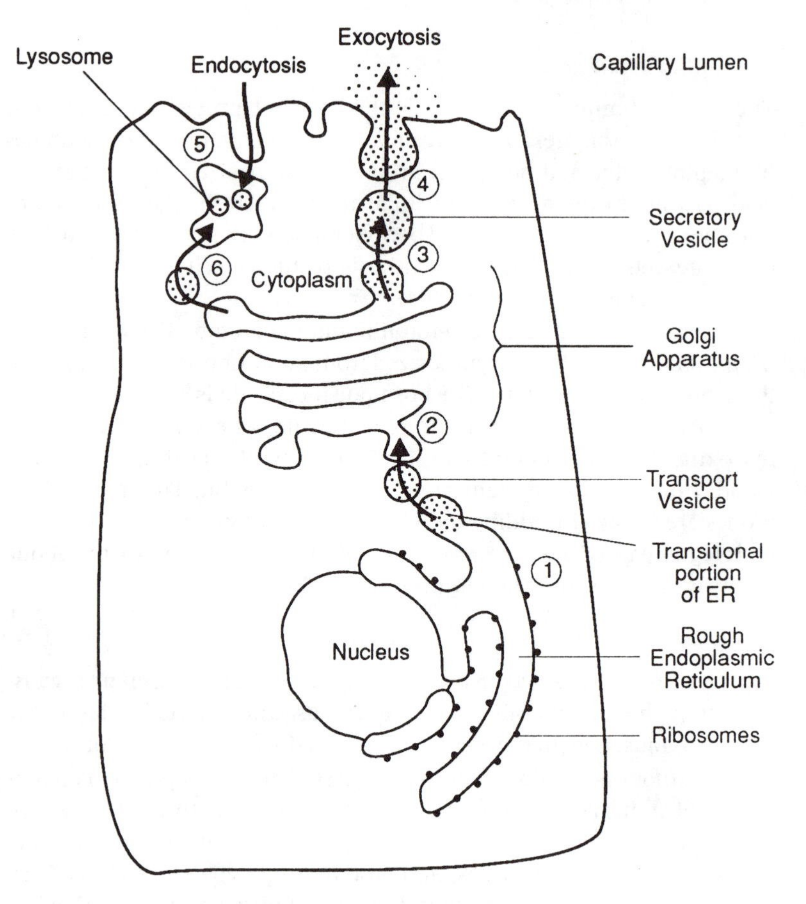

Figure 7.1. Features of peptide hormone synthesis, storage and secretion. (1) Hormone biosynthesis occurs in the ribosomes of the rough endoplasmic reticulum (ER). (2) Transport vesicles containing pre-prohormones bud from the transitional portion of the endoplasmic reticulum and fuse with the Golgi apparatus. (3) The pre-prohormones are cleaved to form prohormones in the Golgi bodies and prohormones or hormones are packaged into secretory vesicles which bud from the Golgi bodies. Hormone synthesis is then completed in the secretory vesicles. (4) Hormones are released into the circulatory system from the secretory vesicles by exocytosis. (5) Reuptake of hormones from the circulation occurs by endocytosis. (6) Excess hormone, whether stored in secretory granules, or taken up by endocytosis, is deactivated by lysosomes. (Redrawn from Alberts *et al.*, 1989.)

and are then packaged into secretory granules in the Golgi apparatus. After their synthesis, most non-steroid hormones are stored in secretory granules (or vesicles) which transport the hormones within the cell and release them into the circulation as shown in Figure 7.1 (Pickering, 1976; Brownstein, Russell and Gainer, 1980; Mains *et al.*, 1987, 1990). Most peptide hormones are synthesized from larger 'pre-prohormones' which are long amino acid chains, manufactured in the ribosomes of the endocrine cell. Proteolytic enzymes in the Golgi bodies then 'clip' these pre-prohormones at specific sites to convert them into shorter amino acid chains which become the 'prohormones'. Some prohormones are enzymatically cleaved within the Golgi bodies to form active hormones. More often, however, the prohormones are packaged in the secretory granules along with the enzymes which convert them to active hormones (Loh, 1987; Sossin, Fisher and Scheller, 1989).

Cleavage of large peptides is the primary mechanism for the production of the smaller peptides as shown in Figure 7.2. Proopiomelanocortin (POMC) is a prohormone which occurs in the adenohypophysis, hypothalamus, gastrointestinal tract and in the placenta. It contains a number of possible cleavage sites and can be converted by proteolytic enzyme action into a number of different hormones, including ACTH, α-MSH and β-endorphin (Figure 7.2A). Oxytocin and vasopressin are cleaved from

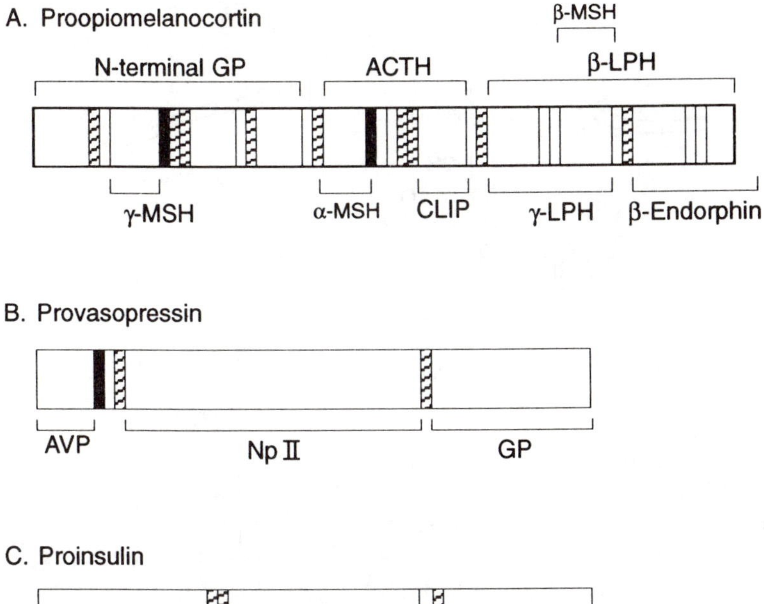

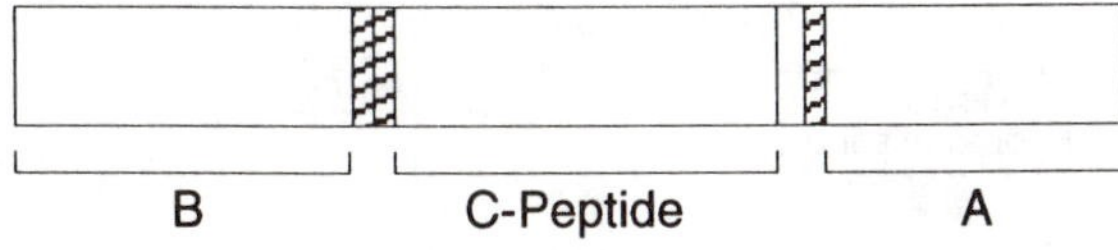

Figure 7.2. The structure of three prohormones, showing the sites of enzyme cleavage sites (hatched bars) of the prohormones into active hormones. (A) Proopiomelanocortin is the prohormone for ACTH and β-lipotropin (β-LPH). ACTH is further cleaved to form α-MSH and corticotropin-like intermediate lobe peptide (CLIP). β-LPH is cleaved to form γ-lipotropin (γ-LPH) and β-endorphin. The N-terminal glycopeptide (GP), which is the residual portion of the prohormone, is cleaved to form γ-MSH. (B) Provasopressin (propressophysin), the prohormone for both arginine vasopressin (AVP) and neurophysin II (NP II), the carrier protein for vasopressin, is cleaved into these two peptides and a glycoprotein (GP) residual. (C) In proinsulin, the A and B amino acid chains of insulin are separated by a C-peptide. Once this residual C-peptide is removed, the A and B chains combine to form the active hormone. (Redrawn from Loh, 1987.)

larger prohormones, prooxytocin and provasopressin (Figure 7.2B). Because these prohormones include both the hormone and neurophysin carrier proteins, they have also been called prooxyphysin and propressophysin (Brownstein *et al.*, 1980; Ramsay, 1991). While small peptides have the same chemical structure in almost all species of vertebrates, the larger peptides are more species specific. For example, human insulin, synthesized from proinsulin (Figure 7.2C), is structurally different from rat insulin.

One might ask why peptide hormones should be synthesized in such a round-about way: first a very large molecule (pre-prohormone) is made, then a piece of this is cut off to make a prohormone, then this is cut up to make a hormone and a residual peptide. It turns out that, in terms of ease of handling during synthesis, prohormones have a number of advantages over active hormones. (a) Prohormones enable the three-dimensional structure of the peptide molecule to be stabilized during synthesis (like carpenters bracing a wall during building). (b) Prohormones are easier to

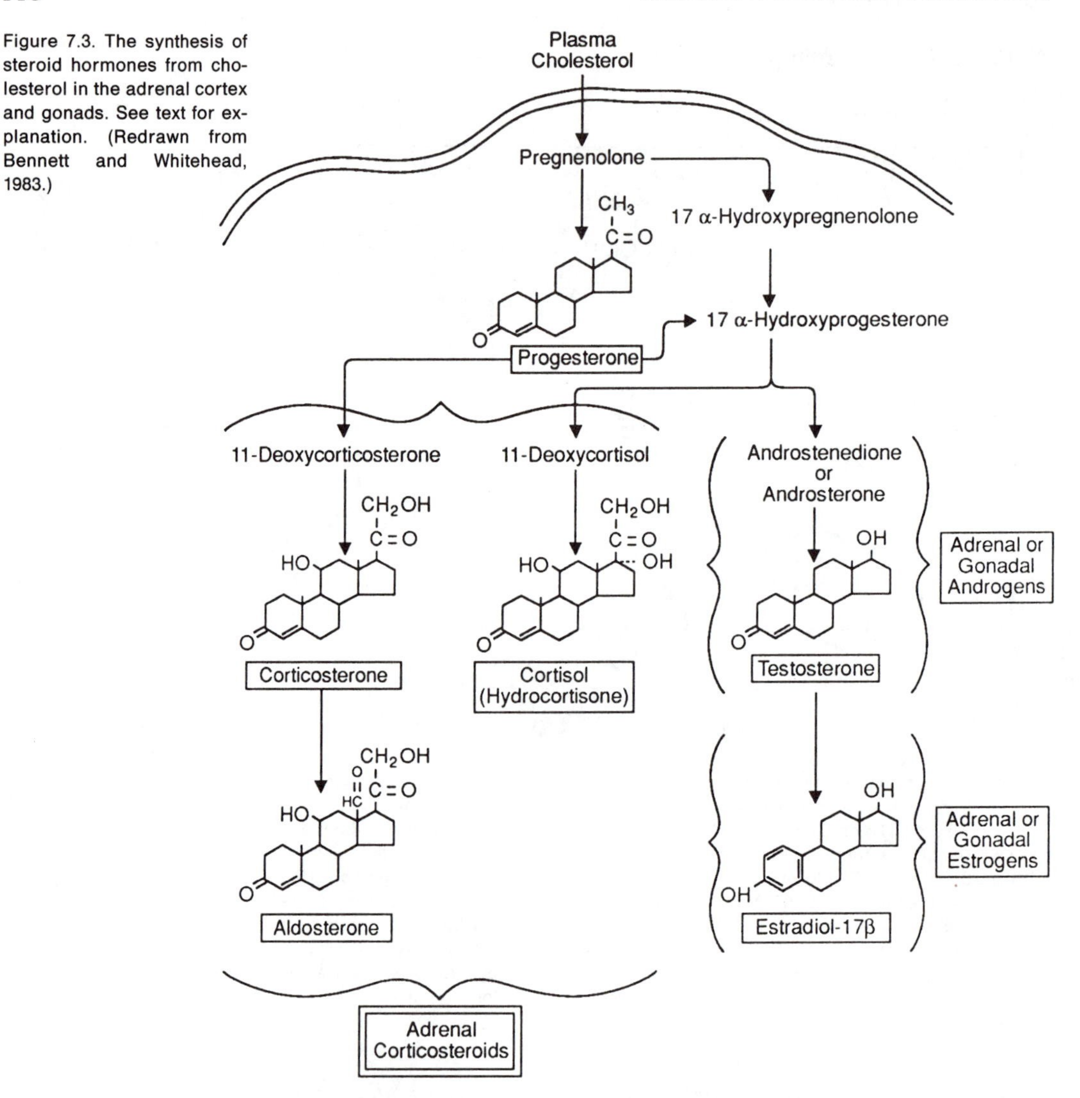

Figure 7.3. The synthesis of steroid hormones from cholesterol in the adrenal cortex and gonads. See text for explanation. (Redrawn from Bennett and Whitehead, 1983.)

transport and package into granules than active hormones. (c) Prohormones act as storage reserves. (d) Prohormones can be used to regulate the amount of hormone released. (e) Prohormones are more resistant to degradation than are active hormones and have longer half-lives (Martin, 1985).

7.2.2 SYNTHESIS OF STEROID HORMONES

Steroid hormones are synthesized from cholesterol in the smooth endoplasmic reticulum of the gonads and adrenal cortex. All steroids have a 4-carbon ring structure similar to that shown in Figure 7.3. Steroid-secreting cells take up cholesterol from the blood and convert it to pregnenolone in the mitochondria of the cell body. This pregnenolone is then converted to progesterone, which can be secreted as a hormone or used as a prohormone for further steroid hormone synthesis, depending on the endocrine gland. Progesterone can be converted to the other gonadal and adrenal steroids through the three different metabolic pathways, as described by Feder (1981) and Martin (1985).

In the testes (and ovaries), 17-α-hydroxypregnenolone and 17-α-hydroxyprogesterone are converted to dihydroepiandrosterone and androstenedione, respectively. These weak androgens are then converted to testosterone. Testosterone is secreted from the testes, and acts as a prohormone for estrogen, which is secreted from the ovaries. For each conversion to occur (i.e. at each arrow in Figure 7.3), specific enzymes are necessary, as described by Miller (1988). If any of these enzymes are missing, the hormones in the next step will not be produced.

In the adrenal cortex, most of the pregnenolone is converted to 17-α-hydroxypregnenolone and then to 17-α-hydroxyprogesterone, whereas the progesterone is converted to 11-deoxycorticosterone (Figure 7.3). The 11-deoxycorticosterone pathway produces corticosterone in the cells of the zona fasciculata of the adrenal cortex and this corticosterone is converted to aldosterone in the cells of the zona glomerulosa of the adrenal cortex (see Figure 2.3). The 17-α-hydroxyprogesterone is converted to 11-deoxycortisol, which leads to the production of cortisol (hydrocortisone). Androgens and estrogens can also be produced in the adrenal cortex through the 17-α-hydroxyprogesterone pathway. Thus, the adrenal cortex synthesizes gonadal as well as adrenal steroid hormones (see Simpson and Waterman, 1983).

7.3 STORAGE AND INTRACELLULAR TRANSPORT OF PEPTIDE HORMONES

Peptide hormones are 'packaged' in secretory granules in the Golgi apparatus as shown in Figure 7.1 (see Alberts *et al.*, 1989). Secretory granules serve a number of functions. (a) They provide intracellular storage and transport for the hormones. (b) In many cases, hormone synthesis is completed in the secretory granule by enzymatic conversion of prohormones (e.g. conversion of provasopressin to vasopressin). (c) The granule protects the hormone from deactivation. (d) Packaging of peptides in granules helps to regulate their rate of synthesis through negative feedback. (e) The secretion of the hormone into the bloodstream is facilitated by the secretory granule, resulting in hormones being released in 'bursts'. In neurosecretory cells, for example, the secretory granules take the hormone from the cell body to the nerve terminal of the axon, where it is released into the circulation (see Figure 4.2). (f) Empty secretory granules may fill up with peptides taken back into the nerve terminal by endocytosis for deactivation or recycling, as shown in Figure 7.1, although this is not as common in peptide-secreting cells as it is in monoaminergic neurons (see Figure 5.3). Thus, secretory granules perform many of the same functions for peptide hormones that vesicles do for neurotransmitters (see Section 5.3).

Steroid hormones are not packaged and stored in secretory granules for release. When they are synthesized, the steroids simply diffuse out of the secreting cells and into the circulation.

7.4. HORMONE RELEASE

Hormone release occurs when the endocrine cell is activated by electrical, neurochemical or hormonal stimulation. Stimulation of peptide hormone-secreting cells causes a change in cell membrane permeability so that the hormone is released into the circulatory system by exocytosis when the secretory granule fuses with the cell membrane (step 4 in Figure 7.1). Depolarization-induced entry of sodium and calcium ions into the cell is essential for peptide secretion. Calcium ion binding is necessary to initiate the fusion of the secretory granule to the cell membrane during exocytosis.

Peptidergic cells release their hormones in bursts or pulses. As each cohort of secretory granules bind to the cell membrane, a burst of hormone is released into the blood, as shown in Figure 6.12. The level of circulating hormones thus rises dramatically and then gradually declines as the hormone is deactivated. Depending upon the cell type and the frequency of stimulation, more granules of hormone are then released, elevating the level of circulating hormone again. Thus, if averaged over 24 hours, the hormone might show a constant level in the blood. But, if blood samples are taken several times per hour, dramatic changes in hormone levels are observed.

This is shown for luteinizing hormone secretion from the pituitary gland in Figure 7.4. This pattern of LH secretion reflects the pulsatile release of LH-RH from the hypothalamic neurosecretory cells (Weick, 1981; Dyer and Robinson, 1989). Oxytocin, LH-RH and other hypothalamic hormones are released in pulses through the action of hypothalamic 'pulse generators' whose electrical activity occurs in bursts as shown in Figure 6.12 (Lincoln *et al.*, 1985). Pulsatile release of hormones appears to be an optimal mechanism for hormonal communication as it prevents down-regulation of receptors on target cells. Continuous release of hormones or abnormal hormone release patterns result in the desensitization of the target cell receptors, which become insensitive to the hormone (Belchetz *et al.*, 1978; Goldbeter, 1988).

Steroid hormone release is controlled by the rate of synthesis (Martin, 1985, p. 584). Steroid hormones are not stored in the secretory cells, since steroids are not packaged in secretory granules. Steroids are, however, secreted in pulses due to the response of the steroid secreting cells to pituitary tropic hormones (ACTH, TSH, LH and FSH), which are released in bursts from secretory granules, as shown in Figure 7.4.

7.5 HORMONE TRANSPORT

Once the hormone is secreted into the blood, it is likely to be rapidly deactivated. To prevent this deactivation, many hormones are bound to protective carrier proteins in the bloodstream. Carrier proteins serve

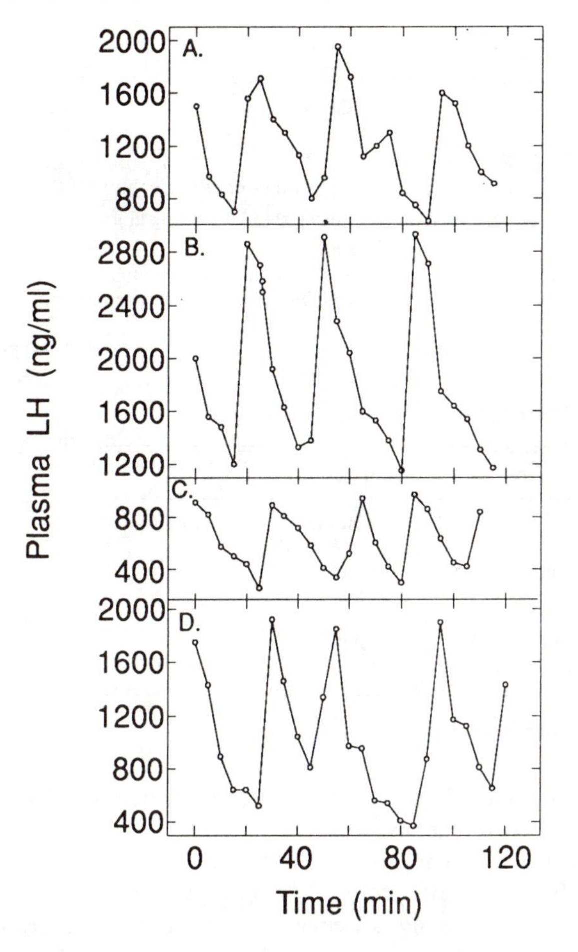

Figure 7.4. The time course of plasma luteinizing hormone (LH) secretion in four different ovariectomized adult female rats (A–D). Blood samples were taken every 5 min through indwelling venous catheters. LH was measured by radioimmunoassay. (Redrawn from Weick, 1981.)

functions similar to those of secretory granules and it is noteworthy that, except for the neurohypophyseal hormones, those hormones which have carrier proteins are primarily those that are not released in secretory granules (i.e. steroid and thyroid hormones).

Plasma albumin acts as a general carrier protein, but most hormones have specific carrier proteins, some of which are listed in Table 7.2. Carrier proteins such as transcortin, thyronine binding globulin and sex hormone binding globulins are glycoproteins produced in the liver, while the neurophysins are produced from prohormones along with oxytocin and vasopressin in the neurohypophysis (Brownstein *et al.*, 1980; Camier, Barre and Cohen, 1985).

Carrier proteins facilitate transportation of the hormones in the blood (steroid hormones would otherwise be insoluble), and prolong the active life of the hormones by protecting them from deactivation. The peptide hormones are not generally bound to carrier proteins and thus have short half-lives in the bloodstream (they are deactivated in a matter of minutes). The steroid and thyroid hormones, which are bound to carrier proteins, have much longer half-lives (they last for hours). While the hormone is bound to the carrier protein, however, it is inactive. Thus,

Table 7.2. *Hormone-specific carrier proteins*

Hormones	Carrier proteins
Oxytocin and Vasopressin	Neurophysins I and II
Thyroxine and triiodothyronine	Thyronine-binding globulin (TBG)
Glucocorticoids and progesterone	Corticosteroid-binding globulin (CBG also called transcortin)
Estrogen and testosterone	Sex-hormone-binding globulin (SHBG also called testosterone-estrogen binding globulin, TeBG)

Table 7.3. *Hormone metabolites in the urine*

Hormone	Urinary metabolite
Serotonin	5-Hydroxyindole-acetic acid (5-HIAA)
Melatonin	6-Hydroxy melatonin
Growth hormone	Many metabolites
Testosterone	17-Ketosteroids
Estrogen	Estrones
Corticosteroids	Many metabolites
Prostaglandins	Various fatty acids

Source: Martin, 1985.

high levels of protein-bound hormones can be held in the blood to act as readily available storage reserves.

Before a hormone can bind to its target cell receptors, it must be uncoupled from its carrier protein. The hormone–carrier protein binding is reversible and this bond is weaker than the hormone–receptor bond. Hormones can, therefore, exist in the blood in two forms, bound or free, and only the free hormone can attach to a receptor at a target cell. Carrier proteins help to regulate the level of free hormone available to the target cells and thus protect the body from excessive hormonal stimulation (Martin, 1985). This 'free hormone hypothesis', while generally accepted, remains unproven, and the actual mechanisms involved in the binding of hormones to carrier proteins and target cell receptors is controversial (Ekins, 1990).

Hormones can only act at cells which have receptors for them and receptors are specific for particular hormones. Thus, for example, estrogen will have no biological effect on a cell unless that cell has an estrogen receptor. Likewise for thyroxine, growth hormone, testosterone, follicle stimulating hormone, etc. The response of target cells to hormonal stimulation is described in Chapters 9 and 10.

7.6 DEACTIVATION OF HORMONES

Most steroid, thyroid and peptide hormones in the circulatory system are deactivated by enzymes in the liver, kidney or blood and their metabolites

Table 7.4. *Methods used to identify cellular sites of hormone synthesis, storage and secretion*

Method	Information gained
Identification of neuroendocrine cells	
Gross anatomy	Anatomical location, size, vascularization and innervation of glands
Light microscopy	Identification of neural and endocrine cells
Electron microscopy	Identification of subcellular structure in neuroendocrine cells
Scanning elecron microscope or confocal microscope	Three-dimensional image of cell structure
Localization of hormones within cells	
Histological staining	Identify cell bodies (nucleus and cytoplasm) and axons
Fluorescence immunocytochemistry	Localize hormones in histological preparations
Immunoenzyme histochemistry	Locate sites of enzyme activity in cells
Examining biosynthetic pathways	
Antibody injection	Detect and isolate specific molecules
Drug injection	Study physiological activity of cells
Radioisotope injection	Measure enzyme activity, trace chemical pathways during synthesis, and determine biochemical responses in a cell
Determining the chemical structure of hormones	
Chromatography	Separates proteins into different fractions
Electrophoresis	Determines the size and structure of proteins
Amino acid sequencers	Determine amino acid sequences of peptides

are then excreted in the urine or feces. Some proteins, however, may be taken up into target cells by endocytosis and deactivated by lysosomes within the target cells (steps 5 and 6 in Figure 7.1; see Martin, 1985). Since hormone metabolites excreted in the urine reflect the levels of these hormones in the circulation, a number of methods for measuring circulating hormone levels in humans and animals involve measuring the hormone metabolites in the urine (see Table 7.3). Hormone metabolites in the urine and feces also provide the basis for some of the pheromones excreted by mammals (see Section 1.4.5).

7.7 METHODOLOGY FOR NEUROENDOCRINE RESEARCH

Many procedures have been developed for studying cellular activity (Alberts *et al.*, 1989, pp. 135–96) and these techniques are also used in neuroendocrinology (Bennett and Whitehead, 1983; Hadley, 1992). Some methods used for the identification of endocrine cells, localizing hormones in tissues, examining biosynthetic pathways, and determining the biochemical structure of hormones are summarized in Table 7.4.

7.8 SUMMARY

This chapter outlines the factors regulating the synthesis, storage, release, transport and deactivation of hormones. Hormones can be divided into two general types: steroids and non-steroids. The non-steroid hormones include modified amino acids, peptides, glycoproteins and prostaglandins. Non-steroid hormones are synthesized from amino acid precursors (pre-prohormones) in the endoplasmic reticulum of the cell and packaged into secretory granules in the Golgi apparatus. Often the prohormones are stored in the secretory granules along with proteolytic enzymes and the final synthesis of the hormone occurs in the secretory granules. As well as the completion of hormone synthesis, secretory granules store hormones, regulate their rate of synthesis and control the release of hormones, serving many of the same functions as secretory vesicles do for neurotransmitters. Steroid hormones are synthesized from cholesterol in the adrenal cortex and gonads and are released directly into the circulation without being packaged in secretory granules. The neurohypophyseal, thyroid, adrenal corticosteroid and gonadal hormones are transported through the blood by specific carrier proteins. These carrier proteins have many functions similar to those of secretory granules. Hormones are deactivated by metabolic enzymes in the kidney, liver and blood and their metabolites are excreted in the urine and feces. Some peptides may also be deactivated by reuptake into neurosecretory cells.

FURTHER READING

Alberts, B., Bray, D., Lewis, J., Raff, M., Roberts, K. and Watson, J. D. (1989). *Molecular Biology of the Cell*, 2nd edn. New York: Garland.

Hadley, M. E. (1992). *Endocrinology*, 3rd edn. Englewood Cliffs, NJ: Prentice-Hall.

Martin, C. R. (1985). *Endocrine Physiology*. New York: Oxford University Press.

REVIEW QUESTIONS

7.1 Name the five steroid hormones

7.2 What is the main precursor of the steroid hormones?

7.3 What are the building blocks of the protein and polypeptide hormones?

7.4 What is the primary mechanism for the synthesis of small peptides?

7.5 In the female, which two endocrine glands can synthesize testosterone?

7.6 Are both steroid and non-steroid hormones stored in secretory granules?

7.7 Hormones are synthesized in the _______ of the cell body.

7.8 The carrier proteins for the neurohypophyseal hormones are _______.

7.9 The contents of a secretory granule are released into the circulation by the process of _______.

7.10 Which three pituitary hormones are synthesized from POMC?
7.11 What controls the rate of synthesis and release of steroid hormones?

ESSAY QUESTIONS

7.1 Compare and contrast the functions of synaptic vesicles and secretory granules.
7.2 Discuss the importance of pre-prohormones and prohormones in the synthesis of peptide hormones.
7.3 Compare the synthesis and secretion of androgens from the testes and adrenal cortex.
7.4 Discuss the importance of the pulsatile release of hormones.
7.5 Discuss the importance of carrier proteins in the transport of hormones in the bloodstream.
7.6 Describe the synthesis of oxytocin and neurophysin I from their prohormone.
7.7 Describe how steroid hormones are deactivated and discuss the advantages and disadvantages of using urinary metabolites of steroids as measures of circulating hormone levels in animals and humans.
7.8 Explain how immunocytochemistry and other methods are used to locate the brain cells which synthesize and release peptides such as CCK.

REFERENCES

Alberts, B., Bray, D., Lewis, J., Raff, M., Roberts, K. and Watson, J. D. (1989). *Molecular Biology of the Cell*, 2nd edn. New York: Garland.

Belchetz, P. E., Plant, T. M., Nakai, Y., Keogh, E. J. and Knobil, E. (1978). Hypophysial responses to continuous and intermittent delivery of hypothalamic gonadotropin-releasing hormone. *Science*, **202**, 631–633.

Bennett, G. W. and Whitehead, S. A. (1983). *Mammalian Neuroendocrinology*. New York: Oxford University Press.

Brownstein, M. J., Russell, J. T. and Gainer, H. (1980). Synthesis, transport, and release of posterior pituitary hormones. *Science*, **207**, 373–378.

Camier, M., Barre, N. and Cohen, P. (1985). Hypothalamic biosynthesis and transport of neurophysins and their precursors to the rat brain stem. *Brain Research*, **334**, 1–8.

Degroot, L. J. and Niepomniszcze, H. (1977). Biosynthesis of thyroid hormone: basic and clinical aspects. *Metabolism*, **26**, 665–718.

Dyer, R. G. and Robinson, J. E. (1989). The LHRH pulse generator. *Journal of Endocrinology*, **123**, 1–2.

Ekins, R. (1990). Measurement of free hormones in the blood. *Endocrine Reviews*, **11**, 5–46.

Feder, H. H. (1981). Essentials of steroid structure, nomenclature, reactions, biosynthesis and measurements. In N. T. Adler (ed.) *Neuroendocrinology of Reproduction*, pp. 19–63. Plenum: New York.

Goldbeter, A. (1988). Periodic signalling as an optimal mode of intercellular communication. *News in Physiological Sciences*, **3**, 103–105.

Goldman, B. D. (1981). Structure of protein and peptide hormones. In N. T. Adler (ed.) *Neuroendocrinology of Reproduction*, pp. 3–12. New York: Plenum.

Hadley, M. E. (1992). *Endocrinology*, 3rd edn. Englewood Cliffs, NJ: Prentice-Hall.

Lincoln, D. W., Fraser, H. M., Lincoln, G. A., Martin, G. B. and McNeilly, A. S. (1985). Hypothalamic pulse generators. *Recent Progress in Hormone Research*, **41**, 369–419.

Loh, Y. P. (1987). Peptide precursor processing enzymes within secretory vesicles. *Annals of the New York Academy of Sciences*, **493**, 292–307.

Mains, R. E., Cullen, E. I., May, V. and Eipper, B. A. (1987). The role of secretory granules in peptide biosynthesis. *Annals of the New York Academy of Sciences*, **493**, 278–291.

Mains, R. E., Dickerson, I. M., May, V., Stoffers, D. A., Perkins, S. N., Ouafik, NL., Husten, E. J. and Eipper, B. A. (1990). Cellular and molecular aspects of peptide hormone biosynthesis. *Frontiers in Neuroendocrinology*, **11**, 52–89.

Martin, C. R. (1985). *Endocrine Physiology*. New York: Oxford University Press.

Miller, W. L. (1988). Molecular biology of steroid hormone synthesis. *Endocrine Reviews*, **9**, 295–318.

Pickering, B. T. (1976). The molecules of neurosecretion: their formation, transport and release. *Progress in Brain Research*, **45**, 161–179.

Ramsay, D. J. (1991). Posterior pituitary gland. In F. S. Greenspan *Basic and Clinical Endocrinology*, 3rd edn, pp. 177–187. Norwalk, CT: Appleton and Lange.

Simpson, E. R. and Waterman, M. R. (1983). Regulation by ACTH of steroid hormone biosynthesis in the adrenal cortex. *Canadian Journal of Biochemistry and Cell Biology*, **61**, 692–707.

Sossin, W. S., Fisher, J. M. and Scheller, R. H. (1989). Cellular and molecular biology of neuropeptide processing and packaging. *Neuron*, **2**, 1407–1417.

Turner, C. D. and Bagnara, J. T. (1976). *General Endocrinology*, 6th edn. Philadelpia: Saunders.

Weick, R. F. (1981). The pulsatile nature of luteinizing hormone secretion. *Canadian Journal of Physiology and Pharmacology*, **59**, 779–785.

8

Regulation of hormone levels in the bloodstream

As was shown in Figures 6.12 and 7.4, hormone levels in the bloodstream fluctuate quite dramatically over short periods of time, yet they are maintained within certain 'average' levels over the long term. The regulation of hormone levels in the bloodstream occurs through the autonomic and central nervous system, non-hormone chemicals in the bloodstream and gastrointestinal tract, and hormonal feedback. In order to understand how these mechanisms regulate the release of hormones, however, assays must be available for measuring hormone levels in the circulation and these assays should be sensitive to small changes in hormone levels. This chapter therefore begins with an examination of the methods for measuring hormone levels in the circulation.

8.1 MEASURING HORMONE LEVELS

The level of a hormone in the bloodstream can be measured directly or estimated by measuring hormone levels in the saliva, urine or feces, measuring urinary metabolites, or by using bioassays.

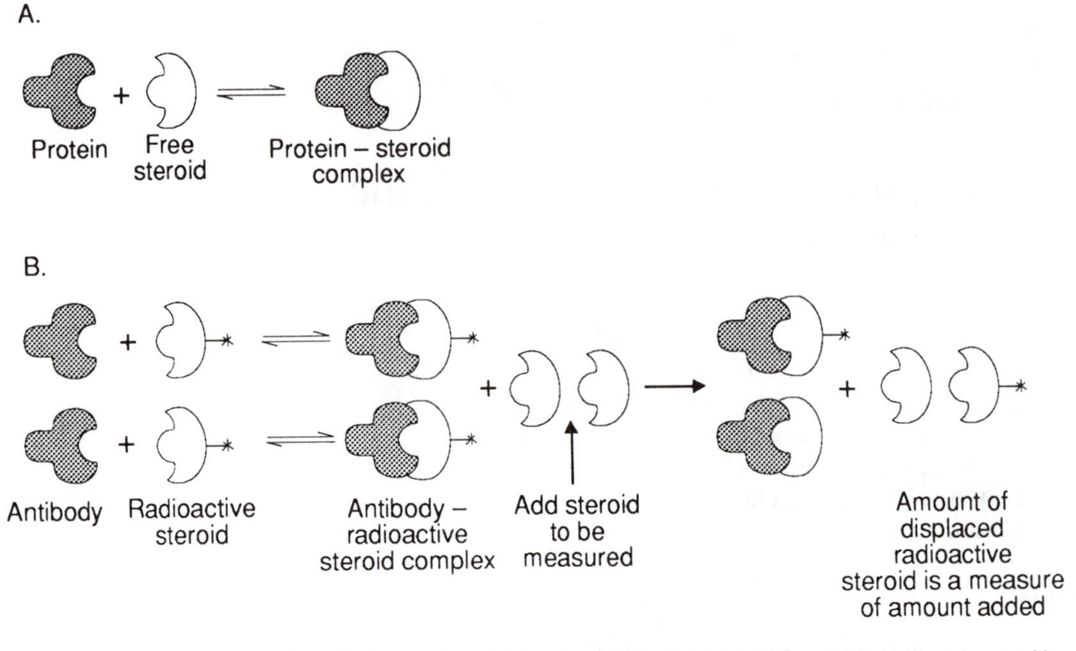

Figure 8.1. The method used for a radioimmunoassay. (A) Reversible binding of a steroid hormone to a binding protein in the plasma to form a protein–steroid complex. (B) Competition for an antibody by radioactive and non-radioactive steroid hormones can be used as an assay. See text for explanation. (Redrawn from Baird, 1972.)

8.1.1 DIRECT MEASUREMENT OF CIRCULATING HORMONES

A number of methods have been developed for the assay of hormone levels in small samples of blood. These methods include gas–liquid chromatography, competitive protein binding assays and radioimmunoassays (Feder, 1981). The most common of these techniques is the radioimmunoassay (RIA), which is based on the fact that a hormone will bind to a specific antibody.

Radioimmunoassays have been developed which are specific for particular hormones, highly accurate, and sensitive to small amounts of hormone (picogram levels = 10^{-12} g). A radioimmunoassay is a competitive binding assay in which serum containing an unknown amount of hormone is added to a known amount of antibody for the hormone of interest (e.g. estrogen) and a small, known amount of radioactively labeled hormone* (* is used to indicate a radioactive isotope). The estrogen in the sample competes with the radioactively labeled estrogen* for binding sites on the antibody (Figure 8.1). The unbound hormone is then removed from the sample and the amount of radioactively labeled hormone bound to the antibody is measured. The more estrogen in the unknown sample, the *less* radioactively labeled estrogen will be bound to the antibody and the lower the radioactivity of the sample, as measured in a scintillation counter. The actual hormone level in the sample can be calculated by comparing the results of the assay to a standard curve (Bennett and Whitehead, 1983; Riad-Fahmy *et al.*,1981).

Although the RIA is the most commonly used method for the analysis of plasma hormone levels, there are some problems with the use of RIAs. RIAs depend on specific antibodies; therefore, if the antibody is not specific, it may detect hormones for which it was not intended. Antibodies may be too specific, detecting only some variants of the hormone, or too general, detecting an entire class of hormones (e.g. all androgens) rather than a single hormone (e.g. testosterone). RIAs are not readily available for all hormones and the RIAs for some hormones (e.g. prolactin) are highly species specific, so can not be used for all species. RIAs can also cross-react with hormone metabolites or other hormones with a similar structure. For example, the RIA for prolactin may cross-react with GH, giving a confounded measure. Some problems associated with the application of RIAs to hormone measurement are discussed by Yalow (1980) and Ekins (1990).

8.1.2 ANALYSIS OF HORMONES IN THE SALIVA, URINE, AND FECES

The collection of saliva, urine, and feces for the analysis of circulating hormone levels is easy, non-invasive, and stress-free, compared to taking blood samples, and can be used to measure hormone levels in free-ranging animals, newborn children, and adults outside of the laboratory or hospital. These methods are particularly useful for obtaining baseline measures of stress-related hormones, such as cortisol. The level of steroid hormones and melatonin in the saliva can be measured by RIA and these salivary hormone levels accurately reflect the level of hormones in the blood serum (Riad-Fahmy *et al.*, 1982; Vakkuri, Leppäluoto and Kauppila, 1985; Dabbs, 1990; Hampl *et al.*, 1990).

Hormones which are eliminated in the urine can be measured by collecting urine samples. Urine analysis gives a measure of the average hormone secretion over a number of hours, but does not permit the minute-by-minute analysis of hormone levels which can be determined from blood samples. Urine samples taken once or twice a day provide consistent measures of hormone release, whereas blood samples taken every few minutes show a great deal of variability (Figure 7.4). Gonadal hormones, as well as FSH, LH, melatonin, and human chorionic gonadotropin are excreted in the urine and detection of these hormones in the urine can provide important information. Urinary assays of LH, for example, are often used to detect the onset of the preovulatory LH surge in clinical settings to time artificial insemination or to collect oocytes for *in vitro* fertilization (Kerin *et al.*, 1980; Paz *et al.*, 1990). LH levels in the urine rise about 6 hours later than in the plasma and show a secretory pattern which accurately reflects that shown in the plasma (Figure 8.2). Detection of the presence of human chorionic gonadotropin in the urine is the basis of pregnancy tests. The analysis of urinary hormones is also used in pediatric medicine to detect endocrine abnormalities in newborn infants

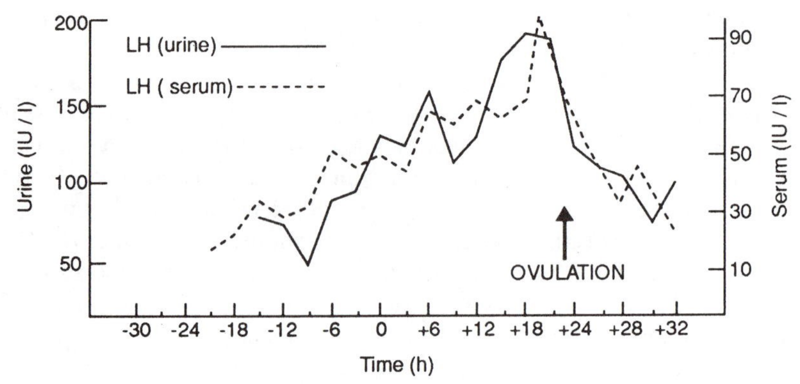

Figure 8.2. A comparison of serum and urinary luteinizing hormone (LH) levels in 3-hour periods near the time of ovulation in a human female. Zero hours was defined as the time when there was a twofold increase in urinary LH concentration above baseline levels. The assessments of serum LH are related to this point. (Redrawn from Kerin *et al.*, 1980.)

(Kivela *et al.*, 1990) and in animal research to measure hormone levels in animals without stressing them by handling, anaesthesia, and venipuncture (Carlshead *et al.*, 1989).

One problem with urine analysis is that many hormones, including oxytocin, vasopressin, TSH, GH, prolactin, insulin, and progestone, are modified or degraded before being excreted in the urine. Thus, urine analysis is not capable of measuring circulating levels of all hormones. In some cases, however, the hormone metabolites excreted in the urine can be measured and estimates of circulating hormone levels made from these assays (see Table 7.3). It is also possible to assay the levels of some hormones and their metabolites from fecal samples (Adlercreutz and Järverpää, 1982).

8.1.3 BIOASSAYS

Simple quantitative bioassays can be conducted by weighing endocrine glands. TSH, for example, stimulates growth of the thyroid gland, so thyroid weight can be used as a bioassay for TSH levels (Figure 8.3). Likewise, gonadal weights can be used as bioassays for gonadotropin levels, and adrenal gland weights as bioassays for ACTH levels (Figure 8.3). Adrenal gland weights can also be used as a bioassay for stress as adrenal activity increases during stress (Christian and Davis, 1964).

Other bioassays measure changes in non-endocrine target tissue. Thus, the size of a cock's comb is a classic bioassay for testosterone, as is the size of a deer's antlers (Figure 8.3). The vaginal smear technique, which involves measurement of cell types in the vaginal epithelium, is a bioassay for estrogen and progesterone levels, and is used to determine the stage of the estrous cycle of female rodents. The earliest pregnancy tests were based on the ability of human chorionic gonadotropin in the urine to induce sperm release in frogs or ovulation in rabbits. The bioassay, therefore, is an important technique for estimating hormone levels in the circulation (Zarrow *et al.*, 1964).

These traditional *in vivo* bioassays, however, are time consuming, require large numbers of experimental animals, are relatively insensitive and lack specificity (Bennett and Whitehead, 1983). More recent bioassays involve *in vitro* techniques, such as isolated tissue preparations or

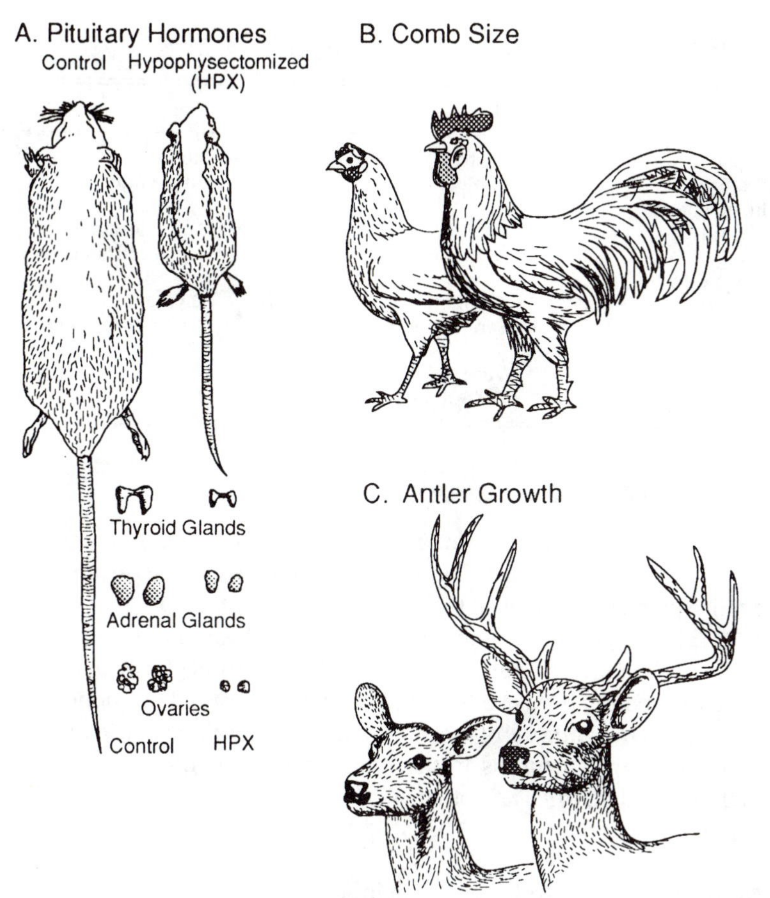

Figure 8.3. Some examples of bioassays. (A) Bioassays for pituitary hormones include weights of the thyroid, adrenal and gonadal target glands. Shown here are two female littermate rats, a control and one which had the pituitary gland removed by hypophysectomy (HPX) at 36 days of age. In adulthood, the control rat was larger than the HPX rat, as were its endocrine glands. Bioassays for testosterone include the measurement of (B) the comb of the white leghorn rooster and (C) antler length of the white-tailed deer. (Redrawn from Zuckerman, 1957.)

cell cultures. For example, the activity of hypothalamic hormones such as CRH can be measured by the amount of pituitary hormone (ACTH) released *in vitro* from pituitary tissue or cell culture, while the action of the pituitary hormones, such as LH, can be assayed by their effect on the release of their target hormones, such as testosterone, from isolated endocrine (e.g. testicular) tissue.

Other bioassays use cultures of specific cells whose growth rate is sensitive to a particular hormone. For example, the NB2 lymphoma cell (a special line of cells) is sensitive to prolactin and the level of prolactin in a blood sample can be determined by the rate of cell division in a culture of these cells (Tanaka *et al.*, 1980; Murphy *et al.*, 1989).

8.2 MECHANISMS REGULATING HORMONE LEVELS

The amount of a hormone in the blood can be regulated by four different mechanisms: (a) the autonomic nervous system, (b) non-hormone chemicals in the blood or gastrointestinal tract, (c) hormonal feedback, and (d) neurotransmitters and neuropeptides in the brain and central nervous system.

Table 8.1. *Non-hormone chemicals regulating hormone levels*

Chemical	Hormone	Gland
Liquids	Gastrin	Stomach
Proteins and fats	Cholecystokinin	Duodenum
Blood glucose		
High	Insulin	Beta cells of pancreas
Low	Glucagon	Alpha cells of pancreas
Blood calcium		
High	Calcitonin	Thyroid
Low	Parathyroid hormone	Parathyroid gland
Blood sodium		
High	Vasopressin	Neurohypophysis
Low	Aldosterone	Adrenal cortex

8.2.1 AUTONOMIC NERVOUS SYSTEM (ANS)

The primary example of hormone regulation by the autonomic nervous system is the adrenal medulla, which is stimulated to release adrenaline by the splanchnic nerves of the sympathetic branch of the ANS (Hadley, 1992). The vagus nerve of the ANS also regulates hormone release from endocrine glands, including the release of insulin and glucagon from the pancreas and the release of some of the gastrointestinal hormones, such as gastrin. The synthesis of melatonin in the pineal gland is also regulated by noradrenergic neurons in the sympathetic branch of the ANS. The thyroid gland, thymus and other endocrine glands are also innervated by nerves of the ANS.

Neuroendocrine reflexes. A neuroendocrine reflex occurs when a stimulus from the central nervous system results in the release of a hormone. Thus, a fear-inducing stimulus causing neurotransmitter release from the sympathetic nerves of the ANS results in the 'reflexive' release of adrenaline from the adrenal medulla. The adrenal medulla is thus a neuroendocrine transducer, converting neural input to hormonal output. Likewise, the magnocellular and parvicellular neurosecretory cells of the hypothalamus are also neuroendocrine transducers which can be induced to release their neurohormones reflexively (e.g. oxytocin, see Figure 6.13, p. 104) in response to external or internal stimuli. The fourth neuroendocrine transducer is the pineal gland.

8.2.2 NON-HORMONAL CHEMICALS

Non-hormonal chemicals and parahormones regulate the secretion of a number of hormones (see Martin, 1985). Drinking fluids, for example, stimulates gastrin secretion from the stomach. Ingestion of proteins and fat in food stimulates cholecystokinin release from the duodenum. When the food is digested, the hormones are no longer stimulated. There are many other cases in which chemicals in the circulation regulate hormone release (Table 8.1). Hormones function to maintain the levels of glucose, calcium and sodium in the circulation at constant (homeostatic) levels

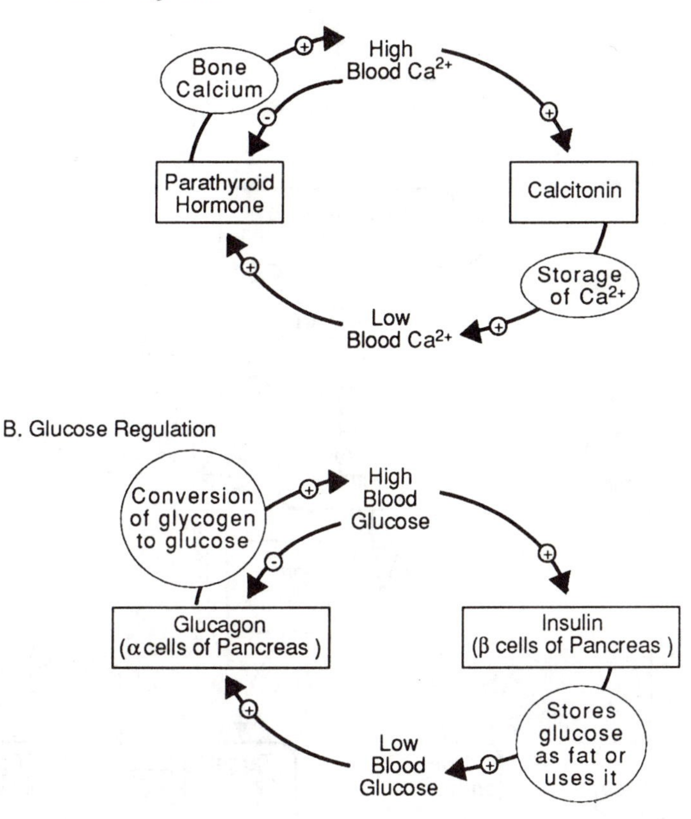

Figure 8.4. (A) The regulation of blood calcium levels by calcitonin and parathyroid hormone secretion and (B) the regulation of blood glucose levels by insulin and glucagon secretion. (Redrawn from Martin, 1985.)

and the levels of the regulatory hormones fluctuate in response to the levels of these chemicals in the blood. Thus, these chemicals also regulate the release of the hormones that regulate them. The endocrine gland releases its hormone in response to an increase in some chemical in the blood (e.g. calcium) and reduces the level of that chemical. When the level of that chemical decreases, stimulation of hormone release declines. Thus, circulating hormone levels are constantly fluctuating by small amounts, increasing or decreasing in response to chemicals in the blood.

In calcium regulation, for example, a decline in blood calcium level stimulates the release of parathyroid hormone, which will cause bone calcium to be released into the blood. The resulting rise in blood calcium levels reduces the stimulation for the release of parathyroid hormone, thus less is secreted. High levels of blood calcium increase calcitonin, which acts to reduce blood calcium levels in a similar feedback fashion (Figure 8.4A). Likewise, high blood glucose levels stimulate insulin release from the β cells of the pancreas and low blood glucose levels stimulate the release of glucagon from the α cells of the pancreas (Figure 8.4B). High blood sodium levels stimulate vasopressin release from the neurohypophysis, and low blood sodium stimulates aldosterone release from the adrenal cortex.

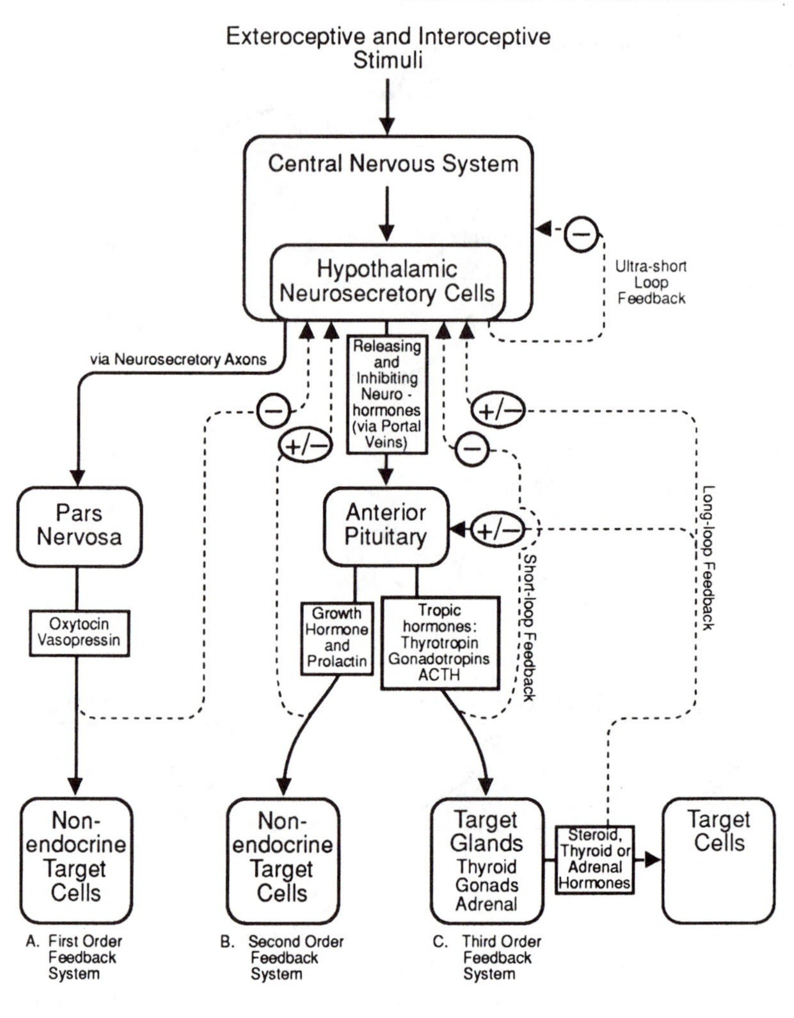

Figure 8.5. Three orders of neuroendocrine feedback loops: (A) first order, (B) second order, and (C) third order. The broken lines indicate negative and positive feedback; either directly upon the anterior pituitary, on the neurosecretory cells of the hypothalamus, or upon both. The third order feedback system has long, short, and ultra-short feedback loops. (Redrawn from Turner and Bagnara, 1976.)

8.2.3 HORMONAL FEEDBACK SYSTEMS

The hypothalamic hormones regulate the synthesis and release of pituitary hormones which act on endocrine and non-endocrine target cells (Figure 8.5). At endocrine target cells, a third hormone is released and all three levels of hormones, hypothalamic, pituitary, and peripheral, act in feedback loops. These feedback loops also involve neurotransmitters and neuropeptides as described in Section 8.2.4, below.

Positive and negative feedback

Feedback is defined as the situation in which the hormones released by the cells of a given system act to regulate the secretion of the hormones which stimulate their release. For example, if hormone A stimulates the release of hormone B, then hormone B can feed back to regulate the release of hormone A. Feedback can be negative or positive. In negative feedback, hormone B feeds back to reduce the secretion of hormone A. In positive feedback, hormone B feeds back to stimulate the release of hormone A.There are three orders of complexity of hormonal feedback loops: first, second and third order feedback systems (Piva, Motta and Martini, 1979).

First order feedback loops. The neurohypophyseal hormones oxytocin and vasopressin, which are released from the hypothalamic neurosecretory cells into the blood from the pars nervosa, act on target cells in the breast and uterus (oxytocin) and in the kidneys (vasopressin), but they do not stimulate pituitary hormones nor any other endocrine target cells. The level of these hormones in the blood is monitored by hormone-sensitive nerve cells in the brain and thus they regulate themselves. These hormones are, therefore, involved in a first order feedback system: there is only one endocrine gland, the hypothalamus, and only one hormone, oxytocin or vasopressin, involved in the feedback mechanism (Figure 8.5A). As oxytocin and vasopressin are released, they feed back to inhibit their own secretion.

Second order feedback loops. Growth hormone, prolactin and melanocyte stimulating hormone are regulated by hypothalamic releasing and inhibiting hormones and do not stimulate other endocrine target cells (Figure 8.5B). These pituitary hormones regulate their own secretion through feedback regulation of the hypothalamic hormones. This constitutes second order feedback (Piva *et al.*, 1979), which is more complex than first order feedback because it involves hypothalamic releasing and inhibiting hormones as well as pituitary hormones. Figure 8.6 illustrates this complexity, using GH as an example. The hypothalamic release of GH-RH stimulates GH from the pituitary and GH provides negative feedback, inhibiting GH-RH release. As GH levels rise, the hypothalamus is stimulated to increase the secretion of GH-RIH (somatostatin), which further inhibits the release of GH from the pituitary. Prolactin and MSH follow the same pattern of feedback control as shown for GH in Figure 8.6.

Third order feedback loops. There are three endocrine glands involved in third order feedback systems: the hypothalamus, pituitary, and a peripheral endocrine gland (Piva *et al.*, 1979). There are three different third order feedback systems: the hypothalamic-pituitary-adrenal, (H-P-A) (Kendall *et al.*, 1975; Keller-Wood and Dallman, 1984); hypothalamic-pituitary-gonadal (H-P-G) (Fink, 1979; Plant, 1986); and hypothalamic-pituitary-thyroid (H-P-T) (Larsen, 1982) feedback systems. Each hormone within a third order feedback system can provide negative feedback to the hypothalamus, so there are three levels of feedback within the third order feedback system: short, long, and ultra-short loop feedback (Figure 8.5C).

Long loop feedback. The hormones from the peripheral endocrine target cell, such as cortisol, testosterone or thyroxine, provide long-loop feedback to the hypothalamus and pituitary. Corticosteroids, for example, inhibit CRH release from the hypothalamus, as well as inhibiting ACTH

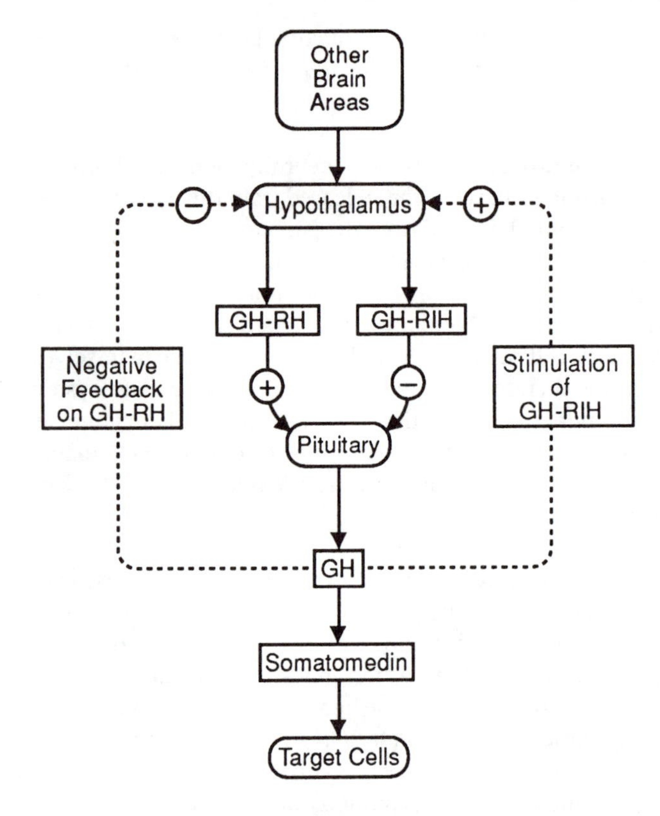

Figure 8.6. An example of the second order feedback system regulating growth hormone secretion. GH-RH = growth hormone releasing hormone; GH-RIH = growth hormone release inhibiting hormone.

release from isolated pituitary glands (Keller-Wood and Dallman, 1984). Thyroid hormones control the release of TSH by direct negative feedback on the pituitary thyrotroph cells (Larsen, 1982). Long loop negative feedback only occurs within a third order feedback system. While the majority of long loop feedback is negative (inhibitory), estrogen has positive feedback on LH secretion. Estrogen facilitates the LH surge by stimulating the release of LH-RH from the preoptic area of the hypothalamus and by promoting the synthesis of LH-RH receptors in pituitary gonadotroph cells (Fink, 1979; Plant, 1986).

Short loop feedback. The pituitary hormones, ACTH, LH, and TSH provide short loop negative feedback to the hypothalamus within a third order feedback system (Figure 8.5C). Short loop feedback within a third order system is analogous to second order feedback, but occurs within the third order feedback system.

Ultra-short loop feedback. The hypothalamic hormones feedback to help to regulate themselves in an ultra-short feedback loop (Figure 8.5C), which is analogous to first order feedback, but is contained within a third order system and thus involves two other hormones (Piva *et al.*, 1979). Thus, the release of LH-RH is inhibited by LH-RH agonists and stimulated by LH-RH antagonists (Valenca *et al.*, 1987; Zanisi, Motta, and Martini, 1987).

Figure 8.7 shows the feedback loops involved in the hypothalamic-

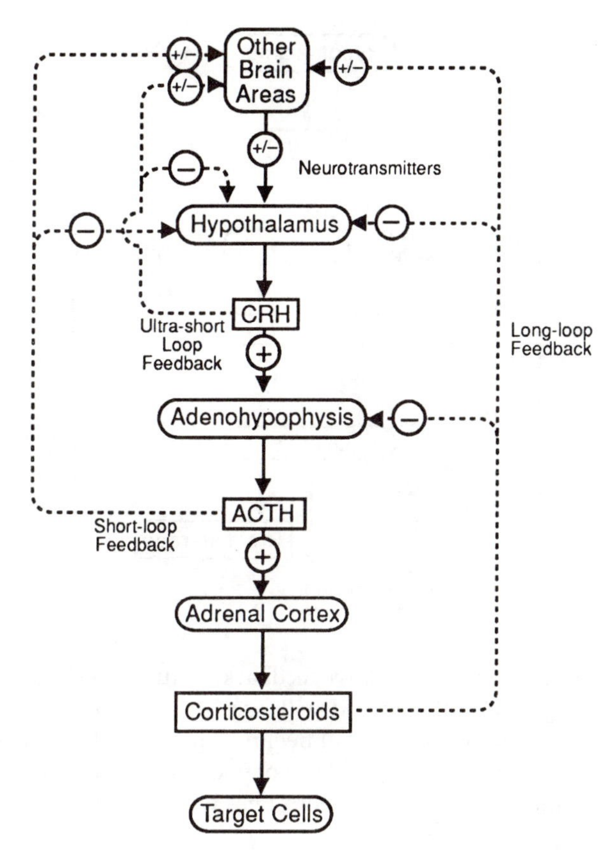

Figure 8.7. An example of the hypothalamic-pituitary-adrenal third order feedback system. This feedback system involves the hormonal regulation of neurotransmitter release from the hypothalamus and other brain regions as well as the neurotransmitter control of hormone release. (Redrawn from Kendall, Tang and Cook, 1975.)

pituitary-adrenal third order feedback system. In this system, CRH provides ultra-short loop negative feedback, ACTH provides short loop negative feedback and the adrenocorticosteroids provide long loop negative feedback. These hormones exert their feedback effects by acting as neuromodulators to regulate the release of the neurotransmitters which control the secretion of the hypothalamic hormones (e.g. CRH) and by direct effects on the endocrine cells in the pituitary gland.

The role of positive and negative feedback in the regulation of the female reproductive cycle is shown in Figure 8.8. The medial basal hypothalamus secretes basal levels of GnRH. As GnRH stimulates the release of FSH and the ova develops, estrogen and inhibin are secreted and provide negative feedback to inhibit FSH secretion. Estrogen also provides positive feedback to the hypothalamus and pituitary to stimulate the pulsatile release of LH-RH and the LH surge, which induces ovulation and stimulates progesterone secretion. Progesterone provides negative feedback to the hypothalamus and pituitary gland, reducing GnRH, FSH and LH secretion. When progesterone levels drop, inhibition of GnRH is reduced and a new reproductive cycle begins.

Effects of altering one hormone in a feedback system

If the level of one hormone is altered in a feedback system, what happens to the level of the other hormones? In a third order feedback system, for example, removal of the testes (castration) will cause testosterone levels

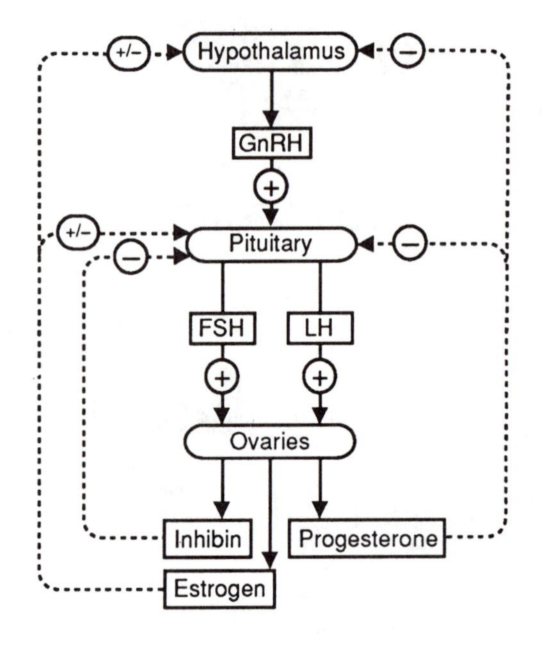

Figure 8.8. Positive and negative feedback in the female hypothalamic-pituitary-gonadal system. For simplicity, only long-loop feedback is shown. Inhibin and estrogen have negative feedback on FSH, while estrogen also has positive feedback, stimulating the pulsatile release of LH. Progesterone has negative feedback on both FSH and LH. GnRH = gonadotropin releasing hormone; FSH = follicle stimulating hormone; LH = luteinizing hormone.

to decline. This removes negative feedback, so that LH-RH and LH levels will no longer be inhibited and will rise. On the other hand, if LH is removed, testosterone levels will decline while LH-RH levels will rise. Thus, by definition, altering the level of one hormone in a feedback loop influences the level of the other hormones within the system.

8.2.4 NEUROTRANSMITTER AND NEUROPEPTIDE CONTROL OF HORMONE LEVELS

Central neurotransmitter regulation of hypothalamic and pituitary hormone secretion was discussed in Chapter 6. Changes in the release of neurotransmitters can alter hormone release at all levels of the neuroendocrine feedback system. For example, increased secretion of the neurotransmitters stimulating the parvicellular release of CRH activates the entire hypothalamic-pituitary-adrenal system, causing not only the release of CRH, but also the release of ACTH and the corticosteroids, which would then stimulate their target cells, as shown in Figure 8.7. Neuropeptides, such as the opiates, also regulate hypothalamic hormone release, as discussed in Chapter 12.

Hypothalamic integration of stimuli influencing neuroendocrine release

As discussed in Chapter 6, the release of hypothalamic hormones is under the control of neurochemical input from sense organs, intero- and extero-receptors, nerve cells, endocrine glands and other chemicals in the blood (Table 8.2). Changes in blood glucose, sodium, fluid and hormone levels are all detected by hypothalamic neurons which then release neurotransmitters to regulate the release of hypothalamic hormones. Similarly, the suckling of an infant at its mother's breast activates the release of oxytocin and prolactin. The complex role of the endocrine

Table 8.2. *Input to the endocrine hypothalamus*

Neural inputs
Sense organs: vision, smell, taste, hearing, touch
Exteroceptors: suckling, coitus, pain, temperature
Interoceptors: uterus, cervix, blood volume, chemoreceptors
Stress: physical, emotional, overcrowding
Time-dependent rhythms and sleep–wake cycles
Cerebrospinal fluid inputs
Hormones
Ions
Other substances, such as peptides
Circulatory system inputs
Hormones: gonadal and adrenal steroids, thyroid hormones, pituitary and hypothalamic peptides, melatonin
Neuropeptides: substance P, bombesin, etc.
Other substances: glucose, amino acids, blood gases
Other properties of blood: e.g. temperature, osmotic pressure

Source: Modified from Hutchinson, 1978.

hypothalamus in the integration of the stimuli controlling neuroendocrine output is discussed by Hutchinson (1978).

8.3 HORMONAL MODULATION OF NEUROTRANSMITTER RELEASE

While neurotransmitters can alter hormone secretion, hormones can also modulate the release of neurotransmitters (Chapter 1). Thus, when hormones feedback to the hypothalamus, they act to modulate neurotransmitter biosynthesis and release. This alteration of neurotransmitter activity regulates the release of the hypothalamic neurohormones. The complete description of a feedback system must therefore include the regulation of neurotransmitters as well as hormones, as shown for the hypothalamic-pituitary-adrenal system in Figure 8.7. Gonadal steroid hormones exert part of their feedback action through the regulation of hypothalamic noradrenaline levels as discussed in Chapter 9 (Nock and Feder, 1981). Likewise, the negative feedback of prolactin on the release of hypothalamic hormones occurs through the regulation of dopamine release. High levels of prolactin stimulate dopamine release from the tuberoinfundibular dopaminergic neurons in the median eminence and this increased dopamine inhibits further prolactin release (Moore, Demarest and Johnston, 1980; Bybee *et al.*, 1983). Hormonal modulation of neurotransmitter synthesis and release also results in emotional, motivational, and behavioral changes as discussed in Chapter 11.

8.4 THE CASCADE OF CHEMICAL MESSENGERS REVISITED

Figure 6.1 showed the cascade of chemical messengers from the release of a neurotransmitter through the stimulation of hormones from the

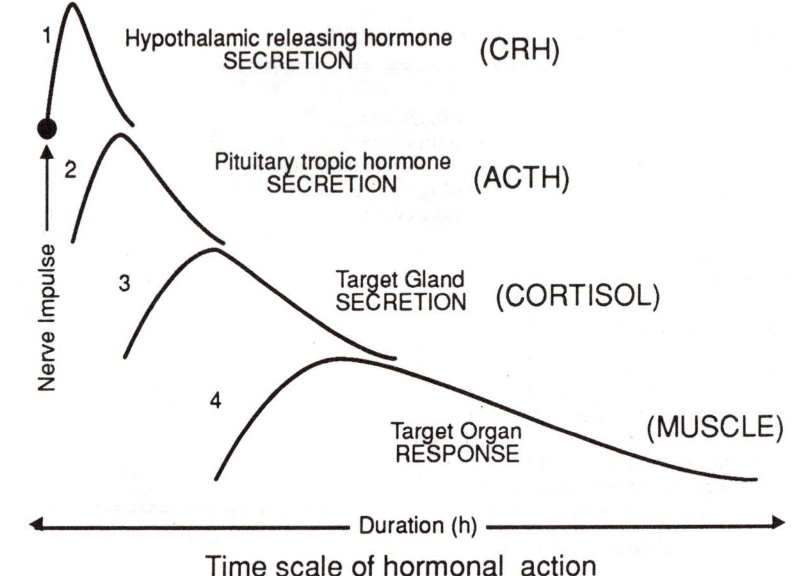

Figure 8.9. The time scale for the action of each element in the cascade of the chemical messengers of the hypothalamic-pituitary-adrenal system. (Redrawn from Mason, 1976.)

hypothalamus, pituitary and peripheral endocrine glands. Figure 8.9 shows the time course of this cascade effect. Whereas the nervous system responds very rapidly to external stimulation, the endocrine system is relatively slow to react. Thus, in response to a stressful stimulus, adrenaline is released almost instantaneously by sympathetic nervous system stimulation of the adrenal medulla, whereas cortisol is released much more slowly from the adrenal cortex because two other hormones, CRH and ACTH, must be released first, each taking some time to reach a peak in the bloodstream. On the other hand, once a hormone is released, it remains, in free and in bound form, for some time in the blood. Thus, the active life of a hormone is much longer than that of a neurotransmitter, especially for hormones bound to carrier proteins. The result is that, after a stimulus occurs, a hormone may take some hours to be secreted into the bloodstream, whereas a neurotransmitter will take only milliseconds to be released into a synapse. As shown in Figure 8.9, however, the hormone will likely stay in the blood for hours, whereas, the transmitter will be deactivated in a few seconds. Likewise, the decline in pituitary hormone levels due to negative feedback is also a slow process (Caminos-Torres, Ma and Snyder, 1977).

The slow release of hormones, coupled with their long active life and slow feedback action results in the fluctuation of hormone levels in the bloodstream. When hormones are released, their levels in the blood are rapidly elevated and then gradually decline, through negative feedback, until another burst is triggered (Figure 7.1, p. 116). Figure 8.10, for example, shows that the pulse rate of LH secretion is very high in a castrated male monkey because no testosterone is present to provide negative feedback. With the resulting high frequency of LH pulses, the average level of LH in the circulation is elevated because the rate of LH secretion is faster than its rate of deactivation. Four days after the

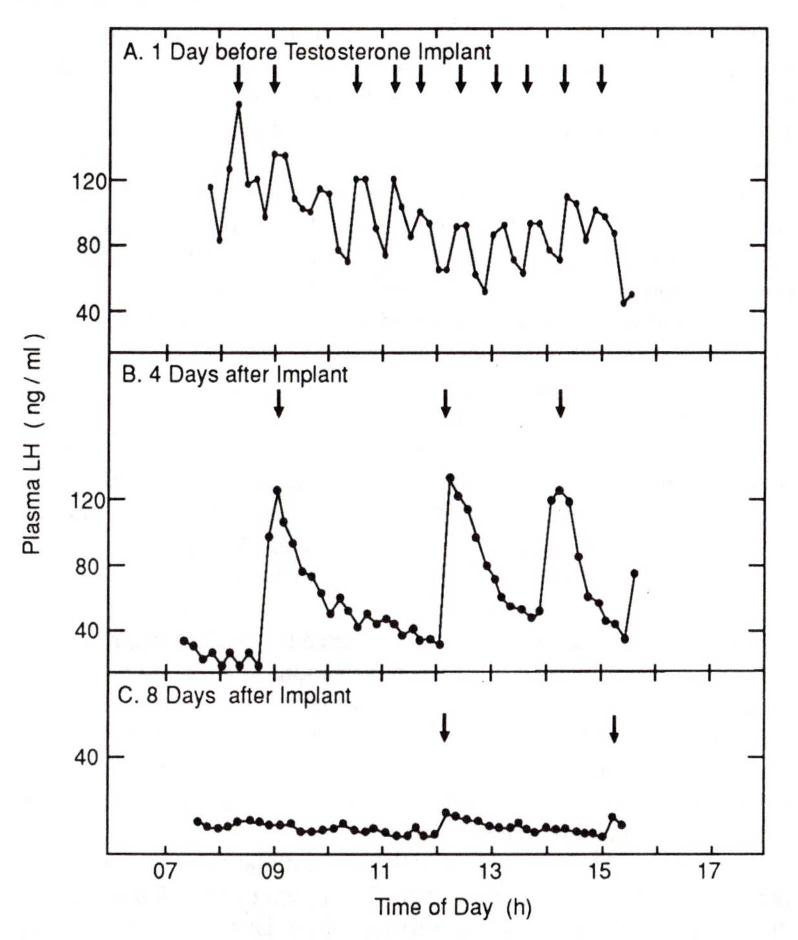

Figure 8.10. Moment-to-moment fluctuations in circulating LH concentrations in a castrated male rhesus monkey (A) 1 day before, (B) 4 days after, and (C) 8 days after the implantation of a silastic capsule containing testosterone which restored the level of this steroid to the normal physiological range. Arrows indicate LH pulses. Note the retardation of LH pulse frequency and, presumably, therefore, the inhibition of the hypothalamic GnRH pulse generator, 4 and 8 days after the initiation of testosterone replacement. (Redrawn from Plant, 1986.)

implantation of a capsule containing testosterone, the LH pulse frequency is greatly reduced and the average level of LH in the blood drops because there is time for it to be degraded between pulses. Eight days after the implantation of testosterone, the pulses of LH are almost completely suppressed. The action of testosterone on LH pulse frequency occurs, presumably, through inhibition of the firing rate of the hypothalamic GnRH pulse generator neurons (Plant, 1986).

8.5 WHEN HORMONE REGULATORY MECHANISMS FAIL

When hormone secretion is reduced (hyposecretion) or elevated (hypersecretion) due to disease or damage to an endocrine gland, the mechanisms regulating hormone levels in the blood may fail to operate properly. Endocrine disorders may occur as a result of abnormal hypothalamic, pituitary or peripheral endocrine gland secretion and a number of diagnostic tests must be performed to determine the site of the abnormality before treatment can occur (see Martin and Reichlin, 1987). Abnormal hormone secretion results in the under- or over-stimulation of target cells, causing a wide range of endocrine-related physiological, behavioral, psychological and psychiatric disorders. The clinical aspects of neuroendocrine disorders are discussed by Jefferson and Marshall (1981) and Martin and Reichlin (1987) and in textbooks on clinical medicine (Greenspan, 1991).

One example of a neuroendocrine disorder is Cushing's syndrome, which occurs as the result of excessive corticosteroid secretion. Most often, this disorder is due to the hypersecretion of ACTH, but it can also be caused by corticosteroid-secreting tumors of the adrenal gland. Elevated corticosteroid secretion can be detected by measurement of corticosteroid levels in the serum, saliva or urine, or by elevated urinary levels of corticosteroid metabolites such as the 17-ketosteroids or 17-hydroxycorticosteroids. Cushing's syndrome patients have a number of physical symptoms, including the rounding of the face, increased weight gain, hypertension, and skin changes such as acne and hirsutism (in women). Patients with Cushing's syndrome may also have psychiatric symptoms, including depression, disturbances of perceptual and cognitive abilities, and, in extreme cases, paranoid states, hallucinations, and confusion (Martin and Reichlin, 1987).

Tumors of the endocrine glands are treated by surgical removal or by chemotherapy. Excess coricosteroid secretion, for example, can be treated by drugs which inhibit corticosterone synthesis by inhibiting the action of the enzyme 11-β-hydroxylase, which converts 11-deoxycorticosterone to corticosterone (see Figure 7.3). Following drug treatment, the hormone levels return to normal and the majority of the symptoms abate (Jefferson and Marshall, 1981).

Feedback regulation in the hypothalamic-pituitary-gonadal system is particularly important for the proper functioning of the female reproductive system and for the stimulation of ovulation. If this feedback system is disrupted, chronic anovulation and reproductive failure can result (Yen, 1978).

8.6 SUMMARY

This chapter examines the methods used for measuring hormone levels in the circulation and the mechanisms through which these hormone levels are regulated. Hormone levels can be measured directly from blood samples using radioimmunoassays (RIAs) or gas–liquid chromatography. Indirect measurements of circulating hormone levels can be obtained from RIAs of hormones in the saliva, urine or feces or by the measurement of hormone metabolites in the urine. Bioassays which examine target cell responses can also be used to estimate the level of hormones in the circulation. The level of hormones in the circulation is regulated by the autonomic nervous system, by non-hormone chemicals such as sodium or glucose in the bloodstream, by feedback from circulating hormones and by the central release of neurotransmitters. Neurohypophyseal hormones are involved in first order feedback loops, prolactin, GH and MSH in second order feedback loops, and ACTH, TSH, LH and FSH in third order feedback loops. Alteration of the level of one hormone in a feedback loop changes the levels of the other hormones in that system, and modulates the neurotransmitters controlling the release of the hypothalamic-pituitary hormones. The hypothalamic neurons regulating the neuroendocrine system must, therefore, be able to integrate stimuli from a wide variety of sources in order to regulate the release of the hypothalamic hormones. These inputs include neurotransmitters, hormonal feedback, and non-hormone chemicals in the circulation, which are detected by osmoreceptors, glucoreceptors and other chemoreceptors. Hormonal feedback can alter the rate of pulsatile release of neurotransmitters and hypothalamic hormones and the responsiveness of the endocrine cells in the pituitary to hypothalamic hormones. Diseases of the hypothalamus, pituitary or peripheral endocrine glands result in hypo- or hyper-secretion of hormones, which can be detected by routine blood or urine analysis. Abnormal hormone levels result in a number of physiological and psychological disorders which can be reversed by surgical intervention or drug treatment. Once the abnormal hormone secretion is treated, the symptoms usually disappear.

FURTHER READING

Greenspan, F. S. (1991). *Basic and Clinical Endocrinology*, 3rd edn. Norwalk, CN: Appleton and Lange.

Hutchinson, J. S. M. (1978). Control of the endocrine hypothalamus. In S. L. Jeffcoate and J. S. M. Hutchinson (eds.) *The Endocrine Hypothalamus*, pp. 75–106. London: Academic Press.

Martin, J. B. and Reichlin, S. (1987). *Clinical Neuroendocrinology*, 2nd edn. Philadelphia: F. A. Davis.

Piva, F., Motta, M. and Martini, L. (1979). Regulation of hypothalamic and pituitary function: long, short, and ultrashort feedback loops. In L. J. DeGroot *et al.*, (eds.) *Endocrinology*, vol. 1, pp. 21–33. New York: Grune and Stratton.

REVIEW QUESTIONS

8.1 What is the difference between short and long loop feedback in a third order feedback system such as the hypothalamic-pituitary-thyroid system?

8.2 What is the effect of removing testosterone on the secretion of LH-RH and LH?

8.3 What stimulates the release of the hormone cholecystokinin and what endocrine gland releases this hormone?

8.4 If I change the substance in A, what happens to the substance in B?

	A	B
(a)	Increase blood glucose	insulin
(b)	Reduce sodium ions (Na^+)	ADH
(c)	Increase thyroxine	TSH-RH
(d)	Increase estrogen	LH

8.5 As prolactin levels increase in the circulation, what effect do they have on dopamine release in the median eminence?

8.6 What is the difference between a second order feedback loop and short loop feedback?

8.7 What is the difference between a bioassay and a radioimmunoassay?
8.8 Give two reasons for using saliva or urine samples for hormone assays rather than blood samples.
8.9 If I use the weight of the thyroid gland as a bioassay, what hormone would I be measuring in the circulation?
8.10 Would you expect to find estrogen receptors in the pituitary gland? Why?

ESSAY QUESTIONS

8.1 Compare the advantages and disadvantages of using radioimmunoassays versus bioassays to measure the levels of a hormone such as prolactin in the circulation.
8.2 Discuss the advantages and disadvantages of assaying hormone levels in blood samples versus saliva, urine or feces samples.
8.3 Discuss the role of the ANS in stimulating the release of pancreatic hormones, with emphasis on how this nervous control interacts with blood sugar levels to maintain homeostatic levels of these hormones in the circulation.
8.4 Discuss the neuroendocrine feedback loops controlling the endogenous menstrual cycle of female primates, noting the changes in neurotransmitters, hypothalamic, pituitary and gonadal hormones over the cycle and how each is inter-related by positive and negative feedback loops.
8.5 Compare the role of dopamine in regulating the release of prolactin and the role of prolactin in regulating dopamine release.
8.6 Discuss the steroid hormone regulation of the hypothalamic GnRH pulse generator.
8.7 Discuss the physiological and psychological aspects of the syndromes associated with hypo- and hyper-secretion of growth hormone (or another pituitary hormone) and the effective treatment of these syndromes.
8.8 Do the thyroid gland and gonads have input from nerves of the CNS? If so, do these endocrine glands qualify as neuroendocrine transducers?

REFERENCES

Adlercreutz, H. and Järvenpää, P. (1982). Assay of estrogens in human feces. *Journal of Steroid Biochemistry*, **17**, 639–645.

Baird, D. T. (1972). Reproductive hormones. In C. R. Austin and R. V. Short (eds.) *Hormones in Reproduction*, pp. 1–28. Cambridge: Cambridge University Press.

Bennett, G. W. and Whitehead, S. A. (1983). *Mammalian Neuroendocrinology*. New York: Oxford University Press.

Bybee, D. E., Nakawatase, C., Szabo, M. and Frohman, L. A. (1983). Inhibitory feedback effects of prolactin on its secretion involve central nervous system dopaminergic mediation. *Neuroendocrinology*, **36**, 27–32.

Caminos-Torres, R., Ma, L. and Snyder, P. J. (1977). Testosterone-induced inhibition of the LH and FSH responses to gonadotropin-releasing hormone occurs slowly. *Journal of Clinical Endocrinology and Metabolism*, **44**, 1142–1153.

Carlstead, K., Wildt, D. E., Monfort, S. L., Killens, R. and Brown, J. (1989). Validation of a urinary cortisol radioimmunoassay for monitoring adrenal activity in domestic and non-domestic cats. Paper presented at the 21st International Ethology Conference, Utrecht, Holland, August 1989.

Christian, J. J. and Davis, D. E. (1964). Endocrines, behavior, and population. *Science*, **146**, 1550–1560.

Dabbs, J. M. Jr (1990). Salivary testosterone measurements: reliability across hours, days, and weeks. *Physiology and Behavior*, **48**, 83–86.

Ekins, R. (1990). Measurement of free hormones in blood. *Endocrine Reviews*, **11**, 5–46.

Feder, H. H. (1981). Essentials of steroid structure, nomenclature, reactions, biosynthesis and measurements. In N. T. Adler (ed.) *Neuroendocrinology of Reproduction*, pp. 19–63. Plenum: New York.

Fink, G. (1979). Feedback actions of target hormones on hypothalamus and pituitary with special reference to gonadal steroids. *Annual Review of Physiology*, **41**, 571–585.

Greenspan, F. S. (1991). *Basic and Clinical Endocrinology*, 3rd edn. Norwalk, CN: Appleton and Lange.

Hadley, M. E. (1992). *Endocrinology*, 3rd edn. New York: Prentice-Hall.

Hampl, R., Foretova, L., Sulcova, J. and Starka, L. (1990). Daily profiles of salivary cortisol in hydrocortisone treated children with congenital adrenal hyperplasia. *European Journal of Pediatrics*, **149**, 232–234.

Hutchinson, J. S. M. (1978). Control of the endocrine hypothalamus. In S. L. Jeffcoate and J. S. M. Hutchinson (eds.) *The Endocrine Hypothalamus*, pp. 75–106. London: Academic Press.

Jefferson, J. W. and Marshall, J. R. (1981). *Neuropsychiatric Features of Medical Disorders*, New York: Plenum Press.

Keller-Wood, M. E. and Dallman, M. F. (1984). Corticosteroid inhibition of ACTH secretion. *Endocrine Reviews*, **5**, 1–24.

Kendall, J. W., Tang, L. and Cook, D. M. (1975). Sites of feedback control in the pituitary-adrenocortical system. In W. E. Stumpf and L. D. Grant (eds.), *Anatomical Neuroendocrinology*, pp. 276–283. Karger: Basel.

Kerin, J. F., Warnes, G. M., Crocker, J., Broom, T. G., Ralph, M. M., Matthews, C. D., Seamark, R. F. and Cox, L. W. (1980). 3-Hour urinary radioimmunoassay for luteinising hormone to detect onset of preovulatory LH surge. *Lancet*, **2** (part 1), 430–431.

Kirelä, A., Kauppila, A., Leppäluoto, J. and Vakkuri, O. (1990). Melatonin in infants and mothers at delivery and in infants during the first week of life. *Clinical Endocrinology*, **32**, 593–598.

Larsen, P. R. (1982). Thyroid-pituitary interaction: Feedback regulation of thyrotropin secretion by thyroid hormones. *New England Journal of Medicine*, **306**, 23–32.

Martin, C. R. (1985). *Endocrine Physiology*. New York: Oxford University Press.

Martin, J. B. and Reichlin, S. (1987). *Clinical Neuroendocrinology*, 2nd edn. Philadelphia: F. A. Davis.

Mason, A. S. (1976). *Hormones and the Body*. Middlesex, England: Penguin.

Moore, K. E., Demarest, K. T. and Johnston, C. A. (1980). Influence of prolactin on dopaminergic neuronal systems in the hypothalamus. *Federation Proceedings*, **39**, 2912–2916.

Murphy, P. R., Friesen, H. G., Brown, R. E. and Moger, W. H. (1989). Verification of NB2 lymphoma cell bioassay for the measurement of plasma and pituitary prolactin in the Mongolian gerbil (*Meriones unguiculatus*). *Life Sciences*, **45**, 301–310.

Nock, B. and Feder, H. H. (1981). Neurotransmitter modulation of steroid action in target cells that mediate reproduction and reproductive behavior. *Neuroscience and Biobehavioral Reviews*, **5**, 437–447.

Paz, G., Yogev, L., Gottreich, A., Rotem, R., Yavetz, H. and Homonnai, Z. T. (1990). Determination of urinary luteinizing hormone for the prediction of ovulation. *Gynecological and Obstetric Investigations*, **29**, 207–210.

Piva, F., Motta, M. and Martini, L. (1979). Regulation of hypothalamic and pituitary function: Long, short, and ultrashort feedback loops. In L. J. de Groot *et al.* (eds.) *Endocrinology*, vol. 1, pp. 21–33. New York: Grune and Statton.

Plant, T. M. (1986). Gonadal regulation of hypothalamic gonadotropin-releasing hormone release in primates. *Endocrine Reviews*, **7**, 75–88.

Riad-Fahmy, D., Read, G. F., Joyce, B. G. and Walker, R. F. (1981). Steroid immunoassays in endocrinology. In A. Voller, A. Bartlett and D. Bidwell (eds.) *Immunoassays for the 80s*, pp. 205–262. Lancaster, England: MTP Press.

Riad-Fahmy, D., Read, G. F., Walker, R. F. and Griffiths, K. (1982). Steroids in saliva for assessing endocrine functions. *Endocrine Reviews*, **3**, 367–395.

Tanaka, T., Shiu, R. P. C., Gout, P. W., Beer, C. T. Noble, R. L. and Friesen, H. G. (1980). A new sensitive and specific assay for lactogenic hormones, measurement of prolactin and growth hormone in human serum. *Journal of Clinical Endocrinology and Metabolism*, **51**, 1058–1063.

Turner, C. D. and Bagnara, J. T. (1976). *General Endocrinology*, 6th edn. Philadelpia: Saunders.

Vakkuri, O., Leppäluoto, J. and Kauppila, A. (1985). Oral administration and distribution of melatonin in human serum, saliva and urine. *Life Sciences*, **37**, 489–495.

Valenca, M. M., Johnston, C. A., Ching, M. and Negro-Vilar, A. (1987). Evidence for a negative ultra-short loop feedback mechanism operating on the luteinizing hormone-releasing hormone neuronal system. *Endocrinology*, **121**, 2256–2259.

Yalow, R. S. (1980). Radioimmunoassay: A major advance but not without problems. *Trends in Pharmacological Sciences*, **1**, 266–268.

Yen, S. S. C. (1978). Chronic anovulation: due to inappropriate feedback system. In S. S. C. Yen and R. B. Jaffe (eds.) *Reproductive Endocrinology*, pp. 297–323. Philadelphia: W. B. Saunders.

Zanisi, M., Messi, E., Motta, M. and Martini, L. (1987). Ultrashort feedback control of luteinizing hormone-releasing hormone secretion *in vitro*. *Endocrinology*, **121**, 2199–2204.

Zarrow, M. X., Yochim, J. M., McCarthy, J. L. and Sanborn, R. C. (1964). *Experimental Endocrinology: A Sourcebook of Basic Techniques*. New York: Academic Press.

Zuckerman, S. (1957). Hormones. *Scientific American*, **196**, 76–87.

9

Steroid and thyroid hormone receptors

In order for a hormone to modulate the biological functions of its target cells that hormone must first bind to a receptor in the target cell. If a cell has no receptor for a given hormone, that hormone can not influence the activity of the cell. Steroid and peptide hormones regulate the activity of their target cells by different receptor mechanisms. The peptide hormones are large, water soluble (hydrophilic) molecules which can not pass through the cell membrane, so bind to receptor proteins on the cell membrane, as described in Chapter 10. The steroid hormones (androgens, estrogens, progesterone, glucocorticoids, and mineralocorticoids) are small lipophilic (lipid soluble) molecules and, if not bound to carrier proteins, they can readily diffuse through the cell membranes into any cell in the body. Thyroid hormones are also able to enter cells by diffusion. Target cells for steroid and thyroid hormones have intracellular receptors in the nucleus which bind to the hormones after they enter the cells.

9.1 THE INTRACELLULAR RECEPTOR SUPERFAMILY

Although steroid and thyroid hormones have different structures and perform very different biological functions, their intracellular receptors

Table 9.1. *The nuclear receptor superfamily*

The following intracellular hormone receptors share structural and functional properties with one another and have been divided into two classes based on their amino acid sequences

Class 1
Androgen receptors
Progesterone (progestin) receptors
Glucocorticoid receptors
Mineralocorticoid receptors
Class 2
Estrogen receptors
Thyroid hormone receptors
Retinoic acid receptors
Vitamin D receptors

have a common structure. These receptors belong to a large superfamily of closely related receptor proteins (Evans, 1988). All of the intracellular receptors have been identified and, on the basis of the similarity of their amino acid sequences, are known to fall into two classes, as summarized in Table 9.1 (Harrison and Lippman, 1989; Funder, 1991). When steroid and thyroid hormones enter their target cells, they bind to these intracellular receptors and form a hormone–receptor complex which regulates specific gene expression in the nucleus of the target cell. A great deal is known about the molecular biology of these receptors, but exactly how they control gene expression in their target cells is still unknown (Beato, 1989; Carson-Jurica, Schrader, and O'Malley, 1990; Weinberger and Bradley, 1990). At least three new hormone receptors have been identified as belonging to this superfamily, suggesting the existence of hormones which have yet to be discovered (Evans and Arriza, 1989).

9.2 HOW ARE HOMONE TARGET CELLS IDENTIFIED?

Target cells for steroid hormones can be identified using several techniques. One of these is called **autoradiography**. This technique can be used to detect steroid target cells anywhere in the body (for example, in the uterus or heart), but is particularly useful for locating target cells in brain tissue (McEwen, 1976; Morrell and Pfaff, 1978, 1981). Knowing the precise anatomical location of these target cells is important for understanding hormone–brain interactions, especially in the control of behavior. To perform autoradiography for estrogen receptors, the female rat is ovariectomized at least 1 week prior to the procedure to reduce circulating levels of endogenous estrogens. Radioactively labeled (tritiated) estrogen is then injected and two hours later the animal is sacrificed, its brain is removed, frozen and cut into thin (6 μm) sections with a cryostat microtome. These sections are then placed on slides coated with photographic emulsion and stored for several weeks or months. The beta particles released from the tritiated estrogen create black dots on the emulsion at the sites of estrogen binding in the brain tissue. The brain tissue is then stained histologically (with cresyl violet or hematoxylin and eosin) to identify the location of the cell bodies.

The autoradiographs can be evaluated by placing the stained tissue slide over the developed photographic emulsion to produce a photomicrograph as shown in Figure 9.1. In this way, the cell bodies containing the labeled estrogen can be identified from the stained tissue. The relative concentration of estrogen in each target cell can be calculated with respect to the level of staining in non-target cells, and the 'sensitivity' of each target cell can be quantified by microscopic analysis of the number of black dots found in the nucleus. Each black dot represents a silver grain in the photographic emulsion 'exposed' by the beta particles in the tritiated estrogen. The number of black dots is proportional to the number of labeled estrogen molecules in the cell. Using this method, target cells for estrogens, androgens, and corticosteroids have been located in the brains of fish, reptiles, birds and a variety of mammals, including rodents, carnivores, and primates (Pfaff and Keiner, 1973; Zigmond, Nottebohm and Pfaff. 1973; Kim *et al.*, 1978; Morrell and Pfaff, 1978).

More recently, immunocytochemical techniques, particularly the use of monoclonal antibodies to receptor proteins, have led to simpler and more accurate methods for the identification of steroid hormone receptors (King and Greene, 1984). *In situ* hybridization techniques have also been used to localize messenger RNA for gonadal and adrenal steroid hormone receptors (Arriza *et al.*, 1988; Simerly *et al.*, 1990).

9.3 HOW ARE STEROID HORMONE TARGET CELLS DIFFERENTIATED FROM NON-TARGET CELLS?

Because steroid hormones can diffuse from the blood into cells and out again, they are found in both target and non-target cells. If there is a steroid hormone receptor in the cell, the hormone will bind to it, forming a hormone–receptor (H–R) complex. The hormone–receptor complex then regulates the biochemical activity of the target cell. The action of estrogen at its target cells will be used as an example of a 'typical' steroid hormone–receptor interaction.

Target cells for estrogen can be discriminated from non-target cells in three ways. First, target cells for estrogen have estrogen receptor proteins in the nucleus of the cell, while non-target cells do not. Second, in a non-target cell, estrogen will occur at the same concentration that it occurs in the blood, because it diffuses into and out of the cell at the same rate. In a target cell, however, the estrogen attaches to a receptor, so the concentration of estrogen increases. Thus, target cells will accumulate estrogen while non-target cells will not. Third, the nuclear hormone–receptor complex stimulates the replication (transcription) of certain information on the DNA by activating messenger RNA (mRNA) synthesis. This mRNA will then promote protein synthesis in the ribosomes of the cell. Thus, estrogen will stimulate mRNA and protein synthesis in a target cell, but not in a non-target cell (Walters, 1985; Blaustein, 1986; Carson-

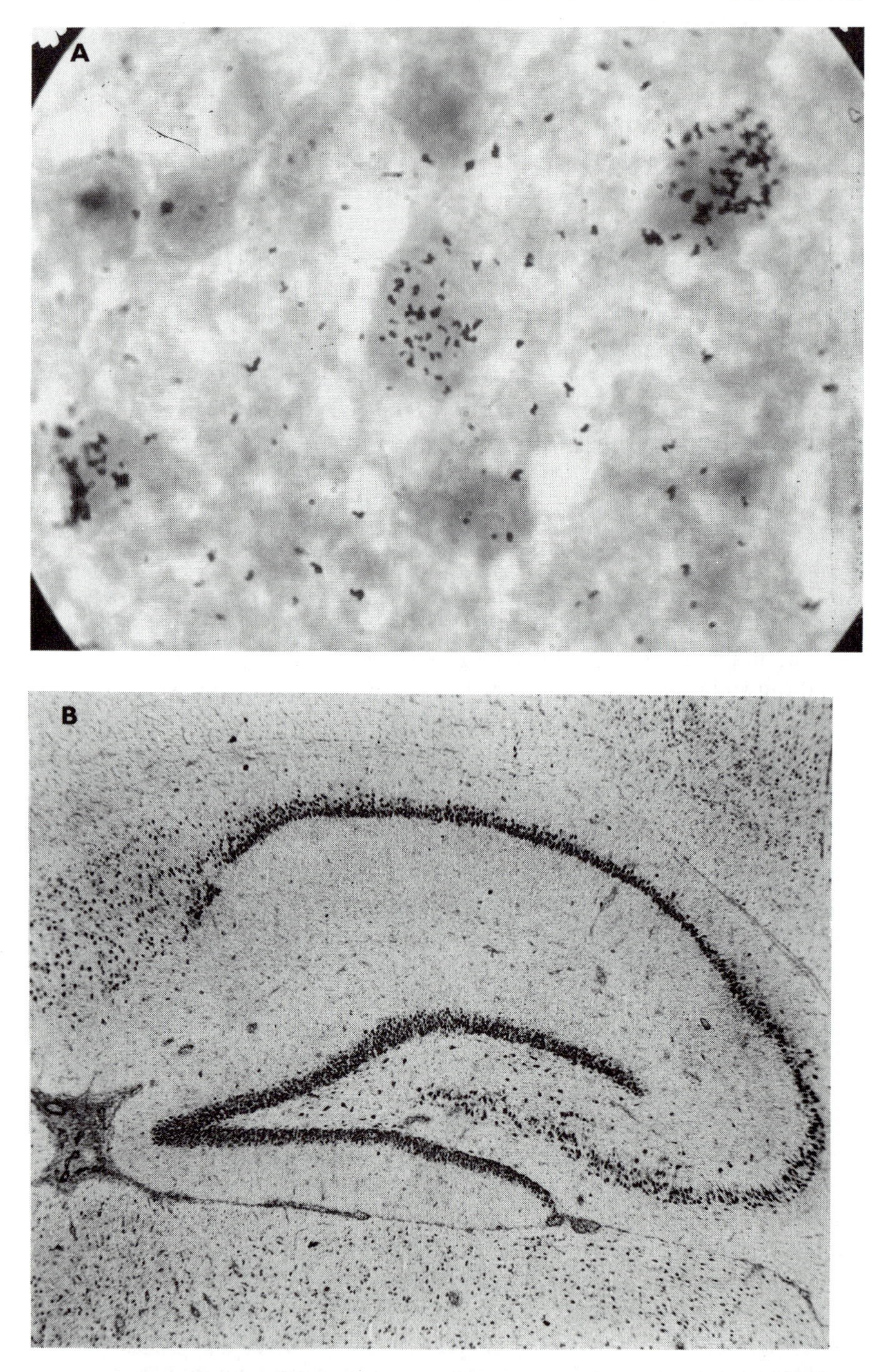

Figure 9.1. (A) An autoradiogram showing the accumulation of radioactive estradiol in the nerve cells of the preoptic area of the hypothalamus of the rat brain. The most heavily labeled regions are the nuclei of the nerve cells, indicating that these are estrogen target cells. This autoradiogram is enlarged 1500 times. (B) A low power magnification of an autoradiogram of the hippocampus of the rat brain after injection of radioactive corticosterone. The nerve cells of the hippocampus are heavily labeled, indicating a high density of corticosteroid target cells. (Unpublished autoradiograms made by John Gerlach and donated by Bruce McEwen.)

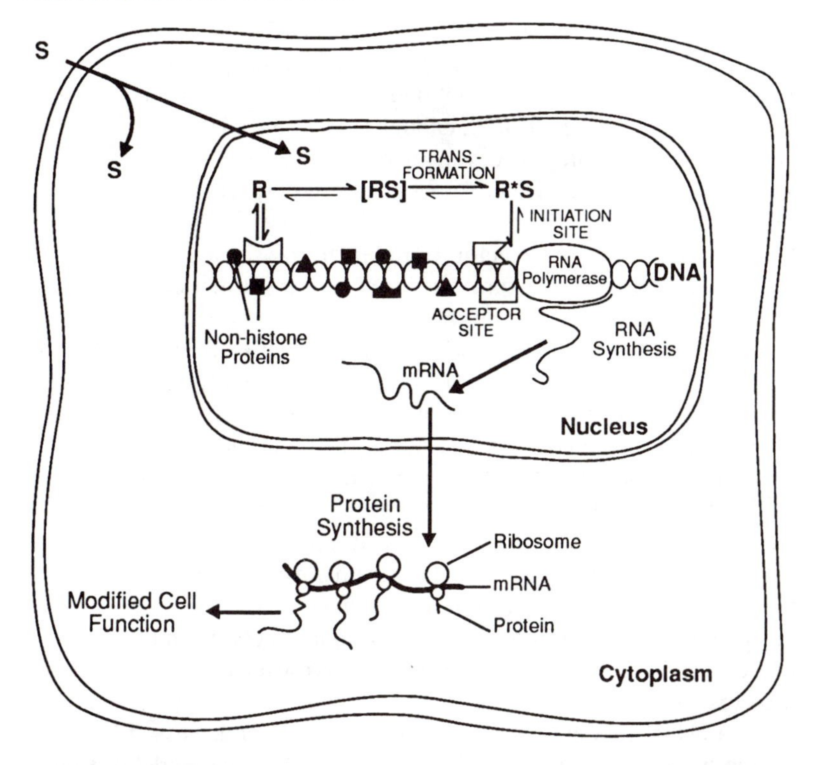

Figure 9.2. The nuclear model of the steroid hormone receptor. The steroid hormones (S) are distributed within both the cytoplasm and the nucleus of the target cell. The unoccupied receptors (R) are believed to be primarily concentrated in the nucleus in a reversible equilibrium binding state with the non-histone proteins of the chromatin. The binding of the steroid hormone to the receptor results in the transformation of the unoccupied receptor (R) into the biologically active hormone–receptor complex (R*S) at the acceptor site. The transformed receptor then opens an initiation site on the DNA, resulting in the induction (or repression) of mRNA synthesis through the action of the enzyme RNA polymerase. The mRNA then enters the cytoplasm where protein synthesis occurs on the ribosomes, resulting in modified cell function. (Redrawn from Walters, 1985.)

Jurica *et al.*, 1990). Estrogen receptors occur in a number of body tissues, including the uterus, vagina, breasts, heart, pituitary gland and brain.

9.4 ACTION OF STEROID HORMONES AT THEIR RECEPTORS

Steroid hormone receptors are proteins with a hormone-binding region, which determines which hormone will bind to that receptor, and a DNA binding region, which attaches to the non-histone proteins in the cell nucleus (Figure 9.2). The nuclear theory of steroid hormone action suggests that the receptor proteins exist in a reversible bond with the non-histone proteins in the chromatin of the cell nucleus (Parker, 1988). Non-histone proteins bind directly to the DNA and define the specific genes to be regulated by the hormone–receptor complex (Weinberger and Bradley, 1990). When no hormone is bound to the receptor, the hormone-binding region inhibits the DNA-binding region. When a steroid hormone binds to its receptor, this inhibition is removed (Evans, 1988). The location at which the steroid hormone–receptor complex binds to the non-histone protein is called an 'acceptor site' (Figure 9.2). Since each cell contains a complete complement of DNA, enough to replicate all the proteins in the body, the acceptor site is necessary to determine the exact segment of the genomic DNA to be replicated for the synthesis of specific proteins in each target cell.

Once the hormone–receptor complex forms an acceptor site, the receptor undergoes a structural alteration or 'transformation', allowing it to open an 'initiation site' on the DNA itself. At the initiation site, the enzyme RNA polymerase transcribes the information from the DNA to a molecule of mRNA (Figure 9.2). The mRNA is then transported to the ribosomes in the cytoplasm where it serves as the template for protein synthesis. The proteins synthesized may be hormones, hormone receptors, carrier proteins, and other proteins in the target cell. Drugs such as anisomycin, which inhibit protein synthesis, will inhibit the action of steroid hormones at their target cells by preventing protein synthesis (Rainbow *et al.*, 1982). The action of steroid hormones at their target cells thus depends on the presence of the enzymes necessary to carry out the biochemical changes initiated by the hormone–receptor complex binding to the acceptor site in the chromatin as well as the presence of the receptor proteins.

The nuclear model of steroid hormone receptors discussed here and shown in Figure 9.2 replaces the 'classical' two-stage model of steroid hormone receptor action (see Gorski, Welshons and Sakai, 1984). This classical model of steroid hormone action at target cells states that the steroid hormones bind to their receptors in the cytoplasm of the cell and, only after binding to the steroid, does the receptor enter the cell nucleus and attach to the non-histone proteins of the chromatin. Although the nuclear model of steroid hormone receptor action has gained wide acceptance, it is not without criticism (see Raam *et al.*, 1988). It is also possible that the nuclear model is correct for estrogen receptors while the two-stage model is correct for glucocorticoid and some other steroid receptors (Harrison and Lippman, 1989).

While the primary action of the steroid hormones is to regulate protein synthesis by attaching to the nuclear receptor site in the target cell (genomic action), steroid hormones can also modulate neural activity by modifying the permeability of nerve cell membranes (non-genomic action), suggesting the existence of membrane receptors for the steroid hormones as well as nuclear receptors (see Section 9.8).

9.5 MEASUREMENT AND REGULATION OF HORMONE RECEPTOR NUMBERS

The number of receptors for a particular hormone can be measured by **receptor binding assays**. In these assays, radioactively labeled hormones are added to cells and then the number of receptor proteins which bind to these hormones are analyzed quantitatively, providing a measure of the number of receptors in each location. Receptor binding assays can also be used to examine receptor specificity for hormones and receptor sensitivity to different hormones, their agonists and antagonists. Receptor proteins can be identified by cell fractionation, chromatographic, and electrophoretic methods (Alberts *et al.*, 1989). Specific methods for the analysis of steroid hormone receptors are described in Agarwal (1983).

The degree of gene activation and subsequent protein synthesis stimulated by a hormone in its target cell is determined by the number of hormone receptors in that target cell as well as by the level of the hormone in the circulation. The number of receptors in a cell is not fixed, but can be altered by the level of hormones in the circulation. For example, the synthesis of progesterone receptors in many target cells is regulated by changes in estrogen levels (Pfaff and McEwen, 1983; Olster and Blaustein, 1990). Hormone receptor numbers also change over the lifespan of an individual. During perinatal development, there is an increase in the number of receptors for many hormones (MacLusky and Náftolin, 1981; Keefer and Holderegger, 1985), and during aging there is a decrease in receptor numbers (Roth, 1979).

Changes in the amount of hormone in the circulation can regulate the number of receptors for that hormone by a process of negative feedback. Prolonged exposure of target cells to a high concentration of circulating hormone results in a reduction in the number of hormone receptors. This is referred to as **down-regulation** of receptors (Martin, 1985). Down-regulation also occurs for receptors of other chemical messengers, i.e. steroid hormones, non-steroid hormones, neurohormones and neurotransmitters. Down-regulation may take some time to occur and is slowly reversed when hormone levels decline.

If, on the other hand, there is a decrease in the amount of a hormone in the circulation, **up-regulation** of receptors occurs, resulting in an increase in the number of receptors, enabling the cell to pick up as much of the hormone as possible. If the hormone is eliminated from the body completely, however, the number of receptors declines because hormones are necessary to stimulate the synthesis of their own receptors. Thus, after a long period of estrogen withdrawal, as occurs after ovariectomy, there are few estrogen receptors, so the target cells must be 'primed' by small doses of estrogen to stimulate production of estrogen receptors before any physiological or behavioral changes can be initiated by estrogen injection.

Because steroid hormones are structurally similar (see Figure 7.3, p. 118), and have receptors belonging to the same superfamily, it is not surprising that two or more steroid hormones have the ability to bind to the same receptor. Androgen receptors, for example, bind to a number of steroid hormones, but have the highest affinity for testosterone and dihydrotestosterone (Sheridan, 1983). Progesterone competes with testosterone for binding at androgen receptors and inhibits the action of androgens (McEwen, 1981). Thus, progesterone acts as an androgen antagonist.

When two different steroid hormone receptors occur in the same nerve cell, the two hormones may act synergistically in the target cells, or antagonistically. For example, estrogen and androgen receptors both occur in the medial preoptic and anterior hypothalamic areas, the median eminence, and cortico-medial amygdala (Sar and Stumpf, 1975; Simerly *et al.*, 1990). Estrogens and androgens (e.g. dihydrotestosterone) may act synergistically to stimulate sexual behavior at these sites (McEwen, 1981). On the other hand, glucocorticoids may inhibit the actions of gonadal steroids at their common target cells in the amygdala, hippocampus and septum (Sar and Stumpf, 1975), and thus inhibit sexual behavior during periods of high stress.

9.6 GONADAL STEROID HORMONE TARGET CELLS IN THE BRAIN

Much of our knowledge about steroid hormone action at receptors comes from studies of the progesterone receptors in the chick oviduct (Schrader *et al.*, 1981) and the estrogen receptors in the mammalian uterus (Gorski *et al.*, 1968; King and Greene, 1984). Much is also known about the actions of steroid hormones on receptors in the cardiovascular system (Stumpf, 1990). The focus of this chapter, however, is on how the steroid hormones are able to modulate neural activity by binding to receptors in target cells in the brain. This section describes the specific neural sites having estrogen, androgen, and progesterone receptors. Section 9.7 examines adrenal steroid hormone receptors in the brain.

9.6.1 ESTROGEN RECEPTORS

Figures 9.3 and 9.4 show the distribution of estrogen receptors in the brain of the rat and mouse from two different views (sagittal and coronal sections). The highest density of estrogen receptors occurs in the medial preoptic area, anterior hypothalamus, ventromedial hypothalamus, the medial and cortical amygdala, the median eminence, and the pituitary gland (Pfaff and Keiner, 1973; Stumpf, Sar and Keefer, 1975; McEwen *et al.*, 1979; Simerly *et al.*, 1990). Moderate densities of estrogen receptors are found in the septum, lateral amygdala, hippocampus, and periventricular area of the hypothalamus, and low densities in a number of other neural sites. Estrogen target cells in the preoptic area, median eminence and pituitary gland provide feedback regulation for the release of GnRH and the pituitary gonadotropic hormones. Estrogen receptors in the ventromedial hypothalamus regulate female sexual behavior and those in the other parts of the hypothalamus and amygdala are involved in regulating parental behavior, aggression, emotionality, hunger, activity levels, and body temperature. One important function of estrogen at target cells in the preoptic area and hypothalamus is to induce the synthesis of progesterone receptors and thus 'prime' these cells for progesterone stimulation (Olster and Blaustein, 1990).

9.6.2 PROGESTERONE RECEPTORS

The highest density of progesterone receptors is found in the medial and lateral preoptic areas, the ventromedial and basomedial hypothalamus,

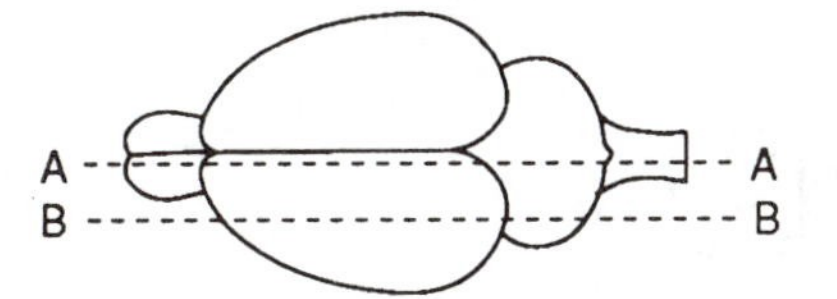

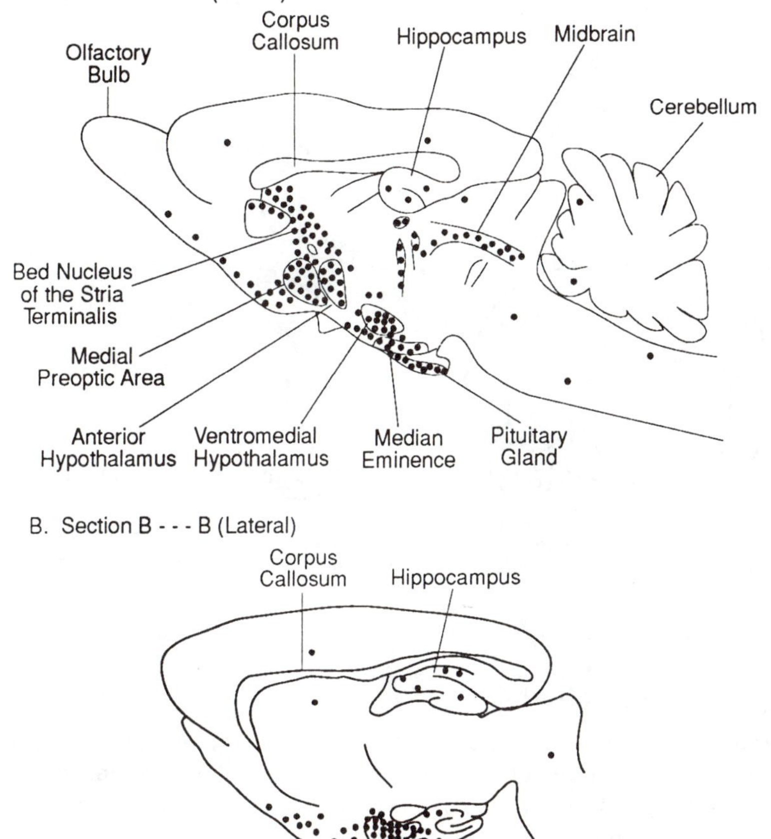

Figure 9.3. Estrogen receptors in the brain. Sagittal sections of the rat brain showing the estrogen sensitive nerve cells concentrated in the bed nucleus of the stria terminalis, medial preoptic area, anterior and ventromedial hypothalamus, median eminence, and amygdala. The dashed lines (A---A and B---B) indicate where the sections were cut from the whole brain. The black dots in each section indicate receptors present in regions in the brain that were made radioactive following injections of radioactive estrogen. (Redrawn from McEwen, 1976, and Pfaff and Keiner, 1973.)

the median eminence and the pituitary gland. Moderate and low levels of progesterone receptors occur in a number of other brain areas (Sar and Stumpf, 1973a; McEwen *et al.*, 1979). There is considerable overlap in the distribution of estrogen and progesterone receptors, with both occurring at high densities in the preoptic area, amygdala, ventromedial nucleus and median eminence of the hypothalamus (McEwen, 1981). Progesterone receptors in the ventromedial hypothalamus act in concert with estrogen to stimulate female sexual behavior, while progesterone receptors in the midbrain (e.g. the substantia nigra) inhibit female sexual behavior (McEwen *et al.*, 1979). Progesterone receptors in the preoptic area and median eminence provide feedback regulation for GnRH release (Pfaff and McEwen, 1983).

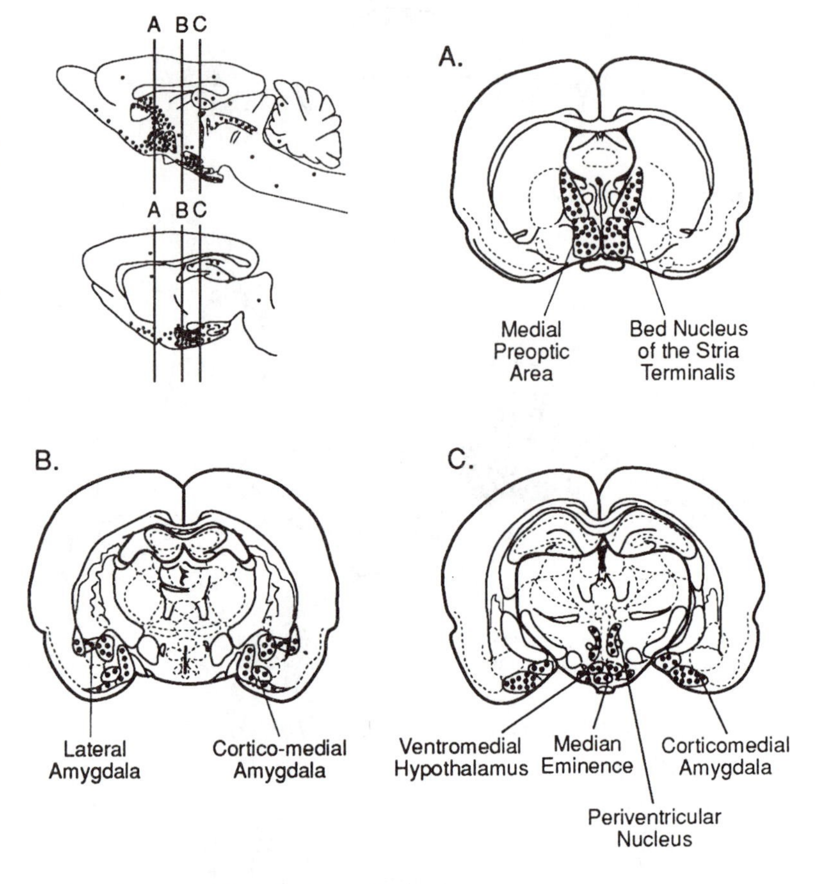

Figure 9.4. Estrogen receptors in the brain. Three coronal (frontal) sections showing the locations of radioactive estrogen uptake in the mouse brain. The lines (A, B, and C) show where the sections were cut with respect to the estrogen receptors shown in the sagittal sections in Figure 9.3. (Redrawn from Stumpf *et al.*, 1975.)

9.6.3 ANDROGEN RECEPTORS

Figure 9.5 shows the distribution of androgen receptors in the rodent brain. Androgen receptors occur at the highest density in the amygdala, septum and hippocampus, bed nucleus of the stria terminalis, the medial preoptic area, anterior and lateral hypothalamus, the median eminence and pituitary gland (Sar and Stumpf, 1975; Selmanoff *et al.*, 1977; Sheridan, 1979; Simerly *et al.*, 1990). Lower densities of androgen receptors occur in the cerebral cortex, thalamus, brain stem and spinal cord. Androgen sensitive neurons in the medial preoptic area and anterior hypothalamus regulate male sexual behavior and provide negative feedback for GnRH secretion, while those in other areas may be responsible for sex differences in a wide variety of behaviors (Beatty, 1979; McEwen *et al.*, 1979).

Although androgen and estrogen receptors have different distribution patterns in the brain, a comparison of Figures 9.4 and 9.5 shows a great deal of overlap in the distribution of these receptors (Simerly *et al.*, 1990). Both androgen and estrogen receptors are found in the medial preoptic area, anterior hypothalamus, ventromedial hypothalamus, and medial and cortical amygdala. Estrogen receptors occur at a higher density than androgen receptors in a number of hypothalamic areas (e.g. the anteroventral and periventricular nuclei), the arcuate nucleus, and the pituitary

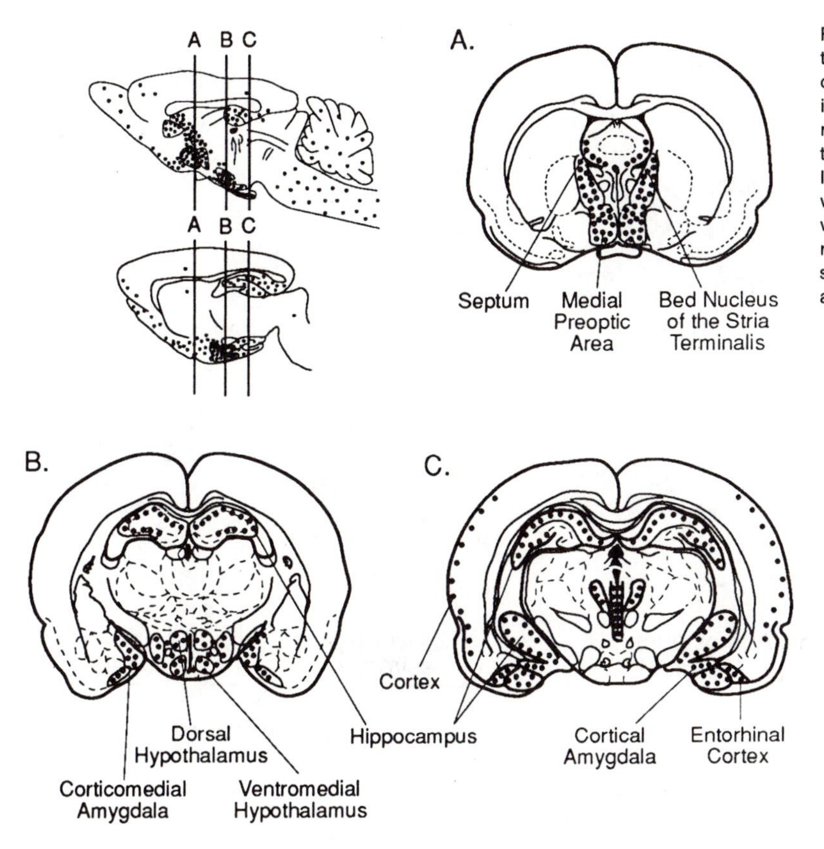

Figure 9.5. Androgen receptors in the brain. Three coronal (frontal) sections showing the locations of radioactive testosterone uptake in the mouse brain. The lines (A, B, and C) show where the sections were cut with respect to testosterone receptors shown in a sagittal section. (Redrawn from Sar and Stumpf, 1975.)

gland. On the other hand, a number of other brain regions have a much higher density of androgen receptors than estrogen receptors. These include the auditory cortex, cerebellum, motor nuclei of the brain stem, nuclei of the brain stem reticular formation, and nuclei of the spinal cord (Simerly *et al.*, 1990).

Although androgen receptors can bind to a number of different steroids, they are most sensitive to testosterone and dihydrotestosterone (Sheridan, 1983; Chang, Kokontis and Liao, 1988). Testosterone can stimulate its target cells by acting directly on androgen receptors or by conversion to dihydrotestosterone, which then acts on the androgen receptors (Figure 9.6). In other target cells, testosterone is converted to estrogen and activates its target cells through the estrogen receptors. Table 9.2 shows some of the androgen-sensitive cells which are stimulated directly by testosterone and some of those which are stimulated through the conversion of testosterone to dihydrotestosterone or estrogen (Mooradian, Morley, and Korenman, 1987).

9.6.4 REDUCTION OF TESTOSTERONE TO 5α-DIHYDROTESTOSTERONE

In certain androgen target cells in the brain, pituitary gland, gonads, and secondary sex organs (including the prostate gland, seminal vesicles and

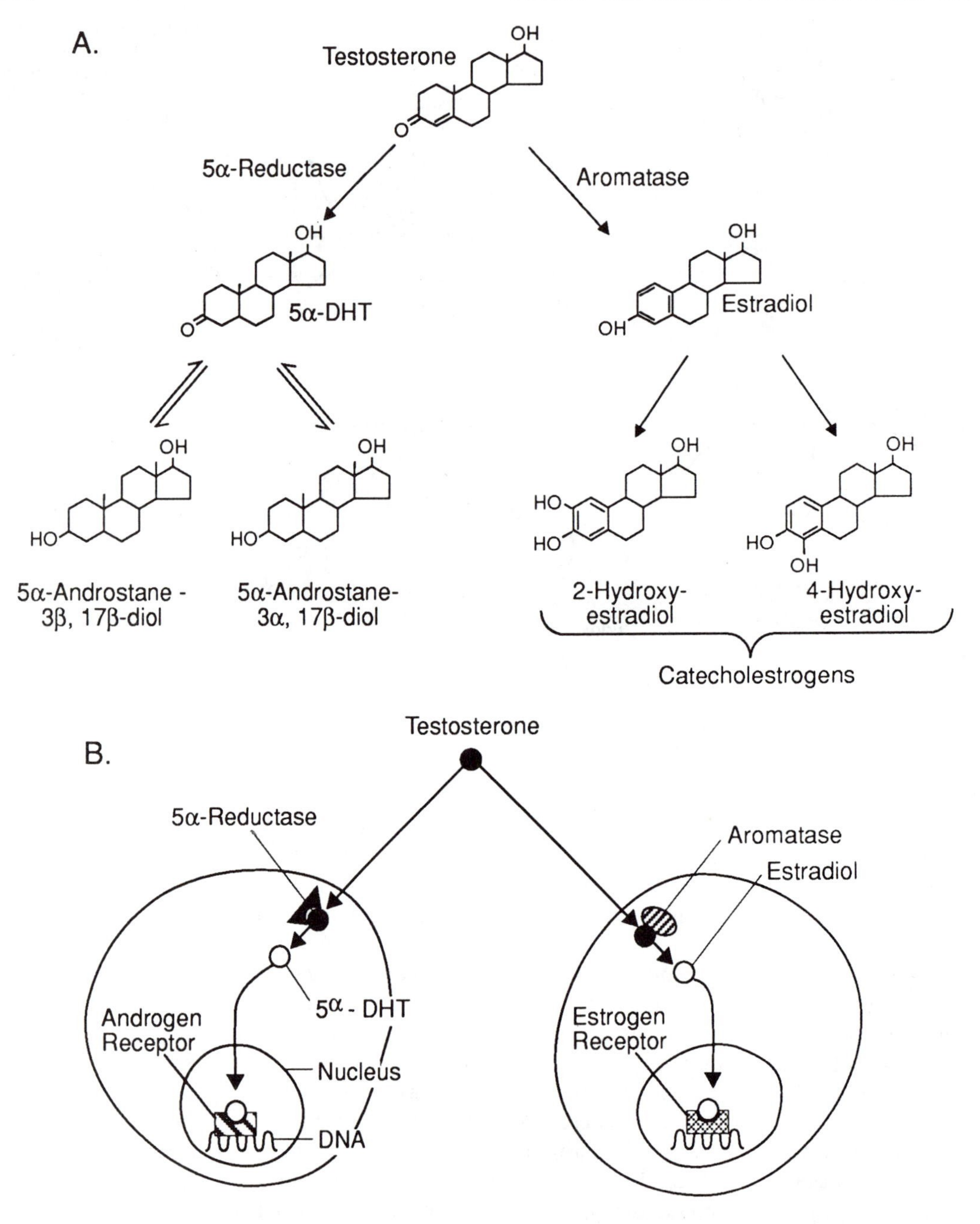

Figure 9.6. The metabolism of testosterone to 5α-dihydrotestosterone (5α-DHT) and estradiol. (A) The major routes of testosterone metabolism in the brain which result in the formation of physiologically active steroids including the catecholestrogens. (Redrawn from MacLusky and Naftolin, 1981.) (B) The reduction of testosterone to 5α-DHT and the aromatization of testosterone to estradiol by enzymes in testosterone target cells. The 5α-DHT acts on androgen receptors in the target cell nucleus, while the estrogen acts on estrodiol receptors (Redrawn from McEwen, 1976.)

Table 9.2. *The location of androgen target cells in the body*

In some target cells, testosterone binds directly to androgen receptors. In other target cells, testosterone is reduced to 5α-dihydrotestosterone (5α-DHT), which then binds to androgen receptors, or is aromatized to estrogen, which binds to estrogen receptors

Testosterone acting on androgen receptors	5α-DHT acting on androgen receptors	Estrogen receptors
Central nervous system[a]	Central nervous system[a]	Central nervous system[a]
	Feedback on GnRH	
Spinal cord nuclei[a]	Spinal cord nuclei[a]	
Testes[a]		Testes[a]
Seminal vesicles[a]	Prostate	Seminal vesicles[a]
Epididymis[a]	Epididymis[a]	
Penis[a]	Penis[a]	
Muscle		
Kidney[a]	Sebaceous glands	
Adipose tissue[a]		Adipose tissue[a]
Immune system	Bone	

Note:
[a] Target cells which are activated by more than one receptor mechanism.

penis), the enzyme 5α-reductase converts testosterone to 5α-dihydrotestosterone (5α-DHT) by a process called 5α-reduction (see Figure 9.6). In the brain, 5α-reductase occurs at high levels in the medial preoptic area, anterior and lateral hypothalamus, median eminence and pituitary gland (Selmanoff *et al.*, 1977). Once formed, 5α-DHT binds to androgen receptors and regulates protein synthesis in these target cells. Androgen receptors bind to both testosterone and 5α-DHT, but appear to have a higher affinity for 5α-DHT (McEwen, 1981; Sheridan, 1983; Mooradian *et al.*, 1987). The function of 5α-DHT at hypothalamic and pituitary androgen receptors appears to be the inhibition of gonadotropic hormone release, while at other neural receptors, 5α-DHT regulates sexual and other behaviors (Sheridan, 1979). In the gonads and sex organs, 5α-DHT is important for sexual differentiation (Callard *et al.*, 1978).

9.6.5 AROMATIZATION OF TESTOSTERONE TO ESTROGEN

The conversion of testosterone to estrogen depends on the action of the aromatase enzyme in the target cells and is referred to as aromatization (Figure 9.6). The ability of these target cells to convert testosterone to estrogen is demonstrated by injection of radioactive testosterone, T* (a * is used to indicate a radioactively labeled hormone), into a castrated male rat and finding radioactive estrogen (E*) in the medial preoptic area, arcuate nucleus, amygdala and ventromedial hypothalamus, indicating that T* has been converted to E* in these cells (Selmanoff *et al.*, 1977; Sheridan, 1979; McEwen, 1981). The aromatase enzyme is not present in the pituitary gland, thus no aromatization of testosterone occurs there (Lieberburg and McEwen, 1977).

The target cells in which testosterone is aromatized to estrogen can be

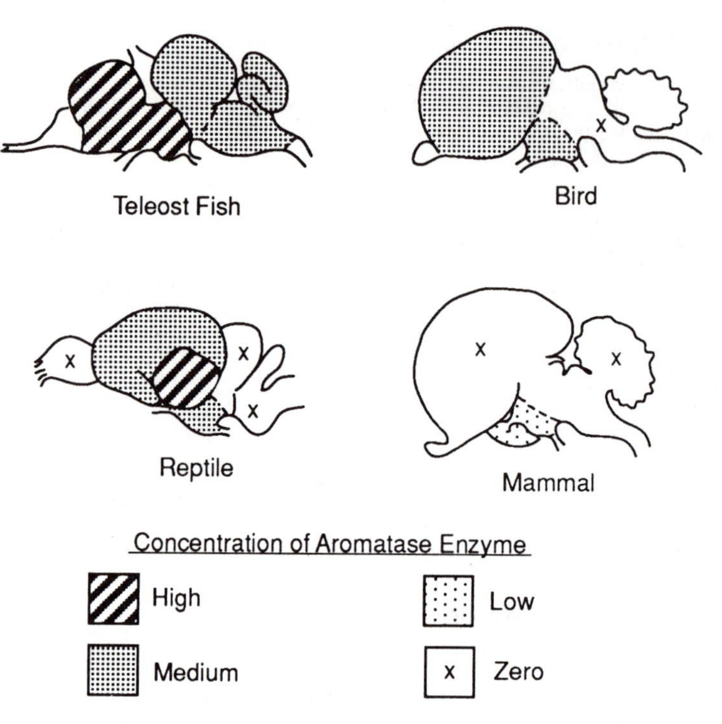

Figure 9.7. Phyletic differences in the neuroanatomical distribution of the aromatase enzyme which converts testosterone to estrogen. Depicted are the brains of a teleost fish (sculpin), reptile (turtle), bird (chicken) and mammal (rat). Notice that the aromatase levels in the mammalian brain are much lower than in the non-mammalian brains and that the distribution of aromatase is restricted to the subcortical areas in the mammals, whereas aromatase levels are high in the cerebral cortex of non-mammalian species. (Redrawn from Callard *et al.*, 1978.)

detected by determining which cells contain the aromatase enzyme (Callard *et al.*, 1978; Callard, 1983). In mammals, aromatase activity occurs primarily in the limbic system (the preoptic area, the hypothalamus, septum, amygdala and hippocampus) and may occur in the same neural locations as 5α-reductase. The aromatase enzyme is far more widely distributed in the brains of non-mammalian vertebrates, occurring in high concentrations in the cerebral cortex and midbrain (see Figure 9.7). In addition to the brain, aromatization of testosterone to estrogen occurs in the gonads, placenta, adrenal cortex, breast tissue, fat cells, kidney, liver and other tissues in both mammalian and non-mammalian vertebrates (Callard *et al.*, 1983).

Aromatization of testosterone to estrogen plays an important function in sexual differentiation, the onset of puberty, adult sexual behavior and the feedback control of gonadotropin secretion in many mammalian and non-mammalian species. Certain sex-specific behaviors are controlled by the aromatization of testosterone to estrogen, while others depend on the stimulation of androgen receptors. There are, however, wide species differences, even within the mammals, in the importance of aromatization in the activation of these behaviors (Adkins-Regan, 1981). For example, only aromatizable androgens will stimulate sexual behavior in castrated male rats but in other species, such as the guinea pig, hamster, rabbit and rhesus monkey, aromatization is not necessary as 5α-DHT and other non-aromatizable androgens act via androgen receptors to stimulate sexual behavior (Sheridan, 1979; Luttge, Hall and Wallis, 1974).

Table 9.3. *Characteristics of type I and type II adrenal steroid receptors in the brain*

Type I Receptors	Type II Receptors
Resemble kidney mineralocorticoid receptors (MR)	Resemble liver glucocorticoid receptors (GR)
Bind to both aldosterone and corticosterone with high affinity; low affinity for dexamethasone	High affinity for dexamethasone, binds to corticosterone with a lower affinity than type I receptors
Do not bind to synthetic glucocorticoid agonists RU26988 and RU28362	Bind to synthetic glucocorticoids RU26988 and RU28362
Highest density in hippocampus, septum, amygdala and dentate gyrus, with low density in other brain areas	Widely distributed in neurons and glial cells of the brain with the highest density in the hypothalamus
Low density in anterior pituitary	High density in anterior pituitary corticotroph cells
Bind to aldosterone to regulate salt appetite	Do not bind to aldosterone
Occupied when corticosteroid levels are low (baseline)	Occupied only when corticosteroid levels are high after stress
Mediate permissive, circadian action of corticosteroids	Terminate stress activation of neuroendocrine pathways

9.7 ADRENAL STEROID HORMONE TARGET CELLS IN THE BRAIN

Like the gonadal steroid hormones, adrenal steroids bind to nuclear receptors to form an activated hormone–receptor complex which binds to the DNA in the cell nucleus and initiates mRNA and protein synthesis (Miesfeld, 1990; Munck *et al.*, 1990). Some reports, however, indicate that glucocorticoid receptors occur in the cytoplasm of the cell as well as the nucleus, supporting the classical two-stage model of steroid hormone receptor action (LaFond *et al.*, 1988).

Using autoradiography (DeKloet and Ruel, 1987; DeKloet *et al.*, 1987) and immunocytochemistry (Evans and Arriza, 1989), two types of receptor for adrenal steroids have been identified: type I (mineralocorticoid) receptors and type II (glucocorticoid) receptors (Table 9.3). Aldosterone binds primarily to the mineralocorticoid receptors in the kidney, where it mediates Na^+ transport. Glucocorticoids bind to receptors in the liver, muscle and fat cells, and almost all other cells in the body. Glucocorticoids regulate glucogen synthesis in the liver, carbohydrate metabolism in the muscles and lipolysis in fat cells. Glucocorticoids also regulate bone growth and the development of lymphocytes in the immune system (see Martin, 1985; Greenspan, 1991; Hadley, 1992 for details). The type I receptors bind to both aldosterone and corticosterone, but because they have a higher affinity for aldosterone and because most plasma corticosteroids are bound to transcortin, the type I receptors in the kidney bind primarily to aldosterone, even though there is much more corticosterone in the circulation. In the kidney, the cells with type I receptors also have an enzyme (11β-hydroxysterone dehydrogenase) which metabolizes cortisol and corticosterone to inactive metabolites (17 ketosteroids), but does not metabolize aldosterone (Funder, 1991).

Both type I and type II adrenal steroid receptors also occur in the brain, where they can be characterized in a number of ways, as shown in Table 9.3. The type I receptors, which are identical to the mineralocorticoid receptors in the kidney, bind to both mineralocorticoids and glucocorticoids. The type II receptors, which resemble liver glucocorticoid receptors, bind only to glucocorticoids (DeKloet and Ruel, 1987; DeKloet *et al.*, 1987). Because glucocorticoids bind to both type I and type II receptors in the brain, it is difficult to distinguish between these two types of receptor using naturally occurring hormones. Synthetic hormones, therefore, have been used to identify the chemical, anatomical and functional differences between these receptor types in the brain. The synthetic glucocorticoids, RU 26988 and RU 28362, for example, bind exclusively to type II receptors and do not bind to type I receptors (Meyer, 1985; McEwen, DeKloet and Rostene, 1986; Funder and Sheppard, 1987; McEwen, 1988). Since these receptors have now been cloned, they can be specifically identified by *in situ* hybridization techniques (Arriza *et al.*, 1988). The distribution of type I and type II receptors in the brain is shown in Figure 9.8. Figure 9.9 shows the details of the distribution of the 'classical' glucocorticoid (type II) receptors.

9.7.1 TYPE I RECEPTORS

Type I receptors have their highest density in the hippocampus, lateral septum, dentate gyrus, brain stem and in some areas of the cerebral cortex, such as the entorhinal cortex and have a lower density in other brain areas (see Figure 9.8). Type I receptors are absent from the hypothalamus and pituitary gland, which have a high concentration of type II receptors (McEwen, 1981, 1976; Arriza *et al.*, 1988). Type I receptors bind equally well to aldosterone and corticosteroids and function to regulate salt appetite, CRH and ACTH secretion, locomotor activity, and mental performance under normal baseline fluctuations in glucocorticoid levels (McEwen *et al.*, 1986; McEwen, 1988). Type I receptors may also modulate behaviors which are influenced by the hippocampus in a circadian rhythm, such as sleep–wake cycles, appetite and mood (Funder, 1991).

Because type I receptors bind to both mineralocorticoids and glucocorticoids, the question arises of whether there are two different forms of type I receptors in the brain, one of which binds to mineralocorticoids and one to glucocorticoids, or whether the same type I receptors bind to glucocorticoids in some circumstances and to mineralocorticoids in others (McEwen *et al.*, 1986). Some evidence suggests that there are type I receptors with different degrees of affinity for aldosterone. Those with high affinity for aldosterone have a low affinity for corticosteroids, while those with low affinity for aldosterone are more likely to bind to corticosteroids. Other evidence suggests that the relative affinity of the type I receptor for aldosterone and corticosteroids may be determined by

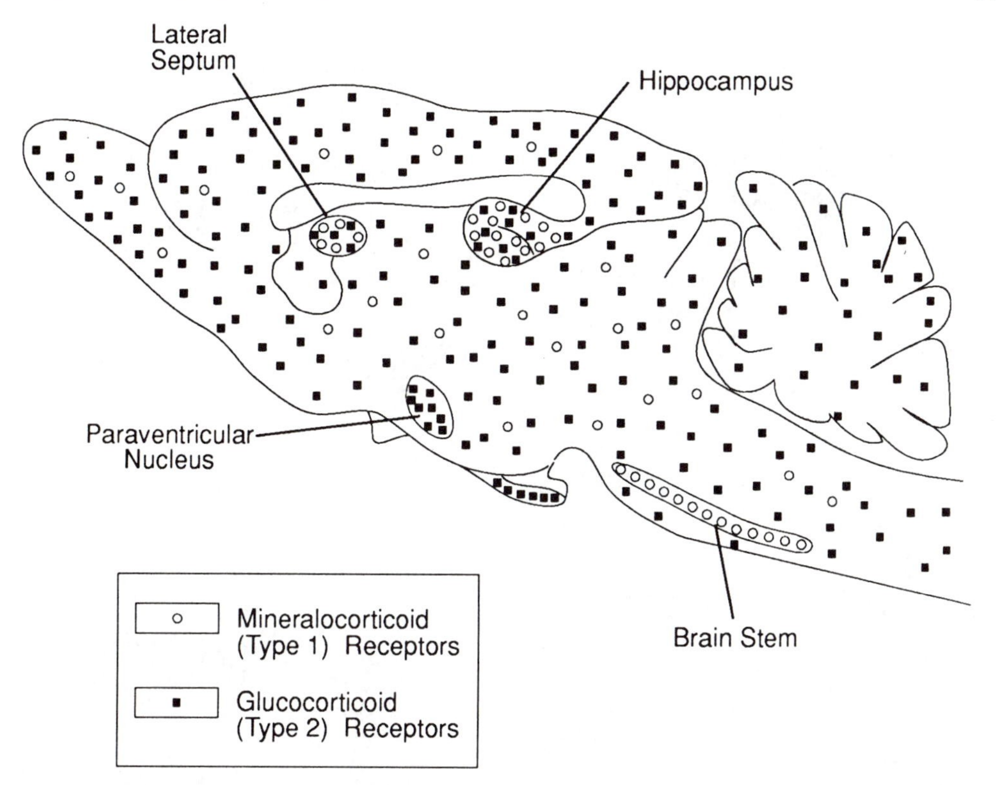

Figure 9.8. The distribution of type I and type II glucocorticoid receptors in the rat brain. Type I receptors occur in high concentrations in the hippocampus, lateral septum and in the brainstem. Type II receptors occur at high levels in the hippocampus, paraventricular nucleus of the hypothalamus, and in the anterior pituitary, and are distributed fairly uniformly throughout the rest of the brain. In the hippocampus, receptors of both types bind glucocorticoids across a range of plasma levels. Outside the hippocampus, the type I receptors function when plasma glucocorticoid levels are low and the type II receptors only when plasma glucocorticoid levels are elevated. (Redrawn from McEwen, 1988.)

the circulating level of corticosteroids and the level of corticosteroid binding globulin (McEwen *et al.*, 1986; Funder and Sheppard, 1987; McEwen, 1988). More recent evidence, however, suggests that the mechanism which determines the affinity of the type I receptors in the brain for aldosterone versus corticosteroids is the presence or absence of the enzyme 11β-hydroxysteroid dehydrogenase, which deactivates corticosteroids, but not aldosterone. As occurs in the kidney, the presence of this enzyme removes corticosteroids from competition with mineralocorticoids, which can then bind to the type I receptors in these cells (Funder *et al.*, 1988). This enzyme is not present in the hippocampal cells, thus the type I receptors in this region of the brain can bind to both mineralocorticoids and glucocorticoids (Evans and Arriza, 1989; Funder, 1991).

9.7.2 TYPE II RECEPTORS

The type II receptors shown in Figures 9.8 and 9.9 are the 'classical' glucocorticoid receptors discovered in the hippocampus and septum

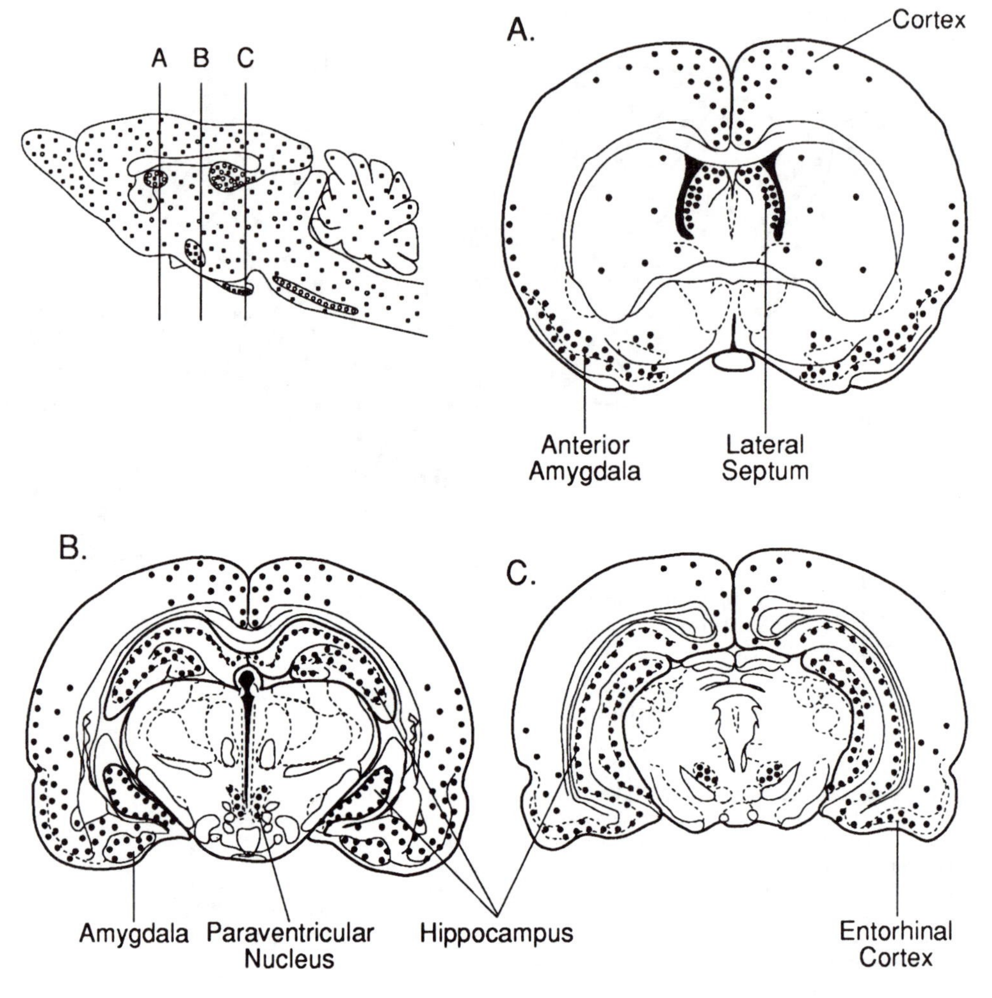

Figure 9.9. Distribution of glucocorticoid type II receptors. Three coronal (frontal) sections showing the locations of radioactive corticosterone uptake by type II receptors in the mouse brain. The lines (A, B, and C) show where the sections were cut with respect to the type II receptors shown in sagittal section in Figure 9.8. (Redrawn from Stumpf and Sar, 1975a.)

(McEwen, Weiss and Schwartz, 1969). Type II receptors are widely distributed throughout the brain, but are concentrated in the hippocampus, septum, amygdala, paraventricular nucleus of the hypothalamus (whose neurons synthesize many peptides, including oxytocin, vasopressin and CRH), certain areas of the cerebral cortex, and in the brain stem, as well as in the corticotroph cells of the anterior pituitary. The type II receptors have a lower affinity for glucocorticoids than the type I receptors and are only activated by the high level of glucocorticoids secreted during stress. They provide negative feedback to terminate stress-activated neural and endocrine (hypothalamic-pituitary-adrenal) activation and regulate behavioral responses to stress (McEwen *et al.*, 1986; McEwen, 1988). The type II receptors thus function to protect the organism from its own stress response (DeKloet *et al.*, 1987). Glucocortioids also function as negative feedback signals to inhibit immune responses (see Chapter 13).

Both type I and type II receptors are down-regulated during chronic stress, in diseases such as diabetes mellitus and diabetes insipidus, and during aging. Type II receptors are down-regulated in response to high corticosteroid levels, whereas type I receptors show a diurnal rhythm, with the highest receptor number in the evening (end of light period) when pituitary-adrenal activity is at its peak. Type I receptor numbers may also be regulated by ACTH, vasopressin and other neuropeptides acting at the hippocampus (DeKloet *et al.*, 1987).

9.7.3 WHY ARE THERE TWO DIFFERENT RECEPTORS FOR GLUCOCORTICOIDS IN THE BRAIN?

Two different glucocorticoid receptors may be necessary in the hippocampus and other brain areas because glucocorticoids serve two different physiological functions in these areas: at low concentrations, they regulate circadian rhythmicity, while at high concentrations, they regulate stress responses. Evans and Arriza (1989) suggest that no single receptor system can respond to the great range of hormone levels shown by glucocorticoids (from 0.5 to 100 nM), whereas two receptors with different sensitivities could accommodate this range. The type I receptor, which has a high affinity for glucocorticoids, responds to low levels of circulating hormone, while the type II receptor, which has a low affinity for glucocorticoids, responds only at high levels of the hormone. Such a binary glucocorticoid response system could allow one cell to respond to both low and high concentrations of glucocorticoids (if it had both type I and type II receptors) or respond only to low or high levels of these hormones (if it had only one type of receptor). Whether or not this hypothesis becomes generally accepted will depend on its ability to explain the disparate actions of adrenal steroids at these two types of receptors (see Funder, 1991).

9.8 GENOMIC AND NON-GENOMIC ACTIONS OF STEROID HORMONES AT NERVE CELLS

Previous chapters have discussed the differences between 'classical' neurons, which have receptors for neurotransmitters and release neurotransmitters when they are stimulated (i.e. neurocrine communication) and neurosecretory cells, or neuroendocrine transducers, which have receptors for neurotransmitters and release neurohormones into the bloodstream when they are stimulated (i.e. neuroendocrine communication). In Chapter 11, I discuss a third type of neuron, which releases neuropeptides when stimulated by neurotransmitters. Each of these types of nerve cell may also have steroid hormone receptors and thus be termed 'steroid hormone-sensitive neurons' (see Figure 9.10).

Steroid hormones are able to modulate messenger RNA and protein synthesis through their actions on nuclear receptors in nerve cells. These genomic or indirect effects of steroids on nerve cells result in hormone-

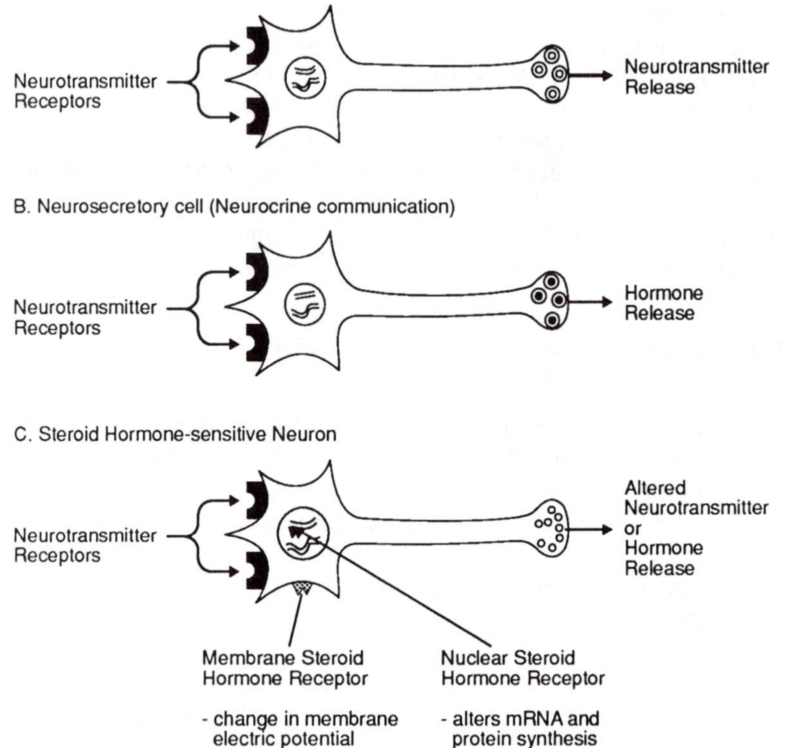

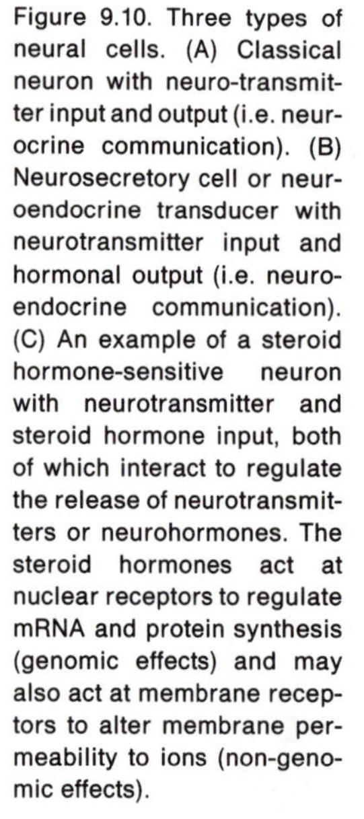
Figure 9.10. Three types of neural cells. (A) Classical neuron with neuro-transmitter input and output (i.e. neurocrine communication). (B) Neurosecretory cell or neuroendocrine transducer with neurotransmitter input and hormonal output (i.e. neuroendocrine communication). (C) An example of a steroid hormone-sensitive neuron with neurotransmitter and steroid hormone input, both of which interact to regulate the release of neurotransmitters or neurohormones. The steroid hormones act at nuclear receptors to regulate mRNA and protein synthesis (genomic effects) and may also act at membrane receptors to alter membrane permeability to ions (non-genomic effects).

induced changes in the number of receptor proteins and the amount of neurotransmitter, neuropeptide, or neurohormone synthesized and stored in the cell. Not all of the effects of steroid hormones on neurons can be accounted for by changes in genomic activity, however. There are also non-genomic or direct effects of steroid hormones on the nerve cell membrane, resulting in changes in electrical potential and the release of neurotransmitters into the synapse. These direct effects of steroids on nerve cells occur rapidly and are of brief duration, whereas the indirect effects, such as protein synthesis and receptor regulation, take longer to occur and are of greater duration (McEwen, 1981). Thus, the ability of steroids to alter the electrophysiological activity of the cell membrane occurs more rapidly than would be possible if it depended on protein synthesis following the activation of a nuclear steroid receptor.

9.8.1 ELECTROPHYSIOLOGICAL RESPONSES OF NEURONS SENSITIVE TO STEROID HORMONES

Intracellular recordings of the electrophysiological activity of steroid hormone-sensitive neurons demonstrate the changes in their firing rate which occur in the presence of steroid hormones (Pfaff, 1981). Hua and Chen (1989), for example, found that within 2 min of bathing an *in vitro* preparation of the guinea pig splanchnic nerve with a glucocorticoid, the resting membrane potential changed from −60 to −70 millivolts (Figure

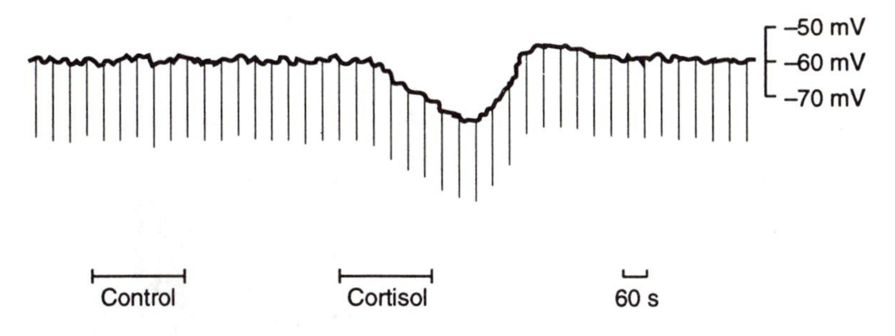

Figure 9.11. The effect of steroid hormones on sensitive neurons *in vitro*. Hyperpolarization of the resting potential of the guinea pig coeliac ganglion occurs about 2 min following the application of a glucocorticoid (cortisol hemisuccinate). The tracing shows the membrane potential in millivolts (mV). Regular injection of a hyperpolarizing current of 100 ms at 20-s intervals caused the downward deflections of the trace. A control solution (Kreb's solution) had no effect on the resting potential of the nerve cell membrane. (Redrawn from Hua and Chen, 1989.)

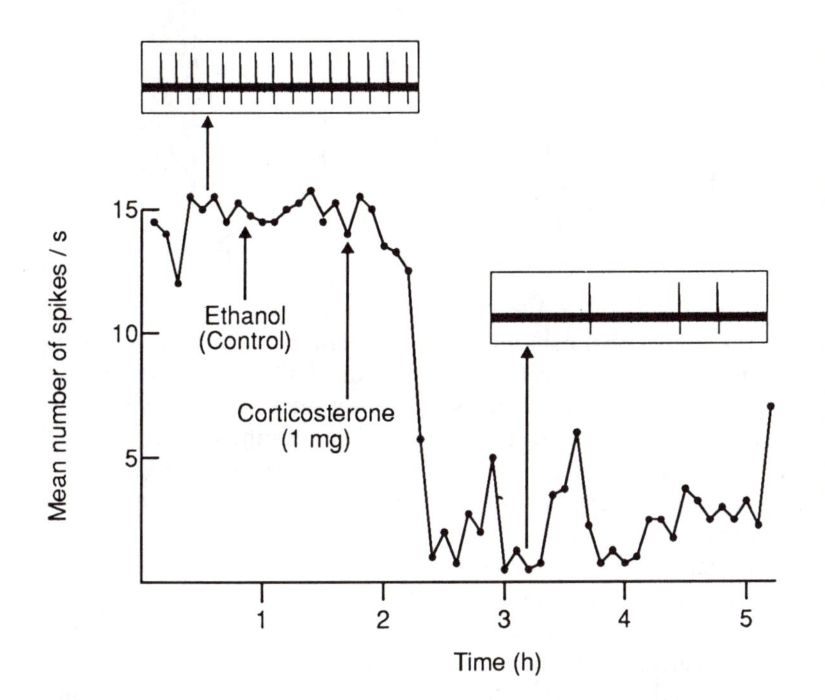

Figure 9.12. The effect of corticosterone (injected intraperitoneally in ethanol) on the electrophysiological activity of a single nerve cell in the hippocampus of a freely moving, hypophysectomized female rat. Samples of 1 s duration from the record show unit activity before (left) and after (right) injection of corticosterone. The control injection of ethanol had no effect on the electrophysiological activity of this neuron, while the corticosteroid injection significantly reduced its firing rate. (Redrawn from Pfaff *et al.*, 1971.)

9.11). No such hyperpolarization of the cell membrane occurred following treatment with a control solution.

When injected into a freely moving rat, glucocorticoids inhibit the electrical activity of neurons in the hippocampus (as shown in Figure 9.12) and other brain areas, including the dorsomedial hypothalamus (Pfaff *et al.*, 1971; Steiner, 1975). The hyperpolarization of the membrane of the *in vitro* neuron observed by Hua and Chen (1989) and shown in Figure 9.11 suggests that glucocorticoids inhibit the firing of neurons *in vivo* (Figure 9.12) by pushing the resting potential further from the threshold necessary for triggering the action potential. (Compare Figure 9.11 with Figure 5.5 (p. 66) to understand how the hyperpolarization of

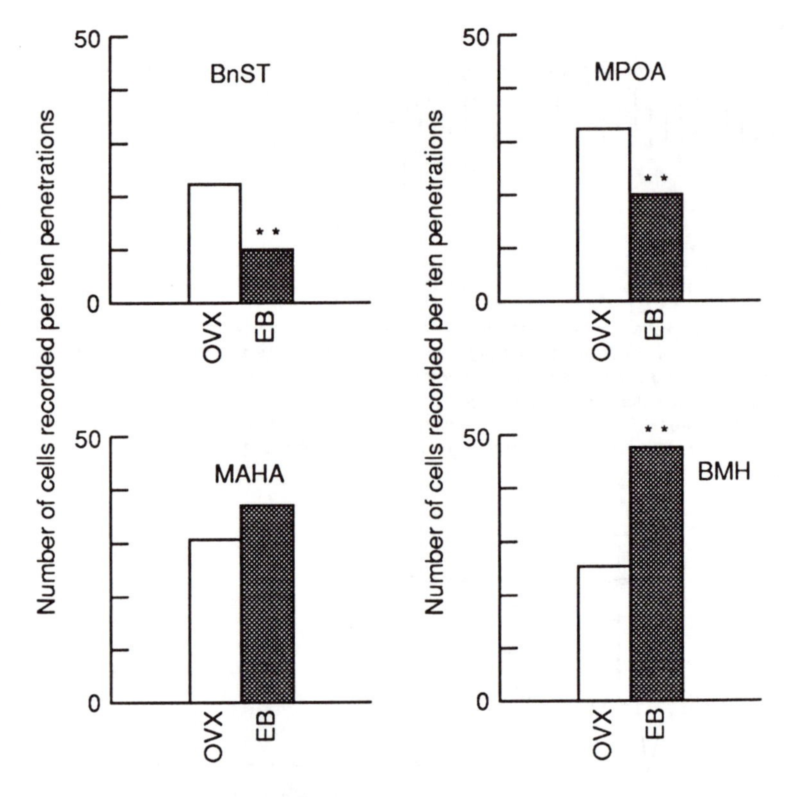

Figure 9.13. The effect of gonadol hormones on neuron activity. Comparison of the number of active cells in different neural areas in estradiol benzoate (EB) treated and untreated ovariectomized (OVX) female rats. The number of neurons with spontaneous electrical activity recorded in ten electrode penetrations through each anatomical structure are plotted. Estradiol depressed baseline electrical activity in the bed nucleus of the stria terminalis (BnST) and medial preoptic area (MPOA) and increased baseline electrical activity in the basomedial hypothalamus (BMH), but had no significant effect on the firing rate of neurons in the medial anterior hypothalamus (MAHA). $^{**}p<0.01$. (Redrawn from Bueno and Pfaff, 1976.)

the resting potential may inhibit the triggering of an action potential.) One way that glucocorticoids may cause this inhibition of electrophysiological activity is by binding to membrane receptors, such as those shown in Figure 5.6, and altering the opening of ion channels in the nerve cell membrane (see Section 9.8.3). Another mechanism is for the glucocorticoid to inhibit the activation of second messengers, such as cyclic AMP, within the postsynaptic cell (Joels and DeKloet, 1989).

Gonadal hormones increase the electrophysiological activity of some neurons and inhibit the activity of other neurons. Estrogen, for example, increases the electrical activity of neurons in the medial basal hypothalamus (arcuate nucleus, dorsomedial and ventromedial hypothalamus) and inhibits the electrical activity of neurons in the medial preoptic area and bed nucleus of the stria terminalis, as shown in Figure 9.13 (Pfaff, 1981; Pfaff and McEwen, 1983). These results correspond with the finding that female sexual behavior is elicited by estrogen *stimulation* of neural activity in the anterior and ventromedial hypothalamus and *inhibition* of the activity of preoptic neurons. In the estrogen-primed female rat, injections of progesterone stimulate electrical activity in cells of the anterior hypothalamus, but inhibit activity in cells of the preoptic area, as shown in Figure 9.14 (Lincoln, 1969). Depending on the site of injection, the dose, and the timing of the injections, progesterone may facilitate or inhibit sexual behavior at these sites (see McEwen *et al.*, 1979; Barfield *et al.*, 1984). Testosterone increases neural activity in the preoptic area and medial forebrain bundle; brain areas which regulate

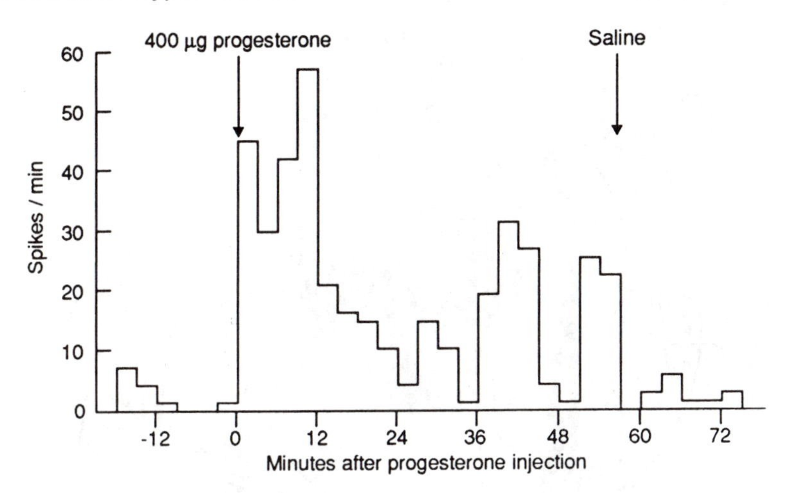

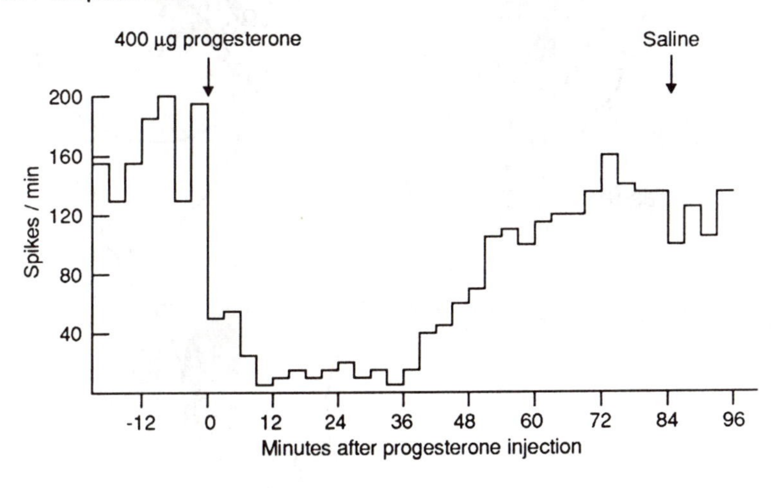

Figure 9.14. Changes in single cell activity following progesterone injection at time zero (0). (A) The increase in firing rate of a single cell in the anterior hypothalamus and (B) the decrease in firing rate of a single cell in the preoptic area after the intravenous infusion of 400 µg progesterone in saline. Control (saline) injections had no effect on the firing rate of these cells. (Redrawn from Lincoln, 1969.)

sexual arousal and sexual behavior in the male rat (Orsini, 1981; Pfaff, 1981).

9.8.2 STEROID HORMONE-INDUCED CHANGES IN NEUROTRANSMITTER RELEASE

When the distribution of steroid hormone receptors in the brain (illustrated in Figures 9.4, 9.5 and 9.9) is compared with the distribution of neurotransmitter pathways (illustrated in Figure 5.8, p. 72) a close relationship is observed between the pattern of steroid hormone receptors and the nerve pathways. The distribution of estrogen and glucocorticoid receptors along the noradrenaline, dopamine and serotonin nerve pathways is shown in Figure 9.15 (Grant and Stumpf, 1975). When a steroid hormone alters the electrical activity of a target neuron along one of these pathways, the amount of neurotransmitter released from that neuron is

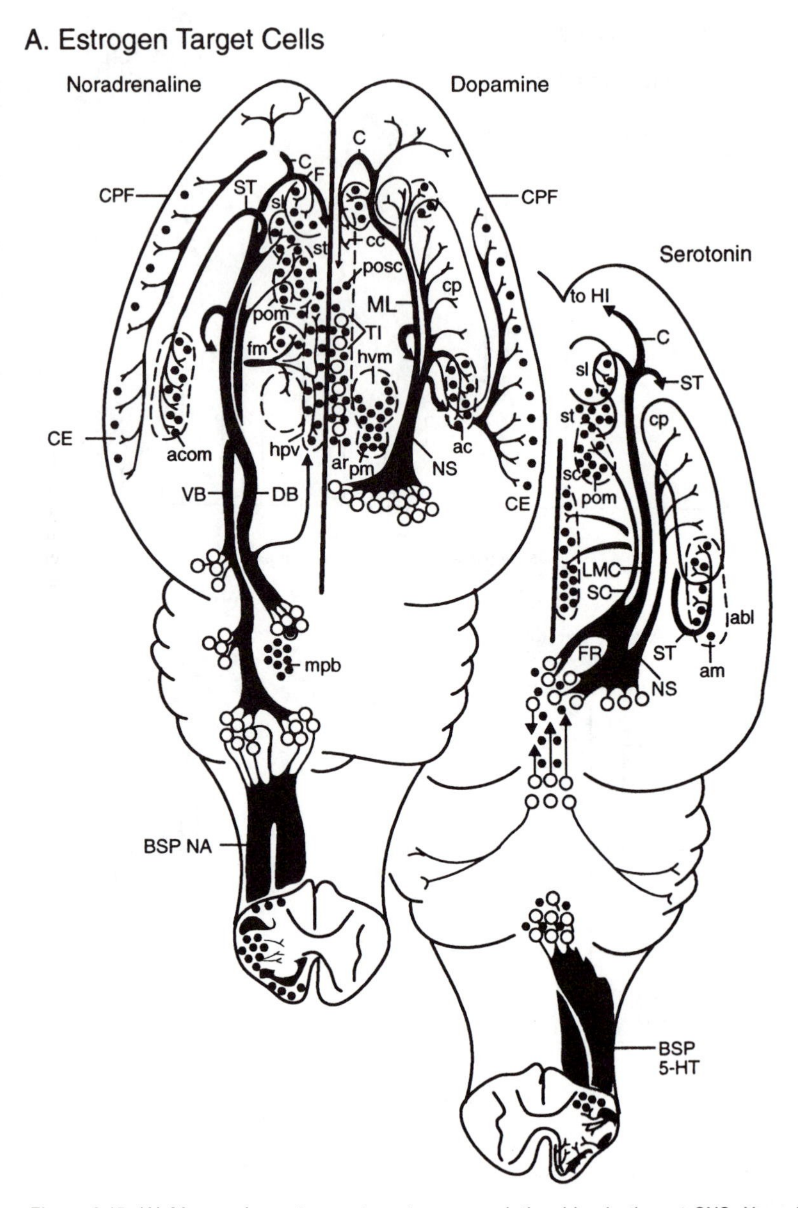

Figure 9.15. (A) Monoamine-estrogen target neuron relationships in the rat CNS. Neural systems identified by histochemical fluorescence methods containing noradrenaline (NA), dopamine (DA), or serotonin (5-HT) are shown from left to right, respectively. Estrogen (and corticosterone in B) target neurons were determined by autoradiography. Dopaminergic neurons displaying nuclear uptake of estradiol-17β have been found among the tuberoinfundibular dopamine neurons (TI). (B) Monoamine-glucocorticoid target neuron relationships in the CNS. Dopamine and serotonin systems are shown as in (A), but with terminal inputs to hippocampus and cingulate cortex sites of corticosteroid target neuron accumulation depicted instead of deeper hypothalamic areas that lack such neurons. The cortico-medial amygdala, lateral septum, hippocampus and several cortical regions stand out as corticosteroid target areas receiving input from ascending monoamine systems. AA = anterior amygdala; abl = basolateral amygdala; abm = basomedial amygdala; ac = central amygdala; aco = cortical amygdala; acom = cortico-medial amygdala; ala = lateral amygdala; alp = posterio-lateral amygdala; am = medial amygdala; ar = arcuate nucleus; BSP 5-HT = bulbospinal serotonin systems; BSP NA = bulbospinal NA system; C = cingulum; CA = anterior commissure; CC = crus cerebri; CE = entorhinal cortex; CF = frontal cortex; cl = claustrum; CO = optic chiasm; cp = caudate putamen; CPF = piriform cortex; DB = dorsal ascending NA bundle; F = fornix; fm = paraventricular nucleus (magnocellular region); FMP = fasciculus medial prosencephali; fp = paraventricular nucleus (parvocellular region);

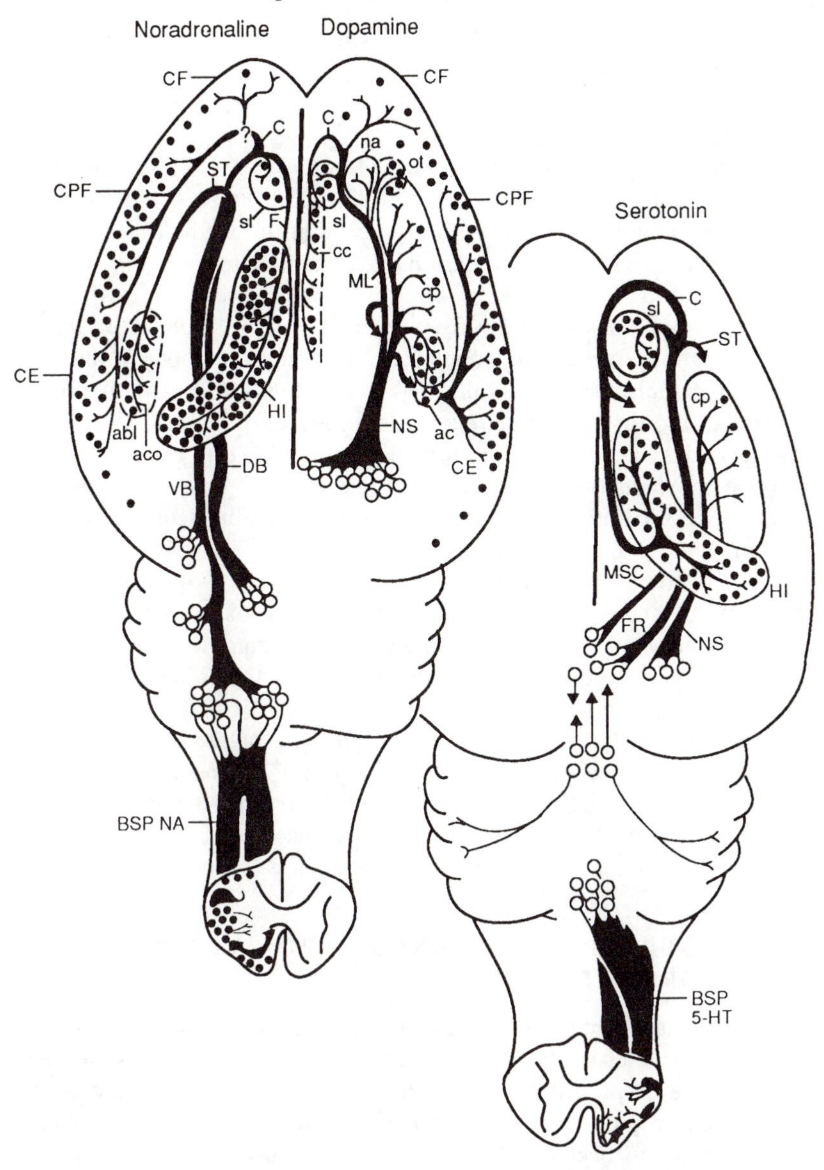

FR = fasciculus retroflexus; gc = griseum centrale; GD = dentate gyrus; hd = dorsomedial hypothalamus; hdm = dorsomedial hypothalamic nucleus; hpv = periventricular nucleus; HI = hippocampus; hvm = ventromedial hypothalamus; IG = indusidium griseum; lh = lateral habenulae; LMC = lateral mesencephalo-cortical serotonin tract; ML = mesolimbic dopamine system; MSC = medial subcortical serotonin tract; na = nucleus accumbens; npb = medio-lateral parabranchial nucleus; NS = nigroneostriatal dopamine system; osc = organum subcommissurale; ot = olfactory tubercule; pf = nucleus parafascicularis; pm = ventral premammillary nucleus; pol = lateral preoptic nucleus; pom = medial preoptic nucleus; posc- = suprachiasmatic nucleus; pv = ventral remammillary nucleus; r = nucleus ruber; RSS = radiotio corpus callosum; S = subiculum; sl = lateral septum; so = supraoptic nucleus; SS = saggital sulcus; ST = stria terminalis; st = nucleus interstitialis stria terminalis; sum- = supramammillary nucleus; sut = subthalamic nucleus; TCC = truncus corpus callosum; TI = tuberoinfundibular dopamine system; TO = optic tract; ts = nucleus triangularis seti; VB = ventral ascending NA bundle; zi = zona incerta.

○ = monoaminergic neurons.

● = steroid target neurons.

(From Grant and Stumpf, 1975.)

also altered. This change in neurotransmitter release along nerve pathways results in motivational, emotional, sensory and behavioral changes and in altered secretion of hypothalamic hormones, depending on the pathway stimulated.

Although the neurotransmitters used in most steroid hormone target neurons remain unidentified, some effects are known. For example, estrogen modulates noradrenaline activity in the medial preoptic area and bed nucleus of the stria terminalis, dopamine release in the arcuate nucleus, and also modulates serotonin (5-HT) release (Crowley and Zemlen, 1981). Estrogen reduces the synthesis of the inhibitory transmitter GABA in the arcuate nucleus, ventromedial hypothalamus and corticomedial amygdala (Wallis and Luttge, 1980). Estrogen can also be metabolized to 'catecholestrogens' (see Figure 9.5) which can bind to adrenergic receptors and thus directly influence the activity of adrenergic nerve pathways (McEwen, 1981). Progesterone elevates noradrenaline release in the anterior hypothalamus and arcuate nucleus of estrogen-primed rats (Crowley and Zemlen, 1981).

Glucocorticoids modulate the synthesis and release of noradrenaline, dopamine, and serotonin by regulating the levels of the enzymes which synthesize or deactivate these transmitters. For example, glucocorticoids acting at type I receptors, stimulate serotonin synthesis in neurons of the midbrain raphe nucleus by increasing tryptophan uptake and tryptophan hydroxylase activity, thus increasing the rate of serotonin synthesis (McEwen *et al.*, 1986; DeKloet and Reul, 1987). Glucocorticoid stimulation of type II receptors promotes the synthesis of the enzyme phenylthonolamine *N*-methyl transferase (PNMT) which is necessary for adrenaline synthesis in adrenergic neurons (see Figure 5.4, p. 64).

9.8.3 ARE THERE MEMBRANE RECEPTORS FOR STEROID HORMONES?

In order to account for the rapid changes in electrophysiological activity of neurons which can be stimulated by steroid hormones, the existence of membrane receptors for steroid hormones in addition to nuclear receptors has been proposed, but the concept of a steroid hormone membrane receptor has not been widely accepted (Szego, 1984). The fact that steroid hormones can stimulate rapid electrophysiological changes in the nerve cell membrane and modulate the release of neurotransmitters suggests that membrane receptors are present, but they have not yet been identified (Towle and Sze, 1983). Szego (1984) postulates that the membrane receptors for steroid hormones act via cyclic AMP activation in the same manner as peptide hormone receptors (see Chapter 10). One of the primary actions of steroid hormones at membrane receptors may be to regulate the opening of ion channels in the cell membrane (see Section 5.5) as suggested by Hua and Chen (1989). Delville (1991) has reviewed the evidence supporting a non-genomic mechanism for the action of progesterone at its neural target cells which facilitates sexual behavior in female rodents. She concludes that, while there is mounting evidence in support of the non-genomic model for progesterone action in the brain, the genomic model can not yet be ruled out.

9.8.4 ARE STEROID HORMONES ('NEUROSTEROIDS') SYNTHESIZED IN THE BRAIN?

The finding that steroid hormones may act via membrane receptors to regulate ion channels in nerve cells has led to the suggestion that the brain may produce its own steroid hormones to regulate neural activity (Lieberman and Prasad, 1990; Robel *et al.*, 1991). These 'neurosteroids' could thus serve a neuromodulatory function, like neuropeptides (Chapter 12). *In vitro* preparations of glial cells cultured from the cerebral cortex of newborn rat brains have the ability to synthesize pregnenolone and progesterone (Jung-Testas *et al.*, 1989). These neurosteroids may bind to GABA receptors and antagonize GABAergic transmission by modulating chloride ion channels in the nerve membranes (Majewska *et al.*, 1986, 1988). In the future, therefore, both the existence of membrane receptors for steroid hormones and the neural synthesis of steroid hormones may become widely accepted findings in neuroendocrinology.

9.8.5 STEROID HORMONE-INDUCED CHANGES IN PROTEIN SYNTHESIS IN NERVE CELLS (GENOMIC EFFECTS)

When a steroid hormone binds to its receptor complex in the cell nucleus, mRNA and protein synthesis is promoted in the target cell (refer to Section 9.4). The proteins synthesized can be neurotransmitters, hormones, neuropeptides, enzymes, neurohormones or the receptors for these chemicl messengers. Estrogen, for example, regulates the synthesis of a number of proteins in target cells and may do so by influencing the activity of the enzymes which control their rate of synthesis or degradation (McEwen, 1981). As mentioned above, estrogen stimulates the synthesis of progesterone receptors in target cells of the preoptic area and ventromedial hypothalamus and in the uterus. Estrogen also regulates the synthesis of oxytocin, vasopressin and enkephalins in the supraoptic nucleus, paraventricular nucleus and ventromedial nucleus of the hypothalamus (McEwen and Pfaff, 1985).

In adult animals, steroid hormone activation of protein synthesis results in temporary changes in neural activity, but during early development, the effects of steroid hormones on protein synthesis result in permanent changes in neural growth and connections (see Section 9.9.5). The effects of steroid hormones on protein synthesis are not immediate and may take hours or days to occur. For example, the activation of sexual behavior in ovariectomized female rats following estrogen injection takes 24–48 hours (McEwen *et al.*, 1978). Steroid hormone stimulation of protein synthesis can be blocked by drugs such as anisomycin which inhibits protein synthesis (see McEwen and Pfaff, 1985). For example, the sexual behavior of female rats treated with estrogen and progesterone can be blocked by injections of anisomycin into the ventromedial hypothalamus, indicating that the stimulation of sexual behavior by gonadal steroids requires protein synthesis in the ventromedial hypothalamus (Rainbow *et al.*, 1982). Blocking protein synthesis also inhibits estrogen stimulation of the pre-ovulatory LH surge

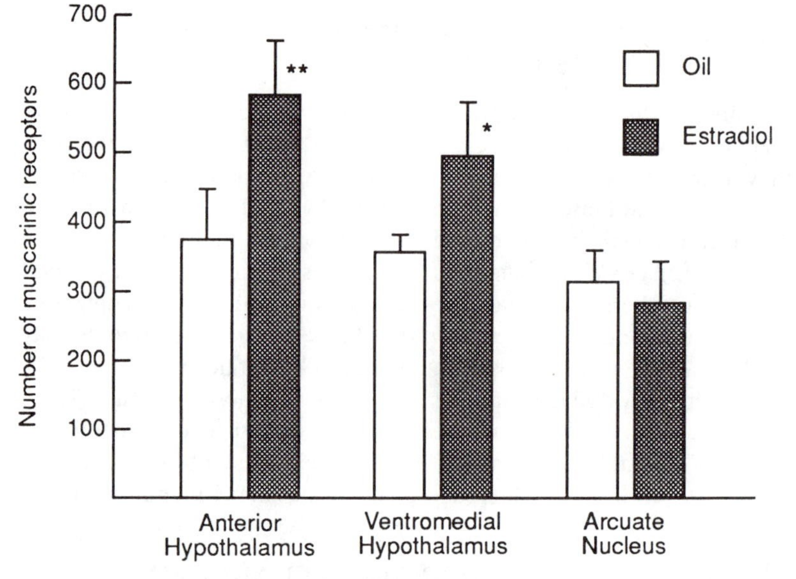

Figure 9.16. Regulation of receptor synthesis by steroid hormones. Effects of estradiol benzoate on the number of muscarinic receptors (femtomoles per milligram of protein) in nerve cells in three regions of the hypothalamus. Ovariectomized female rats were treated for 2 days with either sesame oil or10 micrograms of estradiol benzoate in sesame oil. Estradiol elevated the number of muscarinic receptors in the anterior hypothalamus and the ventromedial nucleus of the hypothalamus, but not in the arcuate nucleus. $^{*}p<0.05$, $^{**}p<0.02$. (Redrawn from Rainbow *et al.*, 1980.)

(positive feedback on LH), the negative feedback effects of corticosteroids on ACTH, and the effects of progesterone on female sexual behavior (McEwen, Krey and Luine, 1978).

Because the proteins synthesized following estrogen stimulation of a target cell may be transported along the cell's axon by axoplasmic flow, the biochemical changes resulting from steroid hormone action on nerve cell bodies in one area of the brain can have their effect in more distant neural sites. Thus, hormonal stimulation of target cells in the locus ceruleus (the site of noradrenaline synthesis; Figure 5.8C) or the midbrain raphe nucleus (the site of serotonin synthesis; Figure 5.8D) may result in changes in neural activity in the hypothalamus, amygdala, cerebral cortex, and other neural sites innervated by axons from these cell bodies (McEwen *et al.*, 1978). Drugs such as colchicine, which inhibit axonal transport, also inhibit the effects of steroid hormones on neural target cells. For example, injection of colchicine into estrogen-primed ovariectomized female rats results in a 2 day delay in the onset of sexual behavior because axonal transport of the newly synthesized proteins is inhibited (Pfaff and McEwen, 1983).

The synthesis of receptors for neurotransmitters is also regulated by steroid hormones in a number of neural sites. For example, the number of dopamine receptors in the anterior pituitary of the female rat changes over the reproductive cycle, with the greatest increase in receptors correlated with the highest estrogen levels (Heiman and Ben-Jonathan, 1982). Estrogen also modulates muscarinic receptor density in some neural cells. About 24 hours after exposure to estrogen, the number of muscarinic receptors in the anterior hypothalamus and ventromedial hypothalamus of the female rat (but not in the male rat) is elevated, as shown in Figure 9.16 (Rainbow *et al.*, 1980; Dohanach *et al.*, 1982). Administration of protein synthesis inhibiting drugs (e.g. anisomycin) prevents this increase in muscarinic receptors and inhibits the activation of sexual behavior via estrogen stimulation of the ventromedial hypothalamus (Rainbow *et al.*, 1984).

Glucocorticoids modulate the levels of serotonergic (5-HT), adrenergic, muscarinic cholinergic and GABA receptors in the hippocampus and other glucocorticoid sensitive nerve cells (DeKloet, 1984; Biegen, Rainbow and McEwen, 1985). In the dorsal hippocampus and dorsal raphe nucleus, for example, glucocorticoids reduce the synthesis of 5-HT receptors and adrenalectomy results in an increase in 5-HT receptors. Chronic exposure to high levels of glucocorticoids results in the down-regulation of 5-HT receptors (McEwen *et al.*, 1986; De Kloet and Reul, 1987) and can result in damage to hippocampal neurons. This neurotoxic effect of glucocorticoids appears to be mediated by type II receptors (Packan and Sapolsky, 1990).

9.9 FUNCTIONS OF STEROID HORMONE MODULATION OF NERVE CELLS

Steroid hormone modulation of nerve cells serves five different functions. These are: (a) the feedback regulation of hypothalamic and pituitary hormone secretion, (b) the regulation of the metabolic rate of neural cells (general metabolic effects), (c) the modulation of emotional, motivational, sensory and behavioral changes, (d) the regulation of adaptive behaviors for coping with stress, and (e) the organization of neural pathways in the brain during perinatal development.

9.9.1 FEEDBACK REGULATION OF HYPOTHALAMIC AND PITUITARY HORMONE SECRETION

As discussed in Chapter 8, steroid hormones have feedback effects on hypothalamic and pituitary hormone release. Steroid hormone modulation of neurotransmitter release in the hypothalamus and other brain regions provides feedback regulation of the release of neurohormones from hypothalamic neurosecretory cells. Steroid hormone feedback can regulate neurohormone secretion by regulating the rate of synthesis, release, and degradation of the hypothalamic hormones, by altering the firing rate of neurons which control neurohormone release from the hypothalamus, and by regulating the number of receptor proteins in the cells of the pituitary gland, thus altering the sensitivity of these cells to the hypothalamic hormones.

The feedback effects of gonadal steroids on the hypothalamic and pituitary hormones are quite complex, as illustrated in Figures 6.6 (p. 96) and 8.8 (p. 138). There are, for example, sex differences in the pattern of GnRH and LH secretion and in the effects of gonadal steroid feedback on the release of these hormones. Adult males have a low-level pulsatile secretion of GnRH, LH and FSH and castration (i.e. removal of testosterone negative feedback) is followed by an increase in the pulsatile secretion of these hormones. Testosterone injections (adding negative feedback) inhibits the release of the gonadotropic hormones, as shown in Figure 8.10 (p. 141) (Moger, 1976). The most likely neural sites for the negative feedback action of testosterone are the medial preoptic area, the anterior hypothalamus and the arcuate nucleus and median eminence of the medial basal hypothalamus (Sar and Stumpf, 1973b).

Adult females have a cyclic release of GnRH, LH and FSH which regulates the reproductive cycle. In ovariectomized female rats, estrogen feedback can either increase or decrease the secretion of these hormones, depending on the dose and treatment schedule. Neurons of the medial basal hypothalamus control the baseline secretion of GnRH and neurons in the preoptic area (of the rat) or medial basal hypothalamus (in primates) provide the stimulus for the ovulatory surge of GnRH. Estrogen injection into the medial basal hypothalamus inhibits LH secretion while estrogen injection into the preoptic area in rats stimulates LH secretion, indicating that the preoptic area is the site of positive feedback of estrogen on GnRH, at least in rats (McEwen *et al.*, 1979). In primates, the site of estrogen positive feedback is the medial basal hypothalamus. In the anterior pituitary gland, estrogen increases the number GnRH receptors and stimulates LH release in females, while testosterone inhibits GnRH and LH release in males. This sex difference is determined by androgens during perinatal development (see Section 9.9.5). Of course, other factors also influence the release of GnRH. During stress, for example, the increased release of glucocorticoids inhibits firing of the hypothalamic neurons responsible for the GnRH rhythm, thus reducing the amount of LH released.

The negative feedback of glucocorticoids on CRH and ACTH involves stimulation of type II receptors in the hypothalamus and anterior pituitary gland as well as type I receptors in the hippocampus, septum and amygdala (Dallman *et al.*, 1987). Glucocorticoids provide negative feedback on the stress-induced synthesis and release of vasopressin and CRH in the paraventricular nucleus and facilitate extinction of stress-induced behaviors via their action on type II receptors (McEwen *et al.*, 1986; DeKloet and Reul, 1987; DeKloet *et al.*, 1987). High glucocorticoid levels also cause down-regulation of glucocorticoid receptors in the hippocampus (DeKloet, 1984). Removal of the adrenal cortex by adrenalectomy is followed by hypersecretion of CRH and ACTH, which can be inhibited by corticosteroid injection.

9.9.2 REGULATION OF METABOLIC RATE

There are two types of metabolic change: anabolic and catabolic. Anabolism, or constructive metabolism, is the use of nutrients (e.g. amino acids, fats, carbohydrates) to build up body tissues. Catabolism, or destructive metabolism, is the breakdown of stored proteins, carbohydrates or fats. Whenever cellular activity increases, there is an increase in metabolic activity above the basal metabolic rate and steroid hormones are able to influence the metabolic rate of their target neurons. The gonadal steroids are anabolic, building up tissues, whereas the glucocorticoids are catabolic, breaking tissues down (except in the liver). Androgens, for example, promote conservation of sodium, chlorine, potassium and calcium and stimulate appetite, thus increasing protein synthesis, resulting in an increase in metabolic rate. Glucocorticoids, on the other hand, inhibit cellular uptake of amino acids and glucose, and thus inhibit protein synthesis, resulting in a decrease in metabolic rate. Thyroid hormones increase basal metabolic rate in their target cells as described in Section 9.10 (Martin, 1985).

9.9.3 MODULATION OF EMOTIONAL, MOTIVATIONAL, SENSORY AND BEHAVIORAL CHANGES

In adult animals, steroid hormones activate the neural pathways which coordinate sexual and parental behaviors as well as a wide range of non-sexual behaviors including locomotor activity, play, aggression, feeding, scent-marking, sensory perception, learning and memory (Beatty, 1979). Male sexual behavior and other male-specific behaviors such as scent-marking are regulated by androgen and estrogen receptors in the medial preoptic area and anterior hypothalamus (McEwen *et al.*, 1979). Female sexual behavior is activated by the actions of estrogen and progesterone in the ventromedial hypothalamus, medial preoptic area, anterior hypothalamus and midbrain central gray (McEwen, 1981).

Figure 9.17 shows a model for the neural and endocrine interactions in the control of sexual behavior in the female rat which can be briefly summarized as follows. The sensory input from the cutaneous stimulation of the somatosensory (touch) receptors on the female's flanks is carried to the spinal cord by the pudendal nerve. The spinal sensory nerves then stimulate the reticular formation in the medulla and the midbrain central gray area of the brain. Estrogen-activated neurons in the ventromedial hypothalamus send axons to the midbrain central gray area, the reticular formation and the medial geniculate body. When activated by the incoming tactile stimulation, these estrogen-primed neurons activate spinal motor neurons in the reticular formation. These spinal motor neurons contract the lateral longissimus dorsi and multifidus muscles, resulting in the lordosis reflex, the sexual posture of the female rat (Pfaff, 1978). If a male approaches a non-estrous female rat (i.e. a female with a low estrogen level), she is aggressive and does not show sexual behavior, but if she is primed with estrogen and progesterone, she responds to his touch by arching her back in the 'lordosis posture' and allowing him to mount.

Female parental behavior is elicited by estrogen acting at the medial preoptic area in the rat (Numan, 1987). The cyclic fluctuations of estrogen and progesterone during the female reproductive cycle also cause cyclic fluctuations in a wide range of behaviors, including aggression, locomotion, food preferences, emotional state, olfactory and visual perception, and cognition (Messent, 1976).

9.9.4 REGULATION OF ADAPTIVE BEHAVIORS FOR COPING WITH STRESS

Glucocorticoids are released in stressful situations, resulting in very high levels of these steroids in the circulation (see Section 9.7). The increased

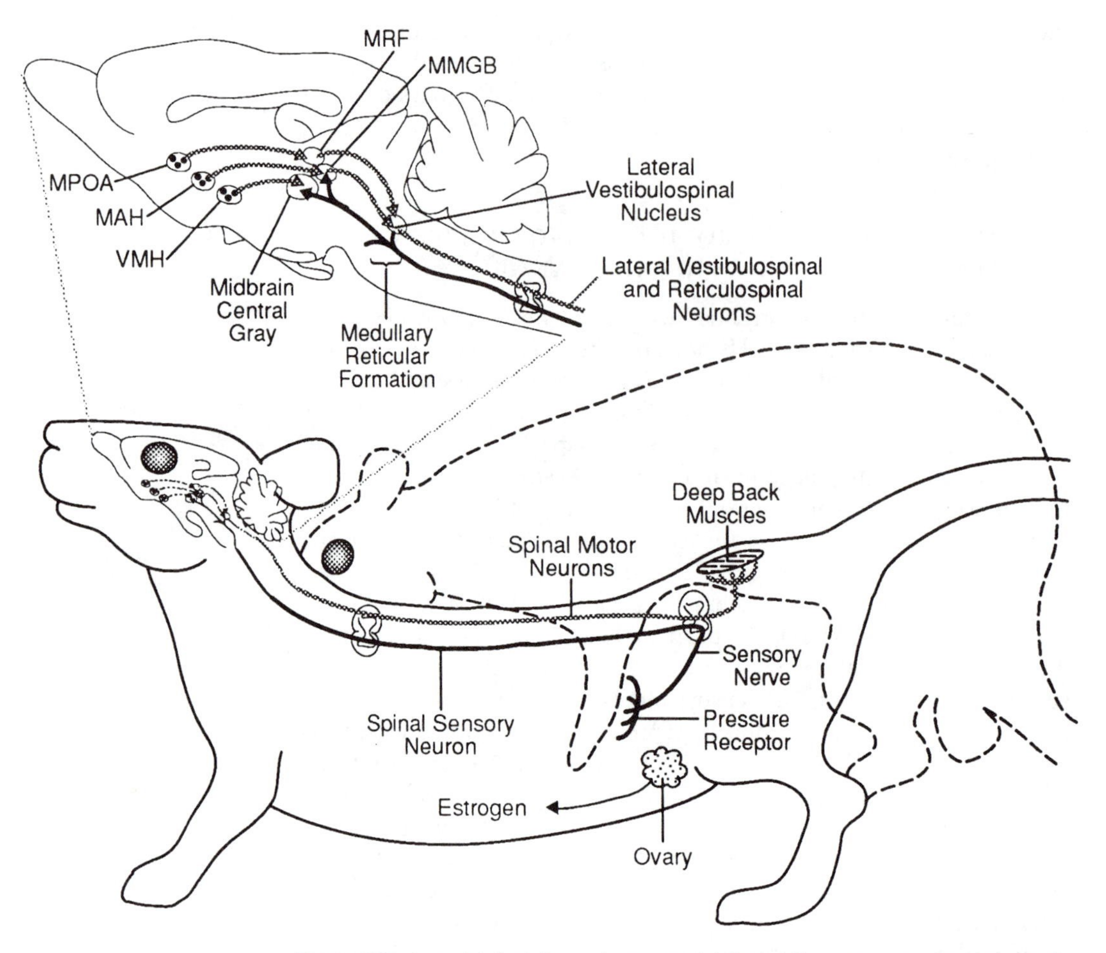

Figure 9.17. A model depicting estrogen modulation of the sensory and motor signals regulating female sexual behavior (lordosis) in rats. The hypothalamic neurons are stimulated both by sensory inputs and the level of estrogen circulating in the blood. When the male stimulates the skin of the female's flanks or rump, the pressure receptors provide the sensory input to the dorsal root ganglion of the spinal cord. This information is then relayed via spinal nerves to the reticular formation in the medulla and the midbrain central gray area. In the presence of estrogen, neurons of the ventromedial hypothalamus (VMH) are activated and those of the medial preoptic area (MPOA) are inhibited. Axons from the VMH descend to the midbrain central gray area, the midbrain reticular formation (MRF) and the medial (magnocellular) region of the medial geniculate body (MMGB) (the peripeduncular nucleus, PPN). The motor neurons for lordosis (the lateral vestibulospinal and the reticulospinal neurons) are then activated. These spinal motor neurons stimulate the lateral longissimus dorsi and the transverso spinalis (multifindus) muscles of the lumbar region which cause dorsiflexion of the rump and tailbase, resulting in the lordosis response. MAH = medial anterior hypothalamus. (Redrawn from Pfaff, 1989.)

glucocorticoid levels during stress provide negative feedback to the brain and immune system and inhibit the over-activity of the stress response, thus protecting the body from its own endocrine and immune responses (Munck, Guyre and Holbrook, 1984). The role of the glucocorticoids in regulating immune responses to stress is described in Section 13.6.2.

The neural responses to glucocorticoids provide behavioral as well as neuroendocrine defenses against excessive stress responses. Glucocorticoids influence learning and memory in aversive situations, such as taste aversion and passive avoidance conditioning and this may be mediated through modulation of noradrenaline and serotonin release in cells with type II glucocorticoid receptors (McEwen *et al.*, 1986; DeKloet and Reul, 1987). Corticosteroids also facilitate exploration of a novel environment, possibly through their action on hippocampal type I glucocorticoid receptors.

Corticosteroids influence mood, sleep, sensory perception, and stress-related behaviors by modulating the release of neurotransmitters such as noradrenaline and serotonin (McEwen, 1981; Stumpf and Sar, 1975a). Prolonged stress-induced hypersecretion of glucocorticoids, however, suppresses neural activity and may cause neural degeneration (cell loss) due to the inhibition of glucose uptake and protein synthesis and the depletion of energy reserves in glucocorticoid target cells (McEwen *et al.*, 1986). Prolonged glucocorticoid secretion may also lead to stress-related diseases such as depression and other affective disorders by altering dopamine levels at glucocorticoid sensitive cells in the hippocampus, thus inhibiting their neural functioning (McEwen *et al.*, 1986; McEwen, 1988). High levels of glucocorticoids may also accelerate brain aging and inhibit mental performance (DeKloet and Reul, 1987). It is, therefore, essential for normal emotional and cognitive functioning to have an efficient mechanism for regulating the adrenocortical stress response. Chronic stress may interfere with this adaptive response (McEwen *et al.*, 1986).

9.9.5 ORGANIZATION OF NEURAL PATHWAYS IN THE BRAIN DURING PERINATAL DEVELOPMENT

During prenatal and early postnatal development, steroid hormones influence the growth and differentiation of nerve cells, the formation of synapses and the synthesis of hormone and neurotransmitter receptors, resulting in permanent changes in the neural structure of the brain (Balazs, Patel and Hajos, 1975; Arnold and Breedlove, 1985; Lauder and Krebs, 1986). During this early organizational period, testosterone, acting through aromatization and reduction pathways, stimulates cell growth and differentiation in the preoptic area, ventromedial hypothalamus, and amygdala, alters synaptic organization in the arcuate nucleus and the amygdala, and influences dendritic branching in the preoptic area of the hypothalamus (Toran-Allerand, 1978; Naftolin *et al.*, 1990). Estrogen receptors mediate the organization of neural development in both males and females and influence feminization through target cells in the hypothalamus, preoptic area and cerebral cortex (Toran-Allerand, 1978). High levels of estrogen receptors are found in the cerebral cortex during the first 2 weeks of postnatal life and estrogen receptors occur in the cerebellum of rats and mice from 22–27 days of age (Shughrue *et al.*, 1990).

Hippocampal glucocorticoid receptors are low in number on the first day of life in rats and reach adult levels at about 4 weeks of age (McEwen *et al.*, 1986). This allows for the normal growth of the brain as high levels of glucocorticoids inhibit neural development. Neonatal glucocorticoid injections, for example, inhibit neural growth and synapse formation in the cerebral cortex, resulting in delayed neural maturation (Meyer, 1985). Thus, although some responses to stressful stimuli may be evoked in newborn animals, the neuroendocrine stress response in rats is not fully developed before 14 days of age, nor is the feedback regulation of CRH and ACTH release. Down-regulation of glucocorticoid receptors in response to high corticosteroid levels begins to develop around 21 days of age, after which the adult negative feedback effects are observed.

The effects of steroid hormones on target cells in the brain (i.e. the stimulation of electrical activity, the altered release of neurotransmitters and the regulation of protein synthesis, as described in Section 9.8) are thus the same, whether hormonal stimulation occurs in adulthood or during fetal development. But, the results of steroid hormone activation of nerve cells during development, when the hormones can permanently alter the pattern of nerve cell growth and differentiation, differ from the results of hormone stimulation in adulthood, when neural development is complete. The organizational effects of steroid hormones during development and their activational effects in adulthood thus involve many of the same neural processes and the age distinction is one of degree (see Arnold and Breedlove, 1985; Naftolin *et al.*, 1990).

9.10 THYROID HORMONE RECEPTORS

Recent evidence indicates that the thyroid hormone receptors belong to the same superfamily as the steroid hormone receptors (Thompson *et al.*, 1987). Thyroid hormone target cells are found in the anterior pituitary, the liver, kidney, heart, and throughout the brain (Oppenheimer, 1979). The highest density of thyroid hormone receptors in the brain occurs in the cerebellum and cerebral cortex, but significant numbers of thyroid hormone receptors also occur in the corpus striatum, hippocampus, amygdala, the ventromedial nucleus and arcuate nucleus of the medial basal hypothalamus, the brain stem and spinal cord (Stumpf and Sar, 1975b; Schwartz and Oppenheimer, 1978; Kaplan *et al.*, 1981; Leonard *et al.*, 1981).

9.10.1 BINDING OF THYROID HORMONES TO THEIR RECEPTORS

As depicted in Figure 9.18, the thyroid hormone receptor is located on a non-histone protein which is attached to a segment of DNA in the cell nucleus (Oppenheimer, 1979, 1985). When activated by binding to T3 at an acceptor site, this receptor opens an initiation site on the DNA and the enzyme RNA polymerase transcribes the genetic information from the DNA to mRNA. The activation of the thyroid hormone target cell involves four steps: (a) the uptake of the thyroid hormones from the plasma; (b) the conversion of thyroxine (T4) to its active derivative, triiodothyronine (T3) by the enzyme 5′-deiodinase; (c) the binding of T3 to its nuclear receptor; and (d) the activation of specific genes by the hormone–receptor

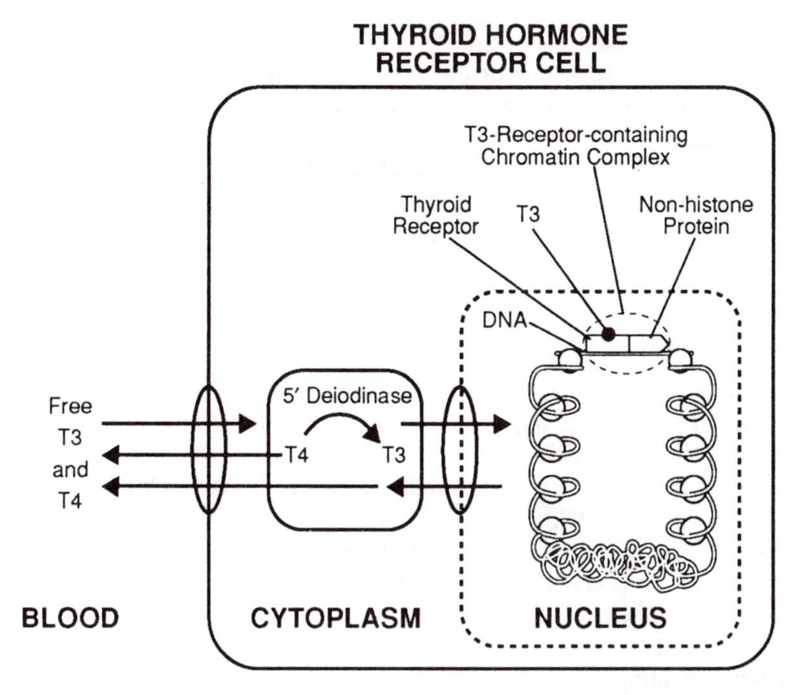

Figure 9.18. The action of the thyroid hormones on their intracellular receptors. Both T3 and T4 circulate in the bloodstream and are able to enter the target cells. Once in the target cell, T4 is converted to T3 by the enzyme 5′ deiodinase. The T3 then enters the nucleus and binds to the receptor at the acceptor site on the non-histone protein of the chromatin. This opens an initiation site on the DNA where genomic information is transcribed to mRNA. (Redrawn from Oppenheimer, 1985.)

complex, resulting in mRNA production and protein synthesis (Nunez, 1984).

The chain of events which takes the thyroid hormone from the bloodstream to the nuclear receptor is shown in Figure 9.18. In the bloodstream, T3 and T4 are bound to thyronine binding globulin (TBG) and must become free hormones before they can enter their target cells (Oppenheimer, 1985). After T4 enters the target cell, it is converted to T3, which is bound more readily to the nuclear receptors than is T4. This conversion depends on the enzyme 5′-deiodinase, which is regulated by the level of thyroid hormones in the circulation (Leonard *et al.*, 1981; Martin, 1985). Once T3 binds to the nuclear receptor, mRNA synthesis occurs and protein synthesis is initiated (Oppenheimer, 1985).

9.10.2 ACTION OF THYROID HORMONES AT TARGET CELLS

Thyroid hormones have many and diverse effects on their target cells. They influence metabolic rate, respiration rate, oxygen consumption, the metabolism of foods, the synthesis of a number of enzymes and provide negative feedback regulation of TRH and TSH secretion (Oppenheimer, 1979). Thyroid hormones act directly on the thyrotrophs of the anterior pituitary to suppress TSH secretion (Larsen, 1982), but the method of TRH regulation is less clear. TRH release may be mediated directly through the actions of thyroid hormones at their receptors in the neurosecretory cells of the paraventricular nucleus (PVN) or indirectly through the modulation of catecholamine pathways which synapse at the PVN (Segerson *et al.*, 1987). Thyroid hormones also regulate many aspects of growth and differentiation and are essential for stimulating the maturation of the brain, heart and lungs during prenatal and early postnatal development. The fetal brain has more T3 receptors than the adult brain and thyroid hormones are essential in infancy for neural growth and synapse formation (synaptogenesis), the synthesis of neurotransmitters and their receptors, the normal development of behavior and the ontogeny of learning (Ford and Cramer, 1977; Murphy and Nagy, 1976; Nunez, 1984; Dussault and Ruel, 1987).

9.11 SUMMARY

Steroid and thyroid hormone receptors belong to a superfamily of nuclear receptor proteins. The target cells which possess these nuclear receptors are identified by autoradiography and by immunocytochemistry. Steroid and thyroid hormones will accumulate in the nucleus of their target cells and will activate mRNA and protein synthesis in these target cells, but not in non-target cells. The amount of protein synthesis in the target cells is directly proportional to the number of hormone–receptor complexes stimulated. The number of nuclear hormone receptors in a cell is not constant and can be regulated by changes in hormone levels, resulting in down-regulation or up-regulation of receptor numbers. Estrogen, progesterone, androgens, glucocorticoids, mineralocorticoids and thyroid hormones all have target cells in the brain and regulate specific neural activity. Estrogen priming induces progesterone receptors and activates female sexual and parental behavior, while androgens stimulate male sexual behavior, aggression and other behaviors. Androgens can act by direct action on androgen receptors, by reduction to 5α-DHT and activation of 5α-DHT receptors, or by aromatization to estrogen and activation of estrogen receptors. Estrogen and progesterone receptors often occur in the same neural areas, as do estrogen and androgen receptors. There are two types of corticosteroid receptor: type I, which bind to both mineralocorticoids and glucocorticoids; and type II, which bind only to glucocorticoids. The neural locations and functions of these two receptors differ, with type II receptors more widely distributed in the brain than are type I receptors and type II receptors involved in more severe stress responses. Types I and II receptors both occur in the hippocampus, where they act to inhibit the activation of the stress response. Steroid hormones are able to alter electrophysiological activity of neural cells, regulate the release of neurotransmitters, and influence protein synthesis. The ability of steroid hormones to alter rapidly the electrophysiological activity of nerve cells has led to the suggestion that nerve cells may release steroids and that these 'neurosteroids' may act on membrane receptors to modulate ion channels in the nerve cell membrane. As well as affecting the electrical activity and release of neurotransmitters from nerve cells, steroid hormones modulate the synthesis and release of hypothalamic and pituitary hormones, regulate the number of receptors for neurotransmitters and hormones, and regulate the metabolic rate of their target cells. Steroid and thyroid hormones also modulate emotional, motivational and behavioral changes, regulate adaptive responses to stress, and organize neural development.

FURTHER READING

Evans, R. M. (1988). The steroid and thyroid hormone receptor superfamily. *Science*, **240**, 889–895.

Harrison, R. W. and Lippman, S. S. (1989). How steroid hormones work. *Hospital Practice*, **24**(9), 63–76.

McEwen, B. S. (1988). Glucocorticoid receptors in the brain. *Hospital Practice*, **23**, 107–121.

McEwen, B. S., Davis, P., Parsons, B. and Pfaff, D. (1979). The brain as a target for steroid hormone action. *Annual Review of Neuroscience*, **2**, 65–112.

Oppenheimer, J. H. (1985). Thyroid hormone action at the nuclear level. *Annals of Internal Medicine*, **102**, 374–384.

Simerly, R.B., Chang, C., Muramatsu, M. and Swanson, L.W. (1990). Distribution of androgen and estrogen receptor mRNA-containing cells in the rat brain: An in situ hybridization study. *Journal of Comparative Neurology*, **294**, 76–95.

REVIEW QUESTIONS

9.1 What is the main difference between a target cell for a hormone and a non-target cell?

9.2 What is the difference between an acceptor site and an initiation site?

9.3 What is the aromatization hypothesis?

9.4 List three ways that testosterone can stimulate target cells, i.e. three types of receptors that can be activated

9.5 Which of the two corticosteroid receptors bind to aldosterone?

9.6 What two inputs interact to regulate the activity of a hormone-sensitive neuron?

9.7 Estrogen stimulates an increase in the number of muscarinic cholinergic receptors in the anterior hypothalamus and lateral ventromedial nucleus. True or false?

9.8 Give four mechanisms through which steroid hormone feedback can act to regulate neurohormone secretion.

9.9 Which nucleus of the hypothalamus stimulates female sexual behavior when activated by estrogen in rats?

9.10 Which thryoid hormone activates nuclear receptors in receptor cells?

9.11 What effect does the prolonged exposure of target cells to a high concentration of hormone have on the receptors for that hormone?

ESSAY QUESTIONS

9.1 Describe the 'superfamily' of nuclear receptors for steroid and thyroid hormones and discuss the evidence that these are in the nucleus and not in the cytoplasm.

9.2 Discuss the factors which influence up-regulation and down-regulation of steroid hormone receptors.

9.3 Discuss the importance of the aromatization of testosterone to estrogen for the activation of male sexual behavior.

9.4 Discuss the role of steroid hormones in the organization of neural pathways during perinatal development.

9.5 Discuss the distinction between type I and type II adrenal steroid receptors with respect to their anatomical distribution and function.

9.6 How do steroid hormones stimulate (or inhibit) the electrical activity of nerve cells if they do not act via synaptic receptors?

9.7 Explain how estrogen and progesterone interact at target cells, i.e. how does estrogen 'prime' a target cell and what effect does this have on the target cell's response to progesterone?

9.8 Explain the role of estrogen in facilitating the sexual behavior of the female rat (i.e. explain Figure 9.17).

9.9 Discuss the action of thyroid hormones at target cells.

REFERENCES

Adkins-Regan, E. (1981). Hormone specificity, androgen metabolism, and social behavior. *American Zoologist*, **21**, 257–271.

Agarwal, M. K. (1983). General considerations in steroid hormone receptor research. In M. K. Agarwal (ed.) *Principles of Receptorology*, pp. 1–67. Berlin: Walter de Gruyter.

Alberts, B., Bray, D., Lewis, J., Raff, M., Roberts, K. and Watson, J. D. (1989). *Molecular Biology of the Cell*. (2nd edn). New York: Garland.

Arnold, A. P. and Breedlove, S. M. (1985). Organizational and activational effects of sex steroids on brain and behavior: a reanalysis. *Hormones and Behavior*, **19**, 469–498.

Arriza, J. L., Simerly, R. B., Swanson, L. W. and Evans, R. M. (1988). The neural mineralocorticoid receptor as a mediator of glucocorticoid response. *Neuron*, **1**, 887–900.

Balazs, R., Patel, A. J. and Hajos, F. (1975). Factors affecting the biochemical maturation of the brain: effects of hormones during early life. *Psychoneuroendocrinology*, **1**, 25–36.

Barfield, R. J., Glaser, J. H., Rubin, B. S. and Etgen, A. M. (1984). Behavioral effects of progestin in the brain. *Psychoneuroendocrinology*, **9**, 217–231.

Beato, M. (1989). Gene regulation by steroid hormones. *Cell*, **56**, 335–344.

Beatty, W. W. (1979). Gonadal hormones and sex differences in nonreproductive behaviors in rodents: organizational and activational influences. *Hormones and Behavior*, **12**, 112–163.

Biegen, A., Rainbow, T. C. and McEwen, B. S. (1985). Corticosterone modulation of neurotransmitter receptors in rat hippocampus: A quantitative autoradiographic study. *Brain Research*, **332**, 309–314.

Blaustein, J. D. (1986). Steroid receptors and hormone action in the brain. *Annals of the New York Academy of Sciences*, **474**, 400–415.

Bueno, J. and Pfaff, D. W. (1976). Single unit recording in hypothalamus and preoptic area of estrogen-treated and untreated ovariectomized female rats. *Brain Research*, **101**, 67–78.

Callard, G. V. (1983). Androgen and estrogen actions in the vertebrate brain. *American Zoologist*, **23**, 607–620.

Callard, G. V., Petro, Z. and Ryan, K. J. (1978). Conversion of androgen to estrogen and other steroids in the vertebrate brain. *American Zoologist*, **18**, 511–523.

Carson-Jurica, M. A., Schrader, W. T. and O'Malley, B. W. (1990). Steroid receptor family: Structure and functions. *Endocrine Review*, **11**, 201–220.

Chang, C., Kokontis, J. and Liao, S. (1988). Structural analysis of complementary DNA and amino acid sequences of human and rat androgen receptors. *Proceedings of the National Academy of Sciences, USA*, **85**, 7211–7215.

Crowley, W. R. and Zemlen, F. P. (1981). The neurochemical control of mating behavior. In N. T. Adler (ed.) *Neuroendocrinology of Reproduction*, pp. 451–484. New York: Plenum.

Dallman, M. F., Akana, S. F., Jacobson, L., Levin, N., Cascio, C. S. and Shinsako, J. (1987). Characterization of corticosterone feedback regulation of ACTH secretion. *Annals of the New York Academy of Sciences*, **512**, 402–414.

de Kloet, E. R. (1984). Adrenal steroids as modulators of nerve cell function. *Journal of Steroid Biochemistry*, **20**, 175–181.

de Kloet, E. R. and Reul, J. M. H. M. (1987). Feedback action and tonic influence of corticosteroids on brain function: A concept arising from the heterogeneity of brain receptor systems. *Psychoneuroendocrinology*, **12**, 83–105.

de Kloet, E. R., Ratka, A., Reul, J. M. H. M., Sutanto, W., and Van Eekelen, J. A. M. (1987). Corticosteroid receptor types in brain: Regulation and putative function. *Annals of the New York Academy of Sciences*, **512**, 351–361.

Delville, Y. (1991). Progesterone-facilitated sexual receptivity: a review of arguments supporting a nongenomic mechanism. *Neuroscience and Biobehavioral Reviews*, **15**, 407–414.

Dohanich, G. P., Witcher, J. A., Weaver, D. R. and Clemens, L. G. (1982). Alteration of muscarinic binding in specific brain areas following estrogen treatment. *Brain Research*, **241**, 347–350.

Dussault, J. H. and Ruel, J. (1987). Thyroid hormones and brain development. *Annual Review of Physiology*, **49**, 321–334.

Evans, R. M. (1988). The steroid and thyroid hormone receptor superfamily. *Science*, **240**, 889–895.

Evans, R. M. and Arriza, J. L. (1989). A molecular framework for the actions of glucocorticoid hormones in the nervous system. *Neuron*, **2**, 1105–1112.

Ford, D. H. and Cramer, E. B. (1977). Developing nervous system in relation to thyroid hormones. In G. D. Grave (ed.) *Thyroid hormones and brain development*, pp. 1–17. New York: Raven Press.

Funder, J. W. (1991). Corticosteroid receptors in the brain. In M. Motta (ed.) *Brain Endocrinology*, 2nd edn. pp. 133–151. New York: Raven Press.

Funder, J. W., Pearce, P. T., Smith, R. and Smith, A. I. (1988). Mineralocorticoid action: target tissue specificity is enzyme, not receptor, mediated. *Science*, **242**, 583–585.

Funder, J. W. and Sheppard, K. (1987). Adrenocortical steroids and the brain. *Annual Review of Physiology*, **49**, 397–411.

Gorski, J., Toft, D., Shyamala, G., Smith, D. and Notides, A. (1968). Hormone receptors: studies on the interaction of estrogen with the uterus. *Recent Progress in Hormone Research*, **24**, 45–80.

Gorski, J., Welshons, W. and Sakai, D. (1984). Remodeling the estrogen receptor model. *Molecular and Cellular Endocrinology*, **36**, 11–15.

Grant, L. D. and Stumpf, W. E. (1975). Hormone uptake sites in relation to

CNS biogenic amine systems. In W. E. Stumpf and L. D. Grant (eds.) *Anatomical Neuroendocrinology*, pp. 445–463. Basel: Karger.

Greenspan, F. S. (1991). Basic and Clinical Endocrinology, 3rd edn. Norwalk, CN: Appleton and Lange.

Hadley, M. E. (1992). Endocrinology, 3rd edn. Englewood Cliffs, NJ: Prentice-Hall.

Harrison, R. W. and Lippman, S. S. (1989). How steroid hormones work. *Hospital Practice*, **24**(9), 63–76.

Heiman, M. L. and Ben-Jonathan, N. (1982). Dopaminergic receptors in the rat anterior pituitary change during the estrous cycle. *Endocrinology*, **111**, 37–41.

Hua, S.-Y. and Chen, Y.-Z. (1989). Membrane receptor-mediated electrophysiological effects of glucocorticoid on mammalian neurons. *Endocrinology*, **124**, 687–691.

Joels, M. and de Kloet, E. R. (1989). Effects of glucocorticoids and norepinephrine on the excitability in the hippocampus. *Science*, **245**, 1502–1505.

Jung-Testas, I., Hu, Z. Y., Baulieu, E. E. and Robel, P. (1989). Neurosteroids: Biosynthesis of pregnenolone and progesterone in primary cultures of rat glial cells. *Endocrinology*, **125**, 2083–2091.

Kaplan, M. M., McCann, U. D., Yaskoski, K. A., Larsen, P. R. and Leonard, J. L. (1981). Anatomical distribution of phenolic and tyrosol ring iodothyronine deiodinase in the nervous system of normal and hypothyroid rats. *Endocrinology*, **109**, 397–402.

Keefer, D. and Holderegger, C. (1985). The ontogeny of estrogen receptors: Brain and pituitary. *Developmental Brain Research*, **19**, 183–194.

Kim, Y. S., Stumpf, W. E., Sar, M., Martinez-Vargas, M. C. (1978). Estrogen and androgen target cells in the brain of fishes, reptiles and birds: Phylogeny and ontogeny. *American Zoologist*, **18**, 425–433.

King, W. J. and Greene, G. L. (1984). Monoclonal antibodies localize oestrogen receptor in the nuclei of target cells. *Nature*, **307**, 745–747.

LaFond, R. E., Kennedy, S. W., Harrison, R. W. and Villee, C. A. (1988). Immunocytochemical localization of glucocorticoid receptors in cells, cytoplasts, and nucleoplasts. *Experimental Cell Research*, **175**, 52–62.

Larsen, P. R. (1982). Thyroid–pituitary interaction: Feedback regulation of thyrotropin secretion by thyroid hormones. *New England Journal of Medicine*, **306**, 23–32.

Lauder, J. M. and Krebs, H. (1986). Do neurotransmitters, neurohumors, and hormones specify critical periods? In W. T. Greenough and J. M. Juraska (eds.) *Developmental Neuropsychobiology*, pp. 119–174. Orlando, FL: Academic Press.

Leonard, J. L., Kaplan, M. M., Visser, T. J., Silva, J. E. and Larsen, P. R. (1981). Cerebral cortex responds rapidly to thyroid hormones. *Science*, **214**, 571–573.

Lieberburg, I. and McEwen, B. S. (1977). Brain cell nuclear retention of testosterone metabolites, 5 α-dihydrotestosterone and estradiol-17β, in adult rats. *Endocrinology*, **100**, 588–597.

Lieberman, S. and Prasad, V. V. K. (1990). Heterodox notions on pathways of steroidogenesis. *Endocrine Reviews*, **11**, 469–493.

Lincoln, D. W. (1969). Effects of progesterone on the electrical activity of the forebrain. *Journal of Endocrinology*, **45**, 585–596.

Luttge, W. G., Hall, N. R. and Wallis, C. J. (1974). Studies on the neuroendocrine, somatic, and behavioral effectiveness of testosterone and its 5α reduced metabolites in Swiss Webster mice. *Physiology and Behavior*, **13**, 553–561.

MacLusky, N. J. and Naftolin, F. (1981). Sexual differentiation of the central nervous system. *Science*, **211**, 1294–1302.

Majewska, M. D., Harrison, N. L., Schwartz, R. D., Barker, J. L. and Paul, S. M. (1986). Steroid hormone metabolites are barbiturate-like modulators of the GABA receptor. *Science*, **232**, 1004–1007.

Majewska, M. D., Mienville, J.-M. and Vicini, S. (1988). Neurosteroid pregenenolone sulfate antagonizes electrophysiological responses to GABA in neurons. *Neuroscience Letters*, **90**, 279–284.

Martin, C. R. (1985). *Endocrine Physiology*. New York: Oxford University Press.

McEwen, B. S. (1976). Interactions between hormones and nerve tissue. *Scientific American*, **235** (July), 48–58.

McEwen, B. S. (1981). Neural gonadal steroid actions. *Science*, **211**, 1303–1311.

McEwen, B. S. (1988). Glucocorticoid receptors in the brain. *Hospital Practice*, **23** (15 August), 107–121.

McEwen, B. S., Davis, P. G., Parsons, B. and Pfaff, D. W. (1979). The brain as a target for steroid hormone action. *Annual Review of Neuroscience*, **2**, 65–112.

McEwen, B. S., de Kloet, E. R. and Rostene, W. (1986). Adrenal steroid receptors and actions in the nervous system. *Physiological Reviews*, **66**, 1121–1188.

McEwen, B. S., Krey, L. C. and Luine, V. N. (1978). Steroid hormone action in the neuroendocrine system: When is the genome involved? In S. Reichlin, R. J. Baldessarini and J. B. Martin (eds.) *The Hypothalamus*, pp. 255–268. New York: Raven Press.

McEwen, B. S. and Pfaff, D. W. (1985). Hormone effects on hypothalamic neurons: Analysing gene expression and neuromodulator action. *Trends in Neuroscience*, **8**, 105–110.

McEwen, B. S., Weiss, J. M. and Schwartz, L. S. (1969). Uptake of corticosterone by rat brain and its concentration by certain limbic structures. *Brain Research*, **16**, 227–241.

Messent, P. R. (1976). Female hormones and behaviour. In B. Lloyd and J. Archer (eds.) *Exploring Sex Differences*, pp. 185–212. New York: Academic Press.

Meyer, J. S. (1985). Biochemical effects of corticosteroids on neural tissues. *Physiological Reviews*, **65**, 946–1012.

Miesfeld, R. L. (1990). Molecular genetics of corticosteroid action. *American Review of Respiratory Disease*, **141**, S11–S17.

Moger, W. H. (1976). Effect of testosterone implants on serum gonadotropin concentrations in the male rat. *Biology of Reproduction*, **14**, 665–669.

Mooradian, A. D., Morley, J. E. and Korenman, S. G. (1987). Biological actions of androgens. *Endocrine Reviews*, **8**, 1–28.

Morrell, J. I. and Pfaff, D. W. (1978). A neuroendocrine approach to brain function: Localization of sex steroid concentrating cells in vertebrate brains. *American Zoologist*, **18**, 447–460.

Morrell, J. I. and Pfaff, D. W. (1981). Autoradiographic technique for steroid hormone localization: Application to the vertebrate brain. In N. T. Adler (ed.) *Neuroendocrinology of Reproduction*, pp. 519–532. Plenum: New York.

Munck, A., Guyre, P. M. and Holbrook, N. J. (1984). Physiological functions of glucocorticoids in stress and their relation to pharmacological actions. *Endocrine Reviews*, **5**, 25–44.

Munck, A., Mendel, D. B., Smith, L. I. and Orti, E. (1990). Glucocorticoid receptors and actions. *American Review of Respiratory Disease*, **141**, S2–S10.

Murphy, J. M. and Nagy, Z. M. (1976). Neonatal thyroxine stimulation accelerates the maturation of both locomotor and memory processes in mice. *Journal of Comparative and Physiological Psychology*, **90**, 1082–1091.

Naftolin, F., Garcia-Segura, L. M., Keefe, D., Leranth, C., Maclusky, N. J. and Brawer, J. R. (1990). Estrogen effects on the synaptology and neural membranes of the rat hypothalamic arucate nucleus. *Biology of Reproduction*, **42**, 21–28.

Numan, M. (1987). Preoptic area neural circuitry relevant to maternal behavior in the rat. In N. A. Krasnegor, E. M. Blass, M. A. Hofer and W. P. Smotherman (eds.) *Perinatal Development: A Psychobiological Perspective*, pp. 275–298. Orlando, FL: Academic Press.

Nunez, J. (1984). Effects of thyroid hormones during brain differentiation. *Molecular and Cellular Endocrinology*, **37**, 125–132.

Olster, D. H. and Blaustein, J. D. (1990). Biochemical and immunocytochemical assessment of neural progestin receptors following estradiol treatments that eliminate the sex difference in progesterone-facilitated lordosis in guinea-pigs. *Journal of Neuroendocrinology*, **2**, 79–86.

Oppenheimer, J. H. (1979). Thyroid hormone action at the cellular level. *Science*, **203**, 971–979.

Oppenheimer, J. H. (1985). Thyroid hormone action at the nuclear level. *Annals of Internal Medicine*, **102**, 374–384.

Orsini, J.-C. (1981). Hypothalamic neurons responsive to increased plasma level of testosterone in the male rat. *Brain Research*, **212**, 489–493.

Packan, D. R. and Sapolsky, R. M. (1990). Glucocorticoid endangerment of the hippocampus: Tissue, steroid and receptor specificity. *Neuroendocrinology*, **51**, 613–618.

Parker, M. G. (1988). The expanding family of nuclear hormone receptors. *Journal of Endocrinology*, **119**, 175–177.

Pfaff, D. W. (1978). Peptide and steroid hormones and the neural mechanisms for female reproductive behavior. In S. Reichlin, R. J. Baldessarini and J. B. Martin (eds.) *The Hypothalamus*, pp. 245–253. New York: Raven Press.

Pfaff, D. W. (1981). Electrophysiological effects of steroid hormones in brain tissue. In N. T. Adler (ed.) *Neuroendocrinology of Reproduction*, pp. 533–544. Plenum: New York.

Pfaff, D. W. (1989). Features of a hormone-driven defined circuit for a model of animal behavior. *Annals of the New York Academy of Sciences*, 563, 131–142.

Pfaff, D. W. and Keiner, M. (1973). Atlas of estradiol-concentrating cells in the central nervous system of the female rat. *Journal of Comparative Neurology*, **151**, 121–158.

Pfaff, D. W. and McEwen, B. S. (1983). Actions of estrogens and progestins on nerve cells. *Science*, **219**, 808–814.

Pfaff, D. W., Silva, M. T. A. and Weiss, J. M. (1971). Telemetered recording of hormone effects on hippocampal neurons. *Science*, **172**, 394–395.

Raam, S., Lauretano, A. R., Vrabel, D. M., Pappas, C. A. and Tamura, H. (1988). Nuclear location of hormone-free estrogen receptors by monoclonal antibodies could be a tissue-fixation dependent artifact. *Steroids*, **51**, 425–439.

Rainbow, T. C., de Groff, V., Luine, V. N. and McEwen, B. S. (1980). Estradiol 17β increases the number of muscarinic receptors in hypothalamic nuclei. *Brain Research*, **198**, 239–243.

Rainbow, T. C., McGinnis, M. Y., Davis, P. and McEwen, B. S. (1982). Application of anisomycin to the lateral ventromedial nucleus of the hypothalamus inhibits the activation of sexual behavior by estradiol and progesterone. *Brain Research*, **233**, 417–423.

Rainbow, T. C., Snyder, L., Berck, D. J. and McEwen, B. S. (1984). Correlation of muscarinic receptor induction in the ventromedial hypothalamic nucleus with the activation of feminine sexual behavior by estradiol. *Neuroendocrinology*, **39**, 476–480.

Robel, P. Akwa, Y. *et al.* (1991). Neurosteroids: biosynthesis and function of pregnenolone and dehydroepiandrosterone in the brain. In M. Motta (ed.) *Brain Endocrinology*, 2nd edn, pp. 105–132. New York: Raven Press.

Roth, G. S. (1979). Hormone receptor changes during adulthood and senescence: Significance for aging research. *Federation Proceedings*, **38**, 1910–1914.

Sar, M. and Stumpf, W. E. (1973a). Neurons of the hypothalamus concentrate [^{3}H] progesterone or its metabolytes. *Science*, **182**, 1266–1268.

Sar, M. and Stumpf, W. E. (1973b). Autoradiographic localization of radioactivity in the rat brain after the injection of 1,2-^{3}H-testosterone. *Endocrinology*, **92**, 251–256.

Sar, M. and Stumpf, W. E. (1975). Distribution of androgen-concentrating neurons in rat brain. In W. E. Stumpf and L. D. Grant (eds.) *Anatomical Neuroendocrinology*, pp. 120–133. Basel: Karger.

Schrader, W. T., Birnbaumer, M. E., Hughes, M. R., Weigel, N. L., Grody, W. W. and O'Malley, B. W. (1981). Studies on the structure and function of the chicken progesterone receptor. *Recent Progress in Hormone Research*, **37**, 583–633.

Schwartz, H. L. and Oppenheimer, J. H. (1978). Nuclear triiodothyronine receptor sites in brain: Probable identity with hepatic receptors and regional distribution. *Endocrinology*, **103**, 267–272.

Segerson, T. P., Kauer, J., Wolfe, H. C., Mobtaker, H., Wu, P., Jackson, I. M. D. and Lechan, R. M. (1987). Thyroid hormone regulates TRH biosynthesis in the paraventricular nucleus of the rat hypothalamus. *Science*, **238**, 78–80.

Selmanoff, M. K., Brodkin, L. D., Weiner, R. I. and Siiteri, P. K. (1977). Aromatization and 5α reduction of androgens in discrete hypothalamic and limbic regions of the male and female rat. *Endocrinology*, **101**, 841–848.

Sheridan, P. J. (1979). The nucleus interstitialis striae terminalis and the nucleus amygdaloideus medialis: prime targets for androgen in the rat forebrain. *Endocrinology*, **104**, 130–136.

Sheridan, P. J. (1983). Androgen receptors in the brain: what are we measuring? *Endocrine Reviews*, **4**, 171–178.

Shughrue, P. J., Stumpf, W. E., MacLusky, N. J., Zielinski, J. E. and Hochberg, R. B. (1990). Developmental changes in estrogen receptors in mouse cerebral cortex between birth and postweaning: studied by autoradiography with 11β-Methoxy-16α-[^{125}I] iodoestradiol. *Endocrinology*, **126**, 1112–1124.

Simerly, R. B., Chang, C., Muramatsu, M. and Swanson, L. W. (1990). Distribution of androgen and estrogen receptor mRNA-containing cells in the rat brain: an *in situ* hybridization study. *Journal of Comparative Neurology*, **294**, 76–95.

Steiner, F. A. (1975). Electrophysiological mapping of brain sites sensitive to corticosteroids, ACTH, and hypothalamic releasing hormones. In W. E. Stumpf and L. D. Grant (eds.) *Anatomical Neuroendocrinology*, pp. 270–275. Basel: Karger.

Stumpf, W. E. (1990). Steroid hormones and the cardiovascular system: direct actions of estradiol, progesterone, testosterone, gluco- and mineralocorticoids, and soltriol (vitamin D) on central nervous regulator and peripheral tissues. *Experientia*, **46**, 13–25.

Stumpf, W. E. and Sar, M. (1975a). Anatomical distribution of corticosterone-concentrating neurons in rat brain. In W. E. Stumpf and L. D. Grant (eds.) *Anatomical Neuroendocrinology*, pp. 254–261. Basel: Karger.

Stumpf, W. E. and Sar, M. (1975b). Localization of thyroid hormone in the mature rat brain and pituitary. In W. E. Stumpf and L. D. Grant (eds.) *Anatomical Neuroendocrinology*, pp. 318–327. Basel: Karger.

Stumpf, W. E., Sar, M. and Keefer, D. A. (1975). Atlas of estrogen target cells in rat brain. In W. E. Stumpf and L. D. Grant (eds.) *Anatomical Neuroendocrinology*, pp. 104–119. Basel: Karger.

Szego, C. M. (1984). Mechanisms of hormone action: Parallels in receptor-mediated signal propagation for steroid and peptide effectors. *Life Sciences*, **35**, 2383–2396.

Thompson, C. C., Weinberger, C., Lebo, R. and Evans, R. M. (1987). Identification of a novel thyroid hormone receptor expressed in the mammalian central nervous system. *Science*, **237**, 1610–1614.

Toran-Allerand, C. D. (1978). Gonadal hormones and brain differentiation: Cellular aspects of sexual differentiation. *American Zoologist*, **18**, 553–565.

Towle, A. C. and Sze, P. Y. (1983). Steroid binding to synaptic plasma membrane: Differential binding of glucocorticoids and gonadal steroids. *Journal of Steroid Biochemistry*, **18**, 135–143.

Wallis, C. J. and Luttge, W. G. (1980). Influence of estrogen and progesterone on glutamic acid decarboxylase activity in discrete regions of rat brain. *Journal of Neurochemistry*, **34**, 609–613.

Walters, M. R. (1985). Steroid hormone receptors and the nucleus. *Endocrine Reviews*, **6**, 512–536.

Weinberger, C. and Bradley, D. J. (1990). Gene regulation by receptors binding lipid-soluble substances. *Annual Review of Physiology*, **52**, 823–840.

Zigmond, R. E., Nottebohm, F. and Pfaff, D. W. (1973). Androgen-concentrating cells in the midbrain of a songbird. *Science*, **179**, 1005–1007.

10

Receptors for peptide hormones, neuropeptides and neurotransmitters

Peptide hormones, neuropeptides, neurotransmitters and other non-steroid chemical messengers stimulate biochemical activity by binding to receptors on the plasma membranes of their target cells. These chemical messengers cannot enter their target cells to stimulate the nucleus in the manner described for steroid and thyroid hormones in Chapter 9. In order to activate biochemical changes within the target cell, they act as first messengers to activate a second messenger, such as cyclic adenosine monophosphate (cyclic AMP), within the cytoplasm of the target cell. The transduction of information from the first to the second messenger is accomplished through the activation of membrane protein transducers (G-proteins) and enzymes, such as adenylate cyclase. This chapter examines the membrane receptors for peptide hormones and neurotransmitters, the mechanisms by which signal transduction across the cell membrane occurs, the role of G-proteins in this signal transduction, the second messenger systems activated, and the actions of the second messengers in the target cells, with special emphasis on neural target cells

10.1 MEMBRANE RECEPTORS

The membrane receptors are complex glycoproteins embedded in the cell membrane. The function of these receptors is to recognize specific ligands in the blood (e.g. peptide hormones, neuropeptides) or in the synapse (e.g. neurotransmitters) and bind to them. Once this binding occurs, signal transduction across the cell membrane occurs as described in Section 10.2 below. Using a number of pharmacological and biochemical techniques employing radioactively labeled ligands, fluorescent dyes, affinity chromatography and immunochemical identification, it is possible to discover the location and structure of the membrane receptors (see Limbird, 1986; Yamamura, Enna and Kuhar, 1990).

Gene cloning and sequencing methods have enabled molecular biologists to develop three-dimensional models of some membrane receptors and four such models are illustrated in Figure 10.1. There are three distinct superfamilies of membrane receptors: ligand-gated ion channel receptors, transmembrane-regulated enzyme receptors, and guanine nucleotide binding protein (G-protein) coupled receptors (Hollenberg, 1991). All of these types of membrane receptor have three regions: an extracellular ligand binding domain, which protrudes from the cell surface and binds to the peptide, neurotransmitter or other ligand molecule; a transmembrane domain, which passes through the bilipid membrane of the cell; and an intracellular domain, which activates the second messenger system. Each of these three domains may differ in complexity as shown by the examples in Figure 10.1. The structure of the receptors for many hypothalamic hormones, pituitary hormones, and other peptides are described by Posner (1985) and Cooke, King, and van der Molin (1988). One of the primary goals of receptor research is to correlate the three-dimensional structure of these receptors with their function (Hollenberg, 1991).

Ligand-gated ion channel receptors. Some receptors (termed ionotropic receptors in Section 5.5) are ligand-gated ion channels. These include the nicotinic acetylcholine receptor, the GABA/benzodiazepine receptor, and the glutamate receptor. The five subunits of the nicotinic receptor, for example, form a transmitter-gated ion channel. One model of this receptor is shown in Figure 10.1A (Popot and Changeux, 1984), but other models have been postulated for this receptor (see Guy and Hucho, 1987). The ligand-activated membrane receptors open ion channels through the cell membrane. Some ion channels open only briefly, allowing a rapid influx of ions which causes a transient change in membrane potential. This occurs, for example, when acetylcholine binds to the nicotinic receptor to open sodium channels in neurons, resulting in an action potential (see Section 5.4). Other ion channels may open for a longer duration, allowing a major influx of ions such as calcium, into the cell.

Transmembrane-regulated enzyme receptors. The receptors may be transmembrane-regulated enzymes, such as the tyrosine kinase

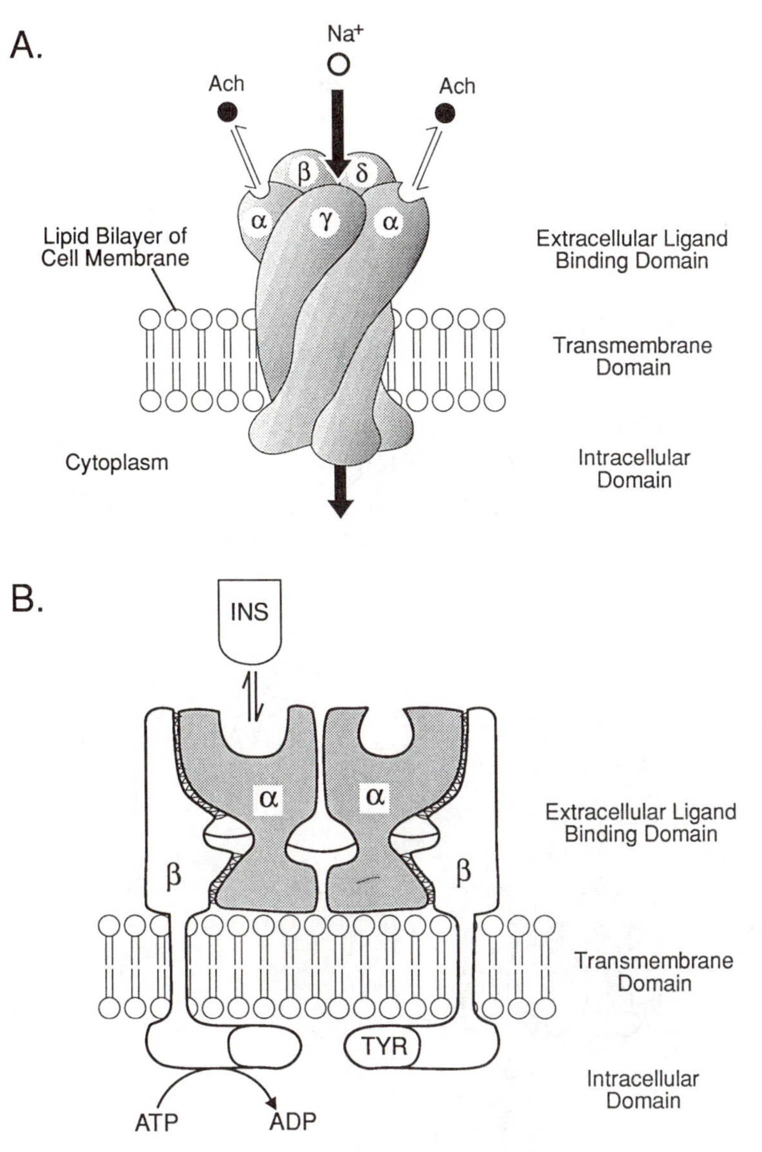

Figure 10.1. Examples of membrane receptors for acetylcholine, insulin, noradrenaline and luteinizing hormone, showing the extracellular ligand binding domains, the transmembrane domains, and the intracellular domains of these glycoproteins. (A) The nicotinic receptor for acetylcholine has five subunits (α, α, β, γ, and δ) which form a transmitter-gated ion channel. When acetylcholine (Ach) binds to the α subunits of the receptor, sodium (Na^+) channels are opened, allowing Na^+ ions into the cell and triggering an action potential. (Redrawn from Popot and Changeux, 1984.) (B) The insulin receptor has four subunits ($\beta\alpha\alpha\beta$). When insulin (INS) binds to the α subunit, the β subunit (tyrosine kinase (TYR)) is disinhibited, and causes phosphorylation of membrane proteins (ATP-ADP). (Redrawn from Anderson, 1989.)

receptors for insulin and epidermal growth factor or the guanylate cyclase-linked receptor for atrial natriuretic factor. The insulin receptor is composed of four subunits, as shown in Figure 10.1B. When there is no insulin bound to the α subunits, they inhibit the activity of the β subunits. When insulin binds to the α subunits, the β subunits are disinhibited and phosphorylate each other, causing an increase in tyrosine kinase activity in the cell membrane (Goldfine, 1987; Anderson, 1989). The elevation of tyrosine kinase leads to the transduction of the signal within the cell as described in Section 10.5.5.

Guanine nucleotide binding protein (G-protein) coupled receptors. Many receptors (termed metabotropic receptors in Section 5.5) are

(C) The β_2-adrenergic receptor has seven transmembrane domains. The ligand binding extracellular domain (N-terminal) binds to noradrenaline and adrenaline. The intracellular domain (C-terminal) activates intracellular activity. (Redrawn from Venter *et al.*, 1989.) (D) The luteinizing hormone (LH) receptor binds to chorionic gonadotropin (CG) as well as LH. The extensive extracellular domain may be necessary for binding these large peptides. The transmembrane domain has seven sections. (Redrawn from McFarland *et al.*, 1989.)

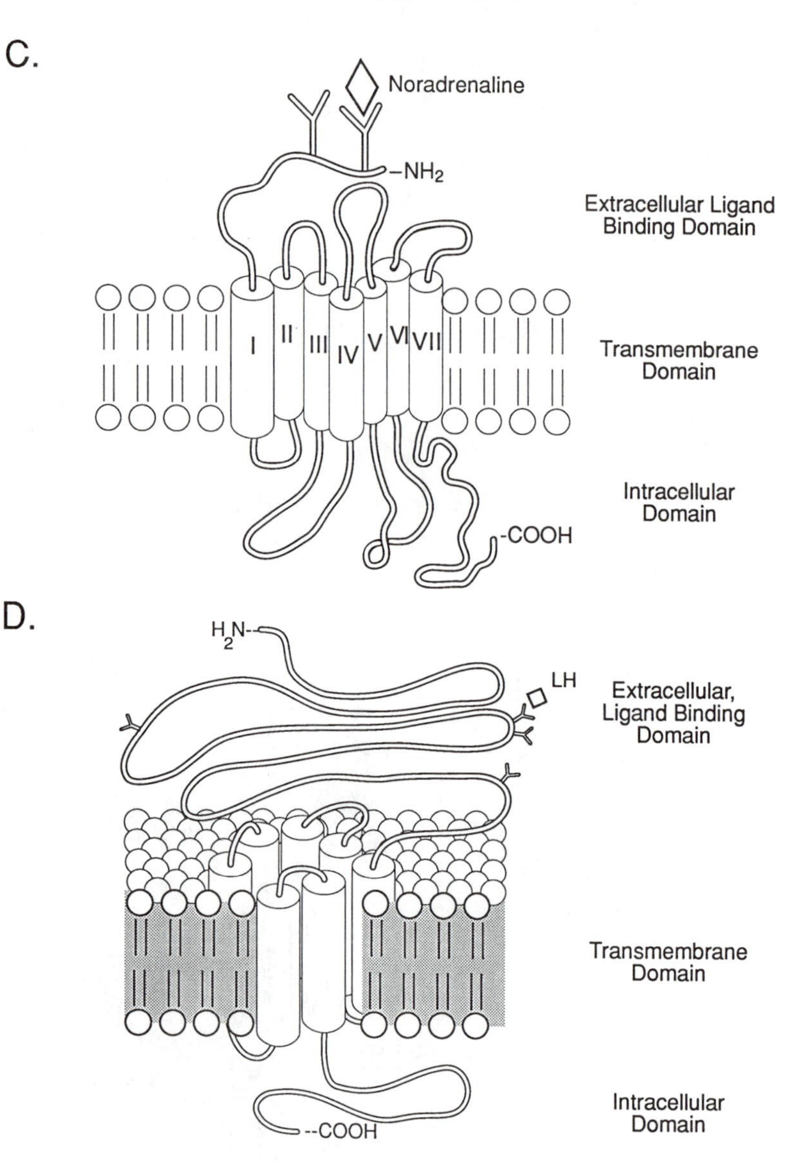

coupled to a G-protein in the cell membrane as occurs in the β_2- adrenergic receptor shown in Figure 10.1C (Ventner *et al.*, 1989), the LH receptor shown in Figure 10.1D (McFarland *et al.*, 1989; Gitelman, 1990), and the TSH receptor, which is very similar in structure to the LH receptor (Parmentier *et al.*, 1989). G-proteins are protein couplers which function in the transduction of the signal from the extracellular ligand–receptor complex to the intracellular second messenger system. They stimulate or inhibit the synthesis of second messengers, such as cyclic AMP, in the target cell by their action on enzymes, such as adenylate cyclase, in the cell membrane.

10.1.1 RECEPTOR REGULATION

The number of receptors on the cell membrane is not fixed, but can be up- or down-regulated in at least four ways. (a) Hormones can regulate the number of their own receptors (homospecific receptor regulation). (b) Hormones can regulate the receptors for other hormones (heterospecific receptor regulation). (c) Non-hormonal factors including viral agents or extracellular chemicals, such as sodium, can regulate receptor numbers. (d) Genetically 'programed' *de novo* receptor synthesis can occur at specific stages of cell differentiation during embryonic development (Hollenberg, 1979, 1981).

There are many examples of peptide hormones which show homospecific receptor regulation. Prolactin, for example, up-regulates its own receptors, while insulin, growth hormone, thyrotropin, and epidermal growth factor (EFG) down-regulate their own receptors (Hollenberg, 1981). Receptor regulation can occur at the genomic (nuclear) level, where the synthesis of new receptor proteins is initiated, or at the membrane level, where the sensitivity of the receptors can be modulated. More detailed information on the mechanisms of receptor regulation is given by Poste and Crooke (1985). The ability of neurotransmitters (Section 5.5) and steroid hormones (Section 9.5) to regulate receptors has already been discussed.

10.1.2 RECEPTOR INTERNALIZATION

Peptide hormones coupled to their receptors may be taken into the cytoplasm of the target cell by receptor-mediated endocytosis. Virtually all peptide hormones (for example, insulin, prolactin, parathyroid hormone, growth hormone, the gonadotropins and the growth factors) are internalized in their target cells by this process (Posner, Khan and Bergeron, 1985; Segaloff and Ascoli, 1988). Why peptides are taken into their target cells has not been determined. The most likely explanation is that the internalization of the peptide–receptor complex results in the deactivation of both the hormone and the receptor by the lysosomes in the cytoplasm. But other functions for receptor internalization are also possible. For example, the hormones and their receptors may be dissociated by internalization. The hormone could then be deactivated and the receptor recycled to the cell membrane. The internalized peptides and/or their receptors may also bind to intracellular receptors and stimulate long-term changes in the target cell, such as cell division or cell differentiation. This mechanism of action has been suggested for nerve growth factor (Johnson and Taniuchi, 1987), but there is little evidence for other hormones acting in this way (Hollenberg and Goren, 1985; Posner *et al.*, 1985; Segaloff and Ascoli, 1988).

10.2 SIGNAL TRANSDUCTION BY G-PROTEINS

The receptors for most of the neurotransmitters, peptide hormones, neuropeptides and other chemical messengers are coupled to G-proteins

Table 10.1. *Some examples of receptors which activate G-proteins, the types of G-protein activated and the target cells in which the second messengers are activated*

First messenger	Receptor	G-protein	Examples of target cells
Adrenaline	β_1-Adrenergic	G_s	Sympathetic synapses
Noradrenaline	β_2-Adrenergic	G_s	Liver, lung
Dopamine	D_1	G_s	Brain
	D_2	G_i	Pituitary lactotrophs (inhibition of prolactin secretion)
Acetylcholine	M_1 and M_2 muscarinic	G?	CNS
		G_s	Heart
ACTH	ACTH	G_s	Adrenal cortex (glucocorticoid secretion)
Opiates	Mu, kappa, delta	G_i	
LH	LH/GC	G_s	Ovaries and testes
TSH	TSH	G_s	Thyroid gland (thyroxine secretion)
CRH	CRH	G_s	Pituitary corticotrophs (ACTH secretion)
GH-RH	GH-RH	G_s	Pituitary somatotrophs (GH secretion)
GnRH	GnRH	G_{pla}	Pituitary gonadotrophs (FSH and LH secretion)
TRH	TRH	G_{plc}	Pituitary thyrotrophs and lactotrophs (TSH and prolactin secretion)
Somoatostatin	Somatostatin	G_i	Pituitary somatotrophs and thyrotrophs (inhibition of GH and TSH secretion)
Vasopressin	Vasopressin	G_{plc}	CNS
		G_i	Liver
		G_s	Kidney
Oxytocin	Oxytocin	G_{plc}	Uterus, CNS
Cholecystokinin	CCK	G_{plc}	Pancreas
Glucagon	Glucagon	G_s	Liver, heart
Neuropeptide Y	NPY	G_k	Heart
Cytokines (TNF, CSF-1, IL-1)	?	G?	Monocytes and lymphocytes

Notes:
G_s = stimulatory G-protein; G_i = inhibitory G-protein; G_{pla} = G-protein activating phospholipase A; G_{plc} = G-protein activating phospholipase C; G_k = G-protein opening K^+ channels.
Source: Birnbaumer and Brown, 1990.

(See Table 10.1 for some examples). As shown in Figure 10.2, these G-proteins are activated when the ligand binds to its membrane receptor. G-proteins stimulate or inhibit the synthesis of second messengers and are, therefore, important in the regulation of the physiological effects of these ligands on their target cells (Gilman, 1987; Birnbaumer *et al.*, 1988; Brown and Birnbaumer, 1989; Birnbaumer and Brown, 1990).

There are at least 12 different types of G-protein in the guanine nucleotide binding protein superfamily. Four of these stimulate the synthesis of second messengers (G_S) and three inhibit second messenger synthesis (G_i). The β-adrenergic and D_1 dopaminergic receptors, for example, are coupled to stimulatory G-proteins which activate cyclic AMP synthesis, while somatostatin and the D_2 dopaminergic receptors are coupled to inhibitory G-proteins which inhibit cyclic AMP synthesis

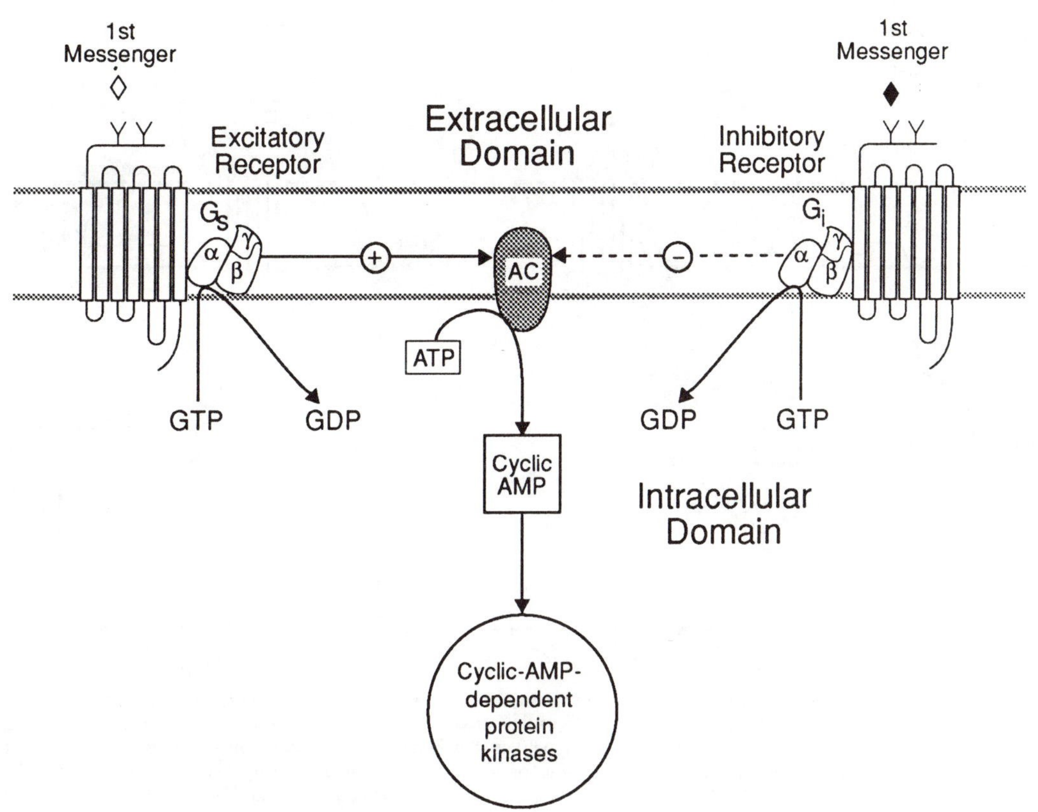

Figure 10.2. The role of G-proteins in the transduction of signals from the ligand-receptor first messenger complex to the second messenger system. The activation of the extracellular ligand binding domain by the peptide hormone, neurotransmitter, growth factor, cytokine or other first messengers can be excitatory or inhibitory, activating stimulatory (G_s) or inhibitory (G_i) G-proteins. The G-proteins comprise three subunits (α, β, and γ) and are activated when guanine diphosphate (GDP) on the α subunit is displaced by guanine triphosphate (GTP). When activated, the G-proteins stimulate (G_s) or inhibit (G_i) enzymes such as adenylate cyclase (AC) in the cell membrane, thus regulating the synthesis of second messengers such as cyclic AMP. (Redrawn from Spiegel, 1989.)

(Spiegel, 1989; Vallar and Meldolesi, 1989). Other G-proteins regulate calcium channels (G_o), potassium channels (G_k) and the phospholipase A (G_{pla}) and phospholipase C (G_{plc}) second messenger systems, which are discussed in Section 10.5 below. The G-proteins have three components (α, β, and γ) which dissociate to activate the enzymes necessary for second messenger synthesis and each of these components varies in structure and function (Johnson and Dhanasekaran, 1989; Taylor, 1990; Simon, Strathmann and Gautam, 1991).

One hormone can have different effects on different target cells if its receptors are coupled to different G-proteins in these target cells. Likewise, one target cell may have receptors for a number of different hormones, each of which activates a different G-protein and, through these G-proteins, different second messenger systems (Taylor, 1990).

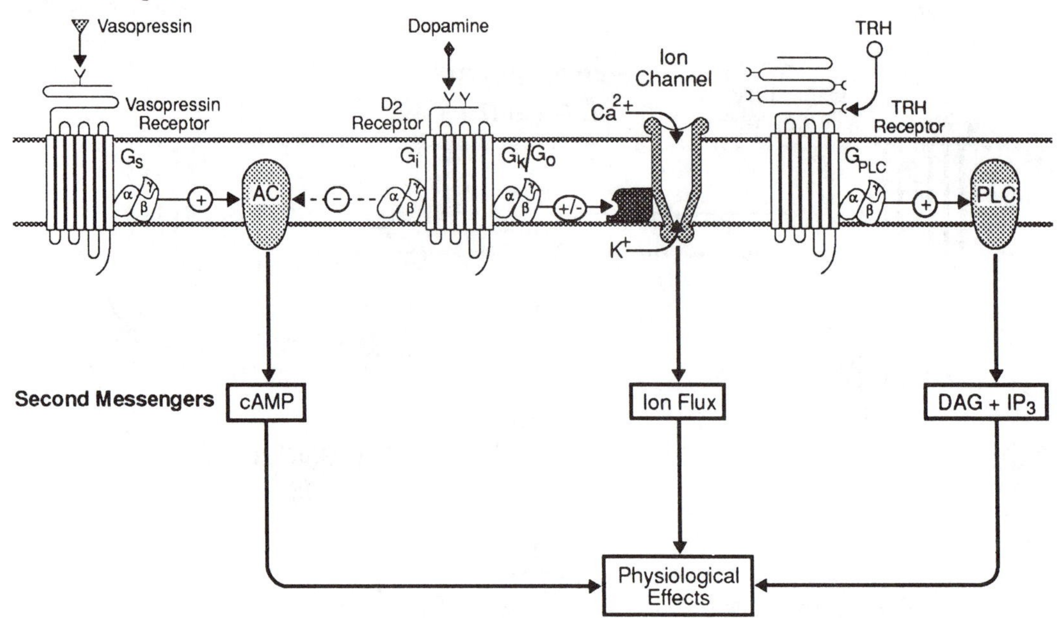

Figure 10.3. An example of how three different first messengers (vasopressin, dopamine and TRH) might act at a single target cell (such as a pituitary lactotroph cell) to regulate physiological changes (such as the synthesis and secretion of prolactin). Each first messenger binds to its receptor on the cell membrane and activates a specific G-protein (G_s, G_i, G_k, G_o or G_{plc}). G_s stimulates adenylate cyclase (AC) and thus stimulates cyclic AMP synthesis, while G_i inhibits adenylate cyclase and thus inhibits cyclic AMP synthesis. G_k and G_o regulate potassium (K^+) and calcium (Ca^{2+}) channels, thus altering the ion flux in the cell. G_{plc} activates phospholipase C (PLC) and the diacylglycerol (DAG) and inositol triphosphate (IP3) second messenger systems. These different second messenger systems then interact to regulate the physiological activity of the cell (and the synthesis and secretion of prolactin). (Redrawn from Spiegel, 1989.)

Figure 10.3 gives an example of how different G-proteins can mediate the intracellular responses of a single cell to three different first messengers. This example shows how prolactin synthesis and secretion may be regulated by vasopressin, dopamine and TRH.

As well as hormonal interactions at target cells, receptor regulation and the negative feedback effects of hormones may also be mediated by G-proteins. The steroid hormones, whose receptors are not coupled to G-proteins, are able to regulate the receptors for neurotransmitters and peptide hormones by regulating the G-proteins coupled to these receptors. Estrogen, for example, alters the sensitivity of dopamine receptors by regulating the amount of inhibitory G-proteins in the cell membrane (Enjalbert, 1989). Disorders of G-proteins may result in physiological disorders (Spiegel, 1989; Birnbaumer and Brown, 1990). Thus, understanding the properties of G-proteins is essential for understanding the mechanisms by which peptide hormones, neuropeptides, and neurotransmitters act on their target cells (see Hollenberg, 1985). The G-proteins are deactivated by the enzyme GTPase (Berridge, 1985b).

Table 10.2. *Some examples of second messenger systems activated by neurotransmitters, peptide hormones and other chemical messengers and their activity in the target cells*

First messenger	Target cell	Activity affected
Increased cyclic AMP		
Catecholamines	β-Adrenergic receptors of heart	Increase heart rate
Dopamine	D_1 receptors in brain	Neural activity
TRH	Pituitary thyrotroph cells	Secretion of TSH
Vasopressin	Kidney cells	Water resorption
TSH	Thyroid gland	Thyroid hormone secretion
ACTH	Adrenal cortex	Glucocorticoid secretion
MSH	Melanophores	Dispersion of pigment granules
LH	Interstitial cells of testes	Testosterone secretion
	Luteal cells of ovary	Progesterone secretion
FSH	Granulosa cells of ovary	Estrogen secretion
Decreased cyclic AMP		
Acetylcholine	Muscarinic receptors of heart	Reduce heart rate
Catecholamines	α_2-Adrenergic receptors of fat cells	Breakdown of triglycerides
Dopamine	D_2 receptors in pituitary lactotrophs	Inhibit prolactin secretion
Opiate peptides	Mu, kappa and sigma receptors	Reduce membrane permeability
Somatostatin	Somatostatin receptors on pituitary somatotrophs	Inhibit GH release
Cyclic GMP		
Atrial natriuretic peptide	ANP receptors	Regulation of blood pressure
Angiotensin II	Heart receptors	Regulates heart contraction
Ca^{2+} – DAG		
Acetylcholine	Muscarinic receptors	Smooth muscle contraction
Oxytocin	Breast	Milk ejection
	Uterus	Smooth muscle contraction
Vasopressin	Liver	Glycogen breakdown
TRH	Pituitary thyrotrophs	TSH release
GnRH	Pituitary gonadotrophs	LH and FSH release
Angiotensin II	Adrenal cortex	Release of aldosterone
Tyrosine kinase		
Insulin	Insulin receptors	Liver, muscle, fat
Epidermal growth factor	EGF receptors	Fibroblast cells

10.3 SECOND MESSENGER SYSTEMS

Once a neurotransmitter, peptide hormone, or neuropeptide binds to its receptor, the specific G-protein coupled to that receptor determines which of a number of intracellular second messenger systems will be activated within the target cell. The second messengers include cyclic AMP, cyclic GMP, calcium, inositol phospholipids, tyrosine kinase and arachidonic acid derivatives (Greengard, 1979; Berridge, 1985b). Examples of the second messenger systems activated by a variety of peptide hormones, neurotransmitters and neuropeptides are shown in Table 10.2. These second messengers stimulate biochemical changes in the cell by activating specific protein kinases (third messengers) which phosphorylate substrate proteins (see Figure 10.4). The phosphorylated substrate proteins (called phosphoproteins) then act as fourth messengers to regulate physiological changes in the cell (Hemmings *et al.*, 1986). One such phosphoprotein is synapsin 1, which is bound to synaptic vesicles in the

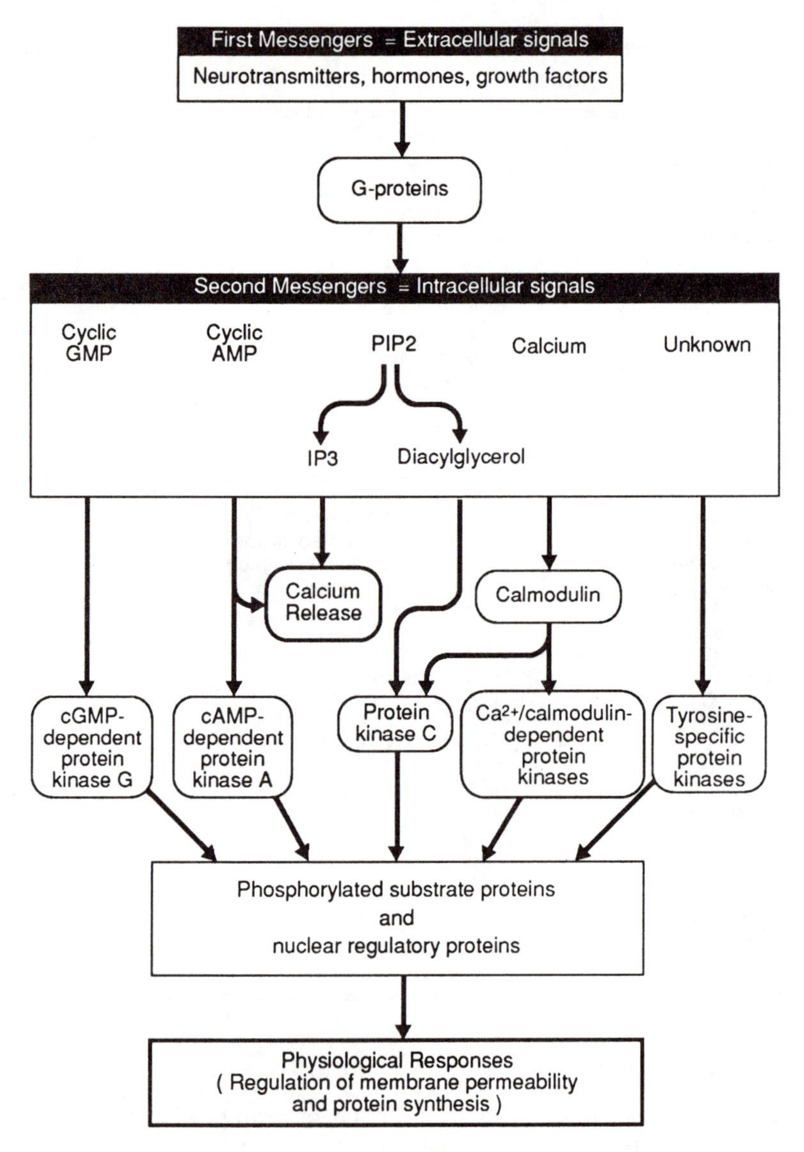

Figure 10.4. The cascade of first, second, third and fourth messengers which interact to alter physiological responses in the target cells of neurotransmitters, peptide hormones, growth factors and other first messengers. Once the first messengers activate their G-proteins, biochemical changes are produced in the target cells via a number of intracellular second messenger systems such as cyclic AMP, cyclic GMP, diacyl-glycerol, and calcium. These second messengers produce their actions through the activation of specific protein kinases, which are often referred to as third messengers. Calcium activates protein kinases in conjunction with calmodulin or diacylglycerol. The protein kinases phosphorylate substrate proteins which act as fourth messengers to stimulate physiological responses, such as changes in membrane permeability and protein synthesis in the cell. (Redrawn from Hemmings, Nairn and Greengard, 1986.)

nerve terminals of axons and functions to regulate neurotransmitter release. A second phosphoprotein is dopamine and cyclic AMP-regulated phosphoprotein (DARPP-32), which functions as a fourth messenger in neurons with D_1 dopaminergic receptors (see Nestler and Greengard, 1983; Hemmings *et al.*, 1986). Protein phosphorylation plays an important role in the regulation of the physiological processes associated with signal transduction because it provides a common pathway through which neurotransmitters, neuropeptides and non-steroid hormones produce their biological effects on their target cells, even though these ligands activate different receptors, G-proteins and second messengers (Hemmings *et al.*, 1989).

10.3.1 THE CYCLIC AMP SECOND MESSENGER SYSTEM

Cyclic adenosine monophosphate (cyclic AMP) was the earliest second messenger to be discovered and has been shown to regulate biochemical changes in a wide variety of target cells (Sutherland, 1972; Aurbach, 1982; Schramm and Selinger, 1984). Many first messengers act to stimulate cyclic AMP synthesis, while others inhibit cyclic AMP synthesis (see examples in Table 10.2). The stimulation of cyclic AMP synthesis occurs via the stimulatory G-proteins (G_s) and the inhibition of cyclic AMP synthesis occurs via the inhibitory G-proteins (G_i), as shown in Figures 10.2 and 10.3 (Gilman, 1984). The dopaminergic D_2 receptor is a good example of a receptor which inhibits cyclic AMP synthesis (Vallar and Meldosi, 1989).

Cyclic AMP has three primary actions once it is activated in a cell. (a) It alters the permeability of the cell membrane, allowing ions to pass in and out of the cell. (b) It alters the activity of enzymes by increasing their degree of phosphorylation. (c) It stimulates protein synthesis through its actions in the cell nucleus. To perform these functions, cyclic AMP activates protein kinase A, which phosphorylates a variety of structural proteins (Figure 10.4). Cyclic AMP may also release calcium from intracellular stores and this calcium then acts as an intracellular messenger (see Section 10.3.4).

The regulation of membrane permeability and protein synthesis by second messenger systems occurs through a 'cascade' of intracellular events. The steps involved in the cyclic AMP cascade are shown in Figure 10.5. When a ligand binds to its membrane receptor, the G-protein activates the adenylate cyclase enzyme which converts adenosine triphosphate (ATP) to cyclic AMP. In the cell membrane, cyclic AMP binds to the regulatory subunit of membrane protein kinase A, allowing the catalytic subunit to phosphorylate the substrate protein, opening an ion channel, and allowing ions to diffuse into or out of the cell.

In the cytoplasm, cyclic AMP binds to the regulatory subunit of soluble protein kinase A, allowing the catalytic subunit to move into the cell nucleus and phosphorylate the non-histone nuclear regulatory protein. The phosphorylation of the nuclear protein defines an acceptor site similar to that defined when steroid hormones bind to their receptors in the cell nucleus (see Figure 9.2, p. 151). The phosphorylated nuclear protein then dissociates from the DNA, allowing the enzyme RNA polymerase to initiate mRNA synthesis at an initiation site. The site on the DNA (i.e. a certain DNA sequence) which is regulated by cyclic AMP-activated protein kinase A has been termed the 'cyclic AMP response element' and the particular non-histone protein which acts as the acceptor site has been termed 'the cyclic AMP response element binding protein' (see Montminy, Gonzalez and Yamamoto, 1990).

When the ligand separates from its receptor, the G-protein is no longer activated, thus production of cyclic AMP stops. Any remaining cyclic AMP in the cytoplasm is deactivated by the enzyme phosphodiesterase, stopping intracellular activity. Another enzyme, phosphoprotein phos-

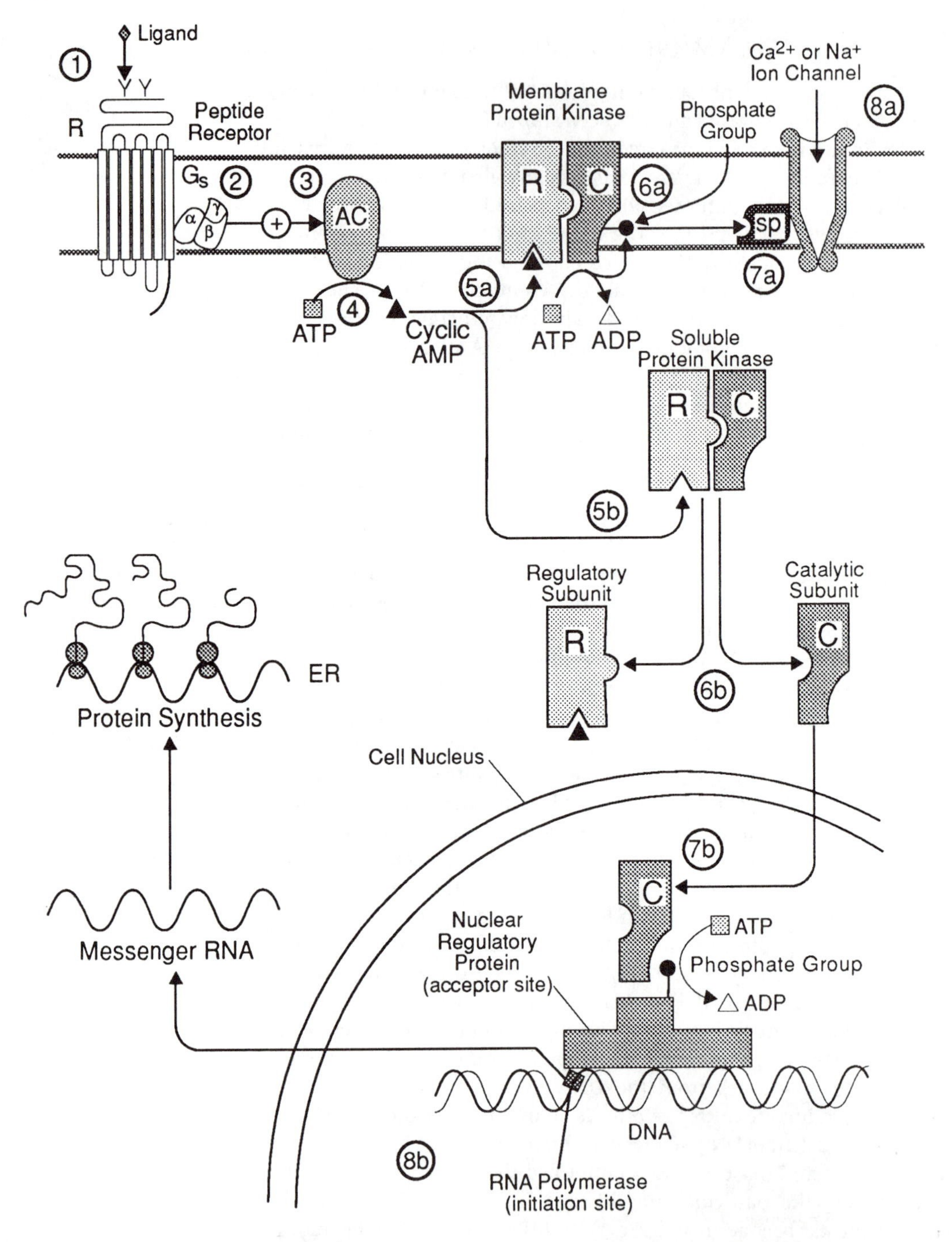

Figure 10.5. A model showing how the cyclic AMP second messenger cascade regulates membrane permeability and protein synthesis in a peptide hormone target cell. When the ligand binds to the receptor on the cell membrane (step 1), the G-protein (G_s) is stimulated (step 2) and activates the adenylate cyclase (AC) enzyme (3) which converts adenosine triphosphate (ATP) to cyclic adenosine monophosphate (cyclic AMP) by the removal of two phosphate ions (4). Once cyclic AMP is activated, two parallel series of events occur; one at the cell membrane, and one in the cytoplasm of the cell. Cyclic AMP binds to the regulatory (R) subunits of the membrane protein kinase (5a), and the soluble (cytoplasmic) protein

phatase, removes the phosphate groups from the substrate proteins in the cell membrane and from the nuclear proteins in the cell nucleus, closing the ion channels and stopping mRNA synthesis (Nathanson and Greengard, 1977; Nathanson, 1977).

10.3.2 THE CYCLIC GMP SECOND MESSENGER SYSTEM

Cyclic guanosine monophosphate (cyclic GMP) is much less prevalent than cyclic AMP, and is distributed unevenly throughout the body, with high levels in the lung, heart, intestine and smooth muscle (Hemmings *et al.*, 1986). Cyclic GMP acts as a second messenger for certain neurotransmitters and peptides (Table 10.2), as well as for the prostaglandins, atrial natriuretic factor and some growth factors (Nathanson, 1977; Berridge, 1985b; Martin, 1985; Garbers, 1989; Hadley, 1992). The activation of the cyclic GMP second messenger system occurs in the same way as described in Figure 10.5 for cyclic AMP. When a ligand binds to its receptor, the G-protein coupled to that receptor activates the enzyme guanylate cyclase, which converts guanosine triphosphate (GTP) to cyclic GMP. The cyclic GMP then activates protein kinase G which phosphorylates membrane or nuclear proteins. Alternatively, the guanylate cyclase can be the intracellular domain of a receptor as occurs in the atrial natriuretic factor receptor. Like cyclic AMP, cyclic GMP is deactivated by phosphodiesterase enzymes.

10.3.3 THE INOSITOL PHOSPHOLIPID (IP)–CALCIUM SECOND MESSENGER SYSTEM

Many peptides and neurotransmitters stimulate their target cells through the complex inositol phospholipid–calcium second messenger system, which is shown in Figure 10.6. For example, receptors in target cells of the anterior pituitary activate the inositol phospholipid second messenger system in response to the hypothalamic hormones. As shown in Table 10.2, the IP system is also involved in the secretion of aldosterone

kinase (5b) causing them to dissociate from the catalytic (C) subunits (6a and 6b). The activated catalytic (C) subunits in the cell membrane transfer a phosphate group from ATP to a substrate protein (SP) (7a). The addition of this phosphate group changes the conformation of the substrate protein, allowing ions to diffuse into or out of the cell (8a). The activated catalytic subunit (C) of the soluble protein kinase enters the cell nucleus, where it transfers a phosphate group from ATP to one of the non-histone nuclear regulatory proteins at an acceptor site (7b). Phosphorylation of this nuclear protein alters its binding properties, causing it to dissociate from the DNA double helix. This exposes an initiation site on the DNA where mRNA synthesis occurs through the action of the enzyme RNA polymerase (8b). The mRNA then enters the cytoplasm where protein synthesis occurs on the ribosomes of the endoplasmic reticulum (ER). Proteins synthesized in this manner include hormone and neurotransmitter receptors, structural proteins, adenylate cyclase, and enzymes involved in hormone and neurotransmitter synthesis and breakdown. When the ligand is released from its receptor, synthesis of cyclic AMP stops and any remaining cyclic AMP in the cytoplasm is deactivated by the enzyme phosphodiesterase. Another enzyme, phosphoprotein phosphatase, removes the phosphate groups from the substrate proteins, closing the ion channels and stopping the influx of ions into or out of the cell. Phosphoprotein phosphatase also removes the phosphate groups from the non-histone proteins in the nucleus and this stops mRNA synthesis. With no mRNA, protein synthesis also ceases. (Redrawn from Nathanson and Greengard, 1977 and Nathanson, 1977.)

from the adrenal cortex in response to angiotensin II stimulation, the response of α_1 adrenergic receptors to catecholamines, and the regulation of cellular growth and cell division in response to growth factors (Berridge, 1985a; Denef, 1988). When one of these ligands binds to its receptor, a specific G-protein (G_{plc}) is activated. This G_{plc}-protein is coupled to the enzyme phospholipase C in the cell membrane and, when activated, phospholipase C cleaves the membrane-bound lipid known as PIP2 (phosphatidylinositol phosphate) into two components: inositol triphosphate (IP3) and diacylglycerol (DAG), each of which has distinct second messenger functions (Guy and Kirk, 1988; Rasmussen, 1989; Hughes *et al.*, 1990). Diacyglycerol remains in the cell membrane and acts in conjunction with Ca^{2+} to activate protein kinase C (calcium-dependent protein kinase), which then phosphorylates the substrate proteins involved in a number of metabolic pathways. In the cell membrane, for example, it can phosphorylate peptide or neurotransmitter receptors, altering their affinity (and thus their sensitivity) to the peptides or neurotransmitters which bind to them (Hanley and Steiner, 1989). IP3 enters the cytoplasm of the cell and activates Ca^{2+} by releasing it from storage in the calcisomes (Figure 10.6). These Ca^{2+} ions then attach to their binding proteins (such as calmodulin) and act as intracellular messengers to activate protein kinase C (Berridge, 1985b; Rasmussen, 1989).

10.3.4 CALCIUM AS A SECOND MESSENGER

Calcium is involved in the second messenger systems for a large number of neurotransmitters, peptides and growth factors, as shown by the examples in Table 10.2 (Khanna *et al.*, 1988; Rasmussen and Barrett, 1988). Calcium is necessary for the action of cyclic AMP, diacylglycerol and other second messengers and it functions as a second messenger itself. Calcium carries out its intracellular action by binding to calcium binding proteins, such as calmodulin. As a second messenger, calcium is important in metabolism, smooth muscle contraction, neurotransmitter release, hormone secretion, and cell growth and proliferation. Noradrenaline, angiotensin II and a number of other ligands which activate the inositol phospholipid second messenger system stimulate the release of stored Ca^{2+} which then acts as a second messenger.

Most cellular functions mediated by cyclic AMP require the presence of calcium and there are many different ways in which calcium and cyclic AMP interact (see Martin, 1985). For example, calcium enhances adenylate cyclase activation of cyclic AMP synthesis following the binding of acetylcholine to cholinergic receptors in the adrenal medulla and following adrenaline stimulation of β-adrenergic receptors in the heart. ACTH stimulation of the adrenal cortex requires calcium bound calmodulin for adenylate cyclase activation of cyclic AMP in order to cause the secretion of glucocorticoids. In this case, Ca^{2+} is the second messenger and cyclic AMP is the third messenger (Martin, 1985). Thus, the biochemical activities of calcium and cyclic AMP often interact to regulate the same protein kinase enzyme (protein kinase A). More often,

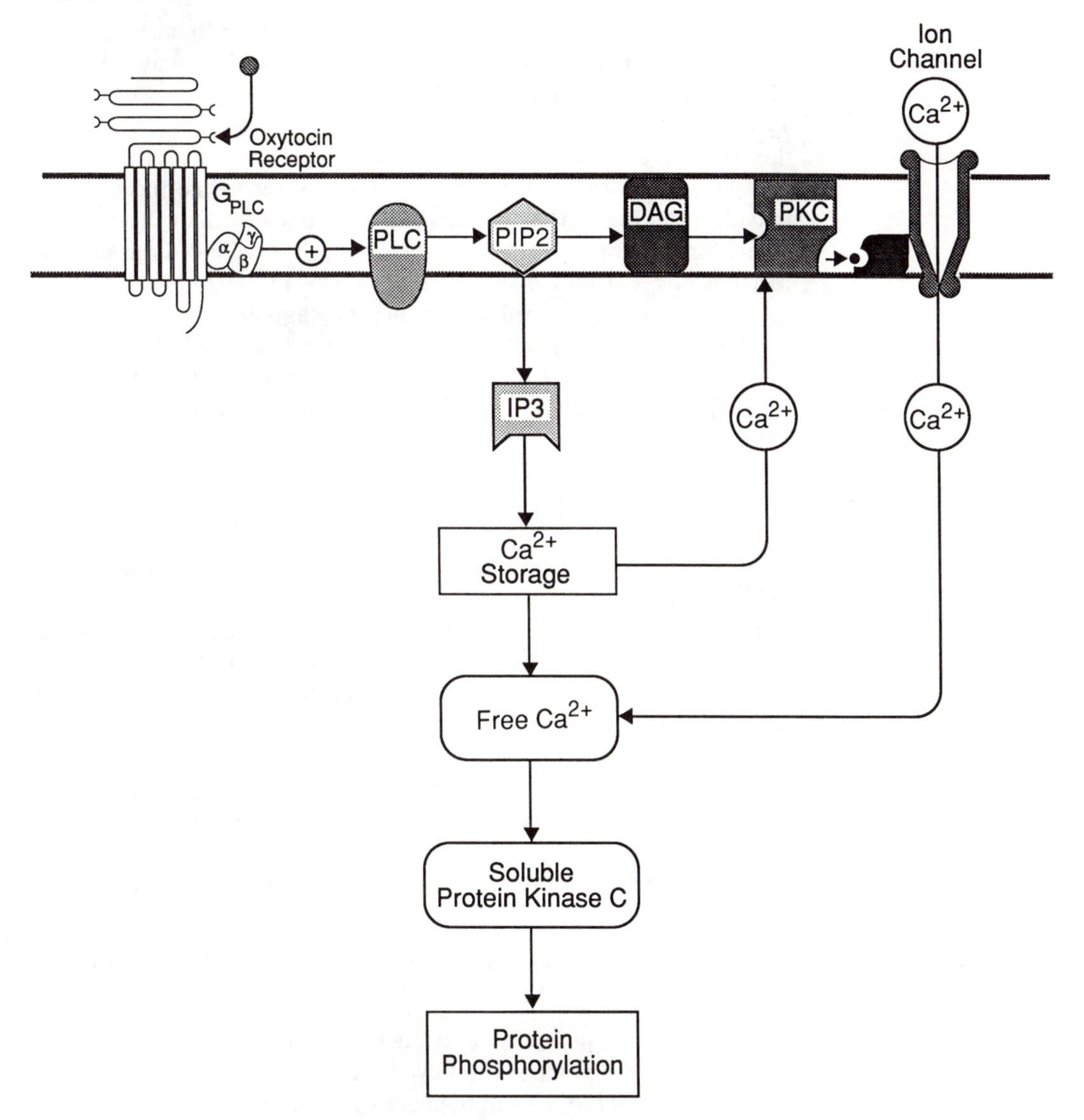

Figure 10.6. The activation of the inositol phospholipid (IP) second messenger system. When a ligand, such as oxytocin, binds to its receptor, a specific G-protein (G_{plc}) is activated. G_{plc} binds to substrate proteins to open calcium (Ca^{2+}) channels, allowing Ca^{2+} into the cell, and activates the enzyme phospholipase C (PLC) in the cell membrane. Phospholipase C converts phosphatidylinositol phosphate (PIP_2) to diacylglycerol (DAG) and inositol triphosphate (IP3). The diacylglycerol acts in the bilipid cell membrane in conjunction with Ca^{2+} to activate protein kinase C (PKC) which then phosphorylates substrate proteins (Sp) which regulate calcium channels. The inositol triphosphate enters the cytoplasm of the cell where it releases Ca^{2+} from storage in calcasomes. This free calcium then acts as a second messenger in the cell to activate soluble protein kinase C, which then phosphorylates nuclear proteins. (Redrawn from Berridge, 1985b; Guy and Kirk, 1988; and Khanna *et al.*, 1988.)

the calcium activated calmodulin activates its own specific protein kinase enzymes (e.g. calcium-calmodulin dependent protein kinase), as shown in Figure 10.4 (Cheung, 1982; Rasmussen, 1989).

The calcium second messenger system is more complex than those for cyclic AMP or cyclic GMP because of the complex mechanism involved in the release of stored calcium (see Figure 10.6) and the wide variety of different calcium binding proteins (Billingsley, Hanbauer and Kuhn, 1985; Khanna *et al.*, 1988; Rasmussen, 1989). Calcium is stored inside the cell in calcium stores (calcisomes) and is pumped out of the cell by a calcium pump. Thus, free Ca^{2+} levels are lower inside the cell than outside, resulting in a calcium imbalance across the cell membrane. The level of free (unbound) calcium inside the cell is regulated through two mechanisms. Calcium levels in the cell can be increased by opening the calcium channels in the cell membrane, allowing calcium to flow into the cell from the extracellular fluid, or by releasing stored calcium. The rise in free calcium is short-lasting, as it is rapidly pumped out of the cell or stored in the calcisomes (Carafoli and Penniston, 1985; Rasmussen, 1989).

10.3.5 TYROSINE KINASE SECOND MESSENGER SYSTEMS

Insulin, insulin-like growth factor (IGF), and epidermal growth factor (EGF) use tyrosine kinase as a second messenger (Goldfine, 1987; Hanley and Steiner, 1989). Insulin target cell tyrosine kinase activity is located in the β subunit of the insulin receptor (see Figure 10.1B). As well as tyrosine kinase, insulin activates the inositol triphosphate–calcium second messenger system (Goldfine, 1987) and may also activate the cyclic AMP second messenger system in some target cells. Because of the multitude of complex effects which insulin has on its target cells and the difficulty in finding a single second messenger which causes all of these effects, Houslay and Wakelam (1988) have suggested a 'multipathway' second messenger system for insulin.

10.3.6 ARACHIDONIC ACID DERIVATIVES AS SECOND MESSENGERS

Prostaglandins are produced from arachidonic acid in the cell membrane and stimulate the production of cyclic AMP or cyclic GMP. While the prostaglandins act in the cell membrane, other products of arachidonic acid act as second messengers within the cell (see Hadley, 1992).

10.4 INTERACTIONS AMONG SECOND MESSENGER SYSTEMS

Each of the second messenger systems described above may occur in the same target cell and these second messengers may interact to regulate cellular activity. As pointed out above, Ca^{2+} interacts with cyclic AMP, cyclic GMP and DAG to regulate protein kinase activity. In many cells, three or more second messenger systems may interact to regulate biochemical activity, especially if the cells have receptors for two or more hormones or neurotransmitters (Berridge, 1985b; Enjalbert, 1989; Hanley and Steiner, 1989). Figure 10.7 gives an example of how CRH and vasopressin activate different second messenger systems which interact to regulate ACTH synthesis and release in the corticotroph cells of the

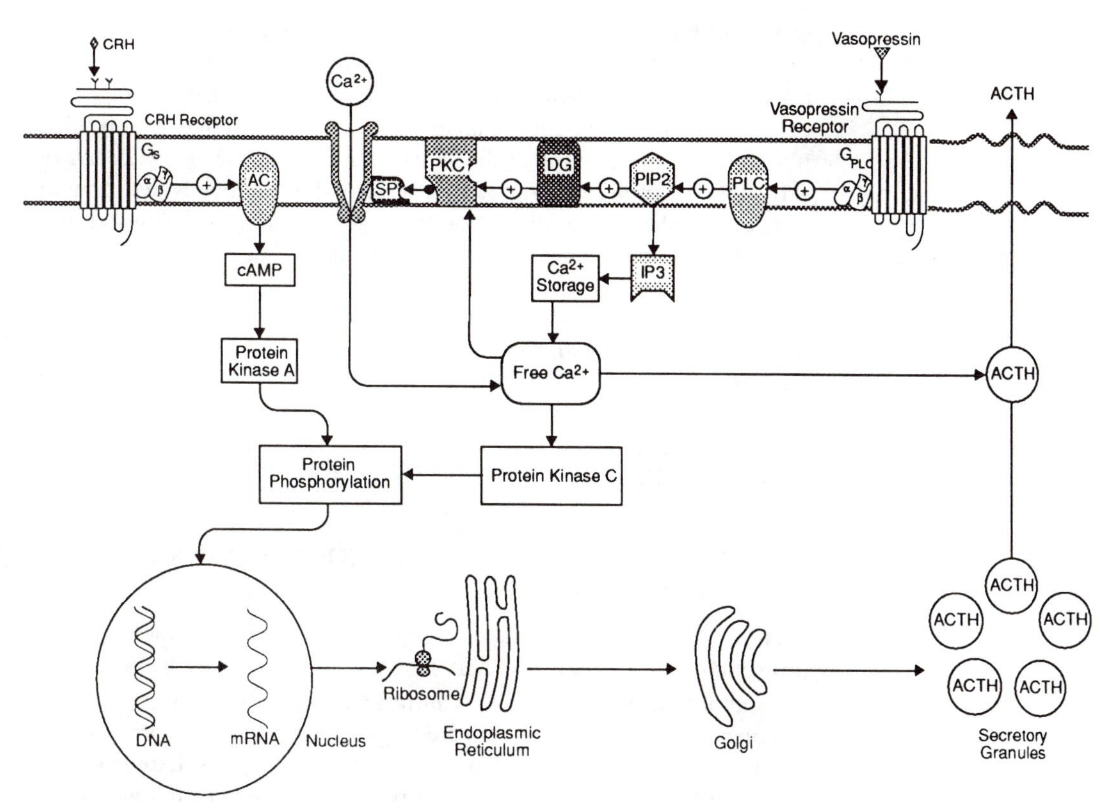

Figure 10.7. An example of how the cyclic AMP and inositol phospholipid second messenger systems interact to regulate the synthesis and release of a hormone such as ACTH. CRH binds to its receptor to activate cyclic AMP synthesis and phosphorylation of proteins by protein kinase A. Vasopressin binds to its receptors to activate the inositol phospholipid second messenger system and open calcium channels. Inositol trisphosphate (IP3) then elevates intracellular calcium levels which act via protein kinase C to phosphorylate proteins. The phosphorylation of nuclear regulatory proteins leads to the replication of ACTH messenger RNA and the synthesis of ACTH in the ribosomes. This ACTH is then packaged into secretory granules in the Golgi bodies and stored before being released. Calcium ions are important in the release of hormones from the secretory granules into the extracellular fluid. (Redrawn from King and Baertschi, 1990.)

anterior pituitary gland (refer back to Figure 6.4, p. 94). In this example, CRH activates cyclic AMP and protein kinase A, while vasopressin activates the IP3–Ca^{2+} second messenger system, elevating intracellular Ca^{2+} levels and activating protein kinase C. The activated protein kinase C in the cell membrane increases the sensitivity of the CRH receptor to CRH, while the increased Ca^{2+} binds to calmodulin and elevates cytoplasmic protein kinase C, which facilitates cyclic AMP phosphorylation of substrate proteins. As a result, ACTH synthesis and release are facilitated (see King and Baertschi, 1990).

10.5 SIGNAL AMPLIFICATION

Although membrane receptors bind only to one ligand at a time, the biochemical activity stimulated in the target cell by that one ligand is amplified and prolonged through the second messenger cascade as shown in Figure 10.8 (Alberts *et al.*, 1989). In the cyclic AMP second messenger system, for example, each ligand–receptor complex activates a number of molecules of G-protein, each of which activates an adenylate cyclase enzyme. Thus, the signal of a single first messenger is amplified into many molecules of cyclic AMP. Each molecule of cyclic AMP activates one molecule of protein kinase A, but each activated protein kinase A molecule can phosphorylate many proteins, leading to long-term changes in ion channel permeability and to the synthesis of many mRNA molecules, each of which can direct the synthesis of many protein molecules. Thus, the second messenger system provides a mechanism for the rapid amplification of a signal from a hormone or neurotransmitter. In this way, a small quantity of the first messenger, which binds only briefly to its receptor, can lead to a high level of biochemical activity within a target cell and this biochemical activity can last for hours after the ligand has been deactivated (Rasmussen, 1989).

10.6 SECOND MESSENGERS IN THE BRAIN AND NERVOUS SYSTEM

Each of the different second messenger systems plays a role in signal transduction in the brain and nervous system. Cyclic AMP is distributed throughout the brain in high concentrations (about 10 times greater than in non-neural tissue) and is activated by a wide range of neuroregulators, as shown in Table 10.2 (Nathanson, 1977). Inositol phosphate/calcium-dependent protein kinase C is also widely distributed throughout the brain. Cyclic GMP is also found in the brain, but at much lower levels than cyclic AMP. The highest concentrations of cyclic GMP in the brain are found in the Purkinje cells of the cerebellum (Greengard, 1979; Nestler and Greengard, 1983; Hemmings *et al.*, 1986, 1989). These second messenger systems have two major modulatory effects in neural cells: they alter membrane permeability and regulate protein synthesis.

Modulation of membrane permeability

When a neurotransmitter binds to its receptor at an **excitatory** synapse on a postsynaptic cell, the cell fires, i.e. there is a rapid change in electrical activity and the cell is depolarized as described in Chapter 5. The ability of the cell to fire depends on the permeability of the cell membrane to Na^+ and K^+ (see Figure 5.5, p. 66). Neuropeptides, peptide hormones and other neuromodulators can alter the firing rate of their neural target cells by altering the sensitivity of their membrane receptors to transmitters and by altering the permeability of their cell membranes to ions. Second messengers can modulate the sensitivity of receptors by altering their binding affinity for their ligands (Figure 10.7) and by up- and down-regulation of receptor numbers (Alkon and Rasmussen, 1988). Second messengers regulate the conductance of Na^+, K^+ and Cl^- channels in the nerve membrane, thus modulating the action potential and the release of neurotransmitters (Billingsley *et al.*, 1985; Hemmings *et al.*, 1986; Levitan, 1988). Changes in membrane permeability are the result of altered phosphorylation of substrate proteins by the protein kinases

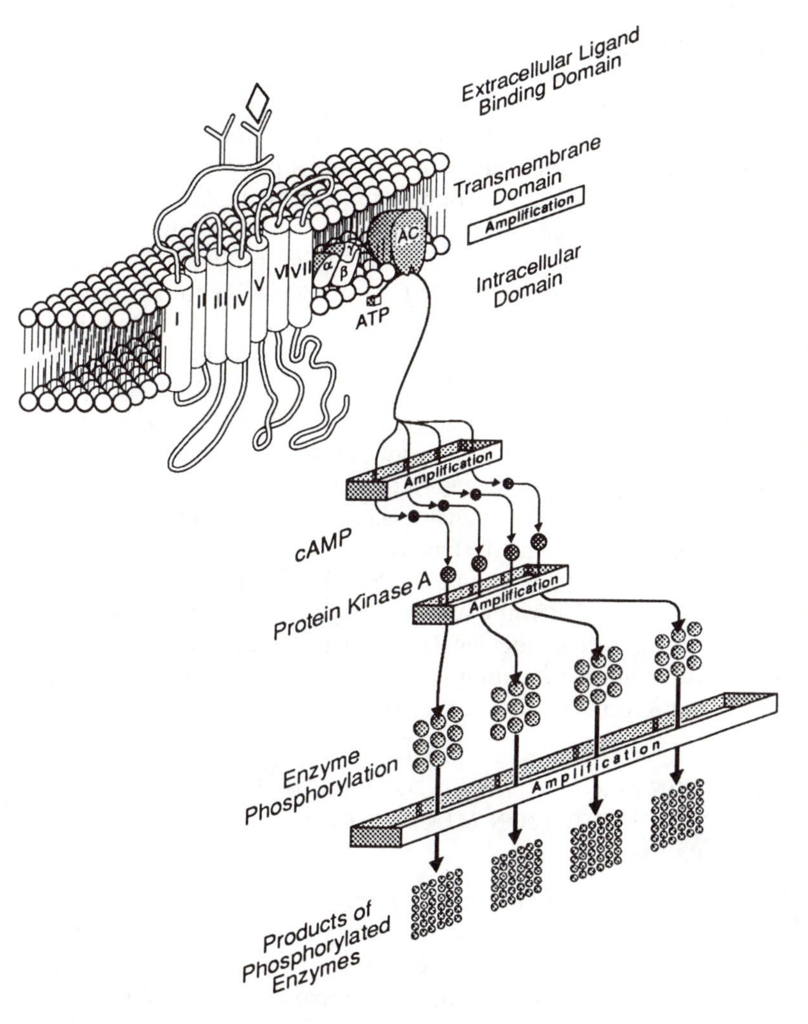

Figure 10.8. The intracellular second messenger cascade and how it results in signal amplification and prolongation of intracellular activity. Amplification of the signal occurs at four places in the second messenger cascade. When a ligand binds to its receptor, that receptor activates many molecules of G-protein, each of which activates one molecule of adenylate cyclase (AC). Each activated adenylate cyclase molecule generates many molecules of cyclic AMP. Each cyclic AMP molecule activates one protein kinase A molecule, and each activated protein kinase A molecule phosphorylates many enzymes or substrate proteins and each of these can stimulate a number of biological actions. (Redrawn from Alberts *et al.*, 1989.)

activated in the second messenger systems as shown in Figures 10.6 and 10.7.

Modulation of protein synthesis

The activation of second messengers results in an increase in protein synthesis in nerve cells. In neurosecretory cells in the hypothalamus, adrenal medulla and pineal gland, for example, the activation of cyclic AMP by a neurotransmitter binding to its receptors stimulates the synthesis and release of neurohormones. Thus, neurotransmitter stimulation of hormone release from neuroendocrine transducers can be very rapid. An example of this is the rapid secretion of adrenaline when acetylcholine is released from the splanchnic nerves and binds to the muscarinic receptors on the adrenal medulla. When a neurotransmitter stimulates hormone synthesis, however, the process takes considerably longer because of the number of biochemical steps involved in the second messenger cascade (see Figure 10.4). An example of this is the stimulation of hormone synthesis and release in the pituitary gland by the hypothalamic hormones. This is a slow process which is regulated by the second messenger cascades, as shown for prolactin and ACTH synthesis in Figures 10.3 and 10.7.

Figure 10.9 summarizes the time scale of the cellular responses which occur when neurotransmitters or peptides bind to their membrane receptors. The muscle contraction stimulated by the firing of nerves at the neuromuscular junction is a very rapid response triggered by ion channel-gated receptors and takes only a few milliseconds to occur. Changes in membrane permeability, which depend on the activation of cyclic AMP and membrane phosphorylation, and which result in prolonged cell firing or in hormone release, take longer to occur (up to 15 min). Long-term changes, such as hormone synthesis, learning, and memory formation, which involve protein synthesis, take much longer to occur (Nathanson and Greengard, 1977; Alkon and Rasmussen, 1988). The amplification of neurotransmitter signals by second messengers thus provides the cellular mechanism for the transformation of rapid synaptic activity into the long-term biochemical changes necessary for neural growth and differentiation and for habituation, sensitization, and other forms of learning and memory to occur (Figure 10.9). The importance of second messenger-induced changes in protein kinase activity for memory formation is discussed by Schwartz and Greenberg (1987) and Alkon and Rasmussen (1988).

10.6.1 CONVERGENCE AND DIVERGENCE OF SIGNAL TRANSDUCTION MECHANISMS IN NEURAL TARGET CELLS

It was pointed out above that one of the primary functions of the receptor mediated activation of second messenger systems is to regulate the permeability of the cell membrane to ions. It was also pointed out that each receptor is coupled to a specific G-protein and second messenger system. Thus, it should be no surprise to find that those neuroregulators which activate the same second messenger system will have the same effects on the neural target cell. Nicoll, Malenka and Kauer (1990) have reviewed the mechanisms for the convergent effects of different neuroregulators on ion channels in the cell membrane. For example, GABA, acetylcholine, enkephalin, noradrenaline and somatostatin all bind to different receptors, but all of these receptors activate the same G-proteins (G_i and G_o) which act to increase the conductance of K^+ channels (see Figure 10.10A). Thus, by being coupled to the same G-protein or second messenger system, all of these different neuroregulators converge to induce the same electrophysiological response in their target neurons.

On the other hand, one neuroregulator may have a number of different receptors (as shown in Table 5.3, p. 68) and each of these different receptors may activate divergent responses in specific target cells, depending on the G-proteins and second messenger systems activated. For example, noradrenaline activates three different receptor types, each of which is coupled to different G-proteins and second messenger systems, resulting in an increase or decrease in K^+ and Ca^{2+} channel conductance, depending on the specific receptors activated (Figure 10.10B). Thus, if a

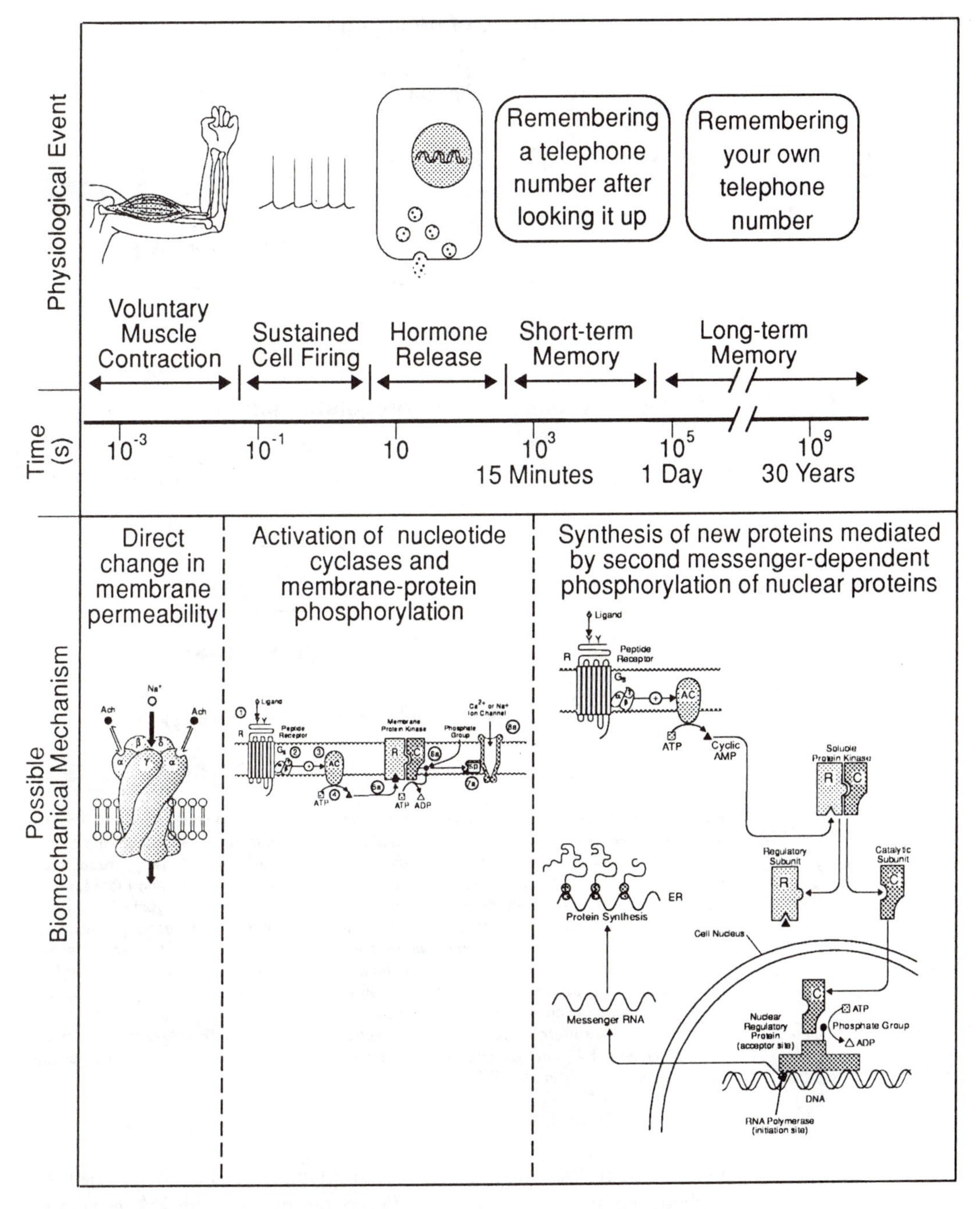

Figure 10.9. The temporal continuum of cellular events in the nervous system from the very brief to the very long lived. At the short end of the continuum are processes such as voluntary muscle contractions that are initiated in the span of a few milliseconds by the direct action of a neurotransmitter, such as acetylcholine, opening an ion channel in the cell membrane. Events lasting from hundreds of milliseconds to minutes are mediated by receptor-induced synthesis of second messengers, such as cyclic AMP, which initiates the phosphorylation of membrane proteins to produce a relatively slow change in membrane permeability. Events ranging from minutes to hours or years in duration, such as short- and long-term memory, may involve the synthesis of new proteins. Such protein synthesis is controlled by the second messenger (cyclic AMP) activation of protein kinase phosphorylation of nuclear regulatory proteins. (Redrawn from Nathanson and Greengard, 1977.)

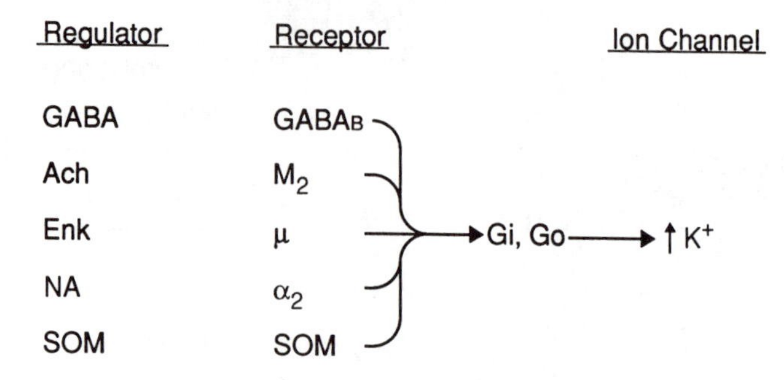

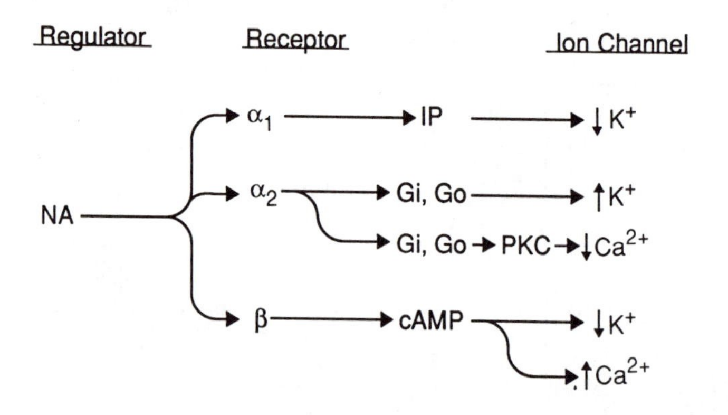

Figure 10.10. Convergence and divergence of neuroregulator action at neural target cells. (A) In this example, GABA acting at $GABA_B$ receptors, acetylcholine (Ach) at muscarinic 2 (M_2) receptors, enkephalin (Enk) at mu (μ) receptors, noradrenaline (NA) at alpha$_2$ (α_2) receptors, and somatostatin (SOM) at its receptors, all activate G_i and G_o proteins and increase K^+ conductance in the cell membrane. (B) One neuroregulator may have divergent effects on its target cells by acting through different receptors, G-proteins and second messenger systems in these target cells. For example, when noradrenaline (NA) binds to α_1-adrenergic receptors, the inositol phospholipid (IP) second messenger system is activated, reducing K^+ conductance. At α_2-adrenergic receptors, NA can act to increase K^+ conductance through G-protein-activated ion channels, or to inhibit Ca^{2+} conductance through activation of protein kinase C. At β-adrenergic receptors, which activate the cyclic AMP (cAMP)second messenger system, K^+ conductance can be decreased and Ca^{2+} conductance increased. (Redrawn from Nicoll *et al.*, 1990.)

neuroregulator has different actions in different parts of the brain, it is probably because it is activating different receptor systems in these brain areas (Nicoll *et al.*, 1990). It is, therefore, understandably difficult to determine the exact effects of particular neurotransmitters in the neuroendocrine system, as was seen in Chapter 6.

10.6.2 DRUG ACTION

Drugs which act as neurotransmitter agonists, such as isoproterenol at β-adrenergic receptors (see Table 5.6, p. 78) for other examples), bind to postsynaptic receptors and act through the same second messenger systems as the neurotransmitters that they are mimicking. These drugs can thus alter cellular activity by their effects on the second messenger cascade or by altering the level of enzymes such as phosphodiesterase, which deactivate second messengers. Through this mechanism, these drugs can alter the synthesis and release of neurohormones from neuroendocrine cells. In the same way, neurotransmitter antagonists (see examples in Table 5.6) can block the activation of the second messenger systems.

10.6.3 RECEPTOR DYSFUNCTION

Disorders of receptors, G-proteins and/or second messenger systems disrupt cellular activity, causing physiological and psychological disorders. For example: insulin receptor disorders occur in obesity and diabetes; disorders of TSH receptors occur in Grave's disease; and allergic rhinitis and asthma have been associated with disorders of the β-adrenergic receptors (Roth *et al.*, 1979; Roth and Taylor, 1982). Depression, schizophrenia, Alzheimer's disease and Parkinson's disease, have also been attributed, at least in part, to defects in receptor regulation (Creese, 1981; Wastek and Yamamura, 1981), but until recently, there were few cases in which such diseases could be causally linked to receptor dysfunction (Snyder and Narahashi, 1990).

Alzheimer's disease may involve changes in second messenger cascades, resulting in reduced protein kinase C levels (Shimohama *et al.*, 1990) and schizophrenia may involve changes in the number of G-proteins (G_i and G_o) in specific brain areas (Okada, Crow and Roberts, 1991). Depression may also involve altered G-protein activity. Lesch and Lerer (1991) suggest that the elevated levels of glucocorticoids in depressed patients alter the levels of G_s and G_i proteins in their serotonergic neuron target cells (refer to Figure 9.15B (p. 171), which shows the type II receptors in the serotonergic nerve pathways). As a result, when serotonin binds to 5-HT_{1A} receptors in these glucocorticoid-sensitive cells, the altered G-protein levels interfere with cyclic AMP synthesis. Their hypothesis is that depression results from glucocorticoid disruption of the serotonin receptor mediated G-protein-activated cyclic AMP second messenger system.

10.7 COMPARISON OF STEROID AND PEPTIDE HORMONE ACTIONS AT THEIR TARGET CELLS

As described in Chapter 9, steroid hormones enter their target cells and bind to nuclear receptors in the chromatin to initiate protein synthesis. Peptide hormones initiate protein synthesis via membrane receptors and the activation of second messenger systems. As shown in Figure 10.11, these two types of hormone both regulate receptor sensitivity, membrane permeability and protein synthesis, but do so through different mechanisms. Szego (1984) has argued that there are more similarities in the actions of peptide and steroid hormones than are commonly acknowledged. Both types of hormone regulate the responses of their neural target

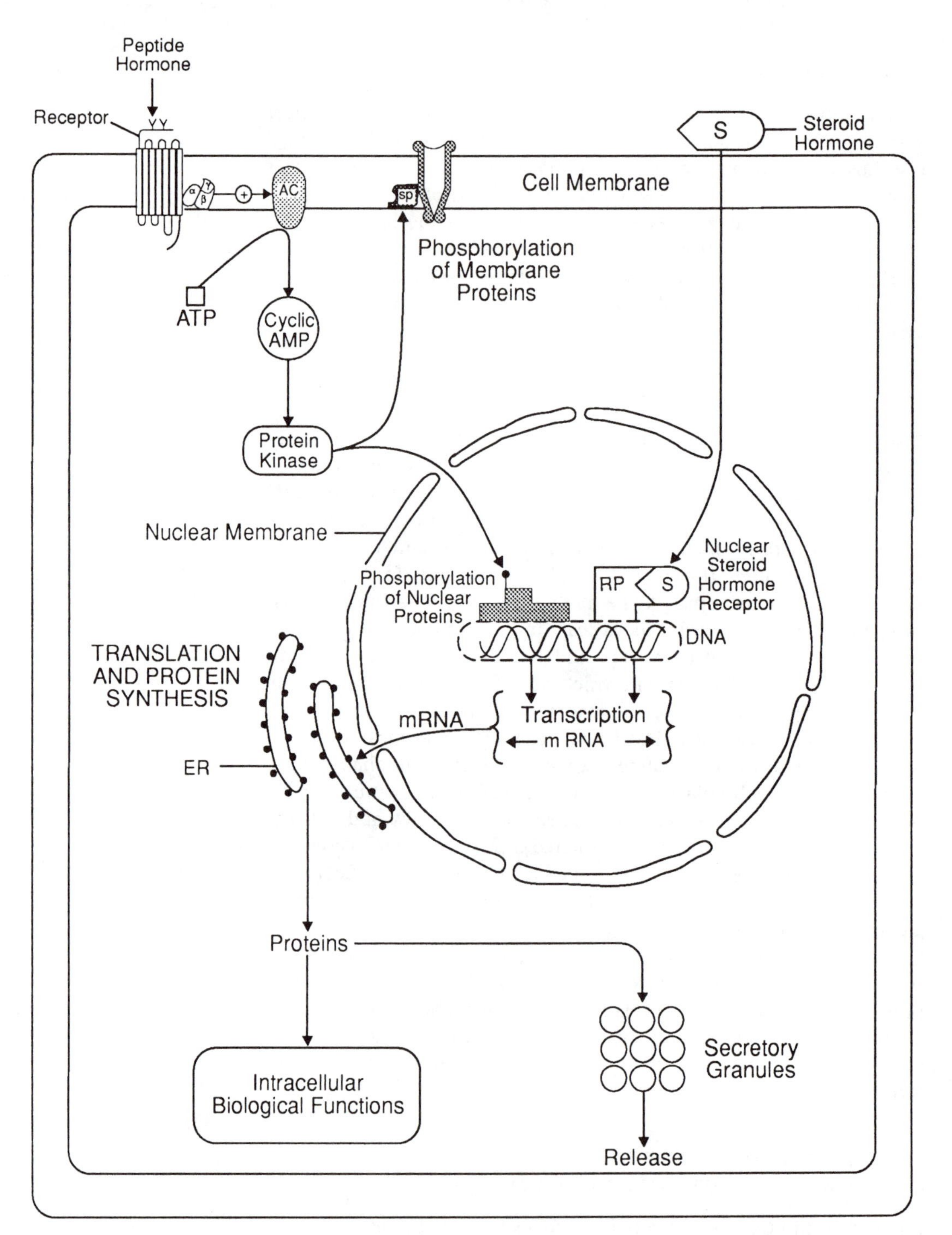

Figure 10.11. A comparison of the actions of steroid and peptide hormones at target cells. Both groups stimulate protein synthesis by activating nuclear regulatory proteins, but do so through different biochemical mechanisms. The steroids bind directly to their nuclear receptors, while the peptides activate second messenger systems. The proteins synthesized may be regulatory proteins such as receptors, G-proteins, enzymes, etc. which serve intracellular functions, or hormones which are stored in secretory granules and then released. The peptide hormones regulate membrane permeability directly through the phosphorylation of substrate proteins. Steroid hormones may alter membrane permeability by regulating the synthesis of receptors, G-proteins, etc. (Redrawn from Turner and Bagnara, 1976.)

cells to neurotransmitters and neuropeptides by regulating the synthesis of receptors, G-proteins and enzymes (genomic effects) or by altering membrane permeability (non-genomic effects) as discussed in Chapter 9. Hadcock and Malbon (1991) have examined the complex interactions of steroid hormone receptors, G-protein coupled receptors and tyrosine kinase receptors in regulating the biochemical activity of their target cells and suggest that these different signalling pathways interact in an 'intracellular network' which regulates the sensitivity, growth and maintenance of the target cell.

10.8 SUMMARY

This chapter has examined the mechanisms by which peptide hormones, neurotransmitters and neuropeptides stimulate biochemical changes in their target cells. These ligands bind to cell membrane receptors which are complex glycoproteins, often having a number of subunits. The number of receptors for each hormone, such as insulin, is not constant but can be regulated homospecifically by insulin levels in the circulation, by other hormones, or by non-hormonal factors. Ligand-bound cell surface receptors can be taken into the cell by endocytosis where they are deactivated or recycled to the cell membrane. Once a ligand binds to its receptor, the signal is transduced by the G-proteins or by enzymes in the cell membrane. The G-proteins activate enzymes which produce various second messengers within the cell. These second messengers amplify the original signal, so that a few first messengers can promote high levels of biochemical activity within the target cell. The best known second messenger is cyclic AMP, but cyclic GMP, diacylglycerol, inositol triphosphate and calcium all act as second messengers, singly or in combination. Calcium ions, for example, whether free or bound to calcium binding proteins, such as calmodulin, interact with the other second messengers to activate protein kinase enzymes in the cell membrane and cytoplasm. These protein kinase enzymes phosphorylate substrate proteins to promote changes in cell membrane permeability and phosphorylate nuclear proteins to activate mRNA and protein synthesis. The enzyme phosphodiesterase then deactivates cyclic AMP and cyclic GMP. The second messenger systems described in this chapter function in the adenohypophyseal target cells of hypothalamic hormones, the endocrine and non-endocrine target cells of pituitary hormones, and the target cells for insulin, glucagon, and other peptide hormones in both the body and the brain. In addition, the neurotransmitters and neuropeptides act at synaptic and non-synaptic receptors through the same second messenger systems to regulate neural growth, behavior changes, emotional and motivational arousal, and short- and long-term memory. Steroid hormones can act to modulate the responses of their target cells to neurotransmitters and peptides by regulating the number of receptors and G-proteins in these cells. Disorders of receptors and second messenger systems result in physiological dysfunction which may underlie disorders such as depression, schizophrenia and Alzheimer's disease.

FURTHER READING

Berridge, M. J. (1985). The molecular basis of communication within the cell. *Scientific American*, **253** (4), 124–134.

Brown, A. M. and Birnbaumer, L. (1989). Ion channels and G-proteins. *Hospital Practice*, **24** (7), 189–204.

Hanley, R. M. and Steiner, A. L. (1989). The second messenger system for peptide hormones. *Hospital Practice*, **24** (8), 59–70.

Hollenberg, M. D. (1991). Structure–activity relationships for transmembrane signalling: The receptor's turn. *Federation of American Societies for Experimental Biology Journal*, **5**, 178–186.

REVIEW QUESTIONS

10.1 What is cyclic AMP and what does it do?

10.2 Both steroid and non-steroid hormones stimulate RNA synthesis, but by different methods. What is the difference?

10.3 What are the two main functions of a membrane receptor for a peptide hormone such as ACTH?

10.4 What does heterospecific receptor regulation mean?

10.5 Which molecules act as transducers of signals in the cell membranes?

10.6 Which enzyme is activated by cyclic AMP to activate protein synthesis in the nucleus of a peptide hormone target cell?

10.7 In order to act as a second messenger, calcium binds to _______ within the cytoplasm of the cell.

10.8 When a neurotransmitter binds to a receptor, three events are triggered. What are they?

10.9 What are the two ways that G-proteins can influence cyclic AMP synthesis?

10.10 How do diacylglycerol and IP3 differ in their function as second messengers?

ESSAY QUESTIONS

10.1 Discuss the components of a typical peptide or neurotransmitter receptor (if there is one) and the function of each component.

10.2 Discuss the role of G-proteins in the transduction of signals in cell membranes.

10.3 Using cyclic AMP or inositol phospholipids as an example, discuss the functions of second messengers in peptide hormone target cells.

10.4 Discuss the functions of calcium and calmodulin as second messengers in peptide hormone target cells.

10.5 Discuss the role of receptors and second messenger systems in one of the following disorders: (a) Alzheimer's disease, (b) Parkinson's disease (c) schizophrenia (d) depression.

10.6 Explain the concept of signal amplification via second messengers using an appropriate example.

10.7 Compare and contrast the mechanisms through which steroid and peptide hormones activate (a) genomic or (b) non-genomic changes in their target cells.

10.8 Explain how the inositol phospholipid second messenger system interacts with the cyclic AMP second messenger system to regulate physiological changes in cells.

10.9 Discuss the role of protein kinase in learning and memory.

REFERENCES

Alberts, B., Bray, D., Lewis, J., Raff, M., Roberts, K. and Watson, J. D. (1989). *Molecular Biology of the Cell.* 2nd edn. New York: Garland.

Alkon, D. L. and Rasmussen, H. (1988). A spatial-temporal model of cell activation. *Science*, **239**, 998–1005.

Anderson, A. S. (1989). Reception and transmission. *Nature*, **337**, 12.

Aurbach, G. D. (1982). Polypeptide and amine hormone regulation of adenylate cyclase. *Annual Review of Physiology*, **44**, 653–666.

Berridge, M. J. (1985a). Inositol triphosphate and diacylglycerol as intracellular second messengers. In G. Poste and S. T. Crooke (eds.) *Mechanisms of Receptor Regulation*, pp. 111–130. New York: Plenum.

Berridge, M. J. (1985b). The molecular basis of communication within the cell. *Scientific American*, **253** (4), 124–134.

Billingsley, M., Hanbauer, I. and Kuhn, D. (1985). Role of calmodulin in the regulation of neuronal function. In A. Lajtha (ed.) *Handbook of Neurochemistry*, 2nd edn, pp. 201–215. New York: Plenum.

Birnbaumer, L. and Brown, A. M. (1990). G proteins and the mechanism of action of hormones, neurotransmitters, and autocrine and paracrine regulatory factors. *American Review of Respiratory Disease*, **141**, S106–S114.

Birnbaumer, L., Codina, J., Mattera, R., Yatani, A. and Brown, A. M. (1988). G proteins and transmembrane signalling. In B. A. Cooke, R. J. B. King and H. J. van der Molen (eds.) *Hormones and their Actions, Part II*, pp. 1–46. Amsterdam: Elsevier.

Brown, A. M. and Birnbaumer, L. (1989). Ion channels and G proteins. *Hospital Practice*, **24** (7), 189–204.

Carafoli, E. and Penniston, J. T. (1985). The calcium signal. *Scientific American*, **253** (5), 70–78.

Cheung, W. Y. (1982). Calmodulin. *Scientific American*, **246** (6), 62–70.

Cooke, B. A., King, R. J. B. and van der Molen, H. J. (eds.) 1988. *Hormones and their Actions, Part II. Specific Actions of Protein Hormones.* Amsterdam: Elsevier.

Creese, I. (1981). Dopamine receptors. In H. I. Yamamura and S. J. Enna (eds.), *Neurotransmitter Receptors, Part II: Biogenic Amines*, pp. 129–183. London: Chapman and Hall.

Denef, C. (1988). Mechanism of action of pituitary hormone releasing and inhibiting factors. In B. A. Cooke, R. J. B. King and H. J. van der Molen (eds.) *Hormones and their Actions, Part II*, pp. 113–134. Amsterdam: Elsevier.

Enjalbert, A. (1989). Multiple transduction mechanisms of dopamine, somatostatin and angiotensin II receptors in anterior pituitary cells. *Hormone Research*, **31**, 6–12.

Garbers, D. L. (1989). Cyclic GMP and the second messenger hypothesis. *Trends in Endocrinology and Metabolism*, **2**, 64–67.

Gilman, A. G. (1984). G proteins and dual control of adenylate cyclase. *Cell*, **36**, 577–579.

Gilman, A. G. (1987). G proteins: transducers of receptor-generated signals. *Annual Review of Biochemistry*, **56**, 615–649.

Gitelman, S. E. (1990). Cloning of the LH/CG receptor. Implications for a unique G-protein-coupled receptor. *Trends in Endocrinology and Metabolism*, **2**, 181–184.

Goldfine, I. D. (1987). The insulin receptor: molecular biology and transmembrane signalling. *Endocrine Reviews*, **8**, 235–255.

Greengard, P. (1979). Cyclic nucleotides, phosphorylated proteins, and the nervous system. *Federation Proceedings*, **38**, 2208–2217.

Guy, H. R. and Hucho, F. (1987). The ion channel of the nicotinic acetylcholine receptor. *Trends in Neuroscience*, **10**, 318–321.

Guy, G. R. and Kirk, C. J. (1988). Inositol phospholipids and cellular signalling. In B. A. Cooke, R. J. B. King and H. J. van der Molen (eds.) *Hormones and their Actions, Part II.*, pp. 47–62. Amsterdam: Elsevier.

Hadcock, J. R. and Malbon, C. C. (1991). Regulation of receptor expression by agonists: transcriptional and post-transcriptional controls. *Trends in Neuroscience*, **14**, 242–247.

Hadley, M. E. (1992). *Endocrinology*. 3rd edn. Englewood Cliffs, NJ: Prentice-Hall.

Hanley, R. M. and Steiner, A. L. (1989). The second-messenger system for peptide hormones. *Hospital Practice*, **24** (8), 59–70.

Hemmings, H. C. Jr, Nairn, A. C. and Greengard, P. (1986). Protein kinases and phosphoproteins in the nervous system. In J. B. Martin and J. D. Barchas (eds.) *Neuropeptides in Neurologic and Psychiatric Disease*, pp. 47–69. New York: Raven Press.

Hemmings, H. C. Jr, Nairn, A. C., McGuinness, T. L., Huganir, R. L. and Greengard, P. (1989). Role of protein phosphorylation in neuronal signal transduction. *Federation of American Societies for Experimental Biology Journal*, **3**, 1583–1592.

Hollenberg, M. D. (1979). Hormone receptor interactions at the cell membrane. *Pharmacological Reviews*, **30**, 393–410.

Hollenberg, M. D. (1981). Membrane receptors and hormone action. I. New trends related to receptor structure and receptor regulation. *Trends in Pharmacological Sciences*, **2**, 320–323.

Hollenberg, M. D. (1985). Receptor models and the action of neurotransmitters and hormones: Some new perspectives. In H. I. Yamamura, S. J. Enna, and M. J. Kuhar (eds.) *Neurotransmitter Receptor Binding*, 2nd edn, pp. 2–39. New York: Raven Press.

Hollenberg, M. D. (1991). Structure–activity relationships for transmembrane signalling: The receptor's turn. *Federation of American Societies for Experimental Biology Journal*, **5**, 178–186.

Hollenberg, M. D. and Goren, H. J. (1985). Ligand-receptor interactions at the cell surface. In G. Poste and S. T. Crooke (eds.) *Mechanisms of Receptor Regulation*, pp. 323–373. New York: Plenum.

Houslay, M. D. and Wakelam, M. J. O. (1988). Structure and function of the receptor for insulin. In B. A. Cooke, R. J. B. King and H. J. van der Molen (eds.) *Hormones and their Actions, Part II*, pp. 321–348. Amsterdam: Elsevier.

Hughes, A. R., Horstman, D. A., Takemura, H. and Putney, J. W. Jr (1990). Inositol phosphate metabolism and signal transduction. *American Review of Respiratory Disease*, **141**, S115–S118.

Johnson, E. M., Jr and Taniuchi, M. (1987). Nerve growth factor (NGF) receptors in the central nervous system. *Biochemical Pharmacology*, **36**, 4189–4195.

Johnson, G. L. and Dhanasekaran, N. (1989). The G-protein family and their interaction with receptors. *Endocrine Reviews*, **10**, 317–331.

Khanna, N. C., Tokuda, M. and Waisman, D. M. (1988). The role of calcium binding proteins in signal transduction. In B. A. Cooke, R. J. B. King and H. J. van der Molen (eds.) *Hormones and their Actions, Part II*, pp. 63–92. Amsterdam: Elsevier.

King, M. S. and Baertschi, A. J. (1990). The role of intracellular messengers in adrenocorticotropin secretion in vitro. *Experientia*, **46**, 26–40.

Lesch, K. P. and Lerer, B. (1991). The 5-HT receptor–G-protein–effector system complex in depression. I. Effect of glucocorticoids. *Journal of Neural Transmission*, **84**, 3–18.

Levitan, I. B. (1988). Modulation of ion channels in neurons and other cells. *Annual Review of Neuroscience*, **11**, 119–136.

Limbird, L. E. (1986). *Cell Surface Receptors: A Short Course on Theory and Methods*. Boston: Nijhoff.

Martin, C. R. (1985). *Endocrine Physiology*. New York: Oxford University. Press.

McFarland, K. C., Sprengel, R., Phillips, H. S., Kohler, M., Rosemblit, N., Nikolics, K., Segaloff, D. L. and Seeburg, P. H. (1989). Lutropin-choriogonadotropin receptor: an unusual member of the G protein-coupled receptor family. *Science*, **245**, 494–499.

Montminy, M. R., Gonzalez, G. A. and Yamamoto, K. K. (1990). Regulation of CAMP-inducible genes by CREB. *Trends in Neuroscience*, **13**, 184–188.

Nathanson, J. A. (1977). Cyclic nucleotides and nervous system function. *Physiological Reviews*, **57**, 157–256.

Nathanson, J. A. and Greengard, P. (1977). 'Second messengers' in the brain. *Scientific American*, **237**, 108–119.

Nestler, E . J. and Greengard, P. (1983). Protein phosphorylation in the brain. *Nature*, **305**, 583–588.

Nicoll, R. A., Malenka, R. C. and Kauer, J. A. (1990). Functional comparison of neurotransmitter receptor subtypes in mammalian central nervous system. *Physiological Reviews*, **70**, 513–566.

Okada, F., Crow, T. J. and Roberts, G. W. (1991). G proteins (Gi, Go) in the medial temporal lobe in schizophrenia: preliminary report of a neurochemical correlate of structural change. *Journal of Neural Transmission*, **84**, 147–153.

Parmentier, M., Liebert, F., Maenhaut, C., Lefort, A., Gerard, C., Perret, J., van Sande, J., Dumont, J. E. and Vassart, G. (1989). Molecular cloning of the thyrotropin receptor. *Science*, **246**, 1620–1622.

Popot, J.-L. and Changeux, J.-P. (1984). Nicotinic receptor of acetylcholine: Structure of an oligomeric integral membrane protein. *Physiological Reviews*, **64**, 1162–1232.

Posner, B. I. (ed.) (1985). *Polypeptide Hormone Receptors*. New York: Marcel Dekker.

Posner, B. I., Khan, M. N. and Bergeron, J. J. M. (1985). Receptor-mediated uptake of peptide hormones and other ligands. In B. I. Posner (ed.) *Polypeptide Hormone Receptors*, pp. 61–90. New York: Marcel Dekker.

Poste, G. and Crooke, S. T. (eds.) (1985). *Mechanisms of Receptor Regulation*.

New York: Plenum.

Rasmussen, H. (1989). The cycling of calcium as an intracellular messenger. *Scientific American*, **261** (4), 66–73.

Rasmussen, H. and Barrett, P. Q. (1988). Mechanism of action of Ca^{2+} dependent hormones. In B. A. Cooke, R. J. B. King and H. J. van der Molen (eds.) *Hormones and their Actions, Part II*, pp. 93–111. Amsterdam: Elsevier.

Roth, J., Lesniak, M. A. *et al.*, (1979). An introduction to receptors and receptor disorders. *Proceedings of the Society for Experimental Biology and Medicine*, **162**, 3–12.

Roth, J. and Taylor, S. I. (1982). Receptors for peptide hormones: alterations in diseases of humans. *Annual Review of Physiology*, **44**, 639–651.

Schramm, M. and Seliner, Z. (1984). Message transmission: Receptor controlled adenylate cyclase system. *Science*, **225**, 1350–1356.

Schwartz, J. H. and Greenberg, S. M. (1987). Molecular mechanisms for memory: Second-messenger induced modifications of protein kinases in nerve cells. *Annual Review of Neuroscience*, **10**, 459–476.

Segaloff, D. L. and Ascoli, M. (1988). Internalization of peptide hormones and hormone receptors. In B. A. Cooke, R. J. B. King and H. J. van der Molen (eds.) *Hormones and their Actions, Part I*, pp. 133–149. Amsterdam: Elsevier.

Shimohama, S., Ninomiya, H., Saitoh, T., Terry, R. D., Fukunaga, R., Taniguchi, T., Fujiwara, M., Kimura, J. and Kameyama, M. (1990). Changes in signal transduction in Alzheimer's disease. *Journal of Neural Transmission*, suppl. **30**, 69–78.

Simon, M. I., Strathmann, M. P. and Gautam, N. (1991). Diversity of G proteins in signal transduction. *Science*, **252**, 802–808.

Spiegel, A. M. (1989). Receptor-effector coupling by G-proteins. Implications for endocrinology. *Trends in Endocrinology and Metabolism*, **2**, 72–76.

Snyder, S. H. and Narahashi, T. (1990). Receptor-channel alterations in disease: Many clues, few causes. *Federation of American Societies for Experimental Biology Journal*, **4**, 2707–2708.

Sutherland, E. W. (1972). Studies on the mechanism of hormone action. *Science*, **177**, 401–408.

Szego, C. M. (1984). Mechanisms of hormone action: parallels in receptor-mediated signal propagation for steroid and peptide effectors. *Life Sciences*, **35**, 2383–2396.

Taylor, C. W. (1990). The role of G proteins in transmembrane signalling. *Biochemical Journal*, **272**, 185–197.

Turner, C. D. and Bagnara, J. T. (1976). *General Endocrinology*, 6th edn. Philadelphia: Saunders.

Vallar, L. and Meldolesi, J. (1989). Mechanisms of signal transduction at the dopamine D_2 receptor. *Trends in Pharmacological Sciences*, **10**, 74–77.

Venter, J. C., Fraser, C. M., Kerlavage, A. R. and Buck, M. A. (1989). Molecular biology of adrenergic and muscarinic cholinegic receptors: a perspective. *Biochemical Pharmacology*, **38**, 1197–1208.

Wastek, G. J. and Yamamura, H. I. (1981). Acetylcholine receptors. In H. I. Yamamura and S. J. Enna (eds.) *Neurotransmitter Receptors, Part 2: Biogenic Amines*, pp. 129–183. London: Chapman and Hall.

Yamamura, H. I., Enna, S. J. and Kuhar, M. J. (eds.) (1990). *Methods in Neurotransmitter Receptor Analysis*. New York, Raven Press.

11

Neuropeptides I: classification, synthesis and colocalization with classical neurotransmitters

Many chemical messengers regulate neural activity, including the neurotransmitters (Chapter 5), the steroid hormones (Chapter 9), and the peptide hormones (Chapter 10). This chapter and Chapter 12 examine how the class of chemical messengers termed 'neuropeptides' regulate neural activity. Because the study of the neuropeptides is a daunting task, this topic is divided into two parts. This chapter examines the classification and synthesis of the neuropeptides and their colocalization with the classical neurotransmitters and Chapter 12 examines the functions of neuropeptides at their target cells.

11.1 CLASSIFICATION OF THE NEUROPEPTIDES

As described in Chapter 1, a neuropeptide is any hormonal or non-hormonal peptide which acts as a neuroregulator, so may have neuromodulator or neurotransmitter action. De Wied (1987, p. 100) has defined neuropeptides as 'endogenous substances, present in nerve cells, which are involved in nervous system function'. Neuropeptides have been localized in the brain using radioimmunoassay, autoradiography,

Table 11.1. *Categories of mammalian brain peptides*

Hypothalamic releasing hormones	*Gastrointestinal peptides*
Thyrotropin releasing hormone	Vasoactive intestinal peptide
Gonadotropin releasing hormone	Cholecystokinin
Somatostatin	Gastrin
Corticotropin releasing hormone	Substance P
Growth hormone releasing hormone	Neurokinin A (substance K)
	Neuropeptide K
Neurohypophyseal peptides	Insulin
Vasopressin	Glucagon
Oxytocin	Secretin
	Motilin
Adenohypophyseal peptides	Pancreatic polypeptide
Adrenocorticotropic hormone	Galanin
α-Melanocyte stimulating hormone	
Prolactin	*Growth factors*
Luteinizing hormone	Nerve growth factor
Growth hormone	Epidermal growth factor
Thyrotropin	Fibroblast growth factor
	Endothelial growth factor
Opioid peptides	
β-Endorphin	*Others*
Enkephalin	Angiotensin II
Dynorphin	Bombesin
Neo-endorphin	Bradykinin
	Calcitonin
	Carnosine
	Delta sleep-inducing peptide
	Neuropeptide Y
	Neurotensin
	Thymosin
	Atrial natriuretic factor

Source: Modified from Krieger, 1986.

immunohistochemistry, and other techniques (Krieger, 1983, 1986; Dockray, 1984) and often have the same chemical structures as peptide hormones, which were previously thought to be secreted only from hypothalamic neurosecretory cells and endocrine glands.

Peptide hormones have traditionally been named after the first function they were known to serve. Thus, the pituitary hormones (ACTH, TSH, FSH, GH, etc.) were named for their functions at their target cells; the hypothalamic hormones (CRH, TRH, GnRH, GH-RH, etc.) were named for their functions at pituitary target cells; and the hormones of the gastrointestinal tract (CCK, VIP, etc.) were named to describe their gastrointestinal functions. This naming process has led to considerable confusion since the discovery that virtually all of the gastrointestinal, pituitary and hypothalamic hormones have functions in the brain which are quite different from those for which they were named. Thus, the name of a neuropeptide (as shown in Table 11.1) may indicate its hormonal function, but may not give any clues as to its function as a neuropeptide (see Kastin *et al.*, 1983; Krieger 1983, 1986). Some of the newly discovered peptides (neuropeptide Y, substance K, etc.) have been given names which have no functional connotation.

The total number of neuropeptides is unknown. While Table 11.1 lists 42 neuropeptides, more than 100 may be active in the nervous system (Lynch and Snyder, 1986; Zadina, Banks and Kastin, 1986). New neuropeptides have been discovered at a rapid rate since the first hypothalamic hormones were isolated in the late 1970s and new peptides continue to be discovered. Scharrer (1987, 1990) and Hökfelt (1991) have presented brief histories of the discovery of the neuropeptides.

11.2 SYNTHESIS, STORAGE, RELEASE AND DEACTIVATION OF NEUROPEPTIDES

This section compares the mechanisms for the synthesis, storage, release, and deactivation of neuropeptides with those for the classical neurotransmitters which were described in Chapter 5.

11.2.1 SYNTHESIS

Figure 11.1 shows the synthesis of neuropeptides and classical neurotransmitters. The enzymes required for the production of the classical transmitters from their amino acid precursors are synthesized on the ribosomes of the endoplasmic reticulum. The neurotransmitter precursor and the enzymes are then packaged into the synaptic vesicles and transported to the nerve ending, where the synthesis of the transmitter is completed. The amount of neurotransmitter at the nerve ending is regulated by the rate of axonal transport of the vesicles, the rate of transmitter synthesis in the vesicles at the nerve ending, the number of vesicles in the storage pool, and the rate of reuptake of the transmitter from the synapse (see also Figures 5.3 and 5.4 on p. 62 and 64).

The synthesis of peptides in neurons follows the same pattern as peptide synthesis in endocrine glands, as described in Chapter 7. After the neuron is stimulated by a transmitter, the second messenger cascade activates the transcription of genetic information from the DNA to messenger RNA in the cell nucleus, as shown in Figure 10.7 (p. 207). The mRNA then moves to the cytoplasm where this information is translated in the ribosomes of the endoplasmic reticulum (see Figure 7.1, p. 116). The ribosomes synthesize large precursor molecules (pre-prohormones) which are long amino acid (polypeptide) chains in which the biologically active peptide is bound by pairs of basic amino acid residues. In the Golgi apparatus, proteolytic enzymes then cut the precursor at these amino acid residues to produce the prohormones, which are packaged into the secretory vesicles along with the enzymes which convert them to biologically active peptides (Krieger, 1983; Loh and Gainer, 1983; White *et al.*, 1985; Lynch and Snyder, 1986). The secretory vesicles are then transported down the axon (Figure 11.1). Thus, unlike the classical neurotransmitters, neuropeptides are not synthesised at the nerve endings. The amount of neuropeptide stored in the nerve terminal is dependent on axonal transport of the secretory vesicles from the cell body (Hökfelt *et al.*, 1980). There is, however, a suggestion that neuropeptides may also be

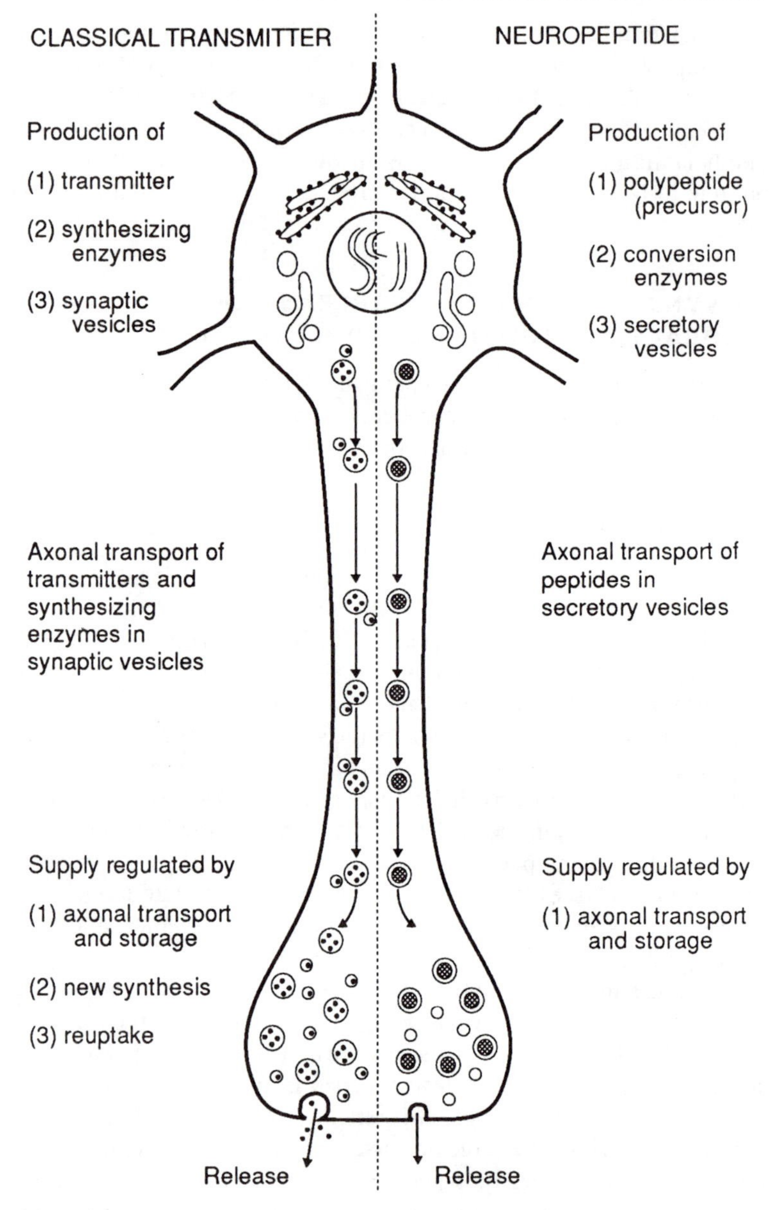

Figure 11.1. A schematic drawing of a neuron comparing the synthesis, storage, release and reuptake of a 'classical' neurotransmitter such as dopamine (left) and a neuropeptide, such as cholecystokinin (CCK) (right). (Redrawn from Hökfelt *et al.*, 1980.)

taken back up into the presynaptic neuron by endocytosis (George and van Loon, 1981), but this is not conclusive (see McKelvy and Blumberg, 1986).

The pre-peptide precursors may contain the amino acid sequences for one or more neuropeptides. Some, such as provasopressin (shown in Figure 7.2, p. 117) contain the amino acid sequence for a single neuropeptide. Others contain the amino acid sequences for many copies of the same peptide or a whole family of peptides. The endogenous opioids provide examples of the complexity of the neuropeptide precursor molecules. There are three families of endogenous opioids: the endorphins, enkephalins, and dynorphins (Akil *et al.*, 1984, 1988). Each

A. PROOPIOMELANOCORTIN

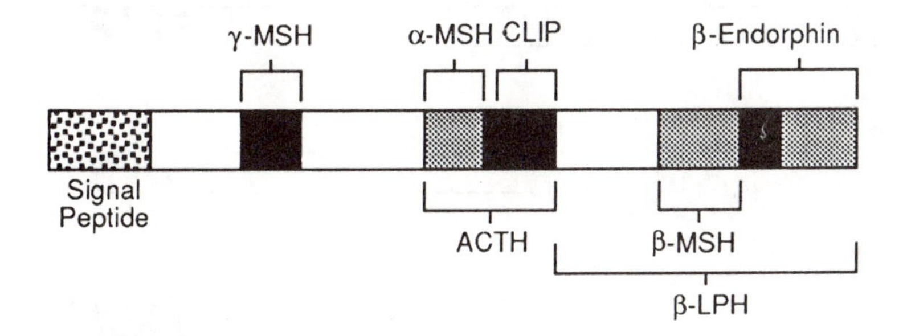

B. PROENKEPHALIN

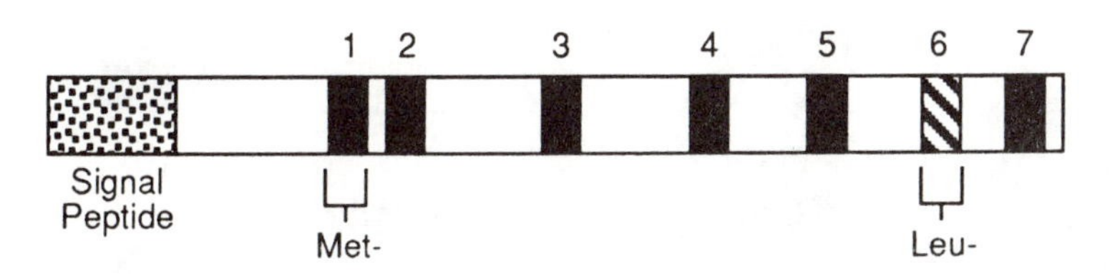

C. PRODYNORPHIN

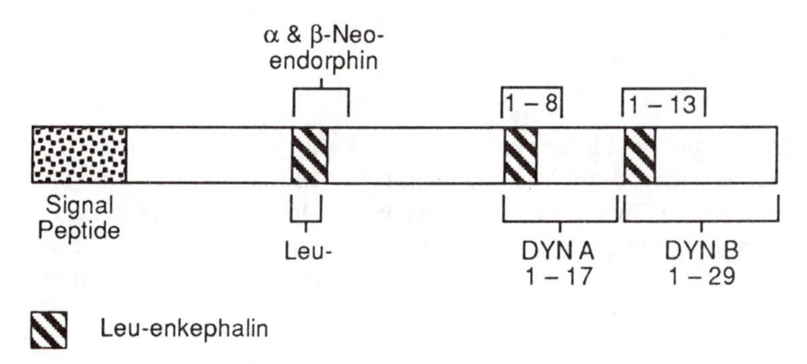

Figure 11.2. Schematic representations of the structure of the three opioid peptide precursors. The amino acid sequence for met-enkephalin is present in both proopiomelanocortin and proenkephalin, while the sequence for leu-enkephalin is common to both proenkephalin and prodynorphin. (A) Proopiomelanocortin is the precursor for β-lipotropin (β-LPH) and β-endorphin as well as ACTH, α-MSH, β-MSH, γ-MSH and corticotropin-like intermediate lobe peptide (CLIP). (B) Proenkephalin is the precursor for both met-enkephalin and leu-enkephalin. (C) Prodynorphin is the precursor for α and β neoendorphin, dynorphin A (DYN A) and dynorphin B (DYN B). (Redrawn from Khachaturian *et al.*, 1985.)

family is derived from a different precursor protein: proopiomelanocortin (POMC), proenkephalin, or prodynorphin (Figure 11.2). ACTH and MSH, as well as β-endorphin, belong to the POMC family (Figure 11.2A). Proenkephalin encodes seven copies of the enkephalins, six of leu-enkephalin and one of met-enkephalin (Figure 11.2B). Prodynorphin encodes dynorphin A, dynorphin B and the neo-endorphins (Figure 11.2C). Pre-procholecystokinin contains the amino acid sequences for the cholecystokinin family, each member of which is synthesized by cleavage from other members of the same family (Figure 11.3).

11.2.2 STORAGE AND RELEASE

Neuropeptides are stored in large secretory vesicles in the nerve endings and released when the nerve is depolarized. Neurotransmitters are stored

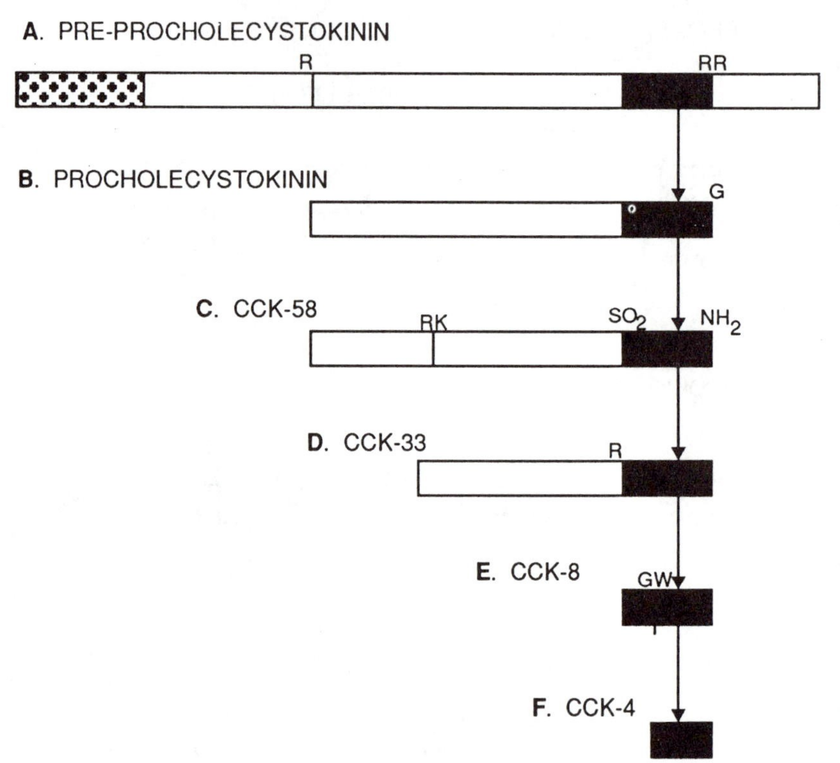

Figure 11.3. The synthesis of the cholecystokinin peptide family. (A) Pre-procholecystokinin contains the sequences of all forms of cholecystokinin. (B) Procholecystokinin is produced by cleavage of the precursor at a single basic amino acid (R) and at the dibasic amino acids (RR). (C) CCK-58 is produced from procholecystokinin by sulfation (SO_2) of a tyrosine and by amidation of the glycine residue (G) to produce the common carboxyl terminus. (D) CCK-33 is produced by cleavage of CCK-58 at dibasic amino acids (RK). (E) CCK-8 is produced by cleavage of CCK-33 at a single basic amino acid (R). (F) CCK-4 is produced from CCK-8 by cleavage between a glycine (G) and a tryptophan (W). (Redrawn from Lynch and Snyder, 1986.)

in both small and large synaptic vesicles and, in many neural cells, the large vesicles contain both monoamine transmitters and neuropeptides, as shown in Figure 11.4A. Thus, when the cell is depolarized, the transmitter, the peptide, or both may be released into the synapse, depending on the nature of the depolarizing stimulus (Hökfelt *et al.*, 1980; Bartfai *et al.*, 1988; Eccles, 1986).

11.2.3 DEACTIVATION

Neurotransmitters are deactivated by a variety of specific enzymes (see Table 5.4, p. 71) and by reuptake into the presynaptic cell. Neuropeptides are deactivated through degradation by peptidase enzymes which exist throughout the brain and nervous system (Lynch and Snyder, 1986). Specific peptidases exist to deactivate the enkephalins (enkephalinase), while other neuropeptides are deactivated by a range of peptidase enzymes. As mentioned above, neuropeptides might also be deactivated by reuptake into the presynaptic cell. The 'deactivation' of neuropeptides often results in the production of small peptide 'fragments' which may have their own neuropeptide actions. For example, CCK is active in many forms (see Figure 11.3). Fragments of TRH and substance P may also have neuropeptide activity (White *et al.*, 1985; McKelvy and Blumberg, 1986; Lynch and Snyder, 1986).

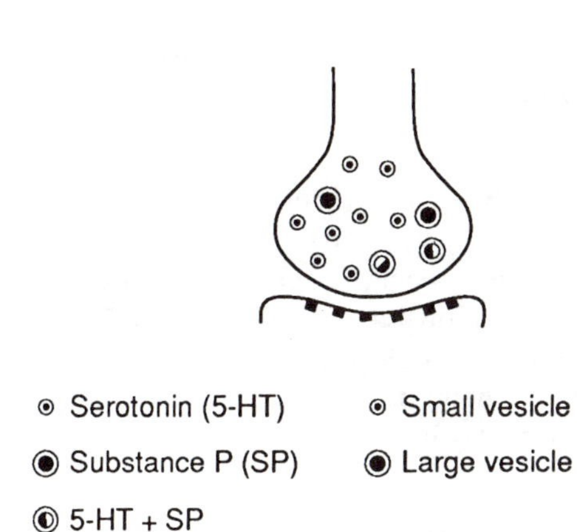

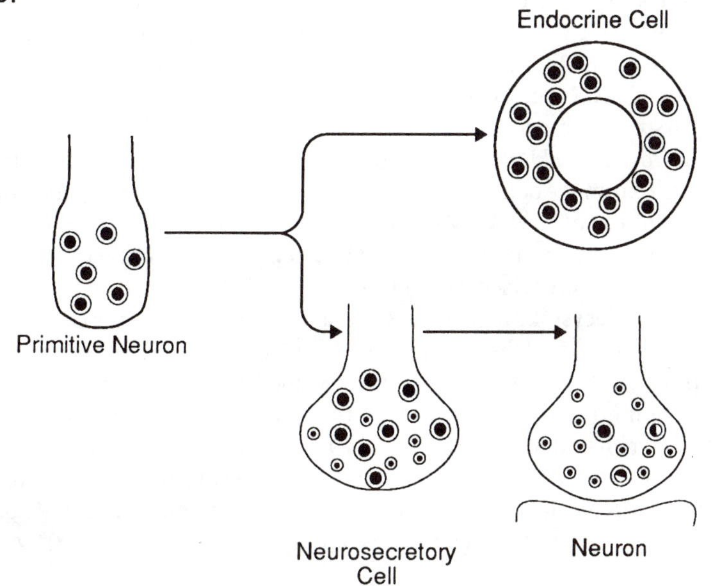

Figure 11.4. The different types of storage granule for neuropeptides and neurotransmitters and their possible evolutionary relationship. (A) Monoamines, such as 5-HT are present in both small (500 angstroms in diameter) and large (1000 angstroms in diameter) vesicles. Neuropeptides, such as substance P are present only in large vesicles. Neuropeptides and monoamines can coexist in the large vesicles, but the small vesicles contain only monoamines. (Redrawn from Hökfelt *et al.*, 1980.) (B) The hypothetical evolution of storage vesicles for neurotransmitters, neuropeptides and hormones. In primitive neurons, one type of large vesicle may have stored both the small monoamines and the neuropeptides as occurs in several neural cells in modern mammalian species. With the demand for faster communication, new types of storage vesicles, the small synaptic vesicles, evolved in neurons for storing and releasing exclusively the classical neurotransmitters and are present in neurons in addition to the larger vesicles storing both classical transmitters and neuropeptide(s). The neurosecretory cells represent an intermediate between endocrine cells and neurons and contain a higher proportion of large vesicles than the neurons which release their messengers at well defined synapses. Endocrine cells may have evolved from the primitive neurons and retained the ability to store peptides and monoamines in the same storage granule. (Redrawn from Hökfelt *et al.*, 1986.)

11.3 EXPLORING THE RELATIONSHIPS BETWEEN NEUROPEPTIDES, NEUROTRANSMITTERS AND HORMONES

Neuropeptides are closely related to both the 'true' hormones and the neurotransmitters and, as noted previously, the terminology used to describe these chemical messengers is often confusing (see Table 1.2, p. 14, for example). The relationships between these different neuroregulators can be examined in three ways: by looking at their common biosynthetic pathways, their common evolutionary pathways, or their common embryological origins (Dockray, 1984, Krieger, 1986).

11.3.1 COMMON BIOSYNTHETIC PATHWAYS

As a result of being encoded by the same genes, peptides often occur in 'families' which are based on common pre-prohormone precursors. For example, ACTH, α-MSH, and β-endorphin belong to the same family, as they are all synthesized from proopiomelanocortin (Figure 11.2). Likewise, the CCK family of peptides is synthesized from the common precursor, procholecystokinin (Figure 11.3); the enkephalins are synthesized from proenkephalin (Figure 11.2) and the tachykinins (substance P, neurokinin A, neuropeptide K) are synthesized from pre-protachykinin (White *et al.*, 1985; Lynch and Snyder, 1986).

The precursor proteins are split at different locations in each biosynthetic pathway to give rise to unique peptides in different cells. For example, pre-procholecystokinin contains the sequences for all forms of CCK. The endocrine glands of the intestine produce the hormone CCK-33, while the neurons of the brain produce CCK-8 and the gastrointestinal nerves innervating the pancreas produce the small CCK-4 fragment (Krieger, 1986; Lynch and Snyder, 1986). The catecholamines, indoleamines and steroid hormones also form families of chemical messengers in which different molecules are synthesized through common biosynthetic pathways in different cells (see Figures 5.4 and 7.3, pp. 64 and 118).

11.3.2 COMMON EVOLUTIONARY PATHWAYS

The same families of hormones, neurotransmitters, peptides and cytokines are found in all vertebrates, and there are vertebrate-like hormones and neurotransmitters in the multicellular and unicellular invertebrates and in the higher plants, as illustrated in Figure 11.5. (Miller *et al.*, 1983; Roth *et al.*, 1982, 1985, 1986; LeRoith and Roth, 1984; LeRoith *et al.*, 1986). For example, substances similar to insulin, glucagon, somatostatin, substance P and ACTH are found in insects, crustaceans and molluscs, and vertebrate-like peptides are found in plants such as alfalfa (TRH), tobacco (somatostatin and interferon), spinach (insulin) and wheat (opioids). The chemicals which act as neurotransmitters and neuropeptides are thus phylogenetically very old, but specific neurons did not exist until the higher invertebrates evolved, and endocrine glands did not appear until the vertebrates evolved (see Figure 11.5). These findings

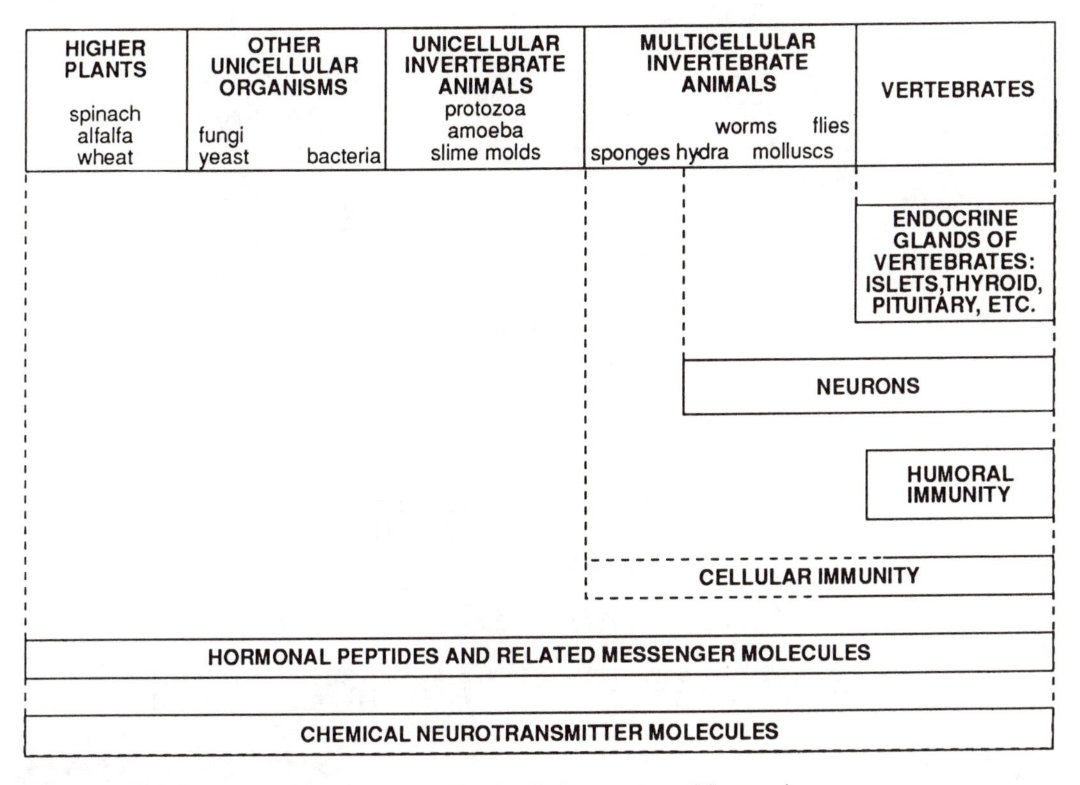

Figure 11.5. The postulated evolutionary origins of the chemical messengers of the neural, endocrine and immune systems. Neurotransmitter and neuropeptide molecules are present in higher plants and unicellular invertebrates. Specialized neurons evolved in the multicellular invertebrates, while endocrine glands did not exist until the vertebrates evolved. Cellular immunity may also have evolved in the multicellular invertebrates and humoral immunity in the vertebrates. (Redrawn from Roth *et al.*, 1985.)

have led to the hypothesis that the neural, endocrine and immune systems have evolved phylogenetically in a Darwinian fashion from a common neuropeptide system which exists in unicellular organisms. Thus, the chemical signals used in mammalian cellular communication may have evolved in different ways, some becoming neurotransmitters and others neuropeptides (Le Roith, Shiloach, and Roth, 1982; Le Roith and Roth, 1984; Roth *et al.*, 1985).

The phylogenetic relationships among the neuropeptides in some families are well known, as shown for the neurohypophyseal hormones in Figure 11.6. Likewise, the GnRH family consists of at least five neuropeptides found in the brains of fish, birds and mammals (Sherwood, 1987) and the opioid precursors, such as POMC, show phylogenetic relationships throughout the vertebrates (Dores *et al.*, 1990). Neuropeptide evolution can be studied by examining the structure of the genes which code for neuropeptides, by analyzing mRNA sequences, or by examining the amino acid sequences of the neuropeptides themselves (Dores *et al.*, 1990). The use of these molecular techniques to examine the structural relationships (homologies) among the chemical messengers in plants and animals indicates that neuropeptide families could have

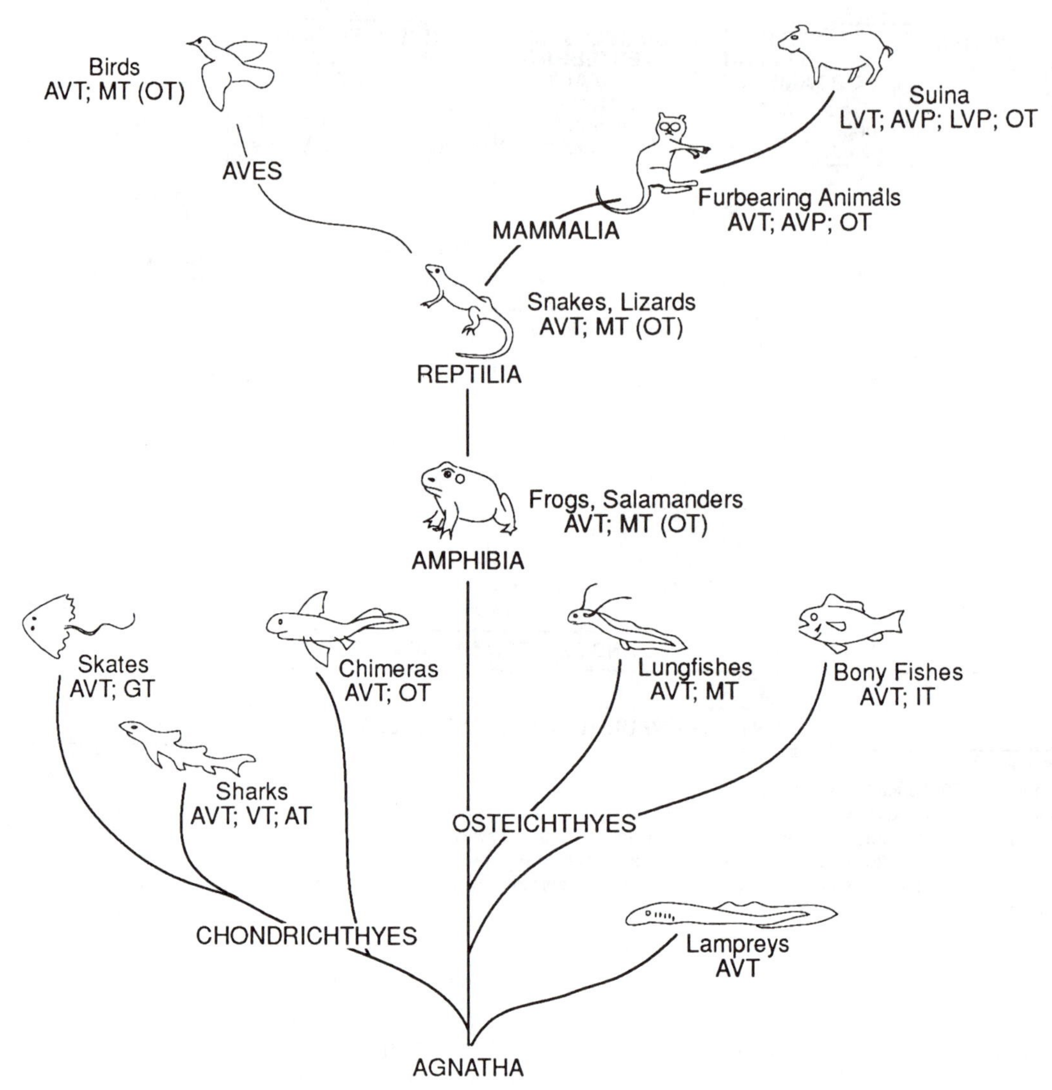

Figure 11.6. The phylogenetic distribution of the neurohypophyseal peptides throughout the vertebrate classes (in capital letters); subclasses, orders, and suborders (in capitals and lower case letters). Hormones separated by semicolons are assumed to be determined by different gene loci; those representing established polymorphisms are separated by commas. A hormone enclosed in parentheses represents an uncertain situation that may be either mistaken identity or polymorphism. AT = aspartocin; AVP = arginine vasopressin; AVT = arginine vasotocin; GT = glumitocin; IT = isotocin; LVP = lysine vasopressin; MT = mesotocin; OT = oxytocin; VT = valitocin. (Redrawn from Turner and Bagnara, 1976.)

evolved from mutations or substitutions in the amino acid sequences of a small number of original biologically active peptides. These changes could have occurred as the neuropeptides were synthesized on the ribosomes after translation of the genetic code via the messenger RNA. The 'gut peptides' (CCK, gastrin, glucagon, secretin and VIP), for example, may all have evolved from a single VIP-like peptide (Dockray, 1984; Krieger, 1986).

There are a number of reasons for thinking that the chemical messengers evolved via changes in their amino acid sequences. First, the evolution of individual hormones can be traced through the vertebrates as illustrated by the evolution of the neurohypophyseal hormones (Figure 11.6). Second, the evolution of families of related peptides such as GH and PRL, the tachykinins and gastrointestinal hormones can be traced through their common amino acid sequences (Stewart and Channabasavaiah, 1979; Miller, Baxter and Eberhardt, 1983). Third, certain amino acid sequences are shared by different peptides. For example, CCK and gastrin share peptide sequences, as do glucagon, secretin and VIP, even though these peptides have different functions. Fourth, different peptides with similar amino acid sequences have similar functions in different species. For example, secretin stimulates pancreatic function in mammals, but not birds, and VIP stimulates pancreatic function in birds, but not mammals (Krieger, 1986). Details of the mechanisms involved in the molecular evolution of the peptides are discussed by Acher (1983).

11.3.3 COMMON EMBRYOLOGICAL ORIGINS

The developing embryo has three layers: ectoderm, mesoderm and endoderm (Figure 11.7A). The brain and spinal cord develop from the 'neural ectoderm', as do the the cells of the neural crest (Figures 11.7B and C). The neural crest cells then differentiate into peripheral sensory neurons, neurons of the sympathetic nervous system, melanocytes, endocrine cells, and other cell types (Landis and Patterson, 1981). Certain embryonic cells from the neural crest develop into neurons which are able to synthesize both neurotransmitters and peptides. These cells, termed the 'neuroendocrine-programed epiblast' cells (Pearse, 1979) give rise to the neuroendocrine and peripheral endocrine cells.

One type of endocrine cell which develops from the neural crest is the chromaffin cell, which secretes the hormones of the adrenal medulla. These chromaffin cells, and certain sympathetic neurons, are derived from a common 'bipotential progenitor cell', which can develop into either a neural or an endocrine cell, depending on the presence of growth factors and hormones in the embryonic environment (Figure 11.8). These cells thus show plasticity during development. If they are stimulated by fibroblast growth factor and nerve growth factor, they develop into neurons, but if they are stimulated by glucocorticoids, they develop into endocrine cells (Anderson, 1989).

The APUD cell concept

Pearse (1969, 1979) originally named the cells derived from the neuroendocrine-programed epiblast cells the 'amine precursor uptake and decarboxylation' (APUD) cells. He believed that, at some stage in their development, these cells are able to take up the amino acid precursors of the amine transmitters (tyrosine and tryptophan) and possess the enzymes (such as tyrosine hydroxylase and DOPA decarboxylase) necessary to convert these precursors to catecholamine transmitters (see Figure 5.4, p. 64). It has been suggested that these common progenitor

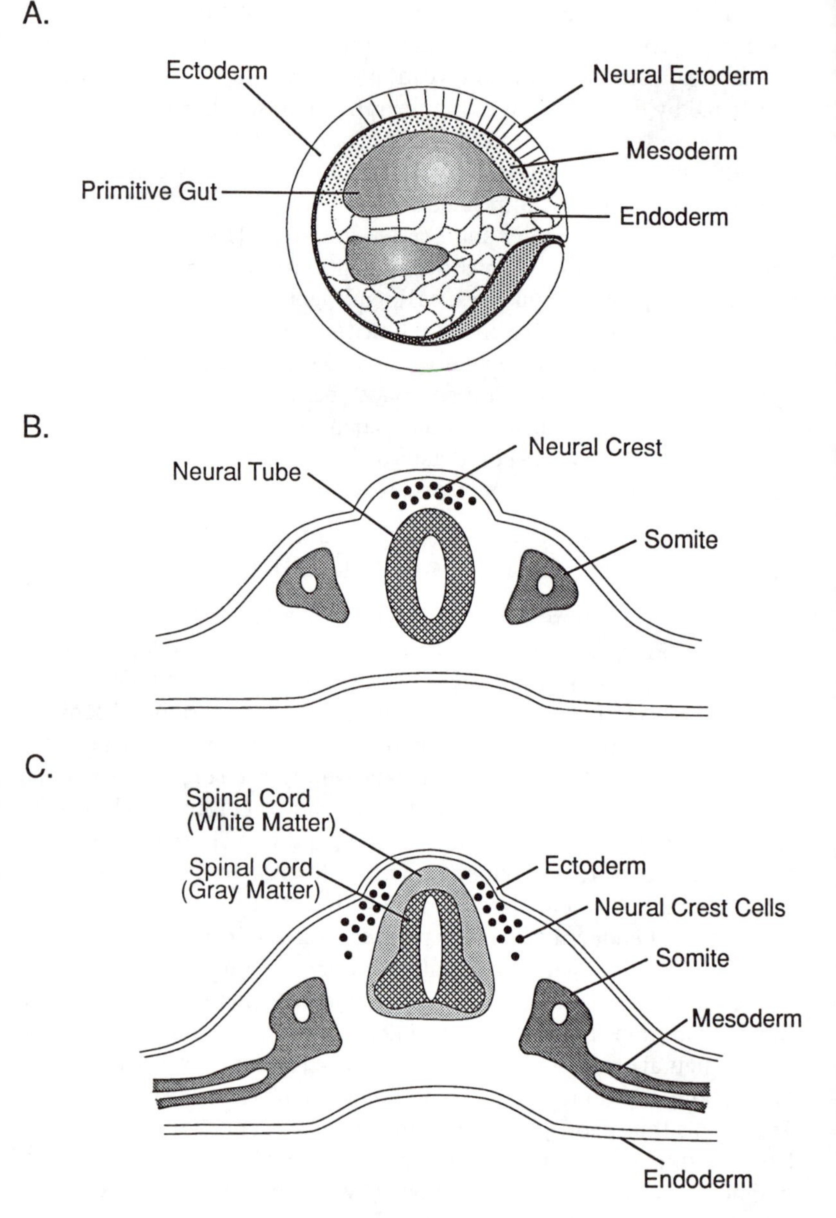

Figure 11.7. The embryological origins of the neuroendocrine system. (A) The three layers of the vertebrate embryo, as shown in an amphibian. The ectoderm gives rise to the epidermis and the nervous system (which arises from the 'neural ectoderm'). The mesoderm gives rise to the muscles, skeleton, and connective tissue. The endoderm gives rise to the gut and related structures. (Redrawn from Purves and Lichtman, 1985.) (B) The development of the peripheral nervous system in a four-week-old human embryo. The neural crest is a group of cells which forms above the neural tube (both of which develop from the neural ectoderm) and gives rise to the peripheral nervous system. The neural tube gives rise to the brain and spinal cord. Somites are blocks of mesoderm which lie along the neural tube and give rise to skeletal muscles, vertebrae, and dermis. (C) The development of the spinal cord and the migration of the cells of the neural crest as the peripheral nervous system develops in the human embryo at about 8 weeks of age. (Redrawn from Cowan, 1979.)

cells provide the embryological origin of more than 40 different peripheral neuroendocrine cell types, including the hypothalamic neuroendocrine cells, the adenohypophyseal cells, the adrenaline secreting chromaffin cells of the adrenal medulla, the calcitonin secreting C cells of the thyroid, and the endocrine cells of the thymus gland, pancreas, gastrointestinal tract and placenta (Pearse and Takor, 1979; Andrew, 1982).

Not all of Pearse's claims about the APUD concept have been confirmed, however (Pictet *et al.*, 1976; Andrew, 1982; Baylin, 1990). For example, some APUD cells do not contain the enzyme dopa decarboxylase so they can not synthesize monoamine transmitters from their amino

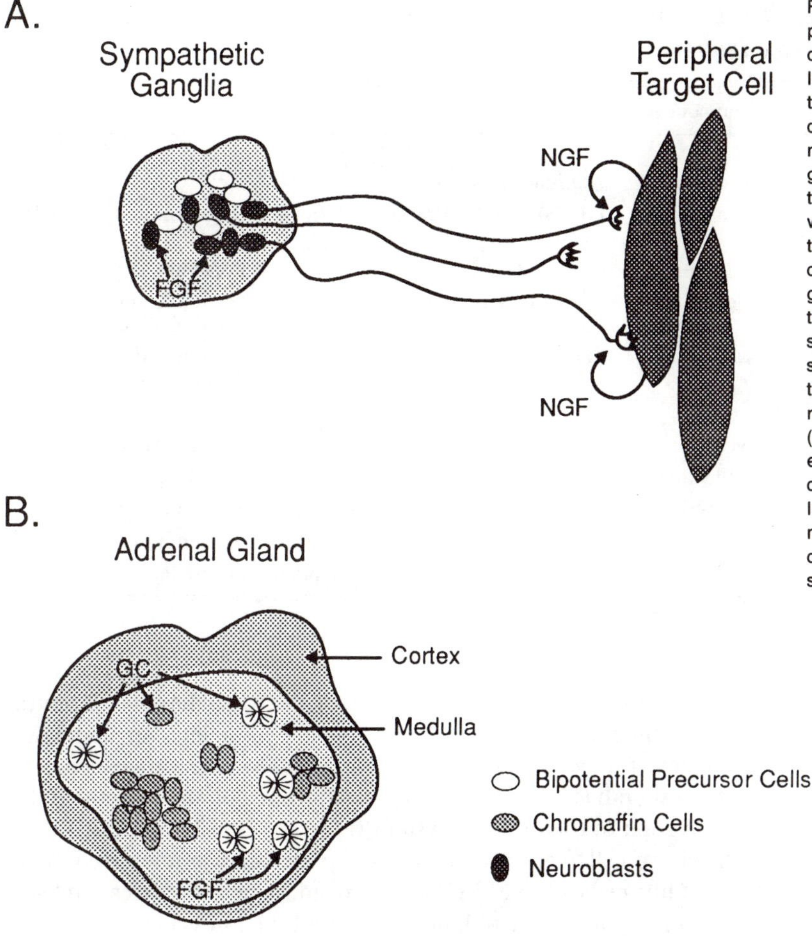

Figure 11.8. The bipotential precursor cells of the neural crest are plastic in their development and may differentiate into neural or endocrine cells. (A) In the sympathetic nerve ganglia, fibroblast growth factor (FGF) induces the development of neurons whose axonal projections extend toward peripheral target cells. The release of nerve growth factor (NGF) from these target cells promotes synapse formation and the survival and maturation of these neurons. (B) In the adrenal cortex, glucocorticoids (GC) inhibit the neural differentiation of the progenitor cells, which are then stimulated by FGF to develop into mature endocrine gland cells. (Redrawn from Anderson, 1989.)

acid precursors (Pearse, 1983). Although Pearse suggested that the APUD cells have a common embryological origin, experimental evidence does not confirm this. APUD cells arise from a variety of embryonic tissues, including the neural crest (thyroid gland and adrenal medulla) and the neuroectoderm (pineal gland and adenohypophysis). The origin of the neuroendocrine cells of the gastrointestinal tract (pancreas, stomach, intestine) is controversial (Andrew, 1982; Krieger, 1986). Despite these criticisms, the APUD theory has provided insights into the embryological origins of the neuroendocrine system and has provided one explanation for the development of the diverse groups of chemical messengers that we call neurotransmitters, hormones, neurohormones and neuropeptides (Baylin, 1990). Cells of the thymus gland also develop from the neural crest, providing a developmental mechanism for the interaction between the neural and immune systems (Roth *et al.*, 1985).

The diffuse neuroendocrine system

Since many APUD cells do not synthesize amines, but do produce peptides, the 'APUD' cells are now considered to be part of the 'diffuse

Table 11.2. *The peptides and amines synthesized by the endocrine and neural components of the central and peripheral divisions of the diffuse neuroendocrine system*

	Component	Peptide(s)	Amine
Central division			
Pineal gland	Endocrine	Arginine vasotocin	Melatonin
Anterior pituitary	Endocrine	LH, FSH, TSH, GH, PRL, ACTH, β-END	Dopamine and serotonin
Intermediate pituitary	Endocrine	α-MSH, β-LPH, β-endorphin	Dopamine and serotonin
Hypothalamic magnocellular neurons	Neural	Oxytocin, dynorphin, vasopressin, ENK	Serotonin, dopamine, noradrenaline
Hypothalamic parvicellular neurons	Neural	Releasing and inhibiting hormones	Serotonin, dopamine, noradrenaline
Peripheral division			
Adrenal medulla	Neural	Enkephalins, dynorphin	Noradrenaline, adrenaline
Thyroid	Endocrine	Calcitonin	Serotonin
Pancreas	Endocrine	Insulin, glucagon, somatostatin	Serotonin, dopamine
Stomach	Endocrine	Substance P	Serotonin
Intestine	Endocrine	Substance P	Serotonin

Source: Pearse, 1983.

neuroendocrine system' (Andrew, 1982). This system has both a central and a peripheral division, each having a neural and an endocrine component as shown in Table 11.2 (Pearse, 1983). The diffuse neuroendocrine system has been referred to as a third (endocrine or neuroendocrine) division of the nervous system (Pearse, 1979; Pearse and Takor, 1979) whose cells can produce both peptides and amines (see Table 11.2). Many cells of the diffuse neuroendocrine system can release peptides in response to neural input and thus function as neuroendocrine 'sensory effectors' or neuroendocrine transducers (as described in Section 4.3). The peptides released can have neurocrine, neuroendocrine, endocrine or paracrine actions (see Figure 1.3, p. 5), depending on the location of the neuroendocrine cells (e.g. pineal gland, hypothalamus, adenohypophysis, thyroid, adrenal medulla, or gastrointestinal tract). These peptides can suppress, amplify, or modulate the activities of the central and autonomic nervous systems, as discussed in Chapter 12.

Pearse (1983) has suggested that, because the cells of the diffuse neuroendocrine system are able to produce both peptides and amine transmitters, this system may be the evolutionary precursor of the more specialized nerve cells and endocrine glands. As the multicellular invertebrates evolved, some neuroendocrine precursor cells may have evolved into neurons which synthesized primarily neurotransmitters (Figure 11.5). In the vertebrates, some neuroendocrine precursor cells may have evolved into neurons and some evolved into the peptide secreting cells of the endocrine system. Still other neuroendocrine cells retained the ability to synthesize both amines and peptides (Le Roith *et al.*, 1982; Pearse, 1983). The hypothesis that the distinct neural and endocrine systems evolved from the diffuse neuroendocrine system may be illustrated by the different types of storage vesicles used in the cells of these systems

(Hökfelt *et al.*, 1986). The storage granules of the primitive neuroendocrine cells may have contained both peptides and amines (Figure 11.4B). Specialized nerve cells retained their ability to synthesize and store amines, while endocrine glands retained their ability to synthesize peptides. Under this hypothesis, the neurosecretory cells represent an intermediary cell type between the primitive neuroendocrine cell and the specialized neuron.

11.4 COEXISTENCE (COLOCALIZATION) OF NEUROTRANSMITTERS AND NEUROPEPTIDES

Pearse (1969) developed the APUD concept from the observation that certain cells of the peripheral neuroendocrine system (the adrenal medulla and gastrointestinal tract) could synthesize both amines and peptides. Since these neuroendocrine cells develop from the neuroectoderm (via the neural crest), other investigators began to look for the coexistence of amine and peptide transmitters in cells of the central and autonomic nervous systems, which have the same embryonic origin. Through the use of radioimmunoassays (Rorstad, 1983), immunocytochemistry (Elde, 1983) and neuroanatomical methods (Palkovits, 1983), many nerve cells in the brain and spinal cord have been found to synthesize and store both a neurotransmitter and a neuropeptide and to release these two chemical messengers to act as cotransmitters in the nervous system.

11.4.1 SYNTHESIS, STORAGE AND RELEASE OF COTRANSMITTERS

In order for a neuron to synthesize and store both neurotransmitters and neuropeptides that cell must have both of the biosynthetic pathways shown in Figure 11.1. In such a cell, peptide synthesis will be completed in the ribosomes and the completed peptides transported down the axons in their secretory vesicles to the nerve terminals. Neurotransmitter synthesis will be completed at the nerve terminals through the action of the enzymes packaged into the synaptic vesicles with the transmitter precursors (Bartfai *et al.*, 1988). The neuron will also have both large and small storage vesicles, as shown in Figure 11.4A. Monoamines are present in both the small and large secretory vesicles, but neuropeptides occur only in the large vesicles; thus the two transmitters coexist in the large vesicles whereas the small vesicles contain only monoamines.

When a peptide and a neurotransmitter occur in the same synaptic vesicle, the neurotransmitter is often referred to as the 'primary transmitter' and the peptide as the 'cotransmitter' (Guidotti *et al.*, 1983). Release of the small vesicles results only in primary transmitter action while release of the large vesicles results in both primary transmitter and cotransmitter action. Whether the small or large vesicles are released depends on the intensity of neural stimulation (Bartfai *et al.*, 1988). The release of neurotransmitters from the small vesicles requires only a low frequency of neural stimulation, whereas the release of both neurotransmitters and neuropeptides from the large vesicles requires a higher frequency of stimulation or a bursting pattern of stimulation (see Section 12.1).

Because the classical neurotransmitters and neuropeptides are colocalized within the nerve terminal, drugs which alter the synthesis, storage and release of the neurotransmitter in a neuron may also influence the synthesis, storage and release of the neuropeptides colocalized in that cell. But, because the amines and peptides are produced in different biosynthetic pathways, the drug may have different effects on the two transmitters (Bartfai *et al.*, 1988). For example, reserpine, which inhibits catecholamine storage (Table 5.5, p. 76) also inhibits the storage of the colocalized peptides in the nerve terminals. Thus, reserpine reduces the amounts of NE and NPY stored in the terminals of the adrenergic nerves innervating the heart and reduces the storage of DA and CCK-8 in the neurons of the nucleus accumbens. On the other hand, certain drugs which alter neurotransmitter synthesis may have no effect on the synthesis of the colocalized peptides. For example, the antipsychotic drugs, haloperidol and chlorpromazine block DA synthesis (Table 5.6, p. 78), but do not block CCK synthesis, so CCK levels increase in the neurons of the midbrain where these transmitters coexist, but DA levels decline (Bartfai *et al.*, 1988). This means that drugs developed to alter catecholamine levels may also alter the levels of enkephalins, CCK and other colocalized neuropeptides and that the behavioral and psychiatric disorders thought to be regulated by classical neurotransmitters may also be influenced by neuropeptides. Likewise, antidepressant drugs may alter TRH levels in the brain (Przegalinski and Jaworska, 1990).

11.4.2 DALE'S PRINCIPLE AND THE PROBLEM OF COLOCALIZATION

The concept that each nerve cell has the ability to synthesize and release only one neurotransmitter has been loosely termed 'Dale's Principle'. The discovery of the colocalization of classical neurotransmitters and neuropeptides in the same neuron has led to a re-examination of this principle (Burnstock, 1976; Osborne, 1981). According to Eccles (1986, p. 3), however, Dale's Principle is stated as 'any one class of nerve cells operates at all of its synapses by the same chemical transmission mechanism. This principle stems from the metabolic unity of a single cell which extends to all of its branches.' By this definition, 'the metabolic unity' of a neuron can accommodate any number of transmitter substances within the same neuron. Thus, Eccles, who coined the term 'Dale's Principle', believes that the colocalization of classical neurotransmitters and neuropeptides within the same neuron does not violate this principle.

11.4.3 WHICH NEUROTRANSMITTERS AND NEUROPEPTIEDES ARE COLOCALIZED?

There are a number of combinations of neurotransmitters and neuropeptides which can be colocalized in the same neuron. A monoamine or an

Table 11.3. *Some examples of the colocalization of neuropeptides and classical neurotransmitters in the same cells in certain brain areas*

Neurotransmitter	*Neuropeptide*	*Neural site*
Acetylcholine (Ach)	Ach and VIP	Neocortex
	Ach and neurotensin	Preganglionic fibers
	Ach and enkephalin	Spinal cord
	Ach and Substance P	Pons
Dopamine (DA)	DA and CCK	Mesolimbic pathways
	DA, CCK and neurotensin	Substantia nigra (A9) and ventral tegmental area (A10)
	DA, dynorphin and GH-RH	Arcuate nucleus
	DA and enkephalin	Arcuate nucleus
	DA and neurotensin	Arcuate nucleus
Noradrenaline (NA)	NA and neuropeptide Y	Medulla oblongata
	NA and enkephalin	Locus ceruleus
	NA and neurotensin	Solitary nucleus
Serotonin (5-HT)	5-HT and substance P	Medulla and spinal cord
	5-HT and TRH	Medulla and spinal cord
	5-HT and CCK	Medulla and spinal cord
	5-HT and enkephalins	Medulla and spinal cord
GABA	GABA and CCK	Cortex and hippocampus
	GABA and somatostatin	Cortex and hippocampus
	GABA and enkephalin	Caudate nucleus
Two or more neuropeptides	Substance P and TRH	Medulla oblongata
	Substance P and CCK	Central gray
	Enkephalins and NPY	Hypothalamus
	NPY and somatostatin	Hypothalamus
	SOM, ENK, gastrin	Hypothalamus
	GH-RH, NPY, NT	Arcuate nucleus
	Vasopressin and DYN	SON and PVN
	Oxytocin and CRH	SON and PVN
Two 'classical' neurotransmitters	Serotonin and GABA	Dorsal raphe nucleus
	Acetylcholine and GABA	Medial septum

Sources: Hökfelt, *et al.*, 1980; Nieuwenhuys, 1985; Müller and Nistico, 1989; Crawley, 1990.

amino acid transmitter can coexist with one or more peptides, as occurs when substance P is colocalized with noradrenaline, enkephalin or CCK with dopamine, VIP with acetylcholine, or CCK with GABA. Two or more peptides can also be colocalized. For example, ACTH and β-endorphin are colocalized in hypothalamic neurons, substance P and TRH in neurons of the medulla oblongata, and vasopressin and dynorphin in cells of the PVN and SON. Two 'classical' neurotransmitters may also be colocalized, as occurs with serotonin and GABA in the dorsal raphe nucleus (Crawley, 1990). Some examples of colocalization are given in Table 11.3. Although no rules have yet been derived to predict which chemical messengers will be colocalized, it appears that monoamines coexist most frequently with the gastrointestinal ('brain–gut') peptides. The peptides derived from proopiomelanocortin are not colocalized with any of the classical neurotransmitters or peptides derived from other precursors, but the peptides derived from proenkephalin and prodynorphin are colocalized with classical neurotransmitters and other peptides (Hökfelt *et al.*, 1980, 1986; Nieuwenhuys, 1985; Kupfermann, 1991).

11.4.4 NEUROPEPTIDES COLOCALIZED WITH PITUITARY HORMONES

Dynorphin, CCK and CRH are colocalized with vasopressin and oxytocin in the neurons of the SON and PVN and are co-released from the neurohypophysis along with these hormones. Dynorphin is co-released with vasopressin and appears to inhibit oxytocin release (see Figure 12.10). CRH is co-released with oxytocin and may act to regulate the release of adenohypophyseal hormones (Bondy *et al.*, 1989). Neuropeptides are also synthesized and released from the endocrine cells of the adenohypophysis along with the 'classical' adenohypophyseal hormones. For example: lactotroph cells (which synthesize prolactin) and thyrotroph (TSH) cells both synthesize VIP; gonadotroph cells (LH and FSH) also synthesize neuropeptide Y, angiotensin II, dynorphin and enkephalin; somatotroph cells (GH) also synthesize neuropeptide Y; and, corticotroph cells (ACTH) synthesize neuropeptide Y as well as the POMC derivatives, β-endorphin and β-lipotropin. Gastrin, CCK, substance P, neurotensin, bombesin, and other neuropeptides, as well as growth factors and cytokines are also synthesized in the adenohypophysis. The functions of these neuropeptides in the pituitary gland have yet to be determined (Houben and Denef, 1990).

11.5 LOCALIZATION OF NEUROPEPTIDE CELL BODIES AND PATHWAYS IN THE BRAIN

Cell bodies containing neuropeptides are located throughout the brain. The use of immunohistochemical methods enables the localization of specific neuropeptide cell bodies and neuropeptide pathways in the brain (Hökfelt *et al.*, 1987). Many neuropeptidergic neurons have cell bodies in more than one neural area, but have short axons, which restrict their distribution. These include VIP, TRH, neurotensin, cholecystokinin and the enkephalins, as well as substance P and somatostatin, whose neural distributions are shown in Figure 11.9. Substance P is colocalized with serotonin neurons in the raphe nuclei of the brain stem and with acetylcholine in the lateral dorsal tegmental nucleus, but most substance P neurons do not appear to have coexisting neurotransmitters. Somatostatin is colocalized with GABA in the cerebral cortex and hippocampus, but the majority of the somatostatin containing neurons have no coexisting neurotransmitters (Hökfelt *et al.*, 1987).

Other neuropeptides, such as the hypothalamic hormones, are synthesized in only a few specific nuclei (as listed in Table 4.3, p. 46), but send axons throughout the brain in neuropeptide pathways, which resemble the pathways for the classical neurotransmitters shown in Figure 5.8 (p. 72). For example, oxytocin is synthesized primarily in the cells of the PVN and SON of the hypothalamus but the axons of these cells extend to

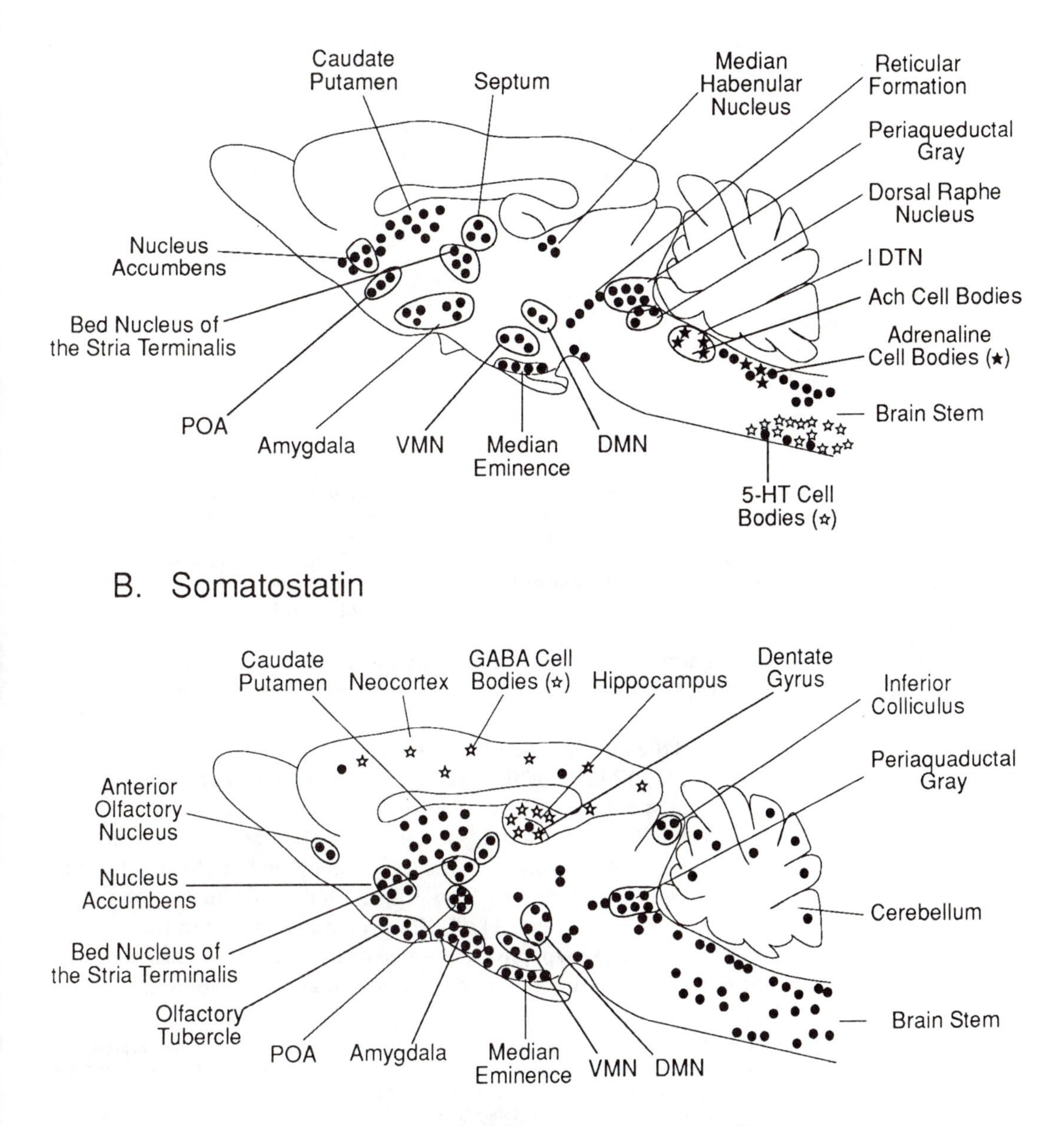

Figure 11.9. A schematic representation of the locations of the major cell groups which synthesize substance P and somatostatin in the rat brain. (A) Substance P is colocalized with acetylcholine (Ach) in cells of the lateral dorsal tegmental nucleus (lDTN) and with adrenaline in cells of the brain stem. In other brain stem cells, as well as in the higher brain areas, substance P is not colocalized with monoamine transmitters. (B) Somatostatin is colocalized with GABA in neurons of the neocortex, hippocampus and dentate gyrus. In other regions of the brain, somatostatin is not colocalized with monoamine transmitters. DMN = dorsomedial nucleus; POA = preoptic area; VMN = ventromedial nucleus. (Redrawn from Hökfelt *et al.*, 1987.)

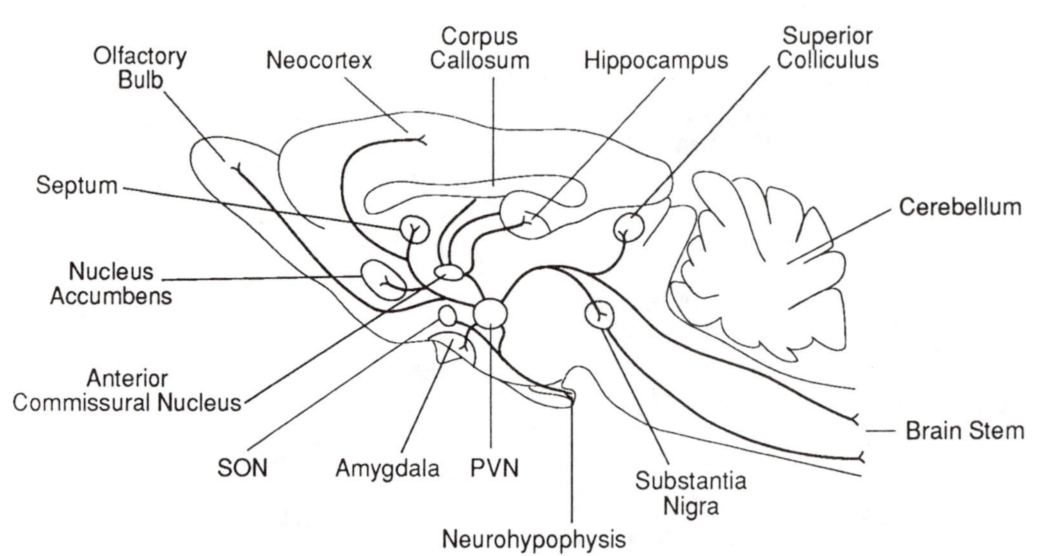

Figure 11.10. A schematic representation of the projections of the oxytocin neurons in the rat brain. The oxytocin cell bodies in the supraoptic nucleus (SON) and paraventricular nucleus (PVN) extend axons to the amygdala, olfactory bulbs, neocortex, nucleus accumbens, hippocampus, superior colliculus, substantia nigra and the pons and medulla of the brain stem, as well as to the neurohypophysis. (Redrawn from Argiolas and Gessa, 1991.)

other brain areas, as shown in Figure 11.10. Vasopressin also has neural pathways throughout the brain (de Vries *et al.*, 1985; Lawrence *et al.*, 1988).

Each of the three endogenous opioid systems has a unique anatomical distribution throughout the brain, spinal cord and peripheral neuroendocrine system (Akil *et al.*, 1984; Khachaturian *et al.*, 1985; Facchinetti, Petraglia and Genazzani, 1987; Evans, Hammond, and Frederickson, 1988). The location of the cell bodies and pathways for the endogenous opioids derived from POMC, proenkephalin and prodynorphin are shown in Figure 11.11. The endorphins are synthesized from POMC in the adenohypophysis, the arcuate nucleus, and in the nucleus of the solitary tract (Figure 11.11A). The enkephalins are synthesized in the

Figure 11.11. The distribution of proopiomelanocortin (POMC), proenkephalin and prodynorphin cell bodies (●) and pathways in the rat brain. (A) The POMC cell bodies are located primarily in the arcuate nucleus and the adenohypophysis. The cell bodies in the arcuate nucleus send axons to other regions of the hypothalamus (POA, PVN, DMN) as well as to the amygdala, bed nucleus of the stria terminalis (BnST), septum, periaqueductal gray, dorsal raphe nucleus and the nuclei of the brain stem. These projections are represented by hatched lines. (B) The proenkephalin cell bodies are located throughout the brain, but are particularly dense in the neocortex, anterior olfactory nucleus, caudate putamen, olfactory tubercule, amygdala, hypothalamus (VMN, PVN,), arcuate nucleus, periaqueductal gray, raphe nuclei, the nucleus of the solitary tract and the dorsal horn of the spinal cord. These cell bodies have both long and short axonal connections to other cells which are represented by hatched lines. (C) The prodynorphin cell bodies are located in the neocortex, caudate putamen, amygdala, hypothalamus (PVN, SON, VMN, DMN), the neurohypophysis and adenohypophysis, and the nucleus of the solitary tract. Prodynorphin-containing neurons also have both long and short axonal connections with other cells. DMN = dorsomedial nucleus; POA = preoptic area; PVN = paraventricular nucleus; SON = supraoptic nucleus; VMN = ventromedial nucleus. (Redrawn from Khachaturian *et al.*, 1985.)

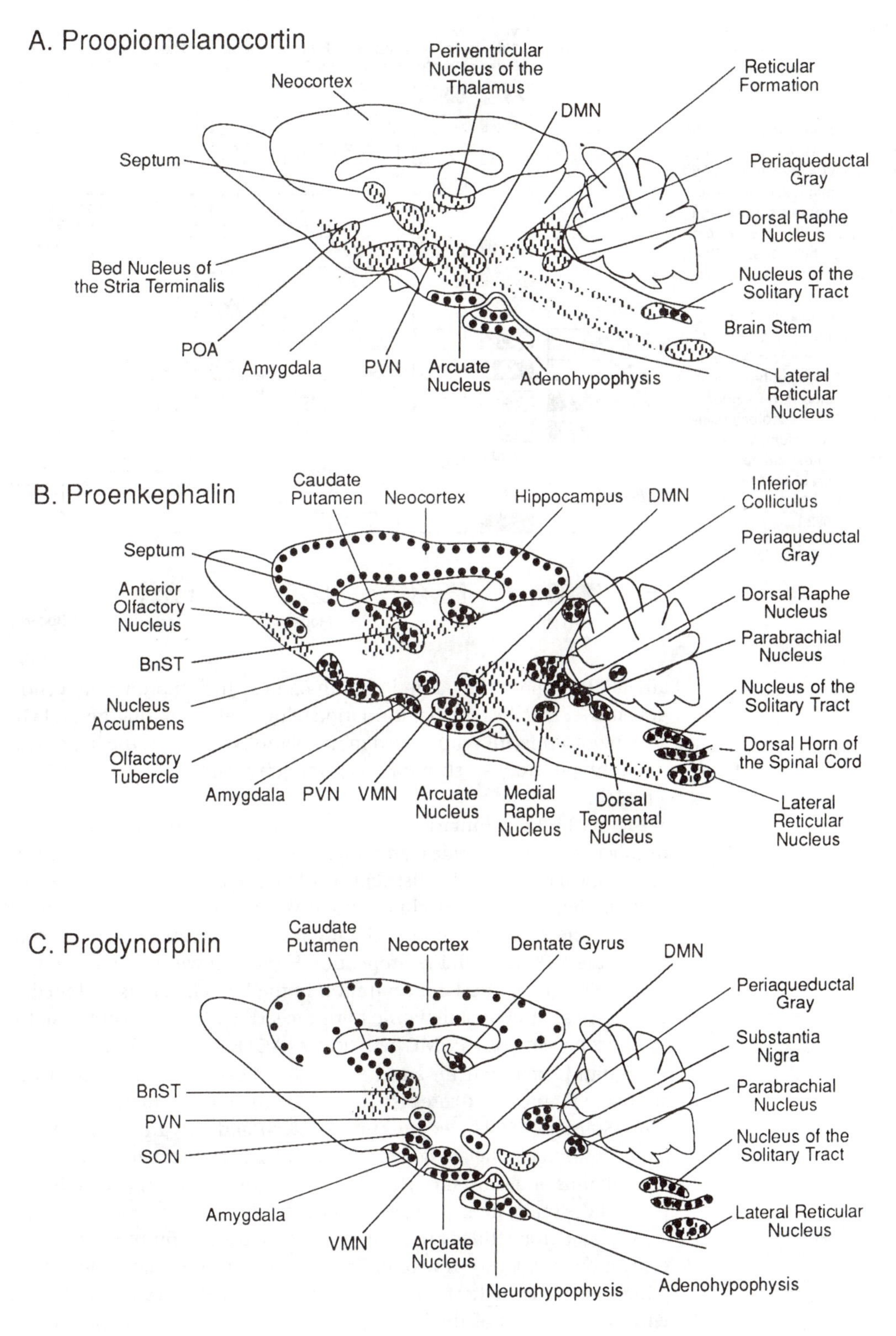
A. Proopiomelanocortin
Neocortex
Periventricular Nucleus of the Thalamus
DMN
Reticular Formation
Septum
Periaqueductal Gray
Dorsal Raphe Nucleus
Bed Nucleus of the Stria Terminalis
Nucleus of the Solitary Tract
Brain Stem
POA
Amygdala
PVN
Arcuate Nucleus
Adenohypophysis
Lateral Reticular Nucleus
B. Proenkephalin
Caudate Putamen
Neocortex
Hippocampus
DMN
Inferior Colliculus
Septum
Periaqueductal Gray
Anterior Olfactory Nucleus
Dorsal Raphe Nucleus
Parabrachial Nucleus
BnST
Nucleus Accumbens
Nucleus of the Solitary Tract
Olfactory Tubercle
Dorsal Horn of the Spinal Cord
Amygdala
PVN
VMN
Arcuate Nucleus
Medial Raphe Nucleus
Dorsal Tegmental Nucleus
Lateral Reticular Nucleus
C. Prodynorphin
Caudate Putamen
Neocortex
Dentate Gyrus
DMN
Periaqueductal Gray
Substantia Nigra
BnST
Parabrachial Nucleus
PVN
SON
Nucleus of the Solitary Tract
Amygdala
Lateral Reticular Nucleus
VMN
Arcuate Nucleus
Neurohypophysis
Adenohypophysis

	Neo-cortex	Hypo-thal-amus	Median emi-nence	Amyg-daloid complex	Hippo-campus	Other limbic areas	Thal-amus	Mesen-ceph-alon	Medulla and pons
LHRH		Moderate, cell bodies	Very High	Low		Low, cell bodies		Low	
TRH	Low	Very High, cell bodies	High	Low		Moderate		Low	Low, cell bodies
SOM	Moderate, cell bodies	Very High, cell bodies	High	Moderate, cell bodies	Low, cell bodies	Moderate, cell bodies	Low	Low	Moderate, cell bodies
ACTH	Low	Very High, cell bodies	High	Moderate	Low	Moderate	Low		Low
α-MSH	Low	Very High, cell bodies	High	Low	Low	Moderate	Moderate	Low	Low
β-LPH	Low	Very High, cell bodies	High	Low		Low		Moderate	Low
β-END		Very High, cell bodies	Moderate	Low	Low	Low	Low	Moderate	Low
ENK		High, cell bodies	Moderate	Moderate, cell bodies	Low	Moderate, cell bodies	Low	Moderate, cell bodies	High, cell bodies
Sub P		High, cell bodies	Low	High, cell bodies	Low	High, cell bodies	Low, cell bodies	High, cell bodies	High, cell bodies
NT		Very High, cell bodies	High	Moderate, cell bodies	Low	Moderate	Low	Moderate	Low
CCK-8	Very High, cell bodies	Low, cell bodies	Low	High	Moderate, cell bodies	High	Low	Low	Moderate
VIP	Very High, cell bodies	Moderate, cell bodies		High, cell bodies	Moderate, cell bodies	Moderate, cell bodies	Low	Low, cell bodies	Low
VP	Low	High, cell bodies	Very High	Low	Low	Low	Low	Low	Low
ANG		Moderate, cell bodies	Moderate	Low			Low	Low	Low
Insulin	Low, cell bodies	Very High				Moderate			

Very High High Moderate Low Cell Bodies

Figure 11.12. Brain areas in which cell bodies containing selected neuropeptides have been demonstrated and the relative concentrations of these neuropeptides in each area. Data were collected by immunocytochemistry and other techniques. LH-RH = luteinizing hormone releasing hormone; TRH = thyrotropin releasing hormone; SOM- = somatostatin; ACTH = adrenocorticotropic hormone; α-MSH = α-melanocyte stimulating hormone; β-LPH = β-lipotropin; β-End = β-endorphin; ENK = enkephalin; Sub P = substance P; NT = neurotensin; CCK8 = cholecystokinin (8 amino acids); VIP = vasoactive intestinal peptide; VP = vasopressin; ANG = angiotensin. (Redrawn from Krieger, 1983.)

amygdala, hippocampus and hypothalamus, the striatum, the central gray matter and the reticular formation of the brain stem (Figure 11.11B). The dynorphins are synthesized in the neurohypophysis, the hypothalamus, hippocampus, striatum, central gray and brain stem (Figure 11.11C).

Figure 11.12 summarizes the location of cell bodies containing a number of neuropeptides and their density in specific brain areas. Anatomical details of the distribution of neuropeptides are given by Elde and Hökfelt (1979), Björklund and Hökfelt (1985) and Nieuwenhuys (1985). The peptides found in the highest concentration in the cerebral cortex are VIP, CCK and neuropeptide Y (not shown in Figure 11.12) . Those with the highest concentration in the hypothalamus and median eminence are the hypothalamic hormones (LH-RH, TRH, somatostatin, and vasopressin), the POMC derivatives (ACTH, α-MSH, β-lipotropin, β-endorphin), neurotensin, and insulin. Bombesin and angiotensin (not shown in Figure 11.10) are also found at high densities in the hypothalamus. Substance P, cholecystokinin (CCK-8) and VIP are found in high concentrations in the amygdala and other limbic system areas, while the enkephalins and substance P are found at high levels in the medulla and pons of the brain stem (Krieger, 1983).

While neuropeptides are found throughout the brain in varying concentrations, and very few brain areas lack neuropeptides, the regions which have the highest concentration of neuropeptides are the neuroendocrine regions of the hypothalamus and median eminence, which contain nearly 40 neuropeptides (Merchenthaler, 1991). The amygdala,

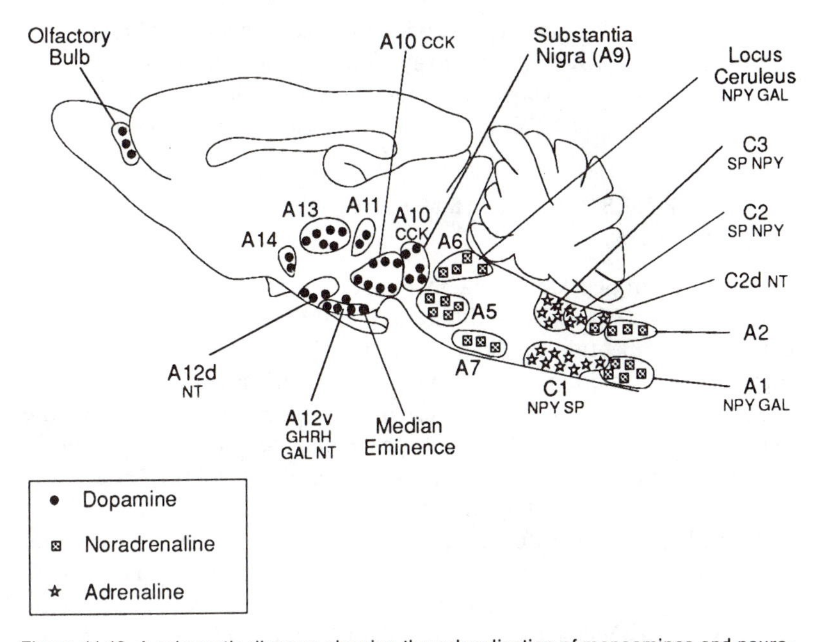

Figure 11.13. A schematic diagram showing the colocalization of monoamines and neuropeptides in specific cell groups in the rat brain. Dopamine is colocalized with cholestokinin (CCK) in the substantia nigra (cell areas A9 and A10) and with growth hormone-releasing hormone (GH-RH), galanin (GAL) and neurotensin (NT) in the arcuate nucleus (A12). Dopamine is not colocalized with neuropeptides in cells of the mesolimbic pathways (A11, A13, A14). Noradrenaline is colocalized with neuropeptide Y and galanin in cells of the locus ceruleus (A6) and the ventral brain stem (A1), but is not colocalized with neuropeptides in the cells of the dorsal brain stem (A2) nor in the medulla (A5 and A7). Adrenaline is colocalized with substance P (SP), neuropeptide Y (NPY) and neurotensin (NT) in the cells of the dorsal motor nucleus of vagus (C2 and C3) and in the ventral brain stem (C1). (Redrawn from Hökfelt *et al.*, 1987.)

hippocampus, other other areas of the limbic system, and the medulla and pons of the brain stem also have high concentrations of neuropeptides. Nieuwenhuys (1985) has pointed out that the brain areas with the highest concentrations of neuropeptides are the same areas which have high densities of steroid hormone receptors (as shown in Figures 9.4, 9.5 and 9.9, pp. 156–64) and he suggests that these neural areas function as a group to regulate the visceral effector mechanisms of the body, maintain homeostasis, and regulate sexual, agonistic and feeding behavior. The neuroendocrine, visceral, cognitive and behavioral functions of the neuropeptides are discussed in Chapter 12.

The colocalization of monoamine neurotransmitters and neuropeptides means that each neuron can be identified on two dimensions: first, by the primary neurotransmitter that it releases; and, second, by the type of coexisting neuropeptide. Figure 11.13 shows that the cell bodies of a number of peptide secreting neurons are closely associated with the cell bodies of the catecholaminergic neurons (Hökfelt *et al.*, 1987). For example, NPY and substance P are colocalized with adrenaline in brain stem neurons (C1), CCK with dopamine in the ventral tegmental nucleus (A10) and neurotensin, galanin and GH-RH with dopamine in the arcuate nucleus. Thus, the neurotransmitter pathways shown in Figures 5.8 and

5.9, p. 72–3 may, therefore, be divided into a number of subtypes. For example, the noradrenergic pathways (Figure 5.8C) may be divided into subtypes depending on whether noradrenaline is colocalized with NPY, neurotensin, the enkephalins, or has no colocalized neuropeptide.

11.5.1 SPECIES AND SEX DIFFERENCES IN NEUROPEPTIDE LOCALIZATION

Although neuropeptides occur in the same general brain areas in all mammals, there are species differences in the exact neural locations of both the neuropeptide cell bodies and the receptors for these neuropeptides. The information given in Figure 11.12 is drawn primarily from studies with rats, but information on the neural localization of the neuropeptides is also available for mice, rabbits, guinea pigs, cats, dogs, pigs, sheep, monkeys and humans (Nieuwenhuys, 1985; Hökfelt *et al.*, 1986, 1987; Mendelsohn *et al.*, 1990).

As well as species differences, there are sex differences in the neural distribution of certain neuropeptides. For example, females have higher levels of oxytocin in the brain than males and have oxytocin in some neural areas in which it is absent in males (Häußler *et al.*, 1990). There are also sex differences in the neural distribution of vasopressin and other neuropeptides. These sex differences may be due to the organizational effects of the gonadal steroids on neural growth and differentiation during prenatal development, as discussed in Section 14.5.3 (see Gorski, 1991).

11.6 NEUROPEPTIDE RECEPTORS AND SECOND MESSENGER SYSTEMS

Neuropeptides bind to G-protein-coupled membrane receptors (as illustrated for the LH receptor in Figure 10.1D, p. 194) to regulate the levels of cyclic AMP and other second messengers in pre- and postsynaptic target cells. Some neuropeptides can bind to synaptic receptors, where they act as neurotransmitters or cotransmitters, while the majority bind to non-synaptic receptors, where they act as neuromodulators. Through the second messenger cascade, neuropeptides regulate receptor levels and alter membrane permeability, increasing or decreasing the sensitivity of the cell to neurotransmitter stimulation as shown in Figure 10.3 (p. 198). Neuropeptides can also regulate protein synthesis in their target cells, altering the synthesis, storage and release of neurotransmitters, neuropeptides, or other proteins, as shown in Figure 10.7 (p. 207) (see Schotman *et al.*, 1985). This is discussed in detail in Chapter 12.

11.6.1 THE IMPORTANCE OF NEUROPEPTIDE AGONISTS AND ANTAGONISTS

The identification and classification of the receptors for the classical neurotransmitters listed in Table 5.3 (p. 68) has relied on the use of pure agonists and antagonist drugs which selectively bind to specific membrane receptor proteins as discussed in Section 5.8. Once the effects of such drugs are known, they can be used clinically to treat disorders of neurotransmitter action. In order to identify the receptors for neuropeptides, to determine their actions on second messenger systems and target cell activity, and to develop clinical treatments for neuropeptide-related disorders, it is necessary to develop synthetic neuropeptide agonists and antagonists. Natural neuropeptides are rapidly metabolized, poorly transported from the blood to the brain, and can not be given orally. Thus, peptide agonists which can be taken orally, are resistant to metabolism by peptidase enzymes, and are easily transported across the blood–brain barrier have been synthesized for a number of neuropeptides (Veber and Friedlinger, 1985). Agonists have been synthesized for somatostatin, substance P, the enkephalins, α-MSH, and GnRH (Veber and Freidinger, 1985; Ehrmann and Rosenfield, 1991). Peptide agonists, however, have many of the same problems as natural peptides, so non-peptide agonists have been developed for a number of neuropeptides, including the opiates and somatostatin. Non-peptide antagonists have been developed which bind to the receptors for CCK, gastrin, angiotensin II, GnRH and the opiates (Friedlinger, 1989).

11.6.2 IDENTIFICATION AND LOCALIZATION OF NEUROPEPTIDE TARGET CELLS IN THE BRAIN

The identification and localization of the neuropeptide cell bodies and pathways is important for understanding the distribution of neuropeptides in the brain, but to understand the functions of the neuropeptides, it is necessary to identify their target cells. Neuropeptide receptors can be localized in the brain by autoradiography, using radioactive peptide ligands, as was described for the identification of steroid hormone receptors in Section 9.2. It is important to identify the location of both the neuropeptide cell bodies and their receptors because the target cells may be located at some distance from the cells which synthesize the peptides. For many neuropeptides, the distribution of receptors does not match the distribution of cell bodies, suggesting that they have neuromodulator, rather than neurotransmitter actions, as discussed in Chapter 12 (see Herkenham, 1987).

As shown in Table 5.3, there are several opioid receptors, each of which has a different neural distribution (Itzhak, 1988; Mansour *et al.*, 1988). Each of the different opioid receptors appears to 'prefer' particular opioid peptides, though there is a great deal of overlap in these preferences. For example, β-endorphin binds most to mu, delta and epsilon receptors; the enkephalins to delta receptors; dynorphins to kappa, delta and mu receptors; and, neoendorphins to kappa and delta receptors (Akil *et al.*, 1988). Mu receptors are most concentrated in the neocortex, thalamus, limbic system and spinal cord, and this distribution corresponds with their function in pain regulation and in sensorimotor integration. Delta receptors are most concentrated in the olfactory bulb, neocortex, caudate putamen, nucleus accumbens, and the amygdala, where they may regulate olfaction, motor integration and cognitive functioning. Kappa receptors are most concentrated in the nucleus accumbens, amygdala, hypothalamus, neurohypophysis, median eminence, and the brain stem and spinal cord. This distribution is consistent with a role in the regulation of water balance and food intake, pain perception and neurohormone release (Akil *et al.*, 1984; Mansour *et al.*, 1988).

Two different tachykinin receptors have been located in the brain and spinal cord of mammals. The NK_1 receptor binds substance P in the septum and other neural areas, while the NK_2 receptor binds substance K in the cerebral cortex, ventral tegmental area and the pineal gland. The NK_3 receptor is found in the body, but not in the brain (Mantyh *et al.*, 1989; Helke *et al.*, 1990). The highest concentration of VIP receptors are found in the neocortex and hypothalamus, which correspond to the areas with the highest concentration of VIP cell bodies and nerve fibers (see Figure 11.12). There are also VIP receptors in the olfactory bulbs, locus ceruleus and pineal gland (Magistretti, 1990). Angiotensin II receptors occur primarily in the hypothalamus (PVN, POA and SCN), the median eminence, and bed nucleus of the stria terminalis (Mendelsohn *et al.*, 1990), so have a close correlation with the locations where the angiotensin-containing cell bodies are found (see Figure 11.12).

Two types of oxytocin and vasopressin receptor are found in the brain: one binds both oxytocin and vasopressin; the other is selective for vasopressin. The oxytocin/vasopressin receptors occur in the ventromedial nucleus, amygdala, bed nucleus of the stria terminalis, and olfactory tubercle. The vasopressin selective receptors occur in the thalamus, A13 dopaminergic nuclei, the suprachiasmatic nucleus and the lateral septal nucleus (Freund-Mercier *et al.*, 1988). Neuropeptide Y receptors are most concentrated in the cerebral cortex and hippocampus, but also occur in the lateral septum, anterior olfactory nucleus, and other neural areas. Only low levels of neuropeptide Y receptors occur in the hypothalamus, where most of the neuropeptide Y cell bodies are located (Quirion *et al.*, 1990).

11.6.3 NEUROPEPTIDE ACTIVATION OF SECOND MESSENGER SYSTEMS

When neuropeptides bind to their receptors, they initiate G-protein-stimulated second messenger synthesis in their target cells, as discussed in Chapter 10 (see Table 10.1, p. 196). Neuropeptides can regulate target cell activity via the cyclic AMP, cyclic GMP, inositol–phospholipid or calcium–calmodulin second messenger cascades (Kaczmarek and Levitan, 1987). In the cyclic AMP system, neuropeptide receptors may be bound to stimulatory (G_s) or inhibitory (G_i) transducer proteins, and thus may stimulate or inhibit cyclic AMP production (see Figure 10.2, p. 197). TRH and VIP, for example, stimulate cyclic AMP synthesis, while somatostatin and the opiates inhibit cyclic AMP production (Table 10.2, p. 199). This is discussed in more detail in Chapter 12.

11.7 NEUROPEPTIDES AND THE BLOOD–BRAIN BARRIER

The walls of the blood vessels in the body have perforations which allow the passage of fluids and chemicals into and out of the capillaries, but the

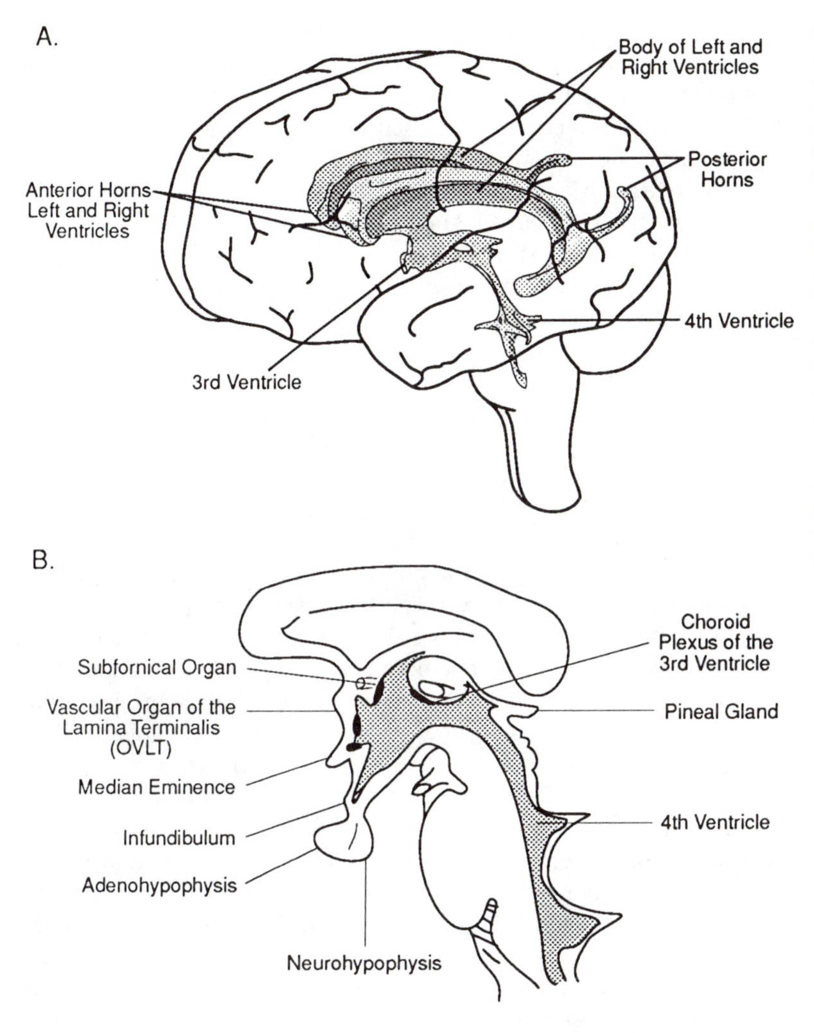

Figure 11.14. The ventricles and the circumventricular organs of the human brain. (A) The locations in the brain of the main body and the anterior and posterior horns of the right and left ventricles are shown , as well as the positions of the third and fourth ventricles. (B) The circumventricular organs of the third ventricle. The capillaries of the choroid plexus are fenestrated and most cerebral spinal fluid is produced in this organ. The pineal gland, subfornical organ, vascular organ of the lamina terminalis (OVLT), median eminence, and neurohypophysis also contain fenestrated capillaries which allow free communication between the blood and the extracellular fluid of the brain. (Redrawn from Reitan and Wolfson, 1985.)

cell walls of the capillaries in the CNS are fused together with tight junctions which prevent the passage of most ions and large molecules from the blood into the brain tissue, and vice versa, thus forming a blood–brain barrier (BBB). There are, however, a few areas of the brain which have fenustrated (permeable) capillaries, which allow the passage of chemicals through them (Broadwell *et al.*, 1987). As shown in Figure 11.14, these permeable capillaries occur in the choroid plexus (where CSF is formed), the circumventricular organs (which include the subfornical organ, the vascular organ of the lamina terminalis, and the area postrema), the median eminence, and the neurohypophysis (Reitan and Wolfson, 1985; Banks and Kastin, 1987, 1988; Merchenthaler, 1991).

Although until recently it was considered impossible for peripherally secreted or injected peptides to cross the BBB, there is now considerable evidence that this can occur via a number of different mechanisms (see Banks and Kastin, 1985). Some peptides are able to cross the BBB by transmembrane diffusion or by non-specific mechanisms. Other peptides, including the opioids (enkephalins and dynorphins), neurohypophyseal hormones, insulin, CCK and others, may be transported across the BBB by specific protein carriers which bind to these peptides and move them across the membrane (see Banks and Kastin, 1987). It is interesting that many of the neuropeptide secreting cells in the brain and the cells with neuropeptide receptors are closely associated with the circumventricular organs, the median eminence, and the neurohypophysis, all of which have a permeable BBB. This will allow the neuropeptides released from the brain to cross the BBB to the body and will allow the passage of peripheral peptides into the CNS, thus enabling the peptide hormones to communicate between the body and brain (McKinley *et al.*, 1989).

11.8 SUMMARY

This chapter describes the different classes of neuropeptides and compares the mechanisms for their synthesis, storage, and release with those for the classical neurotransmitters. Like the peptide hormones, neuropeptides are synthesized from pre-prohormones in the ribosomes and stored in secretory granules. The inter-relationships among the neuropeptides, neurotransmitters, and hormones are examined through their common biosynthetic pathways, common evolutionary pathways, and common embryological origins. Certain peripheral neuroendocrine cells are able to produce both monoamine neurotransmitters and neuropeptides and show plasticity during development, with some becoming neurons and others becoming endocrine cells. The developmental direction of these cells depends on the presence of particular growth factors and steroid hormones in the embryonic environment. In the brain, many neurons synthesize and store monoamine neurotransmitters and neuropeptides in the same large synaptic vesicles and these two transmitters are co-released when the cell is stimulated. Neurotransmitter pathways, therefore, may be subdivided, depending on the type of neuropeptides colocalized with the primary neurotransmitter. While all areas of the brain synthesize neuropeptides, the hypothalamus and median eminence have the most neuropeptides, and other areas, such as the limbic system and the brain stem, have high concentrations of neuropeptides. The areas of maximal neuropeptide receptor concentration may not always correspond to the areas of the highest density of cell bodies. Neuropeptide agonist and antagonist drugs have proved useful in the identification of target cells for neuropeptides and for identifying the second messenger systems activated in these target cells. If the neuropeptides are to be released from the brain to act in the body and the peripheral peptides are to act on neural receptors, these peptides must be able to cross the blood–brain barrier. They can do this in areas such as the median eminence, neurohypophysis and circumventricular organs, where the BBB is permeable. This may be why the neuropeptide secreting cells and neuropeptide receptors are found most frequently in the median eminence and other circumventricular sites.

FURTHER READING

Dores, R. M., McDonald, L. K., Steveson, T. C. and Sei, C. A. (1990). The molecular evolution of neuropeptides: prospects for the '90s. *Brain,*

Behavior and Evolution, **36**, 80–99.

Hökfelt, T., Everitt, B. *et al.* (1986). Neurons with multiple messengers with special reference to neuroendocrine systems. *Recent Progress in Hormone Research*, **42**, 1–70.

Lynch, D. R. and Snyder, S. H. (1986). Neuropeptides: multiple molecular forms, metabolic pathways, and receptors. *Annual Review of Biochemistry*, **55**, 773–799.

Nieuwenhuys, R. (1985). *Chemoarchitecture of the Brain*. Berlin: Springer-Verlag.

Scharrer, B. (1990). The neuropeptide saga. *American Zoologist*, **30**, 887–895.

Zadina, J. E., Banks, W. A. and Kastin, A. J. (1986). Central nervous system effects of peptides, 1980–1985: a cross-listing of peptides and their central actions from the first six years of the journal *Peptides*. *Peptides*, **7**, 497–537.

REVIEW QUESTIONS

11.1 Define the term 'neuropeptide'.

11.2 What is the difference in the sites of the final synthesis of the neurotransmitters and neuropeptides?

11.3 In nerve cells which synthesize both monoamine transmitters and neuropeptides, which are stored in small vesicles and which are stored in large vesicles in the nerve terminal?

11.4 What is the common precursor for substance P and neuropeptide K?

11.5 During evolution, specific endocrine glands first occurred in the ________, while specific neurons first appeared in the ________.

11.6 Certain higher plants produce substances which are chemically similar to TRH, somatostatin and other mammalian neuropeptides: true or false?

11.7 Which layer of the early embryo forms the brain, spinal cord and neural crest cells of the developing embryo?

11.8 What does the acronym 'APUD' stand for?

11.9 If the 'diffuse neuroendocrine system' is considered to be the third division of the nervous system, what are the other two divisions?

11.10 The 'diffuse neuroendocrine system' has a central and a peripheral division, each having a neural and an endocrine component. How would the adrenal medulla be classified along these two dimensions?

11.11 If noradrenaline and neuropeptide Y are colocalized in the same synaptic vesicle, which would be considered the primary transmitter?

11.12 Given the colocalization of dopamine and CCK in neurons of the mesolimbic pathways, what will be the effect of dopamine agonist and antagonist drugs on the release of CCK in these pathways?

11.13 What is the 'correct' version of Dale's Principle, according to Eccles?

11.14 Which neurohypophyseal hormone is colocalized and co-released with dynorphin?

11.15 Which area(s) of the brain have the highest concentration of neuropeptides?

11.16 In what area of the brain are the concentrations of VIP and CCK the highest?

11.17 Why is the BBB important for understanding the action of neuropeptides?

ESSAY QUESTIONS

11.1 Discuss the importance of immunocytochemistry and other techniques for the discovery of new neuropeptides.

11.2 Compare and contrast the mechanisms for the synthesis, storage and release of classical neurotransmitters and neuropeptides.

11.3 Discuss the biosynthesis of the CCK family of peptides, the cells which produce these peptides, and the functions of each member of the family.

11.4 Discuss the theory of the evolution of the neuroendocrine system as proposed in Figure 11.4.

11.5 Discuss what determines whether the bipotential progenitor cells from the neural crest develop into neural or endocrine cells.

11.6 How has the APUD concept aided our understanding of the ontogeny of the neuroendocrine system?

11.7 Discuss the effects of drugs which regulate the synthesis, storage and release of the catecholamines (e.g. reserpine, alpha-methyl-tyrosine, etc.) on the neuropeptides which are colocalized in these neurons.

11.8 Discuss the colocalization of neuropeptides with neurohypophyseal hormones and the functions of these neuropeptides.

11.9 Discuss the hypothesis that the pathways for the monoamine neurotransmitters should be subdivided based on their neuropeptide cotransmitters. How could this increase our understanding of neural function?

11.10 Discuss the advantages of developing neuropeptide agonists and antagonists as therapeutic drugs.

11.11 Discuss the reasons why it is necessary to identify both the neural areas which produce neuropeptides and those with neuropeptide receptors.

11.12 Discuss the mechanisms by which the peptide hormones and neuropeptides can cross the BBB.

REFERENCES FOR CHAPTERS 11 AND 12

Abrams, G. M., Nilaver, G. and Zimmerman, E. A. (1985). VIP-containing neurons. In A. Björklund and T. Hökfelt (eds.) *Handbook of Chemical Neuroanatomy*, vol. 4, *GABA and Neuropeptides in the CNS, Part I*, pp. 335–354. Amsterdam: Elsevier.

Acher, R. (1983). Principles of evolution: the neural hierarchy model. In D. T. Krieger, M. J. Brownstein and J. B. Martin (eds.) *Brain Peptides*, pp. 135–163. New York: John Wiley and Sons.

Akil, H., Bronstein, D. and Mansour, A. (1988). Overview of the endogenous opioid systems: anatomical, biochemical and functional issues. In R. J. Rogers and S. J. Cooper (eds.) *Endorphins, Opiates and Behavioural Processes*, pp. 1–23. Chichester: John Wiley.

Akil, H., Watson, S. J., Young, E., Lewis, M. E., Khachaturian, H. and Walker, J. M. (1984). Endogenous opioids: biology and function. *Annual Review of Neurosicence*, **7**, 223–255.

Anderson, D. J. (1989). Development and plasticity of a neural crest-derived neuroendocrine sublineage. In L. T. Landmesser (ed.) *The Assembly of the Nervous System*, pp. 17–36. New York: Alan R. Liss.

Andrew, A. (1982). The APUD concept: where has it led us? *British Medical Bulletin*, **38**, 221–225.

Anonymous (1981). How does acupuncture work? *British Medical Journal*, **283**, 746–748.

Ansseau, M. and Reynolds, C. F., III. (1988). Neuropeptides and sleep. In C. B. Nemeroff (ed.) *Neuropeptides in Psychiatric and Neurological Disorders*, pp. 178–198. Baltimore: Johns Hopkins University Press.

Arletti, R., Bazzani, C., Castelli, M. and Bertolini, A. (1985). Oxytocin improves male copulatory performance in rats. *Hormones and Behavior*, **19**, 14–20.

Argiolas, A. and Gessa, G. L. (1991). Central functions of oxytocin. *Neuroscience and Biobehavioral Reviews*, **15**, 217–231.

Aronin, N., Carraway, R. E., Difiglia, M. and Leeman, S. E. (1983). Neurotensin. In D. T. Krieger, M. J. Brownstein and J. B. Martin (eds.) *Brain Peptides*, pp. 753–781. Chichester: John Wiley.

Aronin, N., Coslovsky, R. and Leeman, S. E. (1986). Substance P and neurotensin: their roles in the regulation of anterior pituitary function. *Annual Review of Physiology*, **48**, 537–549.

Aronin, N., Difiglia, M. and Leeman, S. E. (1983). Substance P. In D. T. Krieger, M. J. Brownstein and J. B. Martin (eds.) *Brain Peptides*, pp. 783–803. Chichester: John Wiley.

Banks, W. A. and Kastin, A. J. (1985). Permeability of the blood–brain barrier to neuropeptides: the case for penetration. *Psychoneuroendocrinology*, **10**, 385–399.

Banks, W. A. and Kastin, A. J. (1987). Saturable transport of peptides across the blood–brain barrier. *Life Sciences*, **41**, 1319–1338.

Banks, W. A. and Kastin, A. J. (1988). Review: interactions between the blood-brain barrier and endogenous peptides: emerging clinical implications. *American Journal of Medical Sciences*, **295**, 459–465.

Barchas, J. D., Akil, H., Elliott, G. R., Holman, R. B. and Watson, S. J. (1978). Behavioral neurochemistry: neuroregulators and behavioral states. *Science*, **200**, 964–973.

Bartfai, T., Iverfeldt, K., Fisone, G. and Serfözö, P. (1988). Regulation of the release of coexisting neurotransmitters. *Annual Review of Pharmacology and Toxicology*, **28**, 285–310.

Basbaum, A. I. and Fields, H. L. (1984). Endogenous pain control systems: brainstem spinal pathways and endorphin circuitry. *Annual Review of Neuroscience*, **7**, 309–338.

Baskin, D. G., Figlewicz, D. P., Woods, S. C., Porte, D. Jr and Dorsa, D. M. (1987). Insulin in the brain. *Annual Review of Physiology*, **49**, 335–347.

Baylin, S. B. (1990). 'APUD' cells: fact and fiction. *Trends in Endocrinology and Metabolism*, **1**, 198–204.

Beckwith, B. E. and Sandman, C. A. (1978). Behavioral influences of the neuropeptides ACTH and MSH: a methodological review. *Neuroscience and Biobehavioral Reviews*, **2**, 311–338.

Beckwith, B. E., Petros, T. V., Couk, D. I. and Tinius, T. P. (1990). The effects of vasopressin on memory in healthy young adult volunteers. *Annals of the New York Academy of Sciences*, **579**, 215–226.

Beltramino, C. and Taleisnik, S. (1983). Release of LH in the female rat by olfactory stimuli. *Neuroendocrinology*, **36**, 53–58.

Bennett, G. W. and Whitehead, S. A. (1983). *Mammalian Neuroendocrinology*. New York: Oxford University Press.

Benton, D. and Brain, P. F. (1988). The role of opioid mechanisms in social interaction and attachment. In R. J. Rogers and S. J. Cooper (eds.) *Endorphins, Opiates and Behavioural Processes*, pp. 217–235. Chichester: John Wiley.

Bicknell, R. J. (1985). Endogenous opioid peptides and hypothalamic neuroendocrine neurones. *Journal of Endocrinology*, **107**, 437–446.

Björklund, A. and Hökfelt, T. (Eds.) (1985). *Handbook of Chemical Neuroanatomy* vol. 4: *GABA and Neuropeptides in the CNS, Part I*. Amsterdam: Elsevier.

Blaustein, J. D., Nielsen, K. H., Delville, Y., Turcotte, J. C. and Olster, D. H. (1991). Neuroanatomical relationships of substance P and sex steroid hormone-sensitive neurons involved in sexual behavior. *Annals of the New York Academy of Sciences*, **632**, 314–331.

Bloom, F. E. (1988). Neurotransmitters: past, present, and future directions. *Federation of American Societies of Experimental Biology Journal*, **2**, 32–41.

Boer, G. J. and Swaab, D. F. (1985). Neuropeptide effects on brain development to be expected from behavioral teratology. *Peptides*, **6**, (suppl. 2), 21–28.

Bondy, C. A., Gainer, H. and Russell, J. T. (1988). Dynorphin A inhibits and naloxone increases the electrically stimulated release of oxytocin but not vasopressin from the terminals of the neural lobe. *Endocrinology*, **122**, 1321–1327.

Bondy, C. A., Whitnall, M. H., Brady, L. S. and Gainer, H. (1989). Coexisting peptides in hypothalamic neuroendocrine systems: Some functional implications. *Cellular and Molecular Neurobiology*, **9**, 427–446.

Borbely, A. A. and Tobler, I. (1989). Endogenous sleep-promoting substances and sleep regulation. *Physiological Reviews*, **69**, 605–660.

Bozarth, M. A. (1988). Opioid reinforcement processes. In R. J. Rodger and S. J. Cooper (eds.) *Endorphins, Opiates and Behavioural Processes*, pp. 53–75. Chichester: John Wiley.

Bridges, R. S., DiBiase, R., Loundes, D. D. and Doherty, P. C. (1985). Prolactin stimulation of maternal behavior in female rats. *Science*, **227**, 782–784.

Broadwell, R. D., Charlton, H. M., Balin, B. J. and Salcman, M. (1987). Angioarchitecture of the CNS, pituitary gland, and intracerebral grafts revealed with peroxidase cytochemistry. *Journal of Comparative Neurology*, **260**, 47–62.

Brown, D. R. and Miller, R. J. (1982). Neurotensin. *British Medical Bulletin*, **38**, 239–245.

Burnstock, G. (1976). Do some nerve cells release more than one transmitter? *Neuroscience*, **1**, 239–248.

Chronwall, B. M., DiMaggio, D. A., Massari, V. J., Pickel, V. M., Ruggiero, D. A. and O'Donohue, T. L. (1985). The anatomy of neuropeptide-Y-containing neurons in rat brain. *Neuroscience*, **15**, 1159–1181.

Clark, J. T. (1989). A possible role for angiotensin II in the regulation of male sexual behavior in rats. *Physiology and Behavior*, **45**, 221–246.

Cooper, S. J. (1988). Evidence for opioid involvement in controls of drinking and water balance. In R. J. Rogers and S. J. Cooper (eds.) *Endorphins, Opiates and Behavioural Processes*, pp. 187–216. Chichester: Wiley.

Cooper, S. J., Jackson, A., Kirkham, T. C. and Turkish, S. (1988). Endorphins, opiates and food intake. In R. J. Rodgers and S. J. Cooper (eds.) *Endorphins, Opiates and Behavioural Processes*, pp. 143–186. Chichester: Wiley.

Coquelin, A., Clancy, A. N., Macrides, F., Noble, E. P. and Gorski, R. A. (1984). Pheromonally induced release of luteinizing hormone in male mice: involvement of the vomeronasal system. *Journal of Neuroscience*, **4**, 2230–2236.

Cottrell, G. A., Veldhuis, H. D., Rostene, W. H. and de Kloet, E. R. (1984). Behavioural actions of vasoactive intestinal peptide (VIP). *Neuropeptides*, **4**, 331–341.

Cowan, W. M. (1979). The development of the brain. *Scientific American*, **241** (3), 112–133.

Cox, B. M. (1988). Peripheral actions mediated by opioid receptors. In G. W. Pasternak (ed.) *The Opiate Receptors*, pp. 357–422. Clifton, NJ: Humana Press.

Crain, S. M. and Shen, K.-F. (1990). Opioids can evoke direct receptor-mediated excitatory effects on sensory neurons. *Trends in Pharmacological Sciences*, **11**, 77–81.

Crawley, J. N. (1990). Coexistence of neuropeptides and 'classical' neurotransmitters: functional interactions between galanin and acetylcholine. *Annals of the New York Academy of Sciences*, **579**, 233–245.

Cumming, D. C. and Wheeler, G. D. (1987). Opioids in exercise physiology. *Seminars in Reproductive Endocrinology*, **5**, 171–179.

Cunningham, E. T. and Sawchenko, P. E. (1991). Reflex control of magnocellular vasopressin and oxytocin secretion. *Trends in Neurosciences*, **14**, 406–411.

Datta, P. C and King, M. G. (1982). α-Melanocyte-stimulating hormone and behavior. *Neuroscience and Biobehavioral Reviews*, **6**, 297–310.

Dekin, M. S., Richerson, G. B. and Getting, P. A. (1985). Thyrotopin-releasing hormone induces rhythmic bursting in neurons of the nucleus tractus solitarius. *Science*, **229**, 67–69.

Delitala, G. (1991). Opioid peptides and pituitary function: basic and clinical aspects. In M. Motta (ed.) *Brain Endocrinology*, 2nd edn, pp. 217–244. New York: Raven Press.

de Vries, G. J., Buijs, R. M., van Leeuwen, F. W., Caffé, A. R. and Swaab, D. F. (1985). The vasopressinergic innervation of the brain in normal and castrated rats. *Journal of Comparative Neurology*, **233**, 236–254.

de Wied, D. (1987). The neuropeptide concept. *Progress in Brain Research*, **72**, 93–108.

de Wied, D. and Jolles, J. (1982). Neuropeptides derived from pro-opiocortin: behavioral, physiological, and neurochemical effects. *Physiological Reviews*, **62**, 976–1048.

Dismukes, R. K. (1979). New concepts of molecular communication among neurons. *Behavioral and Brain Sciences*, **2**, 409–448.

Dockray, G. J. (1983). Cholecystokinin. In D. T. Kreiger, M. J. Brownstein and J. B. Martin (eds.) *Brain Peptides*, pp. 851–869. Chichester: John

Wiley.

Dockray, G. J. (1984). The chemistry of neuropeptides. *Frontiers in Hormone Research*, **12**, 8–15.

Dores, P. A. (1984). Vasopressin and central integrative processes. *Neuroendocrinology*, **38**, 75–85.

Dores, R. M., McDonald, L. K., Steveson, T. C. and Sei, C. A. (1990). The molecular evolution of neuropeptides: prospects for the '90s. *Brain, Behavior and Evolution*, **36**, 80–99.

Dornan, W. A. and Malsbury, C. W. (1989). Neuropeptides and male sexual behavior. *Neuroscience and Biobehavioral Reviews*, **13**, 1–15.

Drago, F. (1984). Prolactin and sexual behavior: a review. *Neuroscience and Biobehavioral Reviews*, **8**, 433–439.

Dumont, Y., Martel, J.-C., Fournier, A., St-Pierre, S. and Quirion, R. (1992). Neuropeptide Y and neuropeptide Y receptor subtypes in brain and peripheral tissues. *Progress in Neurobiology*, **38**, 125–167.

Dunn, A. J. and Berridge, C. W. (1990). Physiological and behavioral responses to corticotropin-releasing factor administration: is CRF a mediator of anxiety or stress responses? *Brain Research Reviews*, **15**, 71–100.

Eccles, J. C. (1986). Chemical transmission and Dale's principle. *Progress in Brain Research*, **68**, 3–13.

Ehrmann, D. A. and Rosenfield, R. L. (1991). Gonadotropin-releasing hormone agonist testing of pituitary-gonadal function. *Trends in Endocrinology and Metabolism*, **2**, 86–91.

Elde, R. (1983). Immunocytochemistry. In D. J. Krieger, M. J. Brownstein and J. B. Martin (eds.) *Brain Peptides*, pp. 485–494. Chichester: John Wiley.

Elde, R. and Hökfelt, T. (1979). Localization of hypophysiotropic peptides and other biologically active peptides within the brain. *Annual Review of Physiology*, **41**, 587–602.

Emson, P. C. and Hunt, S. P. (1982). Neuropeptides as putative neurotransmitters: endorphins, substance P, cholecystokinin, and vasoactive intestinal polypeptide. In I. R. Brown (ed.) *Molecular Approaches to Neurobiology*, pp. 255–283. New York: Academic Press.

Epelbaum, J. (1986). Somatostatin in the central nervous system: physiology and pathological modifications. *Progress in Neurobiology*, **27**, 63–100.

Epelbaum, J., Agid, F., Agid, Y. *et al.*, (1989). Somatostatin receptors in brain and pituitary. *Hormone Research*, **31**, 45–50.

Evans, C. J., Hammond, D. L. and Frederickson, R. C. A. (1988). The opioid peptides. In G. W. Pasternak (ed.) *The Opiate Receptors*, pp. 23–71. Clifton, NJ: Humana Press.

Facchinetti, F., Petraglia, F. and Genazzani, A. R. (1987). Localization and expression of the three opioid systems. *Seminars in Reproductive Endocrinology*, **5**, 103–113.

Fahrenkrug, J. and Emson, P. C. (1982). Vasoactive intestinal polypeptide: functional aspects. *British Medical Bulletin*, **38**, 265–270.

Fekete, M., Szántó-Fekete, M. and Telegdy, G. (1987a). Action of corticotropin-releasing factor on the nervous system. *Frontiers in Hormone Research*, **15**, 58–78.

Fekete, M., Szipócs, I., Halmai, L., Franczia, P., Kardos, A., Csonka, E., Szántó-Fekete, M. and Telegdy, G. (1987b). Cholecystokinin and the central nervous system. *Frontiers in Hormone Research*, **15**, 175–251.

Figlewicz, D. P., Lacour, F., Sipols, A., Porte, D. Jr and Woods, S. C. (1987). Gastroenteropancreatic peptides and the central nervous system. *Annual Review of Physiology*, **49**, 383–395.

Fisher, L. A. (1989). Corticotropin-releasing factor: endocrine and autonomic integration of responses to stress. *Trends in Pharmacological Science*, **10**, 189–193.

Florey, E. (1967). Neurotransmitters and modulators in the animal kingdom. *Federation Proceedings*, **26**, 1164–1178.

Freidlinger, R. M. (1989). Non-peptide ligands for peptide receptors. *Trends in Pharmacological Sciences*, **10**, 270–274.

Freund-Mercier, M. J., Stoekel, M. E., Dietl, M. M., Palacois, J. M. and Richard, Ph. (1988). Quantitative autoradiographic mapping of neurohypophyseal hormone binding sites in the rat forebrain and pituitary gland. I. Characterization of different types of binding sites and their distribution in the Long-Evans strain. *Neuroscience*, **26**, 261–272.

Fuxe, K., Agnati, L. F. *et al.* (1990). On the role of NPY in information handling in the CNS in normal and physiopathological states. *Annals of the New York Academy of Sciences*, **579**, 28–67.

Gambert, S. R., Garthwaite, T. L., Pontzer, C. H. *et al.*, (1981). Running elevates plasma β-endorphin immunoreactivity and ACTH in untrained human subjects. *Proceedings of the Society for Experimental Biology and Medicine*, **168**, 1–4.

George, S. R. and van Loon, G. R. (1981). Met-enkephalin uptake by a synaptosome-enriched fraction of rat striatum. *Neuroscience Letters*, **26**, 297–300.

Gispen, W. H. and Zwiers, H. (1985). Behavioral and neurochemical effects of ACTH. In A. Lajtha (edn.) *Handbook of Neurochemistry*, 2nd ed., pp. 375–412. New York: Plenum Press.

Gmerek, D. E. (1988). Physiological dependence on opioids. In R. J. Rodgers and S. J. Cooper (eds.) *Endorphins, Opiates and Behavioural Processes*, pp. 25–52. Chichester: John Wiley.

Gorski, R. A. (1991). Sexual differentiation of the endocrine brain and its control. In M. Motta (ed.) *Brain Endocrinology*, 2nd edn, pp. 71–104. New York: Raven Press.

Gray, T. S. and Morley, J. E. (1986). Neuropeptide Y: anatomical distribution and possible function in mammalian nervous system. *Life Sciences*, **38**, 389–401.

Grossman, A. (1987). Opioid peptides and reproductive function. *Seminars in Reproductive Endocrinology*, **5**, 115–124.

Grossman, A. and Rees, L. H. (1983). The neuroendocrinology of opioid peptides. *British Medical Bulletin*, **39**, 83–88.

Guidotti, A., Saiani, L., Wise, B. C. and Costa, E. (1983). Cotransmitters: Pharmacological implications. *Journal of Neural Transmission*, **18**, 213–225.

Hadcock, J. R. and Malbon, C. C. (1991). Regulation of receptor expression by agonists: Transcriptional and post-transcriptional controls. *Trends in Neurosciences*, **14**, 242–247.

Hadley, M. E. (1992). *Endocrinology*, 3rd edn. Englewood Cliffs, NJ: Prentice Hall.

Héußler, H. U., Jirikowski, G. F. and Caldwell, J. D. (1990). Sex differences among oxytocin-immunoreactive neuronal systems in the mouse hypothalamus. *Journal of Chemical Neuroanatomy*, **3**, 271–276.

Hawthorn, J., Ang, V. T. Y. and Jenkins, J. S. (1980). Localization of vasopressin in the rat brain. *Brain Research*, **197**, 75–81.

Helke, C. J., Krause, J. E., Mantyh, P. W., Coutre, R. and Bannon, M. J. (1990). Diversity in mammalian tachykinin peptidergic neurons: Multiple peptides, receptors, and regulatory mechanisms. *Federation of American Societies for Experimental Biology Journal*, **4**, 1606–1615.

Helke, C. J., Sasek, C. A., Niederer, A. J. and Krause, J. E. (1991). Tachykinins in autonomic control systems: The company they keep. *Annals of the New York Academy of Sciences*, **632**, 154–169.

Hemmings, H. C., Jr, Nairn, A. C., McGuinness, T. L., Huganir, R. L. and Greengard, P. (1989). Role of protein phosphorylation in neuronal signal transduction. *Federation of American Societies for Experimental Biology Journal*, **3**, 1583–1592.

Herkenham, M. (1987). Mismatches between neurotransmitter and receptor localizations in brain: observations and implications. *Neuroscience*, **23**, 1–38.

Herman, B . H. and Panksepp, J. (1981). Ascending endorphin inhibition of distress vocalization. *Science*, **211**, 1060–1062.

Hökfelt, T. (1991). Neuropeptides in perspective: the last ten years. *Neuron*, **7**, 867–879.

Hökfelt, T., Everitt, B., Meister, B. *et al.*, (1986). Neurons with multiple messengers with special reference to neuroendocrine systems. *Recent Progress in Hormone Research*, **42**, 1–70.

Hökfelt, T., Johansson, O., Holets, V., Meister, B. and Melander, T. (1987). Distribution of neuropeptides with special reference to their coexistence with classical transmitters. In H. Y. Meltzer (ed.) *Psychopharmacology: The Third Generation of Progress*, pp. 401–416. New York: Raven Press.

Hökfelt, T., Johansson, O., Ljungdahl, Å., Lundberg, J. M. and Schultzberg, M. (1980). Peptidergic neurones. *Nature*, **284**, 515–521.

Hollister, L. E., Davis, K. L. and Davis, B. M. (1980). Hormones in the treatment of psychiatric disorders. In D. T. Krieger and J. C. Hughes (eds.) *Neuroendocrinology*, pp. 167–175. Sunderland, MA: Sinauer.

Houben, H. and Denef, C. (1990). Regulatory peptides produced in the anterior pituitary. *Trends in Endocrinology and Metabolism*, **1**, 398–403.

Imura, H., Kato, Y., Nakai, Y. *et al.*, (1985). Endogenous opioids and related peptides: from molecular biology to clinical medicine. *Journal of Endocrinology*, **107**, 147–157.

Israel, J. M., Brigant, J. L. and Vincent, J. D. (1989). Analysis of the effects of neuropeptides on endocrine cells: an electrophysiological approach. In G. Fink and A. J. Harmar (eds.) *Neuropeptides: A Methodology*, pp. 283–299. Chichester: John Wiley.

Itzhak, Y. (1988). Multiple opioid binding sites. In G. W. Pasernak (ed.) *The Opiate Receptors*, pp. 95–142. Clifton, NJ: Humana Press.

Iversen, S. D. (1982). Behavioural effects of substance P through dopaminergic pathways in the brain. *CIBA Foundation Symposium*, **91**, 307–324.

Jackson, I. M. D. and Lechan, R. M. (1983). Thyrotropin releasing hormone (TRH). In D. T. Krieger, M. J. Brownstein and J. B. Martin (eds.) *Brain Peptides*, pp. 661–685. Chichester: John Wiley.

Jalowiec, J. E., Panksepp, J., Zolovick, A. J., Najam, N. and Herman, B. H. (1981). Opioid modulation of ingestive behavior. *Pharmacology Biochemistry and Behavior*, **15**, 477–484.

Jessell, T. M. and Iversen, L . L. (1977). Opiate analgesics inhibit substance P release from rat trigeminal nucleus. *Nature*, **268**, 549–551.

Jessell, T. M. and Kelley, D. D. (1991). Pain and analgesia. In Kandel, E. R., Schwartz, J. H. and Jessell, T. M. (eds.) *Principles of Neural Science*, 3rd edn, pp. 385–399. Elsevier: New York.

Jolicoeur, F. B., Rioux, F. and St.-Pierre, S. (1985). Neurotensin. In A. Lajtha (ed.) *Handbook of Neurochemistry*, 2nd edn, pp. 93–114. New York: Plenum Press.

Kaczmarek, L. K. and Levitan, I. B. (1987). What is neuromodulation? In L. K. Kaczmarek and I. B. Levitan (eds.), *Neuromodulation: The Biochemical Control of Neuronal Excitability*, pp. 3–17. Oxford: Oxford University Press.

Kalra, S. P., Sahu, A., Kalra, P. S. and Crowley, W. R. (1990). Hypothalamic neuropeptide Y: a circuit in the regulation of gonadotropin secretion and feeding behavior. *Annals of the New York Academy of Sciences*, **611**, 273–283.

Kapás, L., Obál, F., Opp, M. R., Johannsen, L. and Kreuger, J. M. (1991). Intraperitoneal injection of cholecystokinin elicits sleep in rabbits. *Physiology and Behavior*, **50**, 1241–1244.

Kastin, A. J., Banks, W. A., Zadina, J. E. and Graf, M. (1983). Brain peptides: The dangers of constricted nomenclatures. *Life Sciences*, **32**, 295–301.

Katz, R. J. (1988). Endorphins, exploration and activity. In R. J. Rodgers and S. J. Cooper (eds.) *Endorphins, Opiates and Behavioural Processes*, pp. 249–267. Chichester: John Wiley.

Kelley, A. E. (1989). Behavioural models of neuropeptide action. In G. Fink and A. J. Harmar (eds.) *Neuropeptides: A Methodology*, pp. 301–331. Chichester: John Wiley.

Kelly, J. S. and Brooks, P. A. (1984). A review of the electrophysiological evidence for a neurotransmitter role of cholecystokinin. In J. de Belleroche and G. J. Dockray (eds.) *Cholecystokinin (CCK) in the Nervous System*, pp. 84–93. Chichester: Ellis Norwood.

Kendrick, K. M. and Dixson, A. F. (1985). Luteinizing hormone releasing hormone enhances proceptivity in a primate. *Neuroendocrinology*, **41**, 449–453.

Khachaturian, H., Lewis, M. E., Schafer, M. K.-H. and Watson, S. J. (1985). Anatomy of the CNS opioid systems. *Trends in Neurosciences*, **8**, 111–119.

Koob, G. F. and Bloom, F. E. (1983). Behavioural effects of opioid peptides. *British Medical Bulletin*, **39**, 89–94.

Koob, G. F. and Bloom, F. E. (1985). Corticotropin-releasing factor and behavior. *Federation Proceedings*, **44**, 259–263.

Koob, G. F. and Bloom, F. E. (1988). Cellular and molecular mechanisms of drug dependence. *Science*, **242**, 715–723.

Kovács, G. L., Fekete, M., Szabó, G. and Telegdy, G. (1987). Action of ACTH-corticosteroid axis on the central nervous sytem. *Frontiers in Hormone Research*, **15**, 79–127.

Krause, J. E., Hershey, A. D., Dykema, P. E. and Takeda, Y. (1990). Molecular biological studies on the diversity of chemical signalling in tachykinin peptidergic neurons. *Annals of the New York Academy of Sciences*, **579**, 254–272.

Krey, L. C. and Silverman, A. J. (1983). Luteinizing hormone releasing hormone (LHRH). In D. T. Krieger, M. J. Brownstein and J. B. Martin (eds.) *Brain Peptides*, pp. 687–709. Chichester: John Wiley.

Krieger, D. T. (1983). Brain peptides: what, where, and why? *Science*, **222**, 975–985.

Krieger, D. T. (1986). An overview of neuropeptides. In J. B. Martin and J. D. Barchas (eds.) *Neuropeptides in Neurologic and Psychiatric Disease*, pp. 1–33. New York: Raven Press.

Kumakura, K., Karoum, F., Guidotti, A. and Costa, E. (1980). Modulation of nicotinic receptors by opiate receptor agonists in cultured adrenal chromaffin cells. *Nature*, **283**, 489–492.

Kupfermann, I. (1991). Functional studies of cotransmission. *Physiological Reviews*, **71**, 683–732.

Labrie, F., Gigguere, V., Meunier, H., Simard, J., Gossard, F. and Raymond, V. (1987). Multiple factors controlling ACTH secretion at the anterior pituitary level. *Annals of the New York Academy of Sciences*, **512**, 97–114.

Landis, S. C. and Patterson, P. H. (1981). Neural crest cell lineages. *Trends in Neuroscience*, **4**, 172–175.

Lawrence, J. A. M., Poulin, P., Lawrence, D. and Lederis, K. (1988). [3H]Arginine vasopressin binding to rat brain: a homogenate and autoradiographic study. *Brain Research*, **446**, 212–218.

Lembeck, F. (1984). Neuronal versus endocrine effects: substance P as an example. *Frontiers in Hormone Research*, **12**, 118–122.

Le Moal, M., Dantzer, R., Mormede, P. *et al.*, (1984). Behavioral effects of peripheral administration of arginine vasopressin: a review of our search for a mode of action and a hypothesis. *Psychoneuroendocrinology*, **9**, 319–341.

LeRoith, D., Delahunty, G., Wilson, G. L., Roberts, C. T. Jr, Shemer, J., Hart, C., Lesniak, M. A., Shiloach, J. and Roth, J. (1986). Evolutionary aspects of the endocrine and nervous systems. *Recent Progress in Hormone Research*, **42**, 549–587.

LeRoith, D. and Roth, J. (1984). Are messenger-like molecules in unicellular microbes the common phylogenetic ancestors of vertebrate hormones, tissue factors and neurotransmitters? In D. G. Garlick and P. I. Korner (eds.) *Frontiers in Physiological Research*, pp. 87–98. Cambridge: Cambridge University Press.

LeRoith, D., Shiloach, J. and Roth, J. (1982). Is there an earlier phylogenetic precursor that is common to both the nervous and endocrine systems? *Peptides*, **3**, 211–215.

Levine, A. S. and Billington, C. J. (1989). Opioids: are they regulators of feeding? *Annals of the New York Academy of Sciences*, **575**, 209–220.

Loh, Y. P. and Gainer, H. (1983). Biosynthesis and processing of neuropeptides. In D. T. Krieger, M . J. Brownstein and J. B. Martin (eds.) *Brain Peptides*, pp. 79–116. New York: John Wiley.

Lundberg, J. M., Franco-Cereceda, A., Lacroix, J.-S. and Pernow, J. (1990). Neuropeptide Y and sympathetic neurotransmission. *Annals of the New York Academy of Sciences*, **611**, 166–174.

Lundberg, J. M. and Hökfelt, T. (1983). Coexistence of peptides and classical transmitters. *Trends in Neurosciences*, **6**, 325–333.

Lynch, D. R. and Snyder, S. H. (1986). Neuropeptides: multiple molecular forms, metabolic pathways, and receptors. *Annual Review of Biochemistry*, **55**, 773–799.

Magistretti, P. J. (1990). VIP neurons in the cerebral cortex. *Trends in Pharmacological Sciences*, **11**, 250–254.

Magistretti, P. J. and Schorderet, M. (1984). VIP and noradrenaline act synergistically to increase cyclic AMP in cerebral cortex. *Nature*, **308**, 280–282.

Mancillas, J. R., Siggins, G. R. and Bloom, F. L. (1986). Somatostatin selectively enhances acetylcholine-induced excitations in rat hippocampus and cortex. *Proceedings of the National Academy of Sciences, USA*, **83**, 7518–7521.

Mansour, A., Khachaturian, H., Lewis, M. E., Akil, H. and Watson, S. J. (1988). Anatomy of CNS opioid receptors. *Trends in Neurosciences*, **11**, 308–314.

Mantyh, P. W., Gates, T., Mantyh, C. R. and Maggio, J. E. (1989). Autoradiographic localization and characterization of tachykinin receptor binding sites in the rat brain and peripheral tissues. *Journal of Neuroscience*, **9**, 258–279.

Mattson, M. P. (1989). Cellular signalling mechanisms common to the development and degeneration of neuroarchitecture. A review. *Mechanisms of Ageing and Development*, **50**, 103–157.

Mauk, M. D., Olson, G. A., Kastin, A. J. and Olson, R. D. (1980). Behavioral effects of LH-RH. *Neuroscience and Biobehavioral Reviews*, **4**, 1–8.

McCann, S. M. (1991). Neuroregulatory peptides. In M. Motta (ed.) *Brain Endocrinology*, 2nd edn, pp. 1–30. New York: Raven Press.

McKelvy, J. F. and Blumberg, S. (1986). Inactivation and metabolism of neuropeptides. *Annual Review of Neuroscience*, **9**, 415–434.

McKinley, M. J., Allen, A. M., Chai, S. Y., Hards, D. K., Mendelsohn, F. A. O. and Oldfield, B. J. (1989). The lamina terminalis and its neural connections: neural circuitry involved in angiotensin action and fluid and electrolyte homeostasis. *Acta Physiologica Scandinavica*, suppl. **583**, 113–118.

Meisenberg, G. and Simmons, W. H. (1983). Centrally mediated effects of neurohypophyseal hormones. *Neuroscience and Biobehavioral Reviews*, **7**, 263–280.

Meites, J. (1984). Effects of opiates on neuroencordine functions in animals: Overview. In G. Delitala, M. Motta and M. Serio (eds.) *Opioid Modulation of Endocrine Function*, pp. 53–63, New York: Raven Press.

Mendelsohn, F. A. O., Allen, A. M., Chai, S.-Y., McKinley, M. J., Oldfield, B. J. and Paxinos, G. (1990). The brain angiotensin system: insights from mapping its components. *Trends in Endocrinology and Metabolism*, **1**, 189–198.

Merchenthaler, I. (1991). Current status of brain hypophysiotropic factors: morphologic aspects. *Trends in Endocrinology and Metabolism*, **2**, 219–226.

Messing, R. B. (1988). Opioid modulation of learning and memory: Multiple behavioral outcomes. In R. J. Rodgers and S. J. Cooper (eds.) *Endorphins, Opiates and Behavioural Processes*, pp. 269–286. Chichester: John Wiley.

Millan, M. J. (1990). κ-Opioid receptors and analgesia. *Trends in Pharmacological Sciences*, **11**, 70–76.

Millan, M. J. and Herz, A. (1985). The endocrinology of the opioids. *International Review of Neurobiology*, **26**, 1–83.

Miller, W. L., Baxter, J. D. and Eberhardt, N. L. (1983). Peptide hormone genes: Structure and evolution. In D. T. Krieger, M. J. Brownstein and J. B. Martin (eds.), *Brain Peptides*, pp. 15–78. Chichester: John Wiley.

Mitchell, R. and Fleetwood-Walker, S. (1981). Substance P, but not TRH modulates the 5-HT autoreceptor in ventral lumbar spinal cord. *European Journal of Pharmacology*, **76**, 119–120.

Moos, F., Ingram, C. D., Wakerley, J. B., Guerné, Y., Freund-Mercier, M. J. and Richard, Ph. (1991). Oxytocin in the bed nucleus of the stria terminalis and lateral septum facilitates bursting of hypothalamic oxytocin neurons in suckled rats. *Journal of Neuroendocrinology*, **3**, 163–171.

Morley, J. E. (1983). Neuroendocrine effects of endogenous opioid peptides in human subjects: a review. *Psychoneuroendocrinology*, **8**, 361–379.

Morley, J. E. and Flood, J. F. (1990). Neuropeptide Y and memory processing. *Annals of the New York Academy of Sciences*, **611**, 226–231.

Morley, J. E. and Levine, A. S. (1980). Stress-induced eating is mediated through endogenous opiates. *Science*, **209**, 1259–1261.

Morley, J. E. and Levine, A. S. (1985). Cholecystokinin. In A. Lajtha (ed.), *Handbook of Neurochemistry*, 2nd edn., pp. 115–135. New York: Plenum Press.

Mortimer, C. H., McNeilly, A. S., Fisher, R. A., Murray, M. A. F. and Besser, G. M. (1974). Gonadotrophin releasing hormone therapy in hypogonadal males with hypothalamic or pituitary dysfunction. *British Medical Journal*, **4**, 617–621.

Moss, R. L. and Dudley, C. A. (1990). Differential effects of an LHRH antagonist analogue on lordosis behaviour induced by LHRH and the LHRH fragment AcLHRH 5–10. *Neuroendocrinology*, **52**, 138–142.

Myake, M., MacDonald, J. C. and North, R. A. (1989). Single potassium channels opened by opioids in rat locus ceruleus neurons. *Proceedings of the National Academy of Sciences, USA*, **86**, 3419–3422.

Mühlenthaler, M. and Dreifuss, J. J. (1982). Vasopressin excites hippocampal neurones. *Nature*, **296**, 749–750.

Müller, E. E. and Nistico, G. (1989). *Brain Messengers and the Pituitary*. San Diego: Academic Press.

Nakajima, Y., Nakajima, S. and Inoue, M. (1991). Substance P induced inhibition of potassium channels via a pertussis toxin-insensitive G protein. *Annals of the New York Academy of Sciences*, **632**, 103–111.

Needleman, P., Currie, M. G., Geller, D. M., Cole, B. R. and Adams, S. P. (1984). Atriopeptins: potential mediators of an endocrine relationship between heart and kidney. *Trends in Pharmacological Sciences*, **5**, 506–509.

Nemeroff, C. B. (1986). The interaction of neurotensin with dopaminergic pathways in the central nervous system: basic neurobiology and implications for the pathogenesis and treatment of schizophrenia. *Psychoneuroendocrinology*, **11**, 15–37.

Nemeroff, C. B. (ed.) (1988). *Neuropeptides in Psychiatric and Neurological Disorders*. Baltimore: The Johns Hopkins University Press.

Nicoll, R. A., Schenker, C. and Leeman, S. E. (1980). Substance P as a transmitter candidate. *Annual Review of Neuroscience*, **3**, 227–268.

Nieuwenhuys, R. (1985). *Chemoarchitecture of the Brain*. Berlin: Springer-Verlag.

North, R. A. (1986a). Electrophysiological effects of neuropeptides. In J. B. Martin and J. D. Barchas (eds.) *Neuropeptides in Neurological and Psychiatric Disease*, pp. 71–77. New York: Raven Press.

North, R. A. (1986b). Opioid receptor types and membrane ion channels. *Trends in Neurosciences*, **9**, 114–117.

Numan, M. (1985). Brain mechanisms and parental behavior. In N. Adler, D. Pfaff and R. W. Goy (eds.), *Handbook of Behavioral Neurobiology, vol. 7, Reproduction*, pp. 537–605. New York: Plenum Press.

Osborne, N. N. (1981). Communication between neurones: current concepts. *Neurochemistry International*, **3**, 3–16.

Otsuka, M. and Konishi, S. (1983). Substance P – the first peptide neurotransmitter? *Trends in Neurosciences*, **6**, 317–320.

Otsuka, M., Konishi, S., Yanagisawa, M., Tsunoo, A. and Akagi, H. (1982). Role of substance P as a sensory transmitter in spinal cord and sympathetic ganglia. *Ciba Foundation Symposium*, **91**, 13–30.

Ottlecz, A. (1987). Action of gastrointestinal polypeptide hormones on pituitary anterior lobe function. *Frontiers in Hormone Research*, **15**, 282–298.

Palkovits, M. (1983). Neuroanatomical techniques. In D. T. Krieger, M. J. Brownstein, and J. B. Martin (eds.) *Brain Peptides*, pp. 495–545. Chichester: John Wiley.

Panksepp, J., Siviy, S. M. and Normansell, L. A. (1985). Brain opioids and social emotions. In M. Reite and T. Field (eds.) *The Psychobiology of Attachment and Separation*, pp. 3–49. Orlando FL: Academic Press.

Pearse, A. G. E. (1969). The cytochemistry and ultrastructure of polypeptide hormone-producing cells of the APUD series and the embryonic, physiologic and pathologic implications of the concept. *Journal of Histochemistry and Cytochemistry*, **17**, 303–313.

Pearse, A. G. E. (1979). The APUD concept and its relationship to the neuropeptides. In A. M. Gotto, Jr, E. J. Peck, Jr and A. E. Boyd (eds.) *Brain Peptides: A New Endocrinology*, pp. 89–101. Amsterdam: Elsevier.

Pearse, A. G. E. (1983). The neuroendocrine division of the nervous system: APUD cells as neurones or paraneurones. In N. N. Osborne (ed.) *Dale's Principle and Communication Between Neurones*, pp. 37–48. Oxford: Pergamon Press.

Pearse, A. G. E. and Takor, T. T. (1979). Embryology of the diffuse neuroendocrine system and its relationship to the common peptides. *Federation Proceedings*, **38**, 2288–2294.

Pedersen, C. A., Ascher, J. A., Monroe, Y. L. and Prange, A. J., Jr (1982). Oxytocin induces maternal behavior in virgin female rats. *Science*, **216**, 648–650.

Pfaus, J. G. and Gorzalka, B. B. (1987). Opioids and sexual behavior. *Neuroscience and Biobehavioural Reviews*, **11**, 1–34.

Phillips, M. I. (1987). Function of angiotensin in the central nervous system. *Annual Review of Physiology*, **49**, 413–435.

Phillis, J. W. and Kirkpatrick, J. R. (1980). The actions of motilin, luteinizing hormone releasing hormone, cholecystokinin, somatostatin, vasoactive intestinal peptide, and other peptides on rat cerebral cortical neurons. *Canadian Journal of Physiology and Pharmacology*, **58**, 612–623.

Pictet, R. L., Rall, L. B., Phelps, P. and Rutter, W. J. (1976). The neural crest and the origin of the insulin-producing and other gastrointestinal hormone-producing cells. *Science*, **191**, 191–192.

Pittman, Q. J., MacVicar, B. A. and Colmers, W. F. (1987). In vitro preparations for electrophysiological study of peptide neurons and actions. In A. A. Boulton, G. B. Baker and Q. J. Pitman (eds.) *Neuromethods*, vol. 6, *Peptides*, pp. 409–438. Clifton, NJ: Humana Press.

Prange, A. J. Jr, and Utiger, R. D. (1981). What does brain thyrotropin-releasing hormone do? *New England Journal of Medicine*, **305**,

1089–1090.

Prasad, C. (1985). Thyrotropin-releasing hormone. In A. Lajtha (ed.) *Handbook of Neurochemistry*, 2nd edn, pp. 175–200. New York: Plenum Press.

Przegalinski, E. and Jaworska, L. (1990). The effect of repeated administration of anti-depressant drugs on the thyrotropin-releasing hormone (TRH) content of rat brain structures. *Psychoneuroendocrinology*, **15**, 147–153.

Purves, D. and Lichtman, J. W. (1985). *Principles of Neural Development*. Sunderland, MA: Sinauer.

Quirion, R. (1988). Atrial natriuretic factors and the brain: an update. *Trends in Neurosciences*, **11**, 58–62.

Quirion, R., Dam, T.-V. and Guard, S. (1991). Selective neurokinin receptor radioligands. *Annals of the New York Academy of Sciences*, **579**, 137–144.

Quirion, R., Martel, J.-C., Dumont, Y., Cadieux, A., Jolicoeur, F., St-Pierre, S. and Fournier, A. (1990). Neuropeptide Y receptors: autoradiographic distribution in the brain and structure-activity relationships. *Annals of the New York Academy of Sciences*, **611**, 58–72.

Reichlin, S. (1983). Somatostatin. In D. T. Krieger, M. J. Brownstein and J. B. Martin (eds.) *Brain Peptides*, pp. 711–752. Chichester: John Wiley.

Reitan, R. M. and Wolfson, D. (1985). *Neuroanatomy and Neuropathology: A Clinical Guide for Neuropsychologists*. Tucson, AZ, Neuropsychology Press. (Second edition, 1992.)

Renaud, L. P. (1979). Electrophysiology of brain peptides. In A. M. Gotto, Jr, E. J. Peck, Jr and A. E. Boyd (eds.) *Brain Peptides: A New Endocrinology*, pp. 119–139. Amsterdam: Elsevier.

Renaud, L. P., Martin, J. B. and Brazeau, P. (1975). Depressant action of TRH, LH-RH and somatostatin on activity of central neurones. *Nature*, **255**, 233–235.

Richard, P., Moos, F. and Freund-Mercier, M.-J. (1991). Central effects of oxytocin. *Physiological Reviews*, **71**, 331–370.

Rodgers, R. J. and Cooper, S. J. (eds.) (1988). *Endorphins, Opiates and Behavioural Processes*. Chichester: John Wiley.

Rorstad, O. P. (1983). Competitive binding assays. In D. T. Krieger, M. J. Brownstein and J. B. Martin (eds.) *Brain Peptides*, pp. 465–483. Chichester: John Wiley.

Roth, J., LeRoith, D., Collier, E. S., Weaver, N. R., Watkinson, A., Cleland, C. F. and Glick, S. M. (1985). Evolutionary origins of neuropeptides, hormones, and receptors: Possible applications to immunology. *Journal of Immunology*, **135**, 816s–819s.

Roth, J., LeRoith, D., Lesniak, M. A., de Pablo, F., Bassas, L. and Collier, E. S.. (1986). Molecules of intercellular communication in vertebrates, invertebrates and microbes: do they share common origins? *Progress in Brain Research*, **68**, 71–79.

Roth, J., LeRoith, D., Shiloach, J., Rosenzweig, J. L., Lesniak, M. A. and Havrankova, J. (1982). The evolutionary origins of hormones, neurotransmitters, and other extracellular chemical messengers. *New England Journal of Medicine*, **306** (9), 523–527.

Said, S. I. (1991). Vasoactive Intestinal Polypeptide: biologic role in health and disease. *Trends in Endocrinology and Metabolism*, **2**, 107–112.

Sandman, C. A., Barron, J. L., Demet, E. M., Chicz-Demet, A., Rothenberg, S. J. and Zea, F. J. (1990). Opioid peptides and perinatal

development: is β-endorphin a natural teratogen? Clinical implications *Annals of the New York Academy of Sciences*, **579**, 91–108.
Sanger, D. J. (1981). Endorphinergic mechanisms in the control of food and water intake. *Appetite*, **2**, 193–208.
Scharrer, B. (1987). Neurosecretion: beginnings and new directions in neuropeptide research. *Annual Review of Neuroscience*, **10**, 1–17.
Scharrer, B. (1990). The neuropeptide saga. *American Zoologist*, **30**, 887–895.
Schmauss, C. and Emrich, H. M. (1988). Narcotic antagonist and opioid treatment in psychiatry. In R. J. Rodgers and S . J. Cooper (eds.) *Endorphins, Opiates and Behavioural Processes*, pp. 327–351. Chichester: John Wileys.
Schotman, P., Schrama, L. H. and Edwards, P. M. (1985). Peptidergic systems. In A. Lajtha (ed.) *Handbook of Neurochemistry, 2nd edn. vol. 8: Neurochemical Systems*, pp. 243–275. New York: Plenum Press.
Schwartz, J. H. (1991). Chemical messengers: small molecules and peptides. In Kandel, E. R., Schwartz, J. H. and Jessell, T. M. (eds.) *Principles of Neural Science*, 3rd edn, pp. 213–224. New York: Elsevier.
Serra, G., Collu, M. and Gessa, G. L. (1988). Endorphins and sexual behaviour. In R. J. Rodgers and S. J. Cooper (eds.) *Endorphins, Opiates and Behavioural Processes*, pp. 237–247. Chichester: John Wiley.
Sheehan, M. J. and de Belleroche, J. (1984). Central actions of cholecystokinin: behavioural and release studies. In J. de Belleroche and G. J. Dockray (eds.) *Cholecystokinin (CCK) in the Nervous System*, pp. 111–127. Chichester: Ellis Horwood.
Sherwood, N. (1987). The GnRH family of peptides. *Trends in Neurosciences*, **10**, 129–132.
Shults, C. W., Quirion, R., Chronwall, B., Chase, T. N. and O'Donohue, T. L. (1984). A comparison of the anatomical distribution of substance P and substance P receptors in the rat central nervous system. *Peptides*, **5**, 1097–1128.
Siegelbaum, S. A. and Tsien, R. W. (1983). Modulation of gated ion channels as a mode of transmitter action. *Trends in Neurosciences*, **6**, 307–313.
Siggins, G. R., Gruol, D., Aldenhoff, J. and Pittman, Q. (1985). Electrophysiological actions of corticotropin-releasing factor in the central nervous system. *Federation Proceedings*, **44**, 237–242.
Silverman, A.-J. and Zimmerman, E. A. (1983). Magnocellular neurosecretory system. *Annual Review of Neuroscience*, **6**, 357–380.
Sims, K. B., Hoffman, D. L., Said, S. I. and Zimmerman, E. A. (1980). Vasoactive intestinal polypeptide (VIP) in mouse and rat brain: an immunocytochemical study. *Brain Research*, **186**, 165–183.
Sirinathsinghji, D. J. S. (1985). Modulation of lordosis behaviour in the female rat by CRF, β-endorphin and GnRH in the mesencephalic central gray. *Brain Research*, **336**, 45–55.
Smith, A. P., Law, P.-Y. and Loh, H. H. (1988). Role of opioid receptors in narcotic tolerance/dependence. In G. W. Pasternak (ed.) *The Opiate Receptors*, pp. 441–485. Clifton, NJ: Humana Press.
Smith, M. A., Kling, M. A., Whitfield, H. J., Brandt, H. A., Demitrack, M. A., Geracioti, T. D., Chrousos, G. P. and Gold, P. W. (1989). Corticotropin-releasing hormone: from endocrinology to psychobiology. *Hormone Research*, **31**, 66–71.
Snyder, S. H. (1980). Brain peptides as neurotransmitters. *Science*, **209**,

976–983.

Sofroniew, M. V. (1983). Vasopressin and oxytocin in the mammalian brain and spinal cord. *Trends in Neurosciences*, **6**, 467–472.

Spada, A., Reza-Elahi, F., Vallar, L. and Giannattasio, G. (1989). Intracellular signalling in the action of hypothalamic hormones on pituitary cells. In F. F. Casanueva and C. Dieguez (eds.) *Recent Advances in Basic and Clinical Neuroendocrinology*, pp. 87–94. Amsterdam: Excerpta Medica.

Spigelman, I. and Puil, E. (1991). Substance P actions on sensory neurons. *Annals of the New York Academy of Sciences*, **632**, 220–228.

Standaert, D. G., Saper, C . B. and Needleman, P. (1985). Atriopeptin: potent hormone and potential neuromediator. *Trends in Neurosciences*, **8**, 509–511.

Stewart, J. M. and Cannabasavaiah, K. (1979). Evolutionary aspects of some neuropeptides. *Federation Proceedings*, **38**, 2302–2308.

Stoneham, M. D., Everitt, B . J., Hansen, S., Lightman, S. L. and Todd, K. (1985). Oxytocin and sexual behaviour in the male rat and rabbit. *Journal of Endocrinology*, **107**, 97–106.

Strand, F. L., Rose, K. J., Zuccarelli, L. A., Kume, J., Alves, S. E., Antonawich, F. J. and Garrett, L. Y. (1991). Neuropeptide hormones as neurotrophic factors. *Physiological Reviews*, **71**, 1017–1046.

Strupp, B. J. and Levitsky, D. A. (1985). A mnemonic role for vasopressin: the evidence for and against. *Neuroscience and Biobehavioral Reviews*, **9**, 399–411.

Swanson, L. W. and Sawchenko, P. E. (1983). Hypothalamic integration: organization of the paraventricular and supraoptic nucleii. *Annual Review of Neuroscience*, **6**, 269–324.

Taché, Y. and Brown, M. (1982). On the role of bombesin in homeostasis. *Trends in Neurosciences*, **5**, 431–433.

Telegdy, G. (1985). Effects of gastrointestinal peptides on the nevous system. In A. Lajtha (ed.) *Handbook of Neurochemistry*, 2nd edn, pp. 217–241. New York: Plenum Press.

Thody, A. J. and Wilson, C. A. (1983). Melanocyte stimulating hormone and the inhibition of sexual behaviour in the female rat. *Physiology and Behavior*, **31**, 67–72.

Turner, C. D. and Bagnara, J. T. (1976). *General Endocrinology*, 6th edn. Philadelphia: W. B. Saunders.

Vaccarino, F. J., Bloom, F. E., Rivier, J., Vale, W. and Koob, G. F. (1985). Stimulation of food intake in rats by centrally administered hypothalamic growth hormone-releasing factor. *Nature*, **314**, 167–168.

Vaccarino, F. J. and Buckenham, K. E. (1987). Naloxone blockade of growth hormone-releasing factor-induced feeding. *Regulatory Peptides*, **18**, 165–171.

Vale, W. W., Rivier, C., Spiess, J. and Rivier, J. (1983). Corticotropin releasing factor. In D. T. Krieger, M. J. Brownstein and J. B. Martin (eds.) *Brain Peptides*, pp. 962–974. Chichester: John Wiley.

Veber, D. F. and Freidlinger, R. M. (1985). The design of metabolically-stable peptide analogs. *Trends in Neurosciences*, **8**, 392–396.

Vécsei, L., Balázs, M. and Telegdy, G. (1987). Action of somatostatin on the central nervous system. *Frontiers in Hormone Research*, **15**, 36–57.

Voight, M. M., Wang, R. Y. and Westfall, T. C. (1985). The effects of cholecystokinin on the in vivo release of newly synthesized [3H] dopamine from the nucleus accumbens of the rat. *Journal of Neuroscience*, **5**, 2744–2749.

White, R. E., Schonbrunn, A. and Armstrong, D. L. (1991). Somatostatin stimulates Ca^{2+} – activated K^+ channels through protein dephosphorylation. *Nature*, **351**, 570–573.

White, J. D., Stewart, K. D., Krause, J. E. and McKelvy, J. F. (1985). Biochemistry of peptide-secreting neurons. *Physiological Reviews*, **65**, 553–606.

Wood, P. L. and Iyengar, S. (1988). Central actions of opiates and opioid peptides. In G. W. Pasternak (ed.) *The Opiate Receptors*, pp. 307–356. Clifton, NJ: Humana Press.

Zadina, J. E., Banks, W. A. and Kastin, A. J. (1986). Central nervous system effects of peptides, 1980–1985: a cross-listing of peptides and their central actions from the first six years of the journal *Peptides*. *Peptides*, **7**, 497–537.

Zhao, B.-G., Chapman, C. and Bicknell, R. J. (1988). Functional κ-opioid receptors on oxytocin and vasopressin nerve terminals isolated from the rat neurohypophysis. *Brain Research*, **462**, 62–66.

12

Neuropeptides II: neuropeptide function

Neuropeptides are synthesized in a wide variety of neural cells and are often colocalized and co-released with classical neurotransmitters, as discussed in Chapter 11. When they are released, neuropeptides may act either as neurotransmitters or neuromodulators. This chapter examines the neurotransmitter and neuromodulator actions of neuropeptides on the neuroendocrine system, the autonomic nervous system, and the central nervous system. First, however, the definitions of a neurotransmitter and a neuromodulator will be discussed with respect to the neuropeptides.

12.1 NEUROTRANSMITTER AND NEUROMODULATOR ACTIONS OF NEUROPEPTIDES: A DICHOTOMY OR A CONTINUUM?

To be considered as a neurotransmitter, a substance must meet the eight criteria listed in Table 5.1 (p. 58), but demonstrating all of these criteria may be quite difficult. As discussed by Barchas *et al.* (1978), Osborne (1981), and Schwartz (1991), the essential features of a neurotransmitter are that it is:

Table 12.1. *Criteria for defining a neuromodulator in the central nervous system*

1. The substance is not acting as a neurotransmitter, as defined in Table 5.1, in that it does not act transsynaptically.
2. The substance must be present in physiological fluids and have access to the site of potential modulation in physiologically significant concentrations.
3. Alterations in endogenous concentrations of the substance should affect neuronal activity consistently and predictably.
4. Direct application of the substance should mimic the effect of increasing its endogenous concentrations.
5. The substance should have one or more specific sites of action through which it can alter neuronal activity.
6. Inactivating mechanisms should exist which account for the time course of effects of endogenously or exogenously induced changes in concentrations of the substance.
7. Exogenous administration of neuropeptide agonists should have the same effect as increasing the endogenous release of the neuropeptide and exogenous administration of antagonists should have the same effect as decreasing endogenous release.

Source: Barchas *et al.*, 1978.

(1) Synthesized in a neuron
(2) Present in the presynaptic nerve terminal and released into the synapse in amounts sufficient to stimulate the postsynaptic cell when the neuron is stimulated
(3) Whether endogenously released or applied exogenously, a neurotransmitter has exactly the same action on the postsynaptic cell (i.e. it activates the same ion channels or second messenger pathways)
(4) A specific mechanism exists in the synapse for deactivating it.

Concerns about the distinction between neurotransmitters and neuromodulators developed with the discovery of the neuropeptides. At first, neuropeptides were clearly differentiated as modulator substances which had neuroregulatory action, but did not meet the criteria for being neurotransmitters (Florey, 1967). Neuromodulators have been defined by a set of seven criteria (listed in Table 12.1) which differentiate them from neurotransmitters, as defined in Table 5.1 (Barchas *et al.*, 1978). The key criterion which distinguishes neurotransmitters from neuromodulators is that neurotransmitters act via the synapse, whereas neuromodulators do not (Bennett and Whitehead, 1983, pp. 225–8). Before a neuropeptide can be considered to have neurotransmitter activity, therefore, it must be shown to act trans-synaptically.

Whether particular neuropeptides act as neurotransmitters or neuromodulators has been the topic of much debate (Barchas *et al.*, 1978; Dismukes, 1979, and following commentary; Osborne, 1981; Bloom, 1988). As more is learned about the actions of neuropeptides, however, it is clear that they have a continuum of effects on their target cells (Lundberg and Hökfelt, 1983; Kupfermann, 1991). As shown in Figure 12.1, both the 'classical' neurotransmitters and the neuropeptides have multiple postsynaptic and presynaptic actions through their different

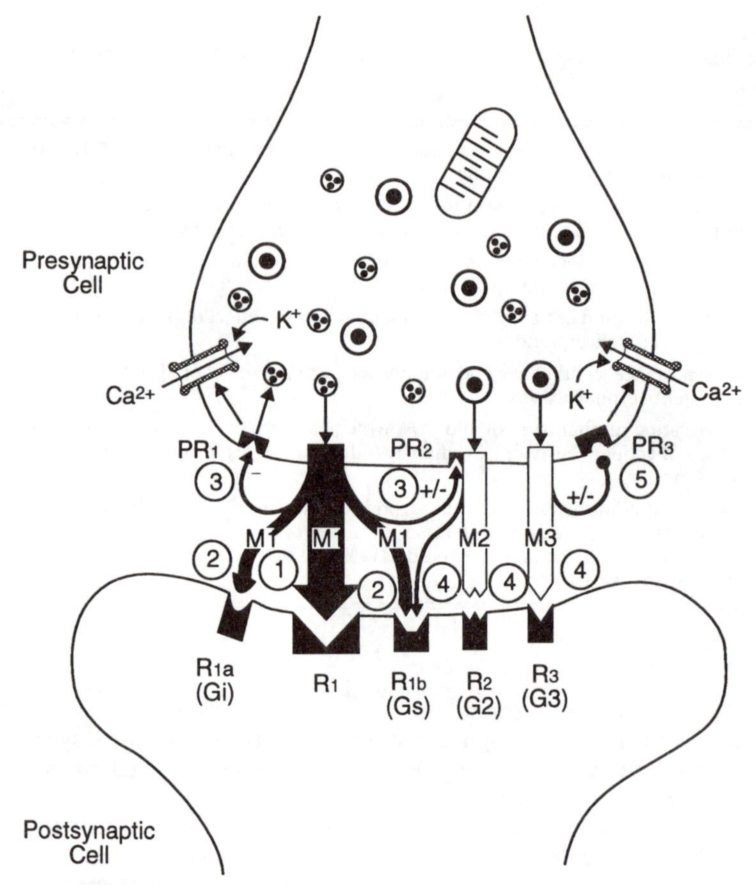

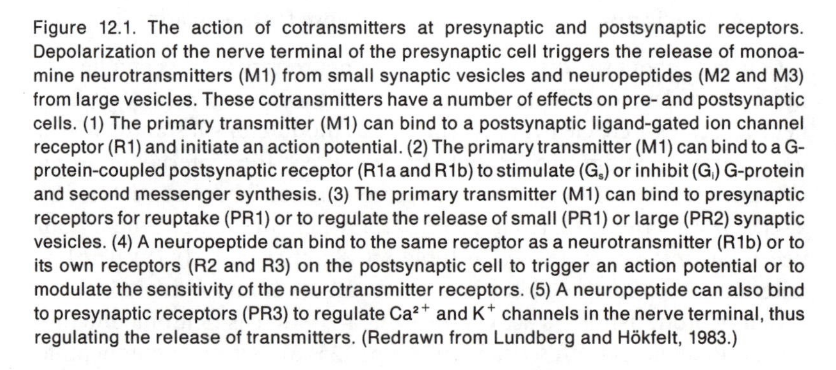
Figure 12.1. The action of cotransmitters at presynaptic and postsynaptic receptors. Depolarization of the nerve terminal of the presynaptic cell triggers the release of monoamine neurotransmitters (M1) from small synaptic vesicles and neuropeptides (M2 and M3) from large vesicles. These cotransmitters have a number of effects on pre- and postsynaptic cells. (1) The primary transmitter (M1) can bind to a postsynaptic ligand-gated ion channel receptor (R1) and initiate an action potential. (2) The primary transmitter (M1) can bind to a G-protein-coupled postsynaptic receptor (R1a and R1b) to stimulate (G_s) or inhibit (G_i) G-protein and second messenger synthesis. (3) The primary transmitter (M1) can bind to presynaptic receptors for reuptake (PR1) or to regulate the release of small (PR1) or large (PR2) synaptic vesicles. (4) A neuropeptide can bind to the same receptor as a neurotransmitter (R1b) or to its own receptors (R2 and R3) on the postsynaptic cell to trigger an action potential or to modulate the sensitivity of the neurotransmitter receptors. (5) A neuropeptide can also bind to presynaptic receptors (PR3) to regulate Ca^{2+} and K^+ channels in the nerve terminal, thus regulating the release of transmitters. (Redrawn from Lundberg and Hökfelt, 1983.)

receptor types. The neuropeptides may act as neurotransmitters at the same postsynaptic receptors as 'classical' neurotransmitters or at their own receptors. At other postsynaptic receptors, the neuropeptides may act as neuromodulators. Both neurotransmitters and neuropeptides also regulate their own release and that of other transmitters by acting at receptors on the presynaptic axon terminals (Figure 12.1).

Figure 12.2 shows four possible receptor mechanisms by which neuropeptides can affect neural activity. Whether or not a cell fires can be determined by the synaptic action of a neurochemical at ligand-gated ion

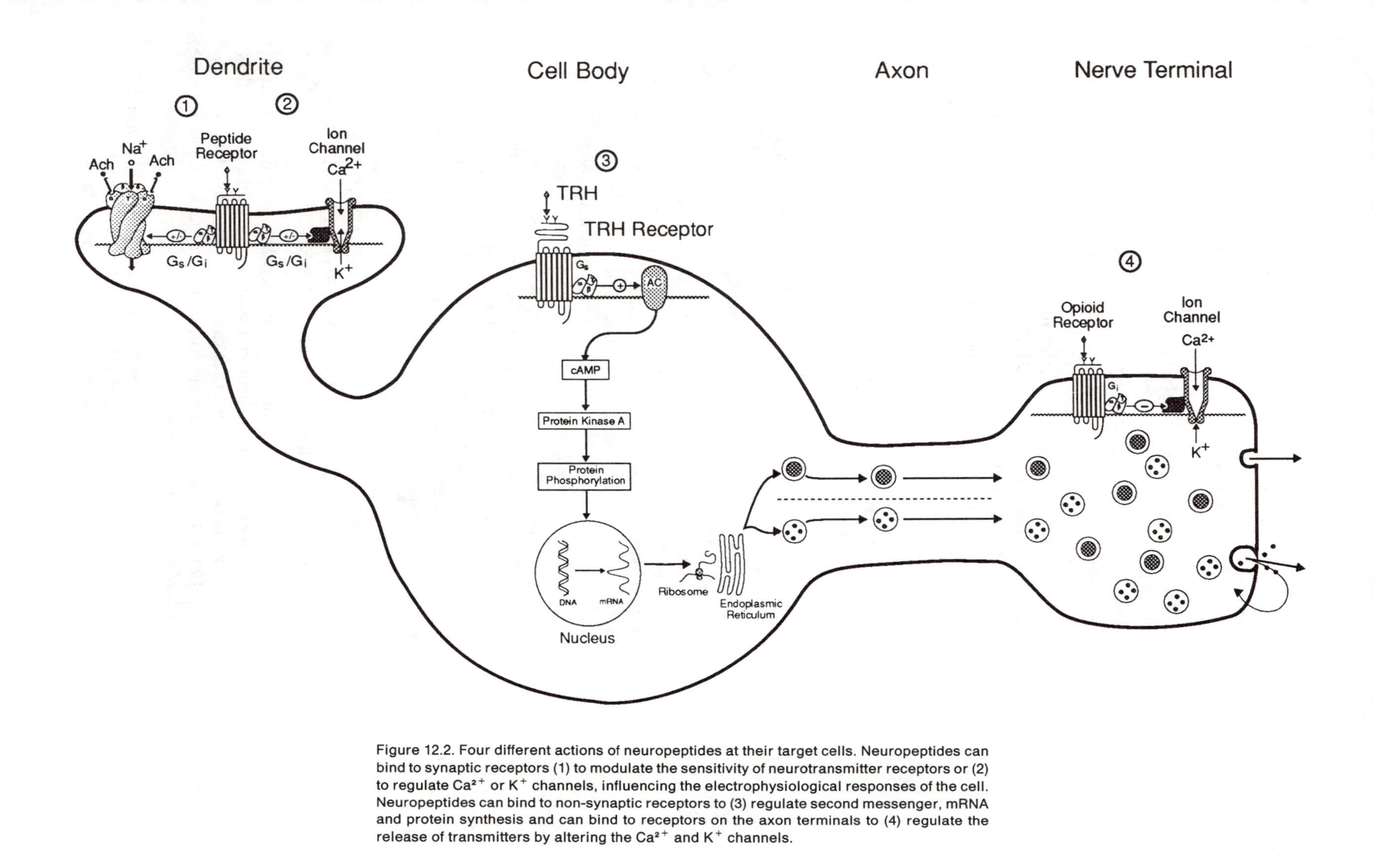

Figure 12.2. Four different actions of neuropeptides at their target cells. Neuropeptides can bind to synaptic receptors (1) to modulate the sensitivity of neurotransmitter receptors or (2) to regulate Ca^{2+} or K^+ channels, influencing the electrophysiological responses of the cell. Neuropeptides can bind to non-synaptic receptors to (3) regulate second messenger, mRNA and protein synthesis and can bind to receptors on the axon terminals to (4) regulate the release of transmitters by altering the Ca^{2+} and K^+ channels.

channel receptors or G-protein-coupled receptors. So far, no neuropeptides are known to bind to ligand-gated ion channels; they all bind to G-protein-coupled receptors. Neuropeptides binding to G-protein-coupled receptors can regulate the sensitivity of other receptors to their ligands, and the opening and closing of non-gated ion channels (actions 1 and 2 in Figure 12.2). These two actions at postsynaptic receptors are the 'neurotransmitter' actions of neuropeptides. The binding of neuropeptides to non-synaptic receptors initiates second messenger cascades which regulate mRNA synthesis in the cell nucleus and thus regulate the synthesis and storage of neurotransmitters, neuropeptides and other proteins in the target cell (action 3 in Figure 12.2). Neuropeptides also bind to presynaptic receptors where they modulate the release of neurotransmitters and neuropeptides by regulating ion channels on the axon terminal (action 4 in Figure 12.2). These last two functions are considered as neuromodulator actions of neuropeptides.

12.2 NEUROTRANSMITTER ACTIONS OF NEUROPEPTIDES

There is a wealth of evidence to suggest that many neuropeptides, including substance P, neurotensin, CCK, VIP, the enkephalins, and others, meet most of the criteria given in Table 5.1 for the definition of a neurotransmitter (Snyder, 1980; Emson and Hunt, 1982; Kelly and Brooks, 1984; Kupfermann, 1991; Schwartz, 1991). Some classical neurotransmitters bind to ligand-gated ion channel receptors, such as the nicotinic cholinergic receptor, and trigger rapid onset, short duration action potentials. Other neurotransmitters act via G-protein-coupled receptors, such as the muscarinic cholinergic and β-adrenergic receptors, to trigger slow onset, long duration action potentials (Siegelbaum and Tsien, 1983). Neuropeptides also bind to G-protein-coupled receptors and trigger slow onset, long duration action potentials, as shown in Figure 12.3. They do this by regulating the ion channels in the membrane of the postsynaptic cell in two ways: by phosphorylating the membrane substrate proteins directly (see Figure 10.3, p. 198) or by activating a second messenger cascade so that protein kinase phosphorylates the membrane substrate proteins (see Figure 10.5, p. 202).

12.2.1 HOW WELL DO NEUROPEPTIDES MEET THE CRITERIA DEFINING A NEUROTRANSMITTER? SUBSTANCE P AS AN EXAMPLE

The first neuropeptide which was considered to have met the criteria given in Table 5.1 for a neurotransmitter was substance P (Nicoll, Schenker and Leeman, 1980; Otsuka and Konishi, 1983; Spigelman and Puil, 1991). This section examines the evidence that substance P has met each of these eight criteria.

1. *The substance must be present in presynaptic elements of neuronal tissue, possibly in an uneven distribution throughout the brain.* Substance P is widely distributed throughout the brain, with high concentrations in

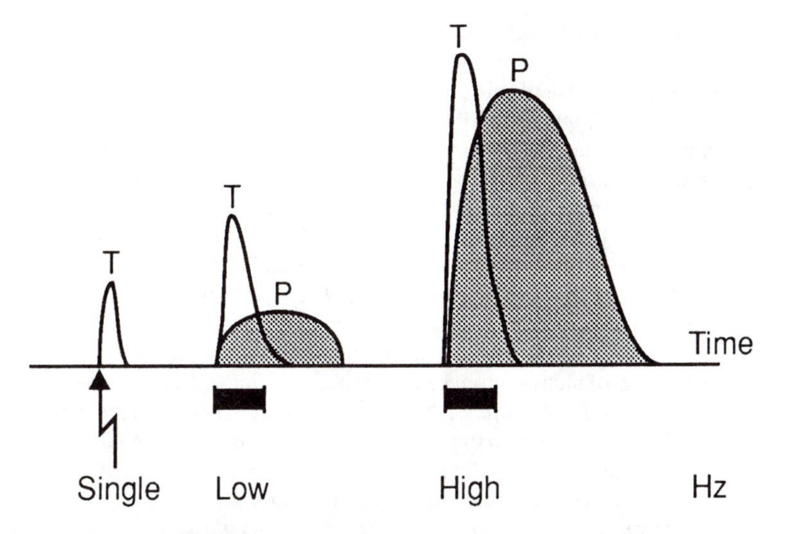

Figure 12.3. Differential release of colocalized classical neurotransmitters and neuropeptides and their effects on the electrophysiological responses of a postsynaptic cell. A single nerve impulse releases a small vesicle containing only the classical neurotransmitter (T), which results in a rapid onset, short duration action potential in the postsynaptic cell. At higher frequencies (bursts) of stimulation, large vesicles containing both monoamines and peptides are released, resulting in a biphasic action potential. The neuropeptides (P) trigger a slow onset, long duration response, which follows the rapid response to the neurotransmitter (T). (Redrawn from Lundberg and Hökfelt, 1983.)

the hypothalamus, amygdala and limbic system, mesencephalon, pons and medulla, and has cell bodies in each of these areas as shown in Figure 11.9A (p. 239) (Shults *et al.*, 1984). Substance P may act as a transmitter in the ascending pain pathways of the dorsal horn of the spinal cord, the striatonigral tract, and the habenulo-interpeduncular tract of the brain (Nicoll *et al.*, 1980). In the spinal cord, substance P is found in the axon terminals of sensory fibers which transmit pain signals (see Figure 12.12). If the axons of these nerves are cut, the amount of substance P in the nerve endings decreases, indicating that substance P is synthesized in the cell body and transported to the dorsal horn of the spinal cord via axonal transport (Nicoll *et al.*, 1980; Helke *et al.*, 1991).

2. *Precursors and synthetic enzymes must be present in the neuron, usually in close proximity to the site of presumed action.* Substance P belongs to the family of tachykinin peptides and is synthesized from preprotachykinin in the dorsal horn of the spinal cord and in the neural areas shown in Figure 11.9A (p. 239) (Helke *et al.*, 1990; Krause *et al.*, 1990). Substance P has been immunohistochemically identified in the nerve terminals of these cells and is often colocalized in synaptic vesicles with serotonin or acetylcholine (see Table 11.3, p. 237). There is, however, no consistent relationship between the density of the substance P cell bodies and nerve fibers and the number of substance P receptors in a given brain region. Both substance P and substance P receptors occur in the dorsal horn of the spinal cord and in the locus ceruleus, but the substantia nigra has high concentrations of substance P and no substance P receptors, while the cerebellum and dentate gyrus have high levels of substance P receptors, but no cell bodies or nerve fibers (Shults *et al.*, 1984).

3. *Stimulation of afferents should cause release of the substance in physiologically significant amounts.* Electrical stimulation of the dorsal root ganglia causes the release of substance P, provided Ca^{2+} is present (Nicoll *et al.*, 1980; Spigelman and Puil, 1991).
4. *Direct application of the substance to the synapse should produce responses which are identical to those of stimulating afferents.* Microinjection of substance P into the spinal cord causes depolarization of the neurons of the dorsal horn and increased firing of these neurons in the same manner as electrical stimulation of the primary afferents (Otsuka and Konishi, 1983; Spigelman and Puil, 1991).
5. *There should be specific receptors present which interact with the substance: these should be in close proximity to presynaptic structures.* Substance P acts at three types of tachykinin receptors (NK_1, NK_2, and NK_3) in the brain and spinal cord (Helke *et al.*, 1990, 1991). Each of these receptors has a different distribution in the nervous system (Krause *et al.*, 1990), but the anatomical relationship between the substance P nerve terminals and these receptors is complex, as mentioned above.
6. *Interaction of the substance with its receptor should induce changes in postsynaptic membrane permeability leading to excitatory or inhibitory postsynaptic potentials.* Substance P has excitatory effects on the neurons of the dorsal horn of the spinal cord. It acts via G-protein-coupled receptors to increase neural excitability by regulating Ca^{2+} or K^+ channels (Spigelman and Puil, 1991; Nakajima *et al.*, 1991).
7. *Specific inactivating mechanisms should exist which stop interactions of the substance with its receptor in a physiologically reasonable time frame.* Substance P is rapidly deactivated by peptidase enzymes in the brain and peripheral nervous system. One of these enzymes, 'substance P-degrading-enzyme' appears to be specific to substance P (Iversen, 1982).
8. *Interventions at postsynaptic sites using agonist drugs should mimic the action of the transmitter and antagonists should block its effects.* Selective agonist and antagonist drugs have been developed to act at each of the three tachykinin receptors (Quirion *et al.*, 1991). With the development of these drugs, the specific locations and functions of each of these receptors is now being discovered.

Summary. All but one of the eight criteria defining a neurotransmitter have been met by substance P. The second criterion, which requires the proximity of substance P cell bodies and receptors, is the only one not met. The mismatch in the neural distribution of cell bodies and receptors characterizes a number of neuropeptides (Herkenham, 1987). Many neuropeptides are now considered to have met the criteria for neurotransmitters (Kupfermann, 1991; Schwartz, 1991) and similar criterion analyses can be done for the enkephalins, cholecystokinin, vasoactive intestinal polypeptide, neurotensin, and other neuropeptides as was done in this section for substance P.

12.3 NEUROMODULATOR ACTIONS OF NEUROPEPTIDES

The traditional principle of neurotransmission, as described in Chapters 5 and 10, is that each neuron releases a single neurotransmitter into a synapse to stimulate a postsynaptic receptor. The finding that two or more chemical messengers may be co-released from a single neuron and interact to alter postsynaptic activity increases the complexity of intercellular communication. There are at least three possible sites where neuropeptides may modulate the actions of the primary neurotransmitters in the postsynaptic cell: the receptor, the transducer, and the second messenger system. As shown in Figure 12.2, the neuromodulator effects of neuropeptides involve: the regulation of receptor sensitivity; the regulation of neurotransmitter synthesis, storage and transport; and the regulation of neurotransmitter release (Florey, 1967; Kaczmarek and Levitan, 1987; Crawley, 1990).

12.3.1 NEUROPEPTIDE REGULATION OF POSTSYNAPTIC RECEPTOR SENSITIVITY

Neuropeptides can modulate the sensitivity of postsynaptic cells to primary transmitters by up- or down-regulating their receptors and G-proteins (Kaczmarek and Levitan, 1987). For example, when opioid peptides bind to their receptors in the adrenal medulla, they down-regulate nicotinic receptors, thus reducing the stimulation of the cell by acetylcholine and inhibiting the release of adrenaline (Kumakura *et al.*, 1980; Guidotti *et al.*, 1983).

12.3.2 MEASURING ELECTROPHYSIOLOGICAL RESPONSES TO NEUROPEPTIDES

When neuropeptides alter the permeability of the ion channels in the postsynaptic cell, they alter the electrophysiological activity of that cell. The effects of neuropeptides on particular ion channels in the postsynaptic cell membrane can thus be experimentally determined by recording the electrophysiological responses of that cell to external stimulation. In many cases the cells can be studied *in vitro*, thus enabling complex pharmacological manipulations (Pittman, MacVicar and Colmers, 1987). The electrophysiological effects of a number of neuropeptides on postsynaptic cells have been determined (Renaud, 1979; Phillis and Kirkpatrick, 1980; North, 1986a; Bloom, 1988). For example, TRH, LH-RH, and somatostatin, regulate the electrophysiological activity of hypothalamic neurons (Renaud *et al.*, 1975); substance P and neuropeptide Y regulate the firing rate of neurons in the spinal cord and sympathetic ganglia (Otsuka and Konishi, 1983; Lundberg *et al.*, 1990) and vasopressin alters the firing rate of hippocampal neurons (Mühlethaler *et al.*, 1982).

As shown in Figure 12.4, neuropeptides can alter the shape of the action potential; the frequency of spontaneous nerve firing; the frequency of neural bursting; and the responsiveness of the cell to stimulation from primary neurotransmitters (Kaczmarek and Levitan, 1987). A change in the shape of the action potential (Figure 12.4A) indicates an alteration in the time that the ion channels stay open. The spontaneous firing rate of neurons (Figure 12.4B) varies from near zero to a high constant activity, even in the absence of stimulation. Some neuropeptides, such as somatostatin, prevent the opening of Ca^{2+} channels, thus inhibit spontaneous

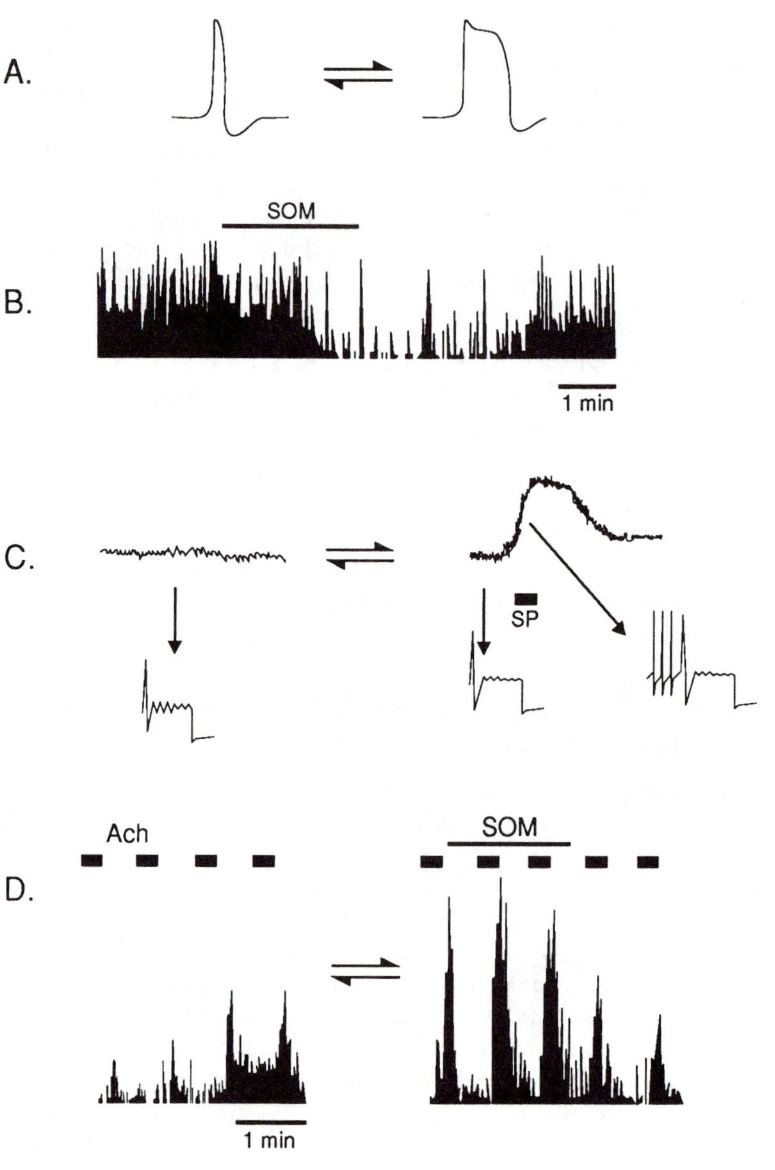

Figure 12.4. Some of the changes that can be observed in the electrical properties of neurons following neuropeptide stimulation. (A) Alteration in the shape of action potentials, as explained in Figure 12.3. (B) Changes in frequency and pattern of spontaneous firing rates. In this example, somatostatin (SOM) depresses the firing rate of a single cortical neuron. (C) Changes in the frequency of neural bursting. In this example, administration of substance P (SP) results in the depolarization of the cell and in repetitive (bursting) discharge when the cell is stimulated. (D) Altered responses to neurotransmitter stimulation. In this example, somatostatin facilitates the firing rate of cells to acetylcholine (Ach) stimulation, even though it depresses the spontaneous firing rate of the same cells, as shown in B. (A redrawn from Kaczmarek and Levitan 1987; B and D redrawn from Mancillas *et al.*, 1986; C redrawn from Spigelman and Puil, 1991.)

nerve firing, while others, such as substance P, close the normally open K^+ channels, thus increasing the depolarization of the cell until the firing threshold is reached (White *et al.*, 1991). The endogenous opioids have complex electrophysiological effects on their target cells. In general, they inhibit neural activity by decreasing Ca^{2+} influx into neurons or by enhancing K^+ conductance, thus hyperpolarizing the receptor cells, and inhibiting the propagation of the action potential along the axon (North, 1986b; Myake, MacDonald and Myake, 1989). Recently, however, opioids have been shown to decrease K^+ conductance, thus facilitating neural activity by prolonging action potentials (Crain and Shen, 1990).

When neuropeptides depolarize the postsynaptic cell, that cell is more likely to show neural bursting when it is stimulated (Figure 12.4C). For

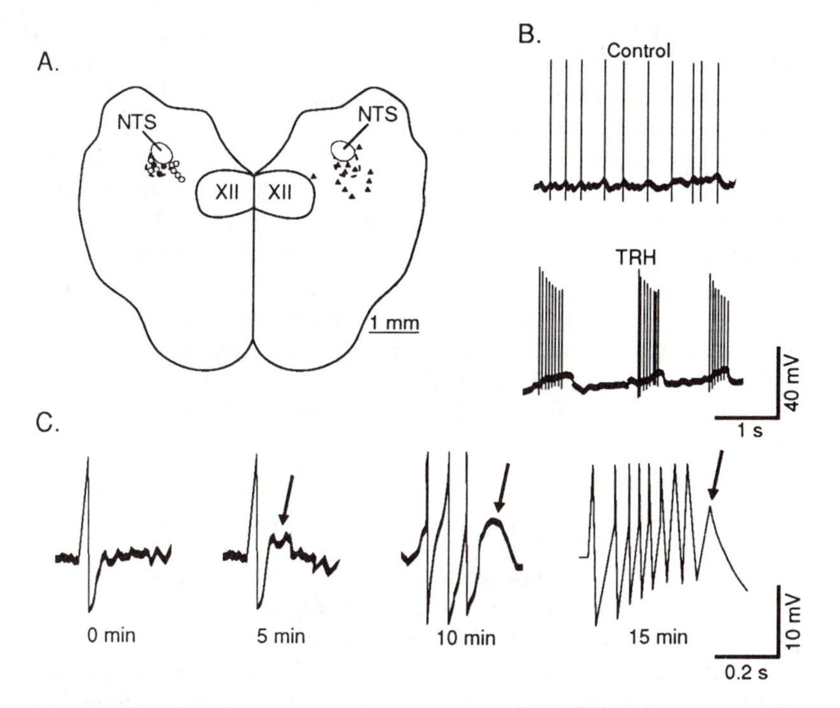

Figure 12.5. Modulation by thyrotropin releasing hormone (TRH) of the discharge pattern of a neuron in the nucleus tractus solitarius of the brain stem. (A) A schematic diagram of a cross-section of the brain stem. The right side shows a map of the dorsal respiratory group of neurons near the nucleus tractus solitarius (NTS). Triangles show the distribution of identified respiratory neurons. The left side indicates the locations of neurons studied in a slice preparation of the nucleus tractus solitarius. Solid circles indicate cells that responded to TRH; open circles show the location of non-responsive cells. (B) The effect of TRH on neuron firing. Unpatterned neural activity occurred prior to TRH injection (control), while phasic bursting was observed during exposure to TRH. (C) A depolarizing after-potential (arrows) and bursting were observed in this neuron during exposure to TRH. (Redrawn from Dekin, Richerson and Getting, 1985.)

example, TRH modulates the pattern of bursting in the respiratory neurons of the nucleus tractus solitarius in the brain stem shown in Figure 12.5. Neuropeptides also modulate the responsiveness of the cell to synaptic stimulation from neurotransmitters. For example, although somatostatin inhibits the spontaneous firing rate of neurons in the rat cortex (Figure 12.4B), it enhances the firing rate of these neurons in response to acetylcholine stimulation (Figure 12.4D).

The pattern of cell firing determines whether small or large synaptic vesicles will be released from the nerve axon. Short duration depolarization stimulates the release of small vesicles (neurotransmitters only) and long duration depolarization or bursts of stimulation release large vesicles containing both neurotransmitters and neuropeptides. Thus, a change in firing pattern may determine whether only the primary transmitter or both the primary transmitter and the peptide cotransmitters are released into the synapse. The ability of neuropeptides to depolarize cell membranes and induce bursts of neural activity makes them important mediators of neuropeptide and neurohormone secretion, as discussed in Section 12.4 below.

The role of oxytocin in the milk ejection reflex provides a good example of the function of the changes in neural activity induced by neuropeptides. Figure 6.13 (p. 104) shows that the suckling of infants at the nipple of a lactating female rat causes a rapid increase in firing rate in the neurons of the PVN. These neurons release oxytocin into the peripheral circulation from the neurohypophysis and from axon terminals in the brain pathways shown in Figure 11.10 (p. 240). Oxytocin has an ultrashort positive feedback loop, stimulating its own release, as shown in Figure 6.10 (p.101). This feedback loop involves oxytocin receptors in cells of the bed nucleus of the stria terminalis and the lateral septum. Oxytocin stimulates these cells, which then increase the firing rate of the neurons in the PVN and SON until a neural burst occurs and sufficient oxytocin is released to stimulate milk ejection (Moos *et al.*, 1991).

12.3.3 COTRANSMITTER ACTION OF NEUROPEPTIDES

Cotransmission results in the combination of a fast neural response, followed by a later slow response as shown in Figure 12.3. Since the electrophysiological effects of each transmitter can be excitatory or inhibitory, the interactions between cotransmitters at their target cells may be additive, subtractive, multiplicative, or divisive (Kupfermann, 1991). Through these operations, cotransmitters fine-tune the sensitivity of the postsynaptic cell to the primary transmitter and thus integrate the information processing properties of the nervous system (Guidotti *et al.*, 1983).

Figure 12.6 shows how a primary neurotransmitter and a neuropeptide cotransmitter may interact additively or multiplicatively to elevate second messenger synthesis in a postsynaptic cell. When VIP binds to its receptor, it activates the G-protein (G_s), increasing cyclic AMP levels. In the cerebral cortex, VIP interacts with noradrenaline, acting at α_1-adrenergic receptors, to potentiate an increase in cyclic AMP production. When the two chemicals act as cotransmitters, cyclic AMP levels are elevated more than when either transmitter alone stimulates the target cell. GABA and VIP also act as cotransmitters to potentiate cyclic AMP production (Magistretti and Schorderet, 1984; Magistretti, 1990).

The complexity of cotransmitter interactions is increased in systems in which a classical neurotransmitter is colocalized with more than one neuropeptide. For example, TRH and substance P are colocalized with serotonin in the medulla oblongata and spinal cord. Both of these neuropeptides are co-released with serotonin and act as cotransmitters, but in different ways. TRH enhances the response of the postsynaptic cell to serotonin by increasing second messenger production, while substance P enhances the presynaptic release of serotonin (Mitchell and Fleetwood-Walker, 1981; Hökfelt *et al.*, 1987). As if this were not complex enough, steroid hormones also interact with neuropeptides to regulate second messenger activity. For example, the glucocorticoids inhibit VIP stimulated increases in cyclic AMP levels in their target cells in the hippocampus, amygdala and septum (Magistretti, 1990).

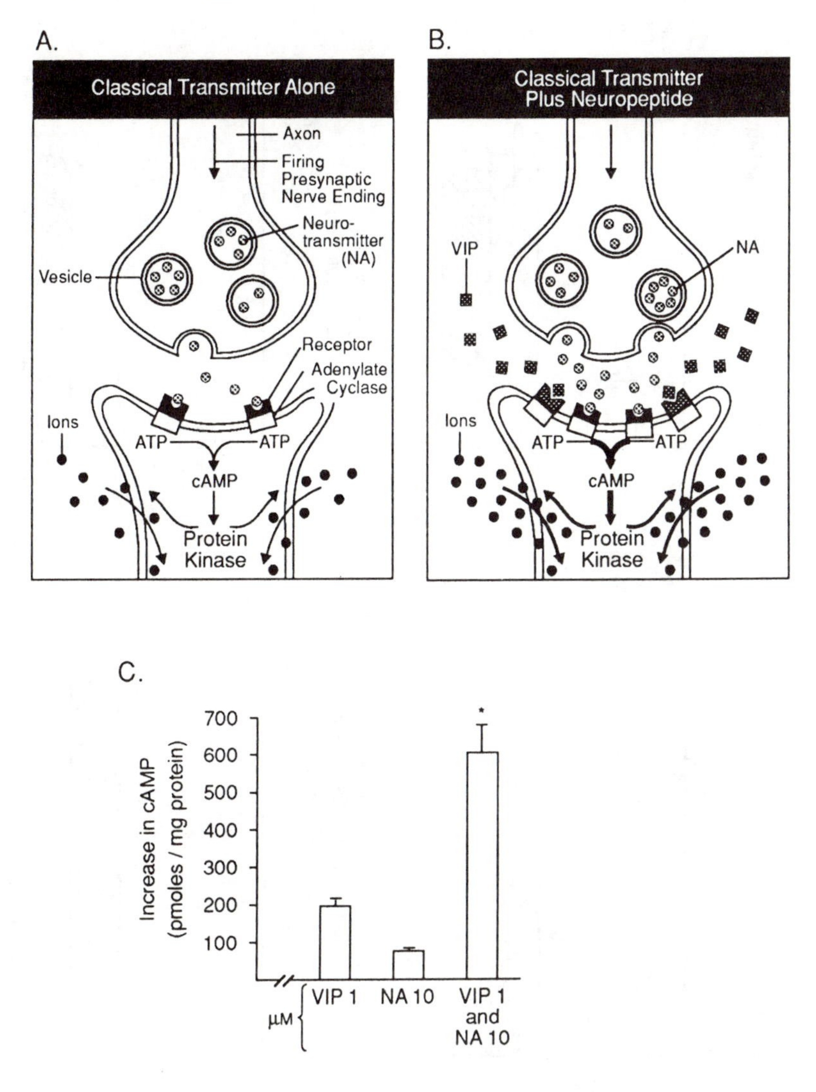

Figure 12.6. Interaction of a primary neurotransmitter and a neuropeptide cotransmitter. One mechanism by which a neuropeptide (such as VIP) and a neurotransmitter (such as noradrenaline (NA)) interact to elevate second messenger activity in a postsynaptic cell. (A) When the presynaptic cell fires, the neurotransmitter is released into the synapse and binds to its receptors on the postsynaptic cell, activating the cyclic AMP second messenger system. The resulting protein kinase phosphorylates the membrane proteins, altering the permeability of the cell to ions such as Ca^{2+} and K^{+}. (B) The binding of the neuropeptide (VIP) to its receptors also increases the production of cyclic AMP, thus enhancing the permeability of the cell membrane to ions. (C) The amount of cyclic AMP stimulated by noradrenaline (10 μM) and VIP (1 μM) alone or in combination has been measured in the mouse cerebral cortex. Both transmitters together stimulate significantly more cyclic AMP synthesis than either transmitter acting alone ($p<0.001$), indicating the facilitative effect of cotransmitter action.(A and B redrawn from Hollister, Davis and Davis, 1980; C redrawn from Magistretti and Schorderet, 1984.)

12.3.4 NEUROPEPTIDE REGULATION OF PROTEIN SYNTHESIS

As discussed in Chapter 10, the activation of second-messenger cascades leads to phosphorylation of both membrane substrate proteins and nuclear regulatory proteins. Phosphorylation of the membrane substrate proteins regulates ion channels in the cell membrane (see Figure 10.5, p. 202). Phosphorylation of the nuclear regulatory proteins regulates mRNA and protein synthesis in the target cells. Thus, the increases in second messenger synthesis shown in Figure 12.6 will increase protein synthesis as well as membrane permeability to ions in the postsynaptic cell.

An example of the action of neuropeptides on protein synthesis is the stimulation of the synthesis and release of the adenohypophyseal hormones by the hypothalamic hormones as discussed in Chapter 4. As shown in Figure 12.7, CRH stimulates the release of ACTH from pituitary

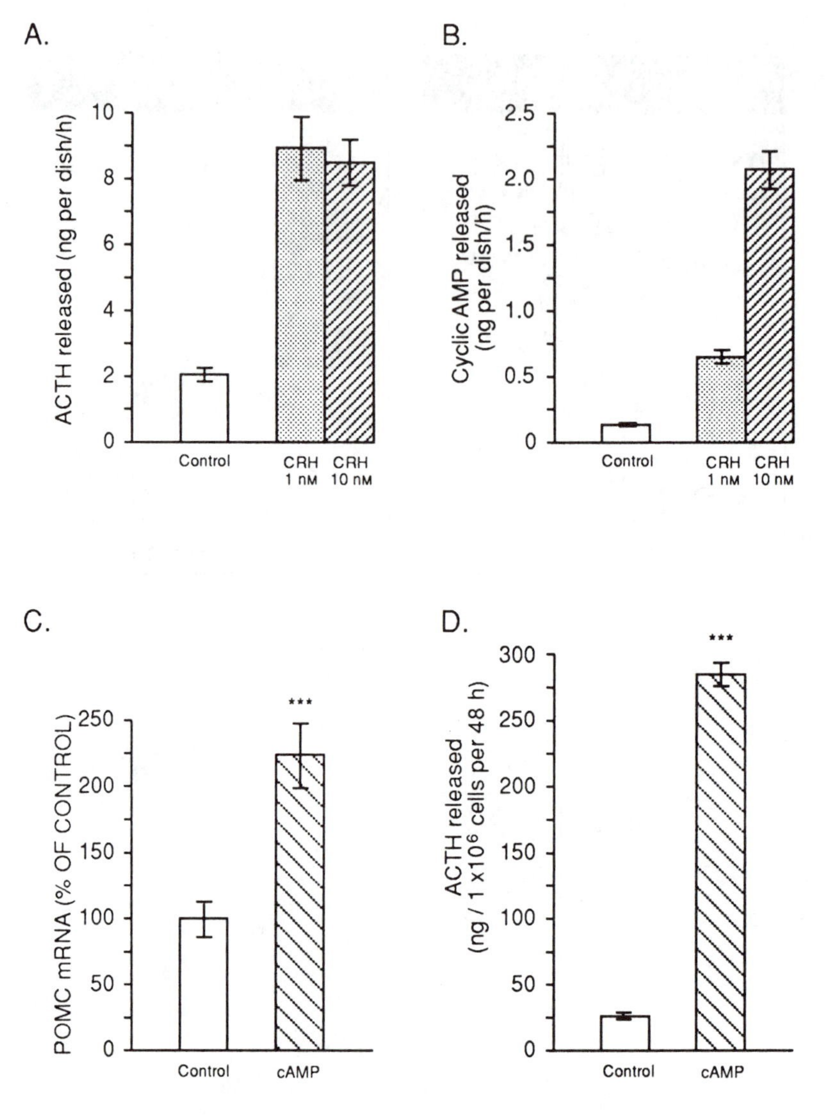

Figure 12.7. The stimulation of ACTH synthesis and release by corticotropin releasing hormone (CRH) is mediated by the cyclic AMP second messenger system. Application of synthetic CRH (1 or 10 nM) to a culture of rat corticotroph cells increases the release of (A) ACTH and (B) cyclic AMP as measured by specific radioimmunoassays. When cyclic AMP is added to a culture of rat anterior pituitary cells, (C) proopiomelanocortin (POMC) mRNA levels and (D) ACTH release are elevated. (Redrawn from Labrie *et al.*, 1987.)

corticotroph cells by activating cyclic AMP production. Infusion of CRH into an *in vitro* culture of pituitary corticotroph cells stimulates cyclic AMP synthesis and ACTH release (see Figures 12.7A and B). Administration of cyclic AMP alone to the corticotroph cell culture stimulates the synthesis of POMC, the prohormone for ACTH, and the release of ACTH (Figures 12.7C and D). Thus, CRH can be shown to elevate cyclic AMP production in its pituitary target cells and the increased cyclic AMP can be shown to stimulate the synthesis and release of ACTH (Labrie *et al.*, 1987). Protein synthesis is stimulated in the target cells of other neuropeptides through different second messenger systems (Spada *et al.*, 1989). The modulation of second messenger systems by neuropeptides also regulates the synthesis of receptor proteins, such as EGF receptors (Hadcock and Malbon, 1991). Neuropeptides also act as neurotropic factors to regulate the growth and development of the brain by stimulating protein synthesis in their target cells (Boer and Swaab, 1985; Mattson, 1989; Strand *et al.*, 1991).

12.3.5 NEUROPEPTIDE MODULATION OF NEUROTRANSMITTER RELEASE FROM PRESYNAPTIC NEURONS

As shown in Figure 12.2, neuropeptides modulate the release of neurotransmitters and neuropeptides from presynaptic cells by regulating Ca^{2+} and K^+ channels in the axon terminals. The mechanism for regulating neurotransmitter release, as discussed in Section 10.3, is the phosphorylation of specific proteins, such as synapsin I, which regulates the binding of the synaptic vesicles to the cell membrane (Hemming *et al.*, 1989). By regulating the phosphorylation of these proteins, neuropeptides can modulate the release of the contents of synaptic vesicles from the axon terminal.

There are many examples of neuropeptides regulating neurotransmitter release at nerve terminals. CCK regulates dopamine release from neurons in the nucleus accumbens in the rat brain (Voight *et al.*, 1985); neuropeptide Y inhibits the release of noradrenaline in the sympathetic nervous system (Lundberg and Hökfelt, 1983); and galanin inhibits the release of acetylcholine in the hippocampus (Crawley, 1990). The opioid peptides inhibit the release of a number of different excitatory neurotransmitters from presynaptic cells, including acetylcholine and substance P in the spinal cord (Jessell and Iversen, 1977) and noradrenaline and dopamine in the hypothalamus (Grossman, 1987).

12.4 EFFECTS OF NEUROPEPTIDES ON THE NEUROENDOCRINE SYSTEM

As discussed in Chapter 11, the highest concentrations of neuropeptide cell bodies in the brain are in the hypothalamus and median eminence, two areas which regulate the neuroendocrine system. Neuropeptides can alter hormone secretion by modulating the release of neurotransmitters which stimulate neuroendocrine cells and by direct action on neuroendocrine and endocrine cells (Israel *et al.*, 1989). When reviewing the effects of neuropeptides on the neuroendocrine system, it is important to know the route of administration and the method of study. Peptides given centrally (by intraventricular or intracerebral injection) often have different effects than if given peripherally (by intravenous injection) because peripherally injected peptides may not cross the blood–brain barrier and because they are rapidly degraded in the bloodstream. Also, studies done *in vivo* often give different results than those done *in vitro* (see Aronin, Coslovsky and Leeman, 1986). *In vivo* preparations have uncontrolled interactions between neuropeptides, neurotransmitters and steroid hormones, while these are absent with *in vitro* preparations. Bearing these problems in mind, this section summarizes some of the effects of neuropeptides on pituitary hormone release (see Table 12.2).

The hypothalamic regulation of the neuroendocrine system is thus

Table 12.2. *Effects of selected neuropeptides on the neuroendocrine system*

	GH	PRL	ACTH	TSH	LH	FSH	OXY	VP
Endogenous opioids	↑	↑	↑ (acute) ↓ (chronic)	↓	↓	↑↓?	↓	↓
Substance P								
Central	↓	0?	0↑?	0	0↓?	0	–	–
Peripheral	↑	↑	0	0	↓	↓	–	–
Neurotensin								
Central	↓	↓	–	↓	↑↓?	0	–	–
Peripheral	↑	↑	–	↑	↑↓?	0	–	–
CCK	↑	↑	↑	↓	↓	↓	–	–
VIP	↑	↑	↑	0	↑↓	0	↑	↑
NPY								
Central	↓	0	↑	–	↑↓*	0	–	–
Angiotensin	↓	↓	↑	–	↑	0	–	–
Bombesin	↑	↑	0	↓	0	0	–	–

Notes:
↑ = stimulation of hormone release; ↓ = inhibition; 0 = no effect; ? = contradictory results; – = not known.
* = ↑ in the presence of estrogen, ↓ without estrogen.
Sources: Ottlecz, 1987; Müller and Nistico, 1989; McCann, 1991; Dumont *et al.*, 1992.

more complex than previously thought. The hypothalamic neuropeptide hormones regulate the release of pituitary hormones (Chapter 4) and regulate their own release through ultra-short feedback loops (Chapter 8). Neurotransmitters regulate the release of the hypothalamic hormones (Chapter 6) and neuropeptides and steroid hormones modulate the effects of the neurotransmitters. For example, somatostatin inhibits the release of GH and TSH from the pituitary gland (see Figure 4.3, p. 51) and, in the presence of estrogen, somatostatin inhibits prolactin release (Epelbaum *et al.*, 1989). The release of somatostatin is inhibited by dopamine and noradrenaline and elevated by GABA (Figure 6.8, p. 99). Finally, VIP and the endogenous opioids decrease somatostatin release, while substance P and neurotensin increase its release (Reichlin, 1983). Thus, VIP and the opioids will increase GH release while substance P and neurotensin inhibit GH release (Table 12.2).

12.4.1 THE ENDOGENOUS OPIOIDS

The endogenous opioids inhibit the release of adrenaline from the adrenal medulla, regulate the secretion of gastrointestinal hormones, and modulate the release of the adenohypophyseal and neurohypophyseal hormones (Bicknell, 1985; Millan and Herz, 1985; Delitala, 1991). Opioids stimulate the release of GH and PRL and inhibit the release of LH and FSH in both rats and humans (Table 12.2). As shown in Figure 12.8, the inhibitory effect of morphine on the hypothalamic-pituitary-gonadal system in the rat is very rapid. Testosterone levels are reduced within 2

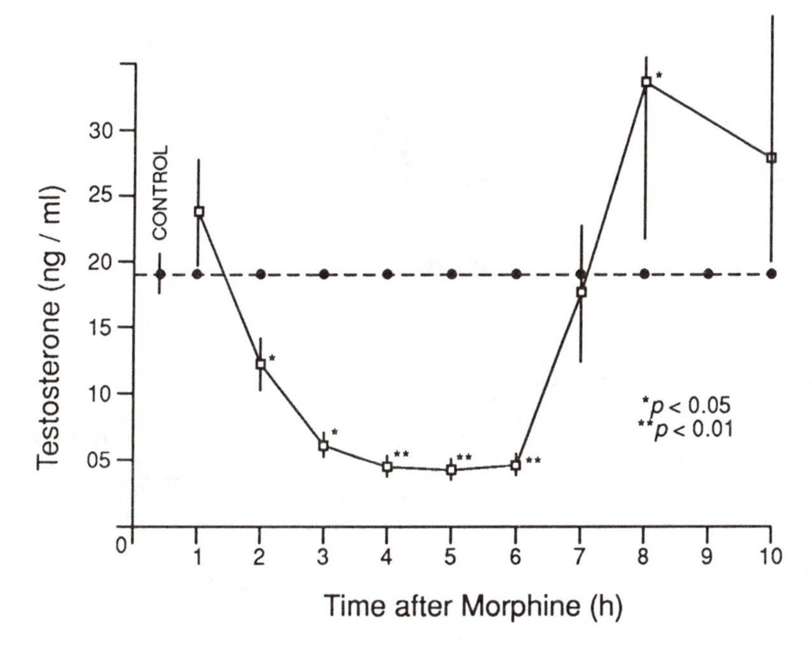

Figure 12.8. The effects of morphine on serum testosterone levels in male rats. Testosterone levels were measured at 1 hour intervals after subcutaneous injection of morphine (10 mg/kg). Values are means ± SEM of 8 to 10 rats. The dashed line represents the mean of all control rats injected with saline. (Redrawn from Cicero 1980.)

hours of morphine injection. Inhibition of the gonadotropic hormones by opioids suppresses the sex drive (see Morley, 1983). This may account for the absence of libido in heroin addicts. The release of TSH is also inhibited by opioids in rats, but not in humans. Acute treatment with opioids elevates ACTH in rats, while chronic treatment inhibits ACTH release in rats and humans. Stress causes the release of β-endorphin and the concomitant release of ACTH, α-MSH, PRL and GH. These hormones modulate two responses to stress: a metabolic adaptive response, in which the increased secretion of ACTH, GH and PRL mobilize physical resources; and a psychological adaptive response in which increased levels of β-endorphin, ACTH and α-MSH mobilize mental resources to reduce stress and increase adaptive behavior.

Opioid regulation of the neuroendocrine system involves interactions with classical neurotransmitters (Meites, 1984; Bicknell, 1985; Grossman, 1987). For example, β-endorphin elevates prolactin release by presynaptic inhibition of dopamine release, as shown in Figure 12.9A. Opioid regulation of gonadotropin secretion is rather more complex (Figure 12.9B). Opioids block ovulation in females and decrease testosterone secretion in males by inhibiting the pulsatile release of pituitary gonadotropic hormones. This is accomplished by β-endorphin and dynorphin inhibiting the release of noradrenaline from hypothalamic neurons. This effectively inhibits secretion of GnRH from the hypothalamus and thus reduces LH release. The release of the opioids is further modulated by the gonadal steroids (Figure 12.9B). Thus, positive feedback of estrogen may stimulate the LH surge at ovulation by inhibiting endogenous opioid release. On the other hand, the negative feedback of estrogen and testosterone probably involves the stimulation of endogenous opioid release (Meites, 1984; Grossman, 1987). Nobody said this was simple.

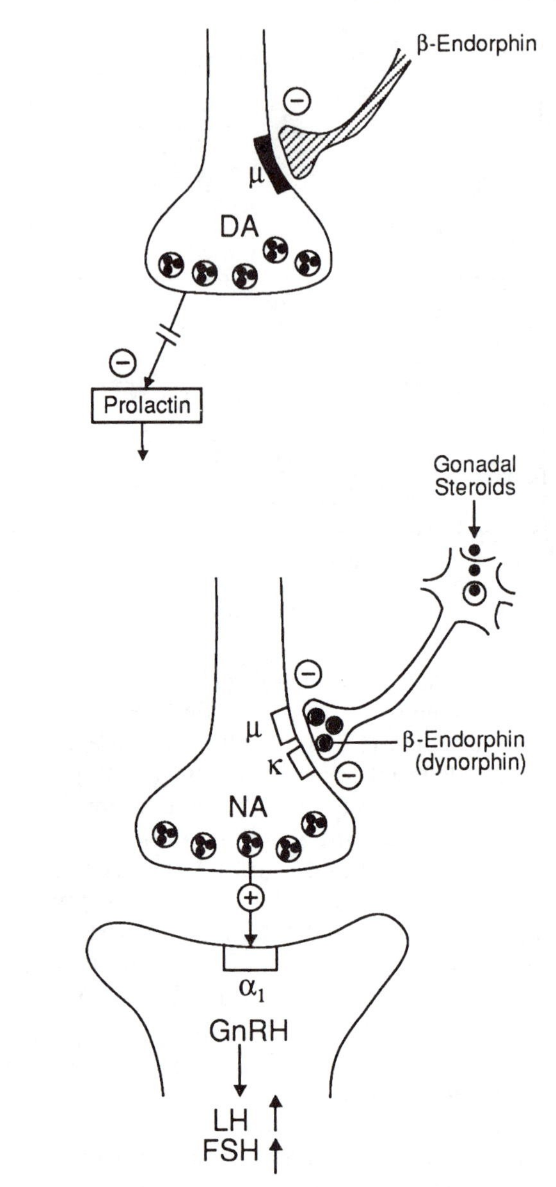

Figure 12.9. Models of two mechanisms of opioid regulation of the neuroendocrine system. (A) A model for the control of prolactin release by endogenous opioids in man. β-Endorphin binding to mu (μ) receptors may directly inhibit dopamine (DA) release, thereby allowing the serum prolactin level to rise. (B) A model of the relationship between endogenous opioid peptides and gonodotropin releasing hormone (GnRH) release in the rat. Most evidence suggests that β-endorphin or dynorphin act at mu and kappa (κ) receptors to modulate the presynaptic release of noradrenaline (NA). Thus, there will be reduced stimulation of the α_1-adrenergic receptors located on GnRH neurosecretory cells, resulting in reduced GnRH release and reduced LH and FSH release from the pituitary gland. The gonadal steroids also modulate the release of the opioids. Thus, the feedback effects of the gonadal steroids on the hypothalamic neurosecretory cells may involve three steps: the steroids regulate the opioids, which regulate NA secretion, which stimulates GnRH release. (Redrawn from Grossman, 1987.)

The secretion of oxytocin and vasopressin from the neurohypophysis is regulated by dynorphin acting at kappa receptors on the nerve terminals of the neurosecretory cells in the posterior pituitary (Bondy *et al.*, 1988; Cox, 1988; Zhao *et. al.*, 1988). Dynorphin is colocalized and co-released with vasopressin and binds to kappa receptors to inhibit oxytocin secretion through its effects on Ca^{2+} or K^+ channels in the nerve terminal (Figure 12.10). Dynorphin may also bind to receptors on the cell bodies in the SON and PVN to inhibit the release of the neurohypophyseal hormones. There is also evidence for opioid control of vasopressin secretion, possibly via presynaptic kappa receptors on the vasopressin nerve terminals (see Figure 12.10). The regulation of oxytocin release by the

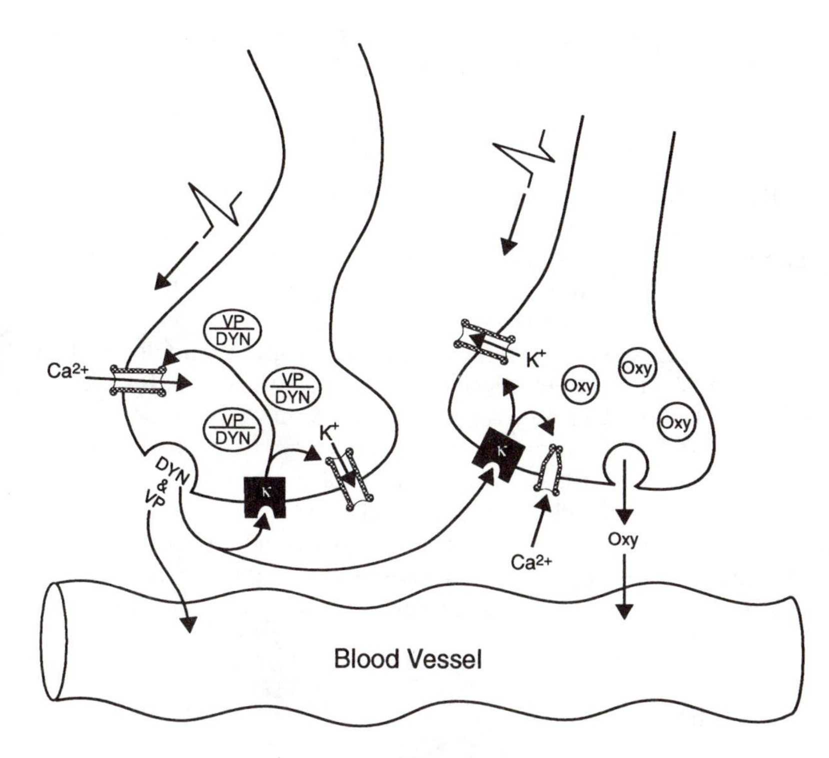

Figure 12.10. A model illustrating the opioid inhibition of oxytocin (OXY) release. In this neurohormonal system, dynorphin (DYN), which is co-released with vasopressin (VP) binds to kappa (κ) receptors on OXY nerve terminals. This decreases Ca^{2+} entry, by direct action on calcium channels, or indirectly, by acting on K^+ channels. The net effect in either case is a reduction in Ca^{2+} entry into the nerve terminal in response to depolarization by the nerve impulse and a reduction in oxytocin secretion. Dynorphin may regulate the release of vasopressin by the same mechanism. (Redrawn from Bondy *et al.*, 1988.)

endogenous opioids could account for the stress-induced inhibition of oxytocin secretion in lactating females (Grossman and Rees, 1983; Delitala, 1991).

12.4.2 SUBSTANCE P

Central injections of substance P suppress GH secretion but have no effect on PRL levels, while peripheral injections of substance P increase the release of both GH and PRL. Neither central, nor peripheral injection of substance P influences TSH release. Central injection of substance P has either no effect (rhesus monkeys) or elevates gonadotropin levels (rats), while peripheral injection inhibits gonadotropin release. Substance P seems to have little effect on ACTH release (Aronin, DiFiglia and Leeman, 1983; Aronin *et al.*, 1986). To complicate matters, the release of substance P is inhibited by thyroxine and stimulated by testosterone (Aronin *et al.*, 1986).

12.4.3 NEUROTENSIN

Central injections of neurotensin inhibit the release of GH and PRL, possibly by increasing the levels of somatostatin in the hypothalamus, while peripheral injection of neurotensin increases GH and PRL release. Peripheral injections of neurotensin also increase TSH release, but the effects of neurotensin on LH are contradictory (Aronin *et al.*, 1986). In humans, intravenous infusion of neurotensin has no effect on levels of GH, PRL, LH or TSH (Brown and Miller, 1982).

12.4.4 GASTROINTESTINAL PEPTIDES

CCK stimulates the release of GH, PRL and ACTH and inhibits the release of TSH, LH and FSH (Dockray, 1983; Ottlecz, 1987). **VIP**, given centrally, stimulates the release of GH, PRL and ACTH and inhibits the pulsatile release of LH through its inhibition of GnRH secretion. VIP appears to have little effect on the secretion of TSH and FSH (Fahrenkrug and Emson, 1982; Ottlecz, 1987). **Neuropeptide Y** given centrally inhibits release of GH and LH, but has no effect on PRL, ACTH, or FSH secretion (Ottlecz, 1987). **Angiotensin II** stimulates ACTH and LH release, inhibits PRL and GH release and has no effect on FSH (Phillips, 1987; Müller and Nistico, 1989). **Bombesin** stimulates the release of PRL and GH and inhibits the release of TSH (Taché and Brown, 1982; Telegdy, 1985). **Gastrin** inhibits the release of PRL, TSH, and LH, and stimulates the release of GH (Telegdy, 1985). **Secretin** elevates PRL when given centrally, but seems to have no effect on other pituitary hormones (Ottlecz, 1987).

12.4.5 SUMMARY

Since neuropeptides are co-released with neurotransmitters and pituitary hormones, they play an influential role in the regulation of the neuroendocrine system. Neuropeptides can alter hormone secretion by regulating neurotransmitter release from nerve axons in the hypothalamus and median eminence, by binding directly to neuroendocrine transducer cells in the hypothalamus, adrenal medulla, and other areas, or by acting on the endocrine cells of the pituitary gland (Millan and Herz, 1985; McCann, 1991).

12.5 VISCERAL, COGNITIVE AND BEHAVIORAL EFFECTS OF NEUROPEPTIDES

As well as their effects on the neuroendocrine system, neuropeptides modulate the visceral functions of the autonomic nervous system and the cognitive and behavioral functions of the central nervous system. All three of these effects are closely related as shown in Figure 12.11. At receptors in the hypothalamus, central amygdala and brain stem nuclei, neuropeptides regulate the visceral functions concerned with basic bodily needs and their 'primary affects'. These include hunger and feeding, fluid regulation and thirst, pain perception and avoidance responses, autonomic and behavioral aspects of temperature regulation and sexual arousal, as well as the functions of the gatrointestinal and circulatory systems. At receptors in the hippocampus, basolateral amygdala and other forebrain and limbic structures, neuropeptides regulate the emotional and motivational states associated with these basic bodily needs and coordinate the

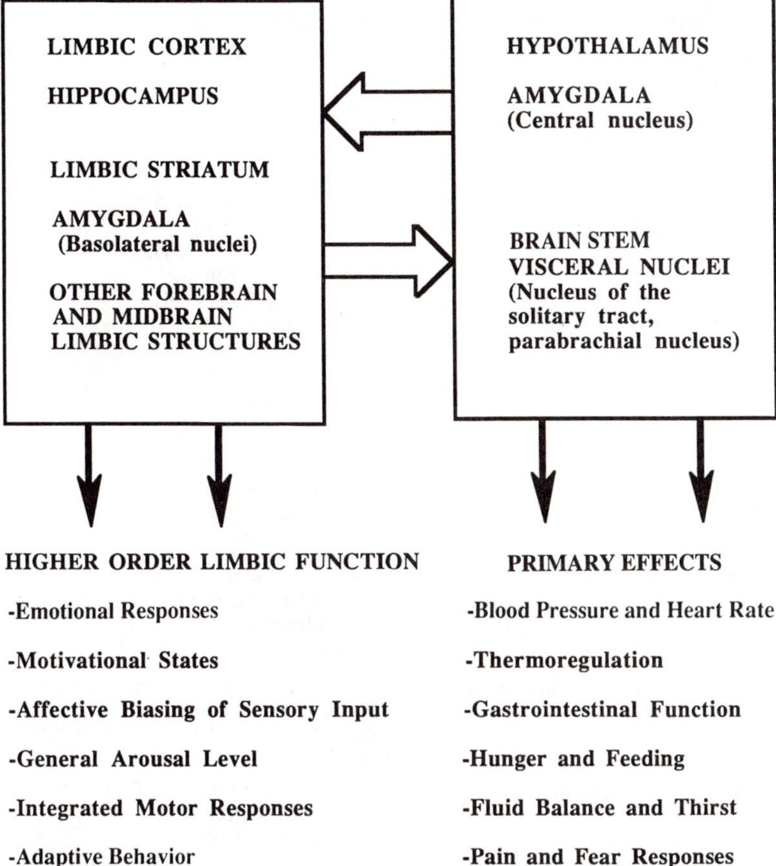

Figure 12.11. A diagram of how the major neuropeptide regions of the brain may interact to coordinate the visceral, cognitive and behavioral responses to bodily needs. The 'lower' neuropeptide receptor nuclei regulate the visceral functions and their primary effects, while the higher order limbic system and cortical areas modulate the cognitive and behavioral correlates of these affective functions. (Redrawn from Kelley, 1989.)

adaptive behavioral responses and general arousal levels correlated with these affective states (Kelley, 1989). Learning and memory are important aspects of adaptive behavior which are regulated by neuropeptide action in the neocortex as well as in the hippocampus and other limbic areas. This section reviews the effects of a number of neuropeptides on these adaptive behaviors as summarized in Table 12.3 (see Zadina *et al.*, 1986, for a cross-indexed reference list of the neural actions of neuropeptides).

12.5.1 THE ENDOGENOUS OPIOIDS

The actions of the endogenous opioid peptides at specific receptors in the brain and spinal cord have been extensively studied. This section reviews the analgesic, visceral, cognitive and behavioral effects of the endogenous opioid peptides and the correlations between opioid functions, receptor types and anatomical sites of action.

Table 12.3. *Some of the visceral, cognitive and behavioral effects of the neuropeptides*

Peptide	Location (site of action)	Function
Endogenous opioids	Spinal cord and brain stem	Induce analgesia; reduce pain perception Lower blood pressure Depress respiration Regulate cardiovascular system
	Hypothalamus and limbic system	Decrease body temperature Increase feeding and drinking Inhibit sexual behavior Reduce isolation-induced stress
	Ventral tegmental area, striatum	Induce euphoria Regulate locomotor behavior
Neurohypophyseal hormones		
Oxytocin	Thalamus and brain stem	Stimulates maternal behavior Modulates sexual behavior Inhibits memory
Vasopressin	Thalamus and limbic system, especially the hippocampus	Regulates blood pressure Facilitates learning and memory
Hypothalamic hypophysiotropic hormones		
Somatostatin	Cerebral cortex and hippocampus	Regulates locomotor activity Regulates body temperature Facilitates learning and memory
GH-RH	Brain	Stimulates feeding
CRH	Cerebral cortex, hippocampus, hypothalamus and other brain areas	Coordinates visceral responses to stress Increases arousal and emotionality Inhibits sexual behavior Influences learning
TRH	Thalamus, olfactory lobe and brain stem	Antidepressant, increases arousal Increases body temperature Elevates blood pressure
LH-RH	Olfactory pathways, limbic system	Elevates sexual behavior and neuroendocrine responses to primer pheromones
Adenohypophyseal hormones		
Prolactin	MPOA-anterior hypothalamus dopamine pathways?	Inhibits male sexual behavior Facilitates female parental behavior
ACTH and α-MSH	Limbic system, hippocampus	Facilitates attention, learning and memory
Gastrointestinal peptides		
Substance P	Brain and spinal cord, hypothalamus	Transmits pain signals Increases arousal and activity Facilitates sexual behavior
VIP	Cerebral cortex and limbic system	Facilitates avoidance learning Inhibits fear-motivated behavior
CCK-8	Cerebral cortex, olfactory bulbs and limbic system	Analgesic; reduces pain Reduces hunger and feeding behavior Induces sleep
Neurotensin	Limbic system and spinal cord	Reduces body temperature, locomotor activity and food intake Suppresses pain
Insulin	Olfactory bulbs, limbic system and hypothalamus	Inhibits hunger and feeding
Other neuropeptides		
Angiotensin	Hypothalamus	Induces thirst and drinking behavior Increases blood pressure Regulates fluid balance

Table 12.3 (*cont.*)

Peptide	Location (site of action)	Function
Bombesin	Hypothalamus and midbrain	Temperature regulation Inhibits feeding Increases blood pressure
Bradykinin	Limbic system and spinal cord	Transmits pain signals
Neuropeptide Y	Hypothalamus and thalamus	Increases feeding and drinking Reduces blood pressure and body temperature Facilitates memory
Delta sleep inducing peptide	Brain, hypothalamus	Induces sleep
Atrial natriuretic factor	Limbic system and brain stem	Reduces blood volume Regulates fluid balance, thirst and drinking.

Analgesic effects

The first known function of the opioids and opiate agonist drugs, such as morphine, was to induce analgesia. Opioids activate a descending inhibitory pain pathway from the thalamus to the spinal cord and inhibit ascending pain pathways in the dorsal horn of the spinal cord as shown in Figure 12.12 (Koob and Bloom, 1983; Basbaum and Fields, 1984). There are two spinothalamic pain pathways: the evolutionarily new neospinothalamic pain pathway, which is concerned with sharp, stabbing pain; and the evolutionarily more primitive paleospinothalamic pain pathway, which is concerned with dull, chronic pain. While the neospinothalamic pain pathway does not appear to have any opioid receptors, the paleospinothalamic pain pathway has a concentration of mu, sigma and kappa opioid receptors in the dorsal horn of the spinal cord, the periaqueductal gray matter, and in the central projections of this pathway in the thalamus and cerebral cortex (Figure 12.12). The opioid peptides have their analgesic effect by inhibiting the firing of neurons along this paleospinothalamic pain pathway.

One of the primary transmitters of pain signals in the dorsal horn of the spinal cord is substance P (see Section 12.2.1). Opioid receptors occur at the primary afferent terminals containing substance P in the dorsal horn of the spinal cord (Figure 12.12) and along the ascending paleospinothalamic pathway (Jessell and Iversen, 1977; Otsuka *et al.*, 1982). Dynorphins acting at kappa receptors (Millan, 1990) and enkephalins acting at mu receptors, inhibit the release of substance P, and possibly other transmitters, from presynaptic neurons in this pathway. They do this by blocking Ca^{2+} channels, hyperpolarizing the substance P neurons and inhibiting their firing (see Jessell and Kelly, 1991).

Because the endogenous opioid peptides reduce pain, factors which stimulate their release decrease pain perception. For example, electrical stimulation of the brain, which causes enkephalin release, results in temporary relief of chronic pain, and acupuncture, which stimulates β-endorphin release, may have its analgesic effect through the opioid inhibition of pain signals (Anonymous, 1981). The increase in the

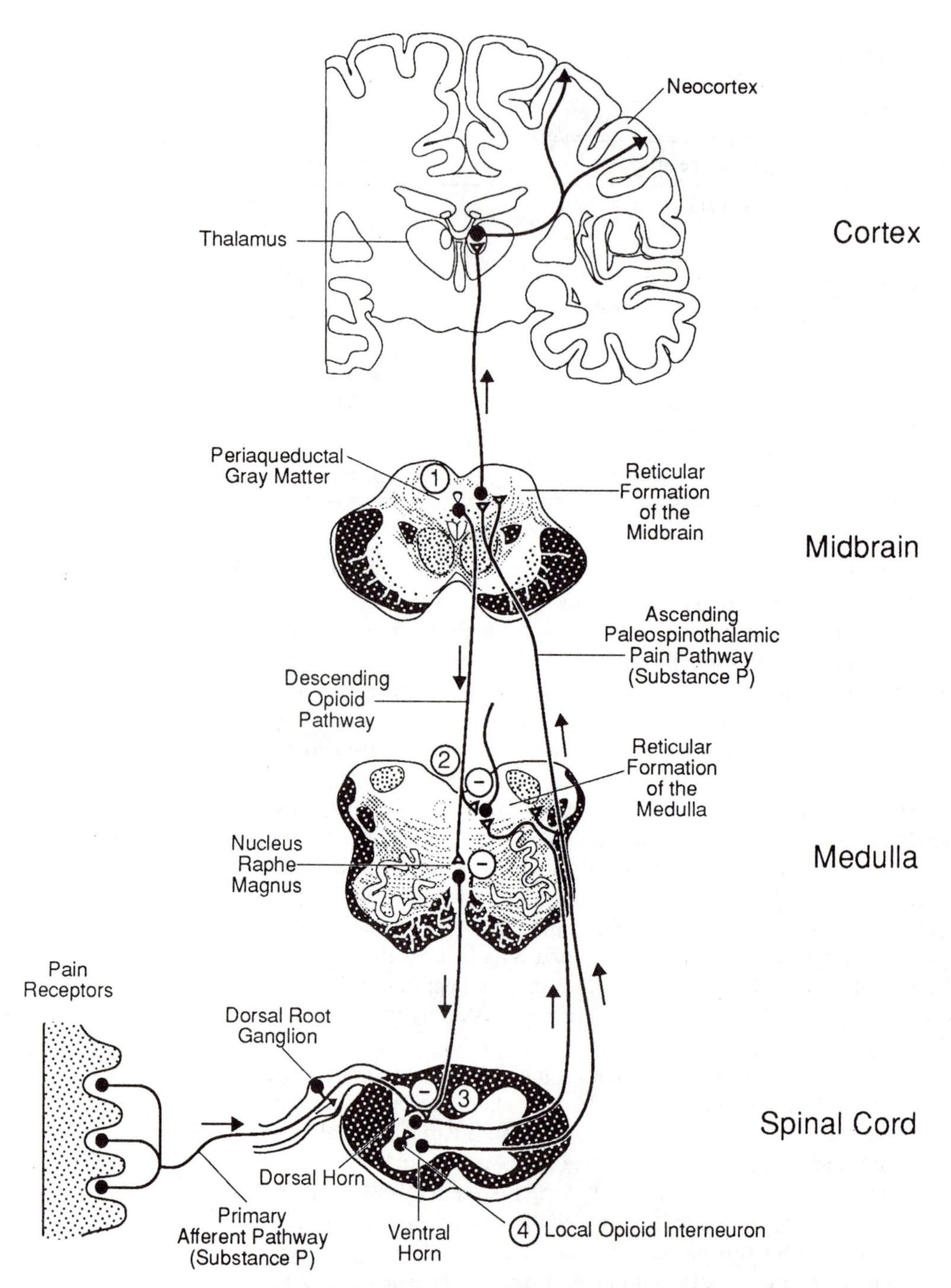

Figure 12.12. The regulation of the paleospinothalamic pain pathway by the endogenous opioid peptides. The pain receptors in the periphery send primary afferent fibers to the dorsal horn of the spinal cord via the dorsal root ganglion. These substance P fibers form an ascending spinothalamic pain pathway through the reticular formation of the medulla and midbrain to the thalamus and neocortex. The endogenous opioid peptides are colocalized with serotonin in the cell bodies in the periaqueductal gray of the midbrain (1). Enkephalins in these cell bodies act via the descending opioid pathways to modulate pain. These pathways make inhibitory connections with the ascending pain pathways in the medulla (2) and in the dorsal horn of the spinal cord (3). In addition to these opioid pathways, there are local opioid interneurons in the dorsal horn of the spinal cord (4) which inhibit the firing of the primary afferent pathways. (Redrawn from Jessell and Kelley, 1991.)

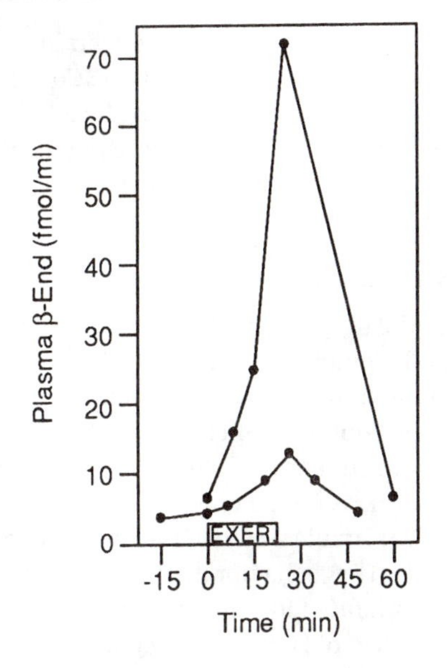

Figure 12.13. The time course of the changes in plasma β-endorphin (β-End) release in two male subjects before, during, and after a short period of moderate exercise (EXER). (Redrawn from Gambert *et al.*, 1981.)

endogenous opioid peptides in the brain, spinal cord, adenohypophysis and adrenal medulla in response to stress results in the phenomenon of stress-induced analgesia. The effects of stress-induced analgesia on pain perception have been compared to morphine injections (Akil *et al.*, 1984). Running also elevates the β-endorphin release from the adenohypophysis (see Figure 12.13), and it has been suggested that the euphoria and analgesia associated with long-distance running may be due to this increased β-endorphin release. The effect of exercise on β-endorphin secretion is well documented, but the physiological and behavioral consequences of this increase in opioid levels are unclear (Gambert *et al.*, 1981). Some postulated effects include changes in cardiovascular and respiratory function, analgesia, euphoria, and eating behavior (Cumming and Wheeler, 1987). Although β-endorphin released from the adenohypophysis may not cross the blood–brain barrier to induce euphoria, exercise may also stimulate the release of β-endorphin or other endogenous opioids in the brain.

Effects on the autonomic nervous system

Within the autonomic nervous system, the endogenous opioid peptides depress respiration, reduce gatrointestinal motility, inhibit catecholamine secretion from the adrenal medulla, modulate heart rate, reduce blood pressure, and lower body temperature. Opioids depress respiration through their action at mu and delta receptors in the respiratory centers of the brain stem (Wood and Iyengar, 1988). The anti-diarrheal effect of opiates occurs through their action at mu receptors in the gastrointestinal tract, where they inhibit gastrointestinal motility and the peristaltic reflex, possibly by inhibiting acetylcholine release (Cox, 1988). Within the cardiovascular system, opioid peptides act at mu, delta and kappa receptors to inhibit sympathetic stimulation of the heart, reducing stress-induced increases in heart rate. The endogenous opioid peptides reduce blood pressure through their action in the brain and by inhibiting noradrenaline release from sympathetic nerves (Cox, 1988). Opioids act at kappa receptors to regulate fluid balance by inhibiting vasopressin release and by their effects on the kidney. Opioids also modulate the activity of the visceral nerves in the pancreas, urogenital tract and other peripheral tissues (Cox, 1988).

Cognitive and behavioral effects

The mu receptors in the limbic system are associated with the regulation of mood states and the emotional aspects of pain and stress reduction. The opioid suppression of neurotransmitter release in these areas produces a feeling of euphoria when stress or pain is inhibited. The opiates may also induce euphoric effects by acting on 'reward' pathways, such as the dopaminergic ventral tegmental pathway, the periventricular gray region, and various limbic system pathways, shown in Figure 5.9 (p. 73) (Bozarth, 1988). The behavioral effects of the opioids include modulation of locomotion, food and water intake, and socio-sexual behavior (Table 12.3). Opioids may also regulate learning and memory and be involved in psychiatric disorders (Imura *et al.*, 1985; see also Rodgers and Cooper, 1988). Opioid stimulation of mu receptors on dopaminergic neurons in the ventral tegmental area facilitates locomotor behavior, while opioids binding to kappa receptors on dopaminergic pathways in the striatum causes sedation and motor rigidity (Mansour *et al.*, 1988).

As shown in Figure 12.14, low doses of opiate agonists, such as morphine, increase food intake, while low doses of opiate antagonists, such as naloxone, decrease food intake (Cooper *et al.*, 1988; Jalowiec *et al.*, 1981). However, the effects of the opioids on feeding are small and they may not be very important regulators of food intake (Levine and Billington, 1989). The elevated levels of endogenous opioids during stress may facilitate stress-induced eating (Morley and Levine, 1980). Exercise-induced stress, which elevates blood opioid levels, may also lead to changes in eating behavior, but the exact nature of these effects is unknown (see Cumming and Wheeler, 1987). As with feeding, opiates have a complex effect on drinking and water balance (Sanger, 1981). Opiate agonists increase thirst and drinking while opiate antagonists reduce drinking, but the mechanism underlying this is unknown. Opioids acting on kappa receptors inhibit vasopressin release, thus stimulating increased urination, while other opioids acting at mu receptors are antidiuretic (Cooper, 1988).

Many types of social behavior, including sexual behavior, mother–infant interaction, affiliative behavior, aggression and submission are influenced by opiates (Panksepp, Siviy and Normansell, 1985; Benton and Brain, 1988). Opiate agonists, such as morphine, inhibit libido and reduce sexual behavior via inhibition of pulsatile GnRH release and inhibition of catecholamine transmitters which control GnRH release and sexual arousal (Pfaus and Gorzalka, 1987; Serra, Collu and Gensa, 1988). Opioid peptides acting at mu receptors appear to increase aggressive

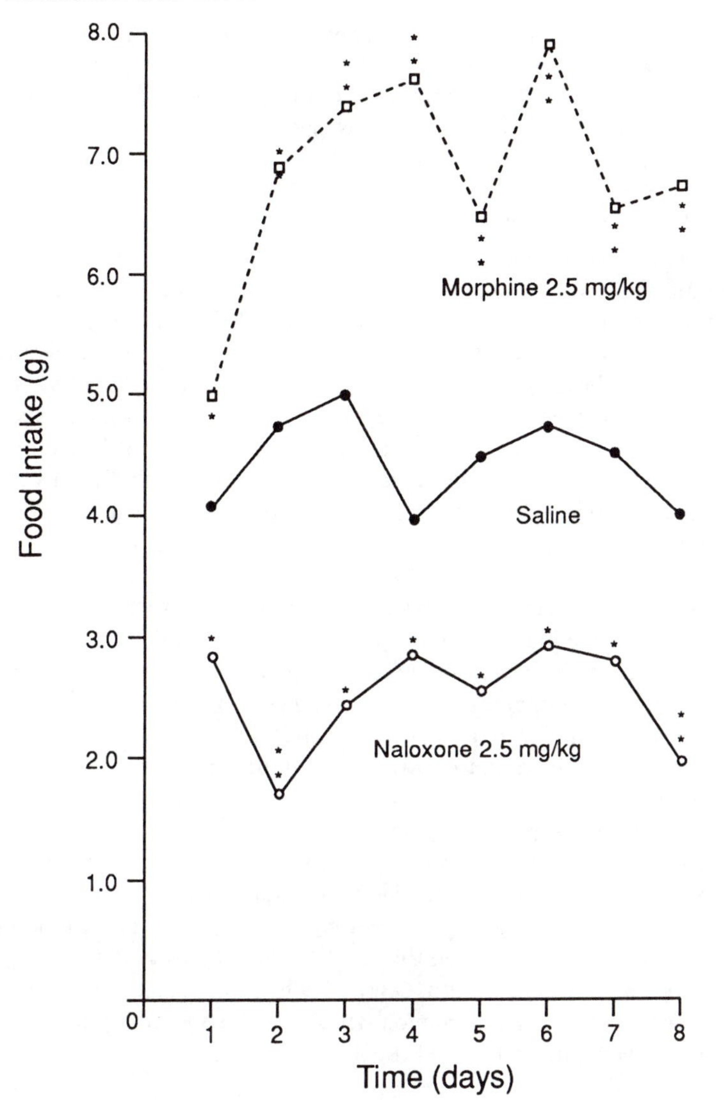

Figure 12.14. The effect of opiates on food intake. Mean 4 h food intake in rats after subcutaneous injections of morphine, naloxone or saline (control). *$p<0.05$, **$p<0.01$, saline versus drug. (Redrawn from Jalowiec *et al.*, 1981.)

behavior in mice, while those acting at kappa receptors elevate defensive behavior (Benton and Brain, 1988).

Social isolation is distressful for many animals and opiates reduce this stress. Opiates reduce the number of 'distress vocalizations' in isolated infants of many species, while naloxone elevates the number of distress vocalizations (Herman and Panksepp, 1981). These and related findings suggest that social attachments, social bonding and imprinting are types of 'opiate addictions' and that the distress which occurs after social separation is comparable to that which occurs after opiate withdrawal (Panksepp *et al.*, 1985). Autistic children are thought to have an imbalance in their endogenous opioids, particularly of β-endorphin, and this may cause their abnormal social behavior (Sandman *et al.*, 1990).

As well as the behaviors discussed above, the endogenous opioid peptides modulate activity levels and exploratory behavior (Katz, 1988), learning and memory (Messing, 1988), and have been implicated in a number of psychiatric disorders, including depression, schizophrenia, and disorders of eating such as anorexia nervosa and bulimia (Nemeroff, 1988; Schmauss and Emrich, 1988).

Addiction: tolerance and dependence

The main problem with using opiate agonists such as morphine is that they are addictive. Addiction, whether to opiates or to other substances, is manifested in terms of tolerance to the drug and physical dependence on the drug. Tolerance refers to the phenomenon that after repeated administration of the drug, higher doses are required to elicit effects that were previously elicited by much smaller doses. Physical dependence exists if termination of drug administration causes severe withdrawal symptoms. In humans, opiate withdrawal results in muscle twitching, yawning, stomach cramps, diarrhea, nausea and vomiting, sleeplessness and nervous excitation, with increased metabolic rate, hyperthermia and a feeling of weakness (Gmerek, 1988).

Opiate addiction may occur because opiate agonists, such as heroin and morphine, cause long-term changes in neural excitability through their modulation of the levels of opiate receptors, G-proteins and second messengers in their target cells. Tolerance may develop because of changes in opiate receptor sensitivity via down-regulation of the receptors, changes in the receptor–G-protein coupling, or changes in the level of second messengers, such as cyclic AMP in target cells in the nucleus accumbens and associated limbic system areas (amygdala, olfactory cortex, etc.). Increased levels of adenylate cyclase and cyclic AMP induced by chronic opiate use lead to a change in cellular activity such that proper functioning of the receptor cell requires opiate stimulation. Removal of the opiate disrupts the new level of cellular activity, leading to withdrawal symptoms (Smith, Law and Loh, 1988). Drug dependence is, however, a complex phenomenon whose cellular and molecular mechanisms are yet to be understood (Koob and Bloom, 1988).

12.5.2 NEUROHYPOPHYSEAL PEPTIDES

Oxytocin and vasopressin have a wide range of central and peripheral effects on visceral, cognitive and behavioral processes (Cunningham and Sawchenko, 1991). These peptides are located primarily in the magnocellular neurons of the PVN and SON, but they are also found elsewhere in the brain because of the extensive projections of the neurons of the PVN and SON to other brain regions (Silverman and Zimmerman, 1983; Swanson and Sawchenko, 1983). Vasopressin is found in the thalamus, medulla, cerebellum, amygdala, substantia nigra and hippocampus (Hawthorn, Ang and Jenkins, 1980), while oxytocin occurs primarily in the midbrain and hindbrain, in the substantia nigra, nucleus of the solitary tract and the dorsal vagal nucleus, as shown in Figure 11.10 (p. 240) (Sofroniew, 1983). The neurohypophyseal hormones are released during stress and modulate adaptive changes in heart rate, respiratory rate and body temperature in response to stressful stimuli (Meisenberg and Simmons, 1983).

Oxytocin. Oxytocin stimulates parturition and initiates milk ejection during lactation (see Hadley, 1992). The visceral functions of oxytocin include the regulation of the cardiovascular and gastrointestinal systems (Richard *et al.*, 1991). Oxytocin also facilitates maternal behavior (Pedersen *et al.*, 1982), regulates grooming and feeding behavior, and has been reported to both stimulate and inhibit sexual behavior (Stoneham *et al.*, 1985; Arletti *et al.*, 1985). Oxytocin also appears to have an amnesic effect, inhibiting memory processes (Meisenberg and Simmons, 1983; Argiolas and Gessa, 1991; Richard *et al.*, 1991).

Vasopressin. Vasopressin regulates blood pressure, blood volume and extracellular fluid levels (Cunningham and Sawchenko, 1991). Through its action in the hippocampus and other limbic areas, vasopressin facilitates avoidance and reward learning and memory consolidation (Meisenberg and Simmons, 1983; Le Moal *et al.*, 1984), and has been tested as a memory-enhancing drug in humans with variable results (Strupp and Levitsky, 1985; Beckwith *et al.*, 1990). The effects of vasopressin on memory may occur through its modulation of noradrenaline and dopamine release in the limbic system (Dores, 1984).

12.5.3 HYPOTHALAMIC HYPOPHYSIOTROPIC PEPTIDES

As well as their effects on the endocrine cells of the adenohypophysis, the hypothalamic hormones have a wide range of central effects through their release from neural axon projections. Some of these effects are summarized here (see Table 12.3).

Somatostatin. Somatostatin receptors are found in the highest concentration in the cerebral cortex and hippocampus and in lower concentrations in other brain areas, as shown in Figure 11.9B (p. 239) (Epelbaum, 1986; Epelbaum *et al.*, 1989). Although the results are contradictory, somatostatin alters activity, causing both motor excitation and sedation, and regulates body temperature, both hyper- and hypothermia being reported (Reichlin, 1983; Epelbaum, 1986). Somatostatin also facilitates memory consolidation and retrieval, most likely through its action in the hippocampus (Vécsei, Balázs, and Telegdy, 1987).

GH-RH. Growth hormone releasing hormone stimulates feeding in rats, possibly via activation of the edogenous opioid peptides (Vaccarino *et al.*, 1985; Vaccarino and Buckenham, 1987).

CRH. Corticotropin releasing hormone nerve pathways extend from the cell bodies in the hypothalamus to the bed nucleus of the stria terminalis, hippocampus, amygdala, cortex and to the brain stem nuclei. CRH modulates the electrical activity of neural cells in the cortex, hippocampus, locus ceruleus and hypothalamus (Siggins *et al.*, 1985) and coordinates the neuroendocrine, visceral, and behavioral adaptive responses to stress (Fisher, 1989; Dunn and Berridge, 1990). CRH elevates noradrenaline levels in the SNS and stimulates the release of adrenaline from the adrenal medulla. CRH also elevates plasma glucose levels, and increases blood pressure and heart rate. CRH increases locomotor activity and arousal, facilitates increased 'emotionality' in stressful situations (Vale *et al.*, 1983; Fekete *et al.*, 1987a), and has been implicated in a number of psychiatric disorders, including anorexia nervosa and depression (Smith *et al.*, 1989). CRH decreases feeding, inhibits sexual behavior and causes animals to freeze in a novel environment, behaviors which have been compared to the effects of depression in humans (Sirinathsinghji, 1985; Smith *et al.*, 1989). CRH can also improve learning in appetitive tasks (Koob and Bloom, 1985). CRH may exert its effects by acting as a co-transmitter with the catecholamines, serotonin, acetylcholine, GABA, or the endogenous opioid peptides (Fisher, 1989; Dunn and Berridge, 1990).

TRH. TRH is found throughout the CNS, with its highest concentrations in the thalamus, hypothalamus, brain stem, olfactory lobe and spinal cord (Jackson and Lechan, 1983). TRH increases activation of the brain and has a general anti-depressant effect, inhibiting sleep, increasing body temperature, elevating blood pressure, and increasing muscle tremor. TRH reduces the symptoms of alcohol and opiate withdrawal, and has been tried as an antidepressant in humans, with varied results (Prange and Utiger, 1981). TRH also decreases food consumption (Prasad, 1985).

LH-RH. LH-RH is widely distributed throughout the brain, especially in the median eminence, the limbic system and along the olfactory pathways (Krey and Silverman, 1983). The primary behavioral effect of LH-RH seems to be a facilitation of sexual behavior in steroid-primed female rodents (Mauk *et al.*, 1980; Moss and Dudley, 1990) and primates (Kendrick and Dixson, 1985). GnRH also increases sexual arousal in human males (Mortimer *et al.*, 1974; Dornan and Malsbury, 1989). Odours from adult rats and mice cause a surge of LH in animals of the opposite sex, which may be mediated by elevation of LH-RH in the accessory olfactory pathways (Beltramino and Taleisnik, 1983; Coquelin *et al.*, 1984).

12.5.4 ADENOHYPOPHYSEAL HORMONES

A number of the adenohypophyseal hormones have been shown to act as neuromodulators. The central effects of prolactin, ACTH and α-MSH are discussed here (see Table 12.3).

Prolactin. Prolactin has two primary effects on behavior: it inhibits sexual behavior (Dornan and Malsbury, 1989) and facilitates parental behavior (Bridges *et al.*, 1985). Prolactin may inhibit male sexual behavior through inhibition of dopaminergic neurons of the substantia nigra and the nucleus accumbens, inhibition of spinal reflexes and neuromuscular connections controlling erection and ejaculation, or the regulation of steroid hormone release from the gonads (Drago, 1984). Parental behavior may be facilitated by prolactin modulation of catecholamine pathways in the medial forebrain bundle or the medial preoptic, anterior hypothalamic area (Numan, 1985).

ACTH and α-MSH. Because the first 13 amino acid residues of ACTH are identical with those of α-MSH, ACTH and α-MSH are often considered together when their effects on behavior are studied (Beckwith and Sandman, 1978; Gispen and Zwiers, 1985). ACTH and α-MSH modulate the release of catecholamines and serotonin through receptors on the raphe–hippocampal pathway (Gispen and Zwiers, 1985; Kovács *et al.*, 1987). The ACTH fragment, $ACTH_{4-10}$ increases attention and arousal, enhances memory of conditioned avoidance responses, elevates scratching and yawning movements, and induces excessive grooming (de Wied and Jolles, 1982; Gispen and Zwiers, 1985). Like ACTH, α-MSH facilitates avoidance learning and appetitive learning and delays extinction (by facilitating memory). α-MSH also facilitates the stretching and yawning movements and modulates sexual behavior (Datta and King, 1982; Thody and Wilson, 1983; Dornan and Malsbury, 1989).

12.5.5 GASTROINTESTINAL PEPTIDES

Most gastrointestinal peptides have been shown to have neuromodulator functions, some of which are summarized here (see Table 12.3).

Substance P. As discussed in Section 12.2, substance P cell bodies and receptors are widely distributed throughout the brain. The primary function of substance P appears to be the transmission of pain signals in the spinal cord, but substance P also functions as a transmitter in the sensory pathways of the gastrointestinal and cardiovascular systems (Helke *et al.*, 1990, 1991). Substance P has a number of behavioral functions which appear to be due to its ability to excite dopaminergic neurons in the substantia nigra. Substance P increases arousal and activity levels (Iversen, 1982; Helke *et al.*, 1990) and facilitates sexual behavior by interacting with steroid-sensitive neurons in the hypothalamus (Dornan and Malsbury, 1989; Blaustein *et al.*, 1991).

Vasoactive intestinal polypeptide (VIP). VIP is found in nerve cells of the cerebral cortex, hypothalamus, amygdala, hippocampus and spinal cord. VIP has a wide range of effects on the cardiovascular, respiratory and gastrointestinal systems and other visceral functions (Said, 1991). VIP may act as an excitatory neurotransmitter as well as a neuromodulator in the CNS (Sims *et al.*, 1980; Fahrenkrug and Emson, 1982; Abrams, Nilaver and Zimmerman, 1985; Magistretti, 1990). VIP facilitates the acquisition of avoidance behavior and may inhibit the expression of fear-motivated behavior (Cottrell *et al.*, 1984).

Cholecystokinin (CCK). Although cholecystokinin is a 33 amino acid peptide, it exists in the brain predominantly as the 8 amino acid fragment, CCK-8, as shown in Figure 11.3 (Morley and Levine, 1985).

CCK-8 is widely distributed throughout the brain, with its greatest concentration of receptors in the cerebral cortex, olfactory bulbs, caudate nucleus, hippocampus and hypothalamus (Dockray, 1983; Morley and Levine 1985). CCK produces analgesia and hypothermia (possibly due to inhibition of TSH release), depresses the activity of the CNS (possibly through the inhibition of dopamine release) and reduces exploratory behavior. As an analgesic, CCK-8 is more potent than morphine, acting in the brain stem, thalamus and cerebral cortex to inhibit pain stimuli, possibly by stimulating opioid receptors (Sheehan and de Belleroche, 1984). CCK also facilitates sleep (Kapás *et al.*, 1991) and reduces food intake in animals and humans. There is, however, a controversy about whether the inhibition of feeding by CCK is due to its central effect on appetite or to peripherally induced feelings of nausea (Sheehan and de Belleroche, 1984; Figlewicz *et al.*, 1987). CCK may inhibit feeding through modulation of noradrenaline, dopamine and GABA secretion in the hypothalamus (Sheehan and de Belleroche, 1984; Morley and Levine, 1985). The colocalization of CCK and dopamine in some neurons has led to the hypothesis that schizophrenic patients may suffer from an overactive CCK system, but this hypothesis has not been proved (Fekete *et al.*, 1987b).

Neurotensin. Neurotensin has its highest concentrations in the stria terminalis, amygdala and hypothalamus, and neurotensin-containing fibers are organized into distinct pathways in the CNS, suggesting that neurotensin may act as a neurotransmitter (Jolicoeur, Rioux and St-Pierre, 1985). Neurotensin reduces body temperature, locomotor activity and food intake in most mammals. Neurotensin modulates the release of dopamine and other transmitters (Brown and Miller, 1982; Jolicoeur *et al.*, 1985), and it has been suggested that neurotensin might be useful in the treatment of psychiatric disorders such as schizophrenia (Nemeroff, 1986).

Insulin. Insulin and insulin receptors are widespread throughout the brain, with their highest concentration in the olfactory bulbs, hypothalamus, and limbic system. Insulin functions as a saturation signal, inhibiting feeding and reducing body weight, leading to the suggestion that insulin might be used to treat obesity in humans (Baskin *et al.*, 1987).

12.5.6 OTHER NEUROPEPTIDES

Many other neuropeptides have neuromodulatory effects in the ANS and CNS, some of which are described here (see Table 12.3).

Angiotensin II. Angiotensin II is synthesized in the spinal cord, brain stem, and amygdala through the action of renin in a mechanism similar to its synthesis in the blood. Central injections of very low doses of angiotensin reliably elicit thirst and drinking behavior by stimulating receptors in the hypothalamus and other neural areas (Phillips, 1987). Angiotensin also increases blood pressure, stimulates vasopressin release and increases SNS activity. Angiotensin levels are influenced by steroid hormones and it plays an important role in the regulation of fluid balance over the menstrual cycle and during pregnancy in human females. Angiotensin may also be involved in learning, memory, and motor control (Phillips, 1987). Angiotensin also appears to inhibit sexual behavior in male rats (Clark, 1989).

Bombesin. Bombesin occurs at high concentrations in the hypothalamus, cerebral cortex and midbrain. Bombesin is the most potent modulator of thermoregulation yet isolated, acting on the temperature regulation sites of the POA-AH region. Bombesin also increases blood pressure, acts as an antidiuretic, and inhibits feeding (Taché and Brown, 1982; Telegdy, 1985).

Bradykinin. Bradykinin is located in the hypothalamus and septum, where it acts to regulate blood pressure, and in the brain stem, where it acts to transmit pain signals. Bradykinin is one of the most potent pain-transmitting substances known (Snyder, 1980).

Neuropeptide Y. Neuropeptide Y occurs in the brain in a higher concentration than any other neuropeptide (Fuxe *et al.*, 1990) and the highest levels of neuropeptide Y are in the hypothalamus and limbic system (Chronwall *et al.*, 1985; Gray and Morley, 1986). Neuropeptide Y has a multitude of effects in the cardiovascular and gastrointestinal systems (Dumont *et al.*, 1992) and influences body temperature regulation. Neuropeptide Y is an extremely potent stimulator of feeding behavior and may be a key regulator of feeding (Kalra *et al.*, 1990). Neuropeptide Y also regulates locomotor behavior, decreases sexual behavior (Gray and Morley, 1986) and facilitates memory (Morley and Flood, 1990). Many of the effects of neuropeptide Y may depend on its modulation of noradrenergic pathways in the brain (Dumont *et al.*, 1992; Fuxe *et al.*, 1990).

Sleep factors. Factor S, delta sleep-inducing peptide and a number of other endogenous sleep-promoting peptides have been identified in the brain (Ansseau and Reynolds, 1988; Borbely and Tobler, 1989).

Atrial Natriuretic Factor (ANF). Atrial natriuretic factor is a hormone synthesized from cells in the atrium of the heart and released into the bloodstream when blood pressure increases. It binds to receptors in the kidney, adrenal cortex and circulatory system, where it acts as a diuretic, reducing blood volume (Needleman *et al.*, 1984). Atrial natriuretic factor is also found in high concentrations in the hypothalamus, limbic system and brain stem, where it regulates fluid balance and cardiovascular function and inhibits the release of vasopressin (Standaert *et al.*, 1985). ANF may also regulate thirst and drinking behavior (Quirion, 1988).

12.5.7 SUMMARY

Neuropeptides play an important role in the regulation of virtually all of the body's autonomic and central nervous system functions. This section

Table 12.4. *Summary of the visceral, cognitive and behavioral effects of neuropeptides*

Physiological or behavioral response	Neuropeptide actions
Pain regulation	Substance P and bradykinin transmit pain signals Opioids, CCK and neurotensin inhibit pain signals
Blood pressure	Angiotensin, CRH, TRH, and bombesin increase blood pressure Opioids and NPY decrease blood pressure
Body temperature	Somatostatin and TRH elevate body temperature Opioids, neurotensin, bombesin, NPY and CCK lower body temperature
Arousal level and motor activity	TRH, CRH, vasopressin and substance P increase arousal and locomotor activity and antagonize sleep Delta sleep inducing peptide, CCK and neurotensin reduce arousal and motor activity and may facilitate sleep
Hunger and feeding behavior	Opioids, GH-RH and neuropeptide Y all stimulate feeding CRH, CCK-8, neurotensin, insulin, TRH and bombesin inhibit feeding
Thirst and drinking behavior	Angiotensin, opioids, and neuropeptide Y increase thirst and drinking Atrial natriuretic factor reduces thirst and drinking
Sexual behavior	LH-RH and substance P facilitate sexual behavior CRH, prolactin, opioids and neuropeptide Y inhibit sexual behavior
Maternal behavior	Oxytocin and prolactin facilitate maternal behavior
Learning and memory	CRH, ACTH, α-MSH, β-endorphin and VIP increase attention and facilitate learning and short-term memory Vasopressin, somatostatin and neuropeptide Y facilitate memory consolidation and long-term memory Oxytocin disrupts memory and learning

has examined the effects of each peptide individually, but many peptides interact to regulate these systems. The influence of various neuropeptides on pain regulation, blood pressure, body temperature, arousal level and motor activity, hunger and thirst, sexual and maternal behavior, and on learning and memory are summarized in Table 12.4.

12.6 SUMMARY

Neuropeptides have both neurotransmitter and neuromodulator actions at their neural target cells. The distinction between these two types of actions has become blurred and there appears to be a continuum of effects rather than a dichotomy. Both neurotransmitters and neuropeptides have multiple actions at pre- and postsynaptic receptors. As neurotransmitters, neuropeptides bind to G-protein-coupled receptors to stimulate slow onset, long duration action potentials. Many neuropeptides meet the majority of the eight criteria used to define neurotransmitters, but the cell bodies and receptors for the neuropeptides are often in different anatomical locations, indicating that they act from a distance to stimulate their receptors. Substance P was the first neuropeptide to be considered as a neurotransmitter, but now a wide variety of neuropeptides are considered to have neurotransmitter activity. As neuromodulators, neuropeptides can regulate receptor sensitivity, G-protein coupling and the activation of second messenger systems in postsynaptic target cells, altering the electrophysiological responses of these cells to synaptic stimulation and regulating protein synthesis. The cotransmitter actions of neuropeptides and neurotransmitters result in additive, subtractive, multiplicative or divisive effects on the amount of second messenger production in the target cell. At presynaptic cells, neuropeptides can alter the release of neurotransmitters and neuropeptides from the nerve terminals. Neuropeptides regulate Ca^{2+} and K^+ channels in their target cells, thus depolarizing or hyperpolarizing the cell, regulating its sensitivity to neurotransmitter stimulation and regulating its release of neurotransmitters. Through their actions on postsynaptic cells, neuropeptides can modulate cell membrane permeability to Ca^{2+} and K^+ ions, electrical excitability of the cell and protein synthesis.

In the neuroendocrine system, neuropeptides regulate the release of neurohypophyseal and adenohypophyseal hormones as well as hormones secreted from peripheral endocrine glands. Neuropeptides may alter hormone release by direct action on the neurotransducer or endocrine cells or by modulating the release of neurotransmitters which stimulate hormone release. The endogenous opioids, substance P, neurotensin, and the gastrointestinal peptides are particularly potent regulators of the neuroendocrine system. Through their action on neurons of the hypothalamus, brain stem and limbic system, neuropeptides also regulate the functions of the ANS and CNS. Neuropeptides regulate integrated adaptive responses to bodily needs such as food, water, pain and temperature regulation and the visceral, affective and behavioral expressions of these bodily needs. Thus, the activation of the endogenous opioid peptides during stress results in analgesia, alteration of gastrointestinal and cardiac function, and adaptive changes in mood, feeding, drinking and sexual behavior. Hypothalamic, pituitary, gastrointestinal and other peptides also have visceral, cognitive and behavioral effects via their receptors in the central and peripheral components of the ANS and CNS. Neuropeptides have been implicated in the etiology and treatment of psychiatric disorders such as depression, schizophrenia and dementia, and in neurological disorders, such as Alzheimer's, Huntington's and Parkinson's diseases, but as yet, few neuropeptides have been found to be useful in the treatment of these neuropsychiatric disorders.

FURTHER READING

Bloom, F. E. (1988). Neurotransmitters: Past, present, and future directions. *Federation of American Societies for Experimental Biology Journal*, **2**, 32–41.

Koob, G. F., Sandman, C. A., and Strand, F. L. (eds.) (1990). A Decade of Neuropeptides. *Annals of the New York Academy of Sciences*, **579**, 1–280.

Motta, M. (ed.) (1991). *Brain Endocrinology*, 2nd edn. New York: Raven Press.

Müller, E. E. and Nistico, G. (1989). *Brain Messengers and the Pituitary*. San Diego, CA: Academic Press.

Rodgers, R. J. and Cooper, S. J. (eds.) (1988). *Endorphins, Opiates and Behavioral Processes*. Chichester: John Wiley.

REVIEW QUESTIONS

12.1 What is the difference in the action potentials triggered by acetylcholine (acting at ligand-gated nicotinic receptors) and substance P acting at its receptors?

12.2 What is the one criterion defining a neurotransmitter that most neuropeptides, including substance P, fail to meet?

12.3 What are the three sites at which neuropeptides can act as cotransmitters to modulate the effects of neurotransmitters at postsynaptic cells?

12.4 If a neuropeptide increases Ca^{2+} channel opening, will it decrease or increase the sensitivity of the cell to a neurotransmitter?

12.5 How do neuropeptides regulate the release of neurotransmitters from presynaptic nerve terminals?

12.6 What is the one main criterion which distinguishes a neurotransmitter from a neuromodulator?

12.7 Enkephalins may function to block pain by inhibiting the release of __________ in the spinal cord.

12.8 Explain why stress-induced analgesia occurs.

12.9 Do the opioid peptides stimulate or inhibit the release of LH and FSH?

12.10 Which neurotransmitter does β-endorphin modulate to increase prolactin release?

12.11 What is the primary behavioral effect of (a) GH-RH and (b) LH-RH?

12.12 Which two neuropeptides elevate body temperature?

ESSAY QUESTIONS

12.1 Discuss the distinction between neurotransmitters and neuromodulators. How have the definitions of these two terms changed since 1967?

12.2 Give the evidence that (a) VIP, (b) CCK, (c) somatostatin, or (d) neurotensin meets each of the eight criteria required for the definition of a neurotransmitter.

12.3 Explain how neuropeptides alter the sensitivity of target cells for 'classical' neurotransmitters by regulating receptor sensitivity and synthesis.

12.4 Discuss the interactive effects of primary transmitters and peptide cotransmitters on the stimulation of second messenger synthesis in postsynaptic cells.

12.5 Discuss the electrophysiological effects of (a) somatostatin, (b) TRH, (c) substance P, (d) neuropeptide Y, or (e) vasopressin on postsynaptic cells and explain what this indicates about the effect of this neuropeptide on the Ca^{2+} and K^+ channels in the postsynaptic cell membrane.

12.6 Discuss the mechanisms through which neuropeptides regulate the presynaptic release of neurotransmitters.

12.7 Discuss the effects of (a) CRH, (b) neuropeptide Y, (c) LH-RH, or (d) VIP on the ANS and CNS.

12.8 Compare and contrast the effects of oxytocin and vasopressin on learning and memory.

12.9 Compare the analgesic effects of the endogenous opioids, CCK and neurotensin.

12.10 Discuss the role of neuropeptides in *one* of the behavioral and/or physiological regulatory systems summarized in Table 12.4: (a) pain, (b) blood pressure, (c) body temperature; (d) arousal level and activity, (e) hunger and feeding, (f) thirst and drinking, (g) sexual behavior (h) maternal behavior (i) learning and memory.

REFERENCES

References for Chapter 12 are listed at the end of Chapter 11.

13

Cytokines and the interaction between the neuroendocrine and immune systems

As pointed out in Chapter 1, the nervous, endocrine and immune systems interact in many ways. Damage to the brain or changes in neurotransmitter and neurohormone release alter immune responses and the chemical messengers released by the cells of the immune system can alter the activity of the nervous and endocrine systems (Smith and Blalock, 1986; Kordon and Bihoreau, 1989). This chapter begins with an overview of the cells of the immune system and their chemical messengers, the cytokines, and then discusses the immune functions of the thymus gland and its hormones. The functions of the cytokines in the immune response to antigens and in the development of blood cells are then summarized and the neuromodulatory effects of cytokines on the brain and neuroendocrine system are examined. This is followed by a discussion of the neural and endocrine regulation of the immune system and the hypothalamic integration of neural, endocrine and immune systems.

13.1 THE CELLS OF THE IMMUNE SYSTEM

The immune system consists of a number of different cell types, including the monocytes and macrophages, T lymphocytes (T cells), B lymphocytes (B cells), granulocytes and natural killer (NK) cells. The role of the immune system is to help maintain homeostasis in the body and its best known function is the protection of the body from foreign invaders such as bacteria and viruses and from abnormal cellular development as occurs in tumor cells. To do this, the immune system must be able to discriminate foreign (non-self) cells from the body's own cells (self). Almost all substances have regions called antigenic determinants or 'epitopes' which can stimulate an immune response. The term **antigen** is used to refer to the molecule as a whole which carries the antigenic determinants. All cells (both foreign and self) carry many antigenic determinants on their surfaces. When a foreign antigen is detected, an immune response is activated.

The immune system has two branches, each of which plays a different, but complementary, role in the protection against infection. The **humoral immune system** protects the body from extracellular antigens (e.g. pollen, bacteria, toxins, etc.). When an antigen is detected in the circulation or other extracellular fluids, the humoral immune response is activated and antibodies (immunoglobulins) are secreted from B lymphocytes to inactivate the foreign material. The **cell-mediated immune system** protects the body from intracellular antigens (e.g. viruses), as well as from foreign (transplanted) tissue, tumor cells and protozoa. If one of the body's cells is infected by a virus or a tumor develops, then the cell-mediated immune response is activated and the infected cell or tumor cell is destroyed by cytotoxic T lymphocytes or NK cells.

13.1.1 MONOCYTES AND MACROPHAGES

Monocytes develop from precursor cells in the bone marrow and travel through the bloodstream to various tissues where they mature into macrophages. Macrophages ingest (phagocytose) and break down foreign antigens and present epitopes of these antigens to T cells. Macrophages also synthesize and release the cytokine, interleukin 1.

13.1.2 T LYMPHOCYTES (T CELLS)

The precursor cells for the T lymphocytes are produced in the bone marrow and migrate to the thymus gland through the circulatory system. In the thymus gland, the T cell precursors differentiate to form helper T cells (T_H), cytotoxic T cells (T_C) and suppressor T cells (T_S). Cytotoxic T cells destroy body cells that have been infected by viruses, tumor cells, and foreign cells (e.g. tissue transplants). Helper T cells release cytokines which stimulate the immune functions of other lymphocytes in a number of ways as discussed in Section 13.3. Suppressor T cells are able to suppress immune responses by inhibiting the actions of T_H and T_C cells. The maturation, differentiation and proliferation of the T cells in the thymus gland is regulated by thymic hormones and the thymic microenvironment (see Section 13.2). After the T cells mature in the thymus gland, they are stored in the secondary lymph organs, e.g. the spleen,

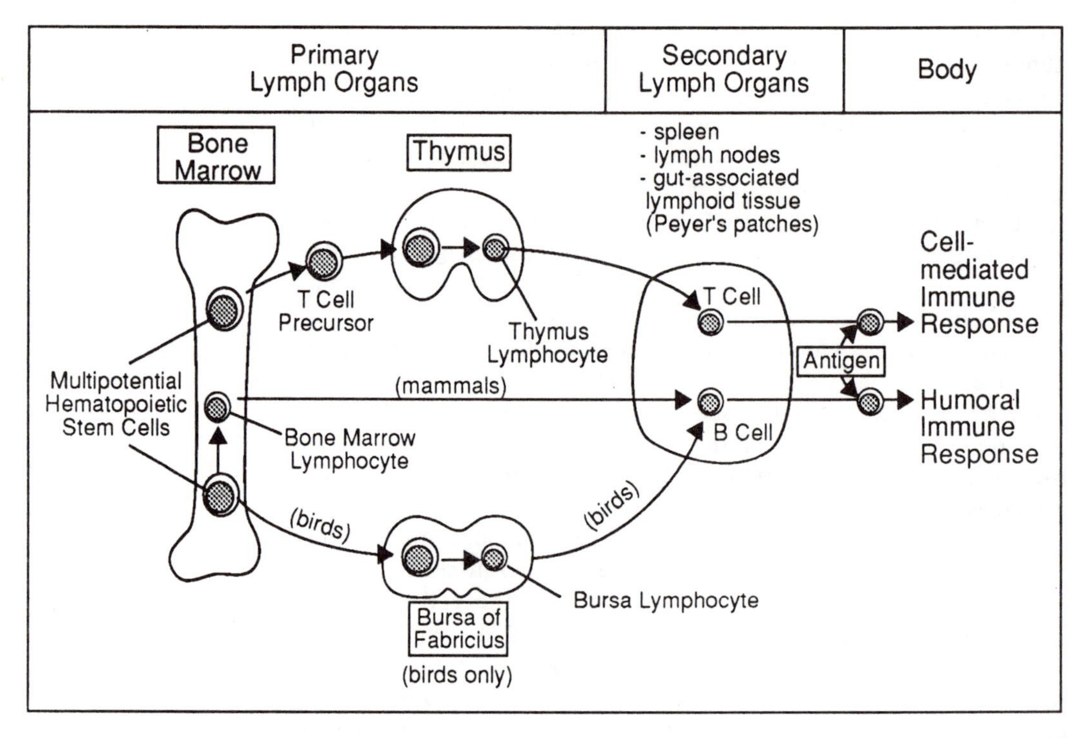

Figure 13.1. The development of T and B lymphocytes. Multipotential hematopoietic stem cells in the bone marrow produce the precursors of both T and B lymphocytes. In both mammals and birds, the precursors of the T lymphocytes migrate from the bone marrow through the bloodstream to the thymus gland, where they differentiate into thymus lymphocytes. When the thymus lymphocytes mature, they migrate to secondary lymph organs to become thymus-derived lymphocytes (T cells) which regulate immune responses. In mammals, the precursor cells destined to become B cells differentiate into lymphocytes in the hematopoietic tissue of the bone marrow and then migrate to secondary lymph organs to become B cells, which produce antibodies for the humoral immune responses. In birds, B cell precursors migrate to the bursa of Fabricius, where they differentiate into bursa lymphocytes. Many of these lymphocytes die, but others migrate to secondary lymph organs to become bursa-derived lymphocytes (B cells). The stage at which precursor cells become committed to develop into T or B lymphocytes is uncertain. (Redrawn from Alberts *et al.*, 1989.)

lymph nodes and Peyer's patches (the gut-associated lymph tissue) before being released to circulate through the body (see Figure 13.1). If the thymus gland is absent, the T cells fail to mature and cell-mediated immunity can not occur. Because the cytokines released by the T_H cells mediate antibody production by B cells, the lack of T cells also inhibits humoral immunity.

13.1.3 B LYMPHOCYTES (B CELLS)

As with T cells, the precursors of the B lymphocytes are produced in the bone marrow. In mammals, the B cells mature in the bone marrow and then migrate to the secondary lymph organs, where they are stored and released into the circulation. In birds, B cells mature in the bursa of Fabricius before being stored in the secondary lymph organs (Figure

13.1). The committed B cells are antigen specific and when they detect *their* antigen in the body, they produce antigen-specific antibodies (immunoglobulins) which bind to the antigens and inactivate them. As well as antibody-producing B cells, there are also 'memory' B cells, which remember which antibodies the body has created. B cells also produce cytokines, which play a role in immunoregulation.

13.1.4 NATURAL KILLER (NK) CELLS

Natural killer cells are large lymphocytes which develop in the bone marrow. They are stored in the secondary lymph organs, from which they are released to circulate in the bloodstream. NK cells are able to spontaneously kill virus-infected cells and tumor cells. NK cells also release interferon γ.

13.1.5 GRANULOCYTES

Granulocytes are mature granular leukocytes (white blood cells). There are three types of granulocyte: neutrophils, eosinophils and basophils. Neutrophils destroy dead cells, bacteria and other foreign cells by phagocytosis. Eosinophils destroy parasites and modulate inflammatory responses in damaged tissues. Basophils destroy parasites by phagocytosis and are involved in immediate hypersensitivity (i.e. allergic) reactions through their secretion of histamine.

It is beyond the scope of this chapter to discuss the details of immune system activity. Brief overviews of the cells involved in the immune response are given by Tonegawa (1985); Schneider, Cohn and Bulloch (1987); Blalock (1989); and Dunn (1989). The functions of the immune system are summarized by Alberts *et al.* (1989) and given in detail in textbooks of immunology (e.g. Coleman *et al.*, 1989; Golub and Green, 1991).

13.2 THE THYMUS GLAND AND ITS HORMONES

The thymus is a two-lobed gland which is located in the chest, above the heart (see Figure 2.1, p. 20). It has two regions: one in which the T lymphocytes mature, and one containing epithelial cells, which secrete the thymic hormones. The thymus gland produces at least ten peptide hormones, which are collectively referred to as the thymic hormones. These include thymosin a, thymosin b, thymosin fraction 5, thymopoietin, thymic humoral factor and serum thymic factor (Trainin, Pecht and Handzel, 1983; Low and Goldstein, 1984; Oates and Goldstein, 1984; Norman and Litwack, 1987).

The thymus has been called 'the master gland of the immune system' because the thymic hormones are essential for both cell-mediated and humoral immune responses (Oates and Goldstein, 1984; Deschaux and Rouabhia, 1987). The thymic hormones facilitate the production of T cell precursors in the bone marrow, regulate the differentiation of precursor T cells into helper, cytotoxic and suppressor T cells in the thymus gland and

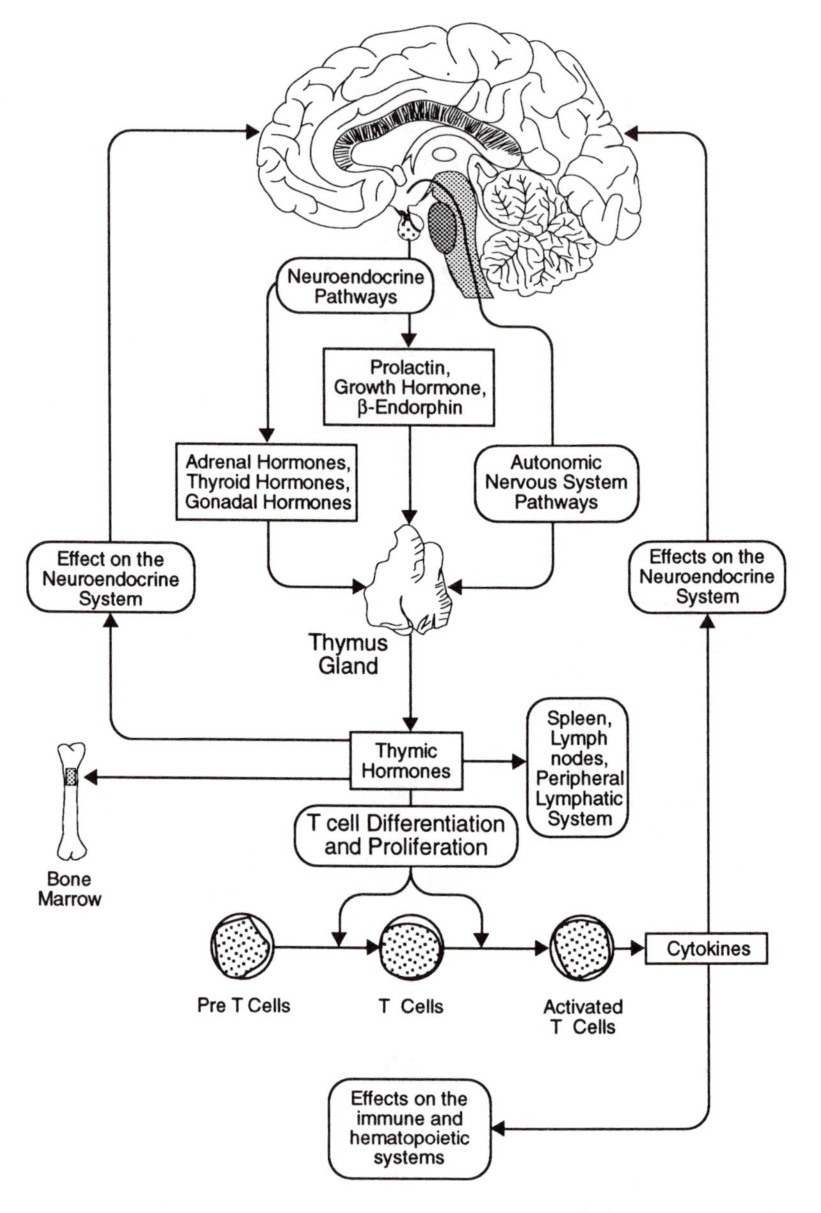

Figure 13.2. The interactions between the thymus gland and the neuroendocrine system. Thymus gland development and thymic hormone release is regulated by hormones from the pituitary, adrenal gland and gonads, and by the autonomic nervous system. The thymic hormones regulate T cell precursor development in the bone marrow, the differentiation and proliferation of T cells in the thymus gland and the maturation of the T cells in the secondary lymph organs. Both the thymic hormones and the cytokines released from activated T cells have effects on the brain and neuroendocrine system as well as on the immune and hematopoietic systems.

activate mature T cells in the spleen and lymph nodes (Figure 13.2). Thymic hormones also mobilize natural killer cells and stimulate the release of cytokines from macrophages and helper T cells (Trainin *et al.*, 1983; Low and Goldstein, 1984; Hall *et al.*, 1985).

The activity of the thymus gland is mediated by autonomic nerve pathways as well as pituitary, adrenal, thyroid and gonadal hormones, as shown in Figure 13.2. The thymus gland is particularly sensitive to adrenal and gonadal steroids, which inhibit its growth and function. In turn, the thymic hormones act on both neural and endocrine cells to regulate their activity, as described in Section 13.5.3 (Hall *et al.*, 1985).

The thymus gland is essential for the maturation and differentiation of T cells and the development of cell-mediated immune responses in

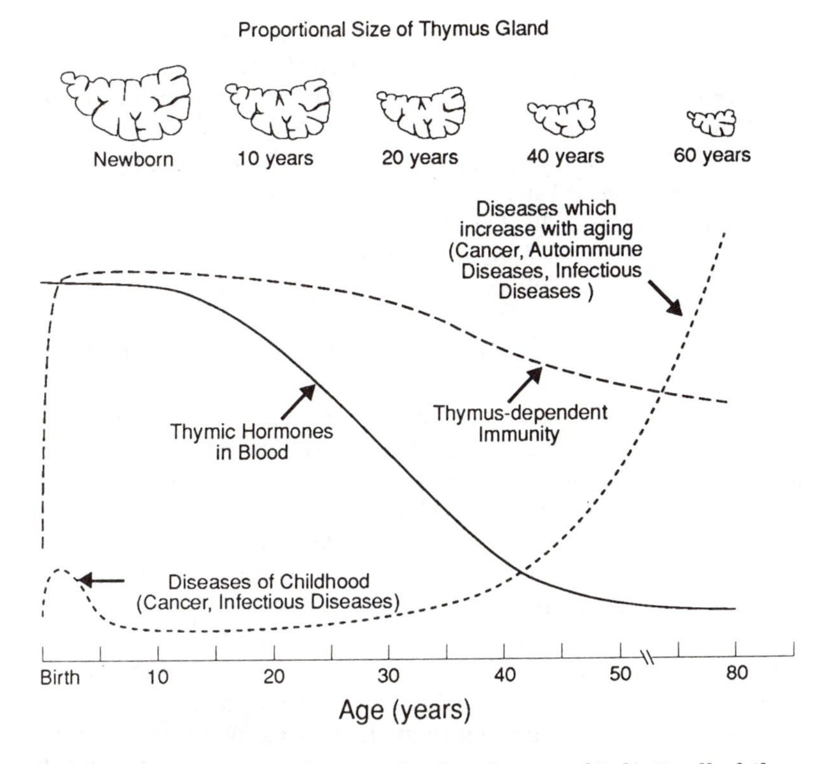

Figure 13.3. The decrease in the size of the thymus gland in proportion to body weight in humans as aging occurs. Thymic hormones in the blood and thymus-dependent immunity (cellular immune responses) also decline as aging occurs. The frequency of diseases associated with cellular immunity is high in newborns, as the T cells in the thymus must develop neonatally. During infancy and early adulthood there are high levels of thymic hormones in the circulation, but these decline with aging as the thymus shrinks, giving rise to an increase in diseases with aging. (Redrawn from Low and Goldstein, 1984.)

newborn animals. If the thymus gland is absent at birth, T cells fail to develop and the animal lacks the ability to cope with viral and bacterial infections. In neonatal rats and mice, for example, thymus gland removal impairs growth and leads to 'wasting syndrome'. Such animals have reduced growth, a hunched posture and difficulty walking, suffer from diarrhea, are dirty and ungroomed and show delayed puberty. These symptoms are caused by overwhelming infections with micro-organisms due to the absence of mature T cells (Pierpaoli and Sorkin, 1972).

In most mammals, the absolute size of the thymus gland increases until puberty, after which it begins to decline in both size and function (Norman and Litwack, 1987). In proportion to body weight, the thymus gland is largest in infancy and declines in size as aging occurs (see Figure 13.3). The decline in thymus gland size after puberty may be due to the immunosuppressive action of adrenal corticosteroids. As the size of the thymus gland declines, so does the level of thymic hormones in the blood and the ability to exhibit thymus-dependent (T cell) immunity (Low and Goldstein, 1984). As a result, there is an increase in diseases associated with failure of the cell-mediated immune system (e.g. cancer and infectious diseases) as aging occurs (Makinodan and Kay, 1980; Weigle, 1989). There are also a number of diseases of childhood which occur before the immune system matures.

13.3 CYTOKINES: THE MESSENGERS OF THE IMMUNE SYSTEM

Cytokines are protein messengers which are produced and released by macrophages, T cells, B cells, and other cells of the immune system when these cells are activated by antigen presenting cells, neural stimulation or hormones. Cytokines communicate between the cells of the immune system and are essential for the activation of immune responses and the development of blood cells (haematopoiesis). Cytokines may be considered as hormones (in the broadest definition of the term), since they are produced in one cell and may travel through the blood to act at distant cells, as defined for endocrine communication in Chapter 1. Cytokines also have paracrine and autocrine actions.

When first discovered, cytokines were named according to their function (e.g. T cell growth factor) and classified as lymphokines or monokines. However, it soon became clear that each cytokine had a number of different functions and that two or more cytokines often served the same function. To reduce the confusion, individual cytokines were named interleukins, based on their ability to communicate between leukocytes (Aarden *et al.*, 1979). As the structure and function of individual cytokines was identified, they were reclassified as shown in Table 13.1. The identifed cytokines include at least ten interleukins, three interferons and a number of other factors, not all of which are listed in Table 13.1 (see Dinarello and Mier, 1987; Hamblin, 1988).

Each cytokine has its own specific receptors on the surface of its target cells and the synthesis of these receptors is usually activated at the same time as the synthesis of the cytokines. The receptors for some cytokines have been cloned (see Whitacre, 1990). Like the receptors for the peptide hormones, cytokine receptors are coupled to G-proteins and activate second messenger systems within the target cell (see Chapter 10). These second messengers include cyclic AMP, calcium and the other second messengers shown in Figure 10.3 (see Hamblin 1988; Whitacre, 1990). The different effects of cytokines on their target cells are mediated through different second messenger systems. After the cytokine binds to its receptor, the entire complex is taken into the cell (internalization) and degraded by lysosomes, thus deactivating both the cytokine and its receptor. When the stimulus for cytokine synthesis is removed, down-regulation of receptors also occurs.

It is difficult to develop specific bioassays for cytokines because several cytokines have similar biological effects (see Table 13.1). Furthermore, one cytokine may stimulate the synthesis and release of a second cytokine, whose biological effects may be ascribed to the first cytokine! Since each cytokine may have several functions, and each bioassay usually detects only one of these, it is difficult to characterize all of the different functions of each cytokine. Some of the best known functions of each cytokine are given below and summarized in Table 13.1.

13.3.1 INTERFERON γ (IFNγ)

Interferon γ is produced by T lymphocytes following stimulation by specific antigens or mitogens (substances which induce DNA synthesis and cell proliferation). Interferon γ is an antiviral protein which activates macrophages and stimulates B cell proliferation and differentiation. Interferon γ can also increase the synthesis of interleukin 2 and its

Table 13.1. *Examples of some cytokines, their cellular source and functions*

Cytokine	Cellular source	Function
Interferon γ (IFNγ)	Activated T cells	Activates macrophages Stimulates B cell proliferation and differentiation Stimulates IL-2 synthesis
Interleukin 1 (IL-1α and β) (Lymphocyte activating factor)	Macrophages, T cells, B cells, NK cells, fibroblasts, etc.	Enhances T and B cell activation Promotes proliferation of T and B cells Stimulates prostaglandin synthesis Causes inflammation and fever Stimulates synthesis of IFNγ, IL-2, IL-3 and IL-4
Interleukin 2 (IL-2) (T cell growth factor)	Activated T cells	Stimulates T cell growth Activates cytotoxic T cells Stimulates B cell growth and proliferation Stimulates growth of NK cells
Interleukin 3 (IL-3) (hematopoietic cell growth factor)	Macrophages, activated T cells	Stimulates growth and differentiation of cells produced by the hematopoietic stem cells
Interleukin 4 (IL-4) (B cell stimulation factor 1)	Macrophages, activated T cells	Promotes T cell growth Stimulates B cell proliferation and differentiation Activates macrophages
Interleukin 5 (IL-5) (B cell growth factor 2)	Activated T cells	Promotes B cell proliferation Stimulates antibody production
Interleukin 6 (IL-6) (B cell stimulation factor 2)	Monocytes, activated T and B cells, fibroblasts	Promotes B cell proliferation and differentiation Activates T cells Induces fever and inflammation
Tumor necrosis factor (TNF)	Macrophages, T cells, B cells	Destruction of tumor cells Mediates endotoxic shock reaction Inhibits feeding Causes cachexia (wasting)
Lymphotoxin (LT) (tumor necrosis factor β)	Activated T cells	Destruction of tumor cells
Colony stimulating factors (CSFs)	Fibroblasts, monocytes	Stimulate growth and differentiation of hematopoietic stem cells in bone marrow

receptors in helper T cells. Two other interferons (α and β) are produced in other cells of the immune system and act to inhibit the spread of viruses, bacteria, protozoa and other antigens (Hamblin, 1988; Whitacre, 1990).

13.3.2 THE INTERLEUKINS

As of 1990, ten interleukins have been identified. Some of their main functions are summarized in this section (see Table 13.1).

Interleukin 1 (IL-1)

Interleukin 1 exists in two forms (IL-1α and IL-1β) which are synthesized by monocytes, macrophages, T lymphocytes, fibroblasts and other immune cells which have been activated by endotoxins or other noxious factors. Once released, IL-1 has many roles in immunity and inflamma-

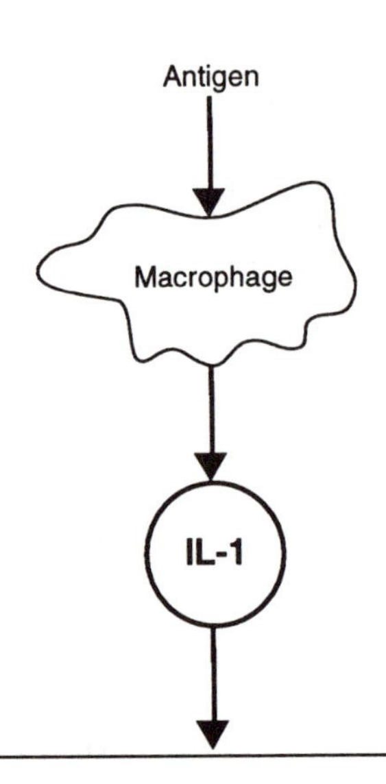

1. BRAIN: prostaglandin synthesis, sleep, anorexia, fever.
2. NEUROENDOCRINE SYSTEM: modulates release of hormones.
3. LYMPHOCYTES: promotes T cell and B cell proliferation, and synthesis of other cytokines.
4. BONE MARROW: hematopoiesis.
5. LIVER: acute phase proteins.
6. MUSCLE: protein synthesis.
7. BONE AND CARTILAGE: synthesis of prostaglandins and collinases.
8. ENDOTHELIUM AND EPITHELIUM: local inflammation and wound healing.

Figure 13.4. Some of the actions of interleukin 1 (IL-1). As well as its actions on the lymphocytes and bone marrow, IL-1 has receptors in the brain, liver, muscle, skin and bone and has a number of actions on these non-immune cells.

tion. IL-1 promotes T cell and B cell differentiation and the growth and activity of macrophages, neutrophils, and NK cells during inflammatory responses. IL-1 causes an increase in arachidonic acid metabolism and thus promotes prostaglandin synthesis during the inflammatory response. IL-1 also stimulates the synthesis of IL-2, IL-3, IL-4 and interferon γ by helper T cells and the synthesis of IL-2 receptors on helper T cells (Mizel, 1989).

There are IL-1 receptors on non-immune cells, including liver, bone, muscle and brain cells (see Figure 13.4). During immune responses, IL-1 acts on the brain to induce fever and sleep, decrease locomotor activity and suppress food intake (Plata-Salaman, 1989, 1991). IL-1 also regulates glial cell growth and proliferation, modulates the release of hypothalamic and pituitary hormones, and may modulate the activity of endogenous opioid peptide and noradrenergic neurotransmitter systems. The link between IL-1 and the neuroendocrine system is discussed in Section 13.5. IL-1 activity may also play a significant role in certain autoimmune diseases such as rheumatoid arthritis (see Durum, Schmidt and Oppenheim, 1985). The corticosteroid hormones and some immunosuppressive drugs, such as cyclosporine, inhibit IL-1 production and may be used to treat autoimmune disorders. Aspirin neutralizes some of the effects of IL-1 by inhibiting the synthesis of the prostaglandins (see Hamblin, 1988; Coleman *et al.*, 1989; Whitacre, 1990).

Interleukin 2 (IL-2)

Interleukin 2 was originally named T cell growth factor. It is produced by helper T cells following activation by antigen presenting cells (macrophages) and acts to promote T cell growth and B cell differentiation. IL-2 can also induce macrophages to become more cytotoxic (Mizel, 1989). IL-2 synthesis is promoted by IL-1 and a number of neuroendocrine factors (see Table 13.2). IL-2 modulates acetylcholine release in the CNS and neurohormone release from the magnocellular neurosecretory cells of the hypothalamus (Plata-Salaman, 1989). Some immunosuppressive agents, such as the corticosteroids and cyclosporine, inhibit IL-2 synthesis (Whitacre, 1990).

Interleukin 3 (IL-3)

Interleukin 3 was originally named hematopoietic cell growth factor (Table 13.1). It is produced by activated helper T cells and macrophages and stimulates the growth and differentiation of cells produced by multipotential hematopoietic stem cells in the bone marrow (Whitacre, 1990).

Interleukin 4 (IL-4)

Interleukin 4 is produced by activated helper T cells and macrophages and is best known for stimulating the proliferation and differentiation of B cells. It also acts as a growth factor for T cells, mast cells and bone marrow hematopoietic cells and may stimulate other cells of the immune system (Mizel, 1989).

Interleukin 5 (IL-5)

Interleukin 5, which is also known as B cell growth factor 2, is produced by activated T cells and stimulates the proliferation and differentiation of antigen activated B cells. IL-5 also functions to up-regulate IL-2 receptors on T cells and to enhance the differentiation of cytotoxic T cells (Mizel, 1989).

Interleukin 6 (IL-6)

Interleukin 6 is produced in a number of cell types, including helper T cells, B cells, fibroblasts and macrophages. IL-6 has a variety of biological effects, including the up-regulation of IL-2 receptors on T cells and the differentiation of B cells (Kishimoto, 1989). IL-6 often acts synergistically with IL-1 and has many functions in common with IL-1. IL-6 is also secreted by the anterior pituitary gland (Spangelo and MacLeod, 1990; Plata-Salaman, 1991).

Other interleukins

Interleukin 7 and interleukin 8 were discovered in 1988 (Whitacre, 1990). IL-7 is produced in the bone marrow, thymus, and spleen and acts in the bone marrow to stimulate the production of the B cell and T cell precursors. IL-8 is produced by monocytes, endothelial cells and fibroblasts and has a number of biological functions, including the chemotaxis of neutrophils and T cells. Interleukin 9, discovered in 1990, is a T cell growth factor which stimulates the proliferation of bone marrow-derived mast cells and stimulates hematopoietic stem cells in the bone marrow (Donahue, Yang and Clark, 1990; Hültner *et al.*, 1990). Interleukin 10 is cytokine synthesis inhibitory factor which is produced in helper T cells in mice and inhibits the synthesis of interferon γ by activated helper T cells (Hsu *et al.*, 1990).

13.3.3 CYTOTOXIC CYTOKINES

Tumor necrosis factor (TNF), and lymphotoxin (LT), which is also called tumor necrosis factor β, are closely related cytotoxins. TNF is produced by activated macrophages, T cells and B cells, while LT is produced primarily by T cells (Table 13.1). Both TNF and LT destroy tumor cells. TNF also stimulates the release of hypothalamic hormones (e.g. CRH) and may act directly on the brain to suppress feeding (Plata-Salaman, 1989; 1991). TNF also mediates endotoxic shock, causes cachexia (wasting) and induces fever, both by acting directly on the hypothalamus and by stimulating the release of IL-1 (Beutler, 1990).

13.3.4 COLONY STIMULATING FACTORS

There are four different colony stimulating factors which regulate blood cell formation (haematopoiesis), as shown in Figure 13.7 (Metcalf, 1991). Granulocyte-macrophage colony stimulating factor (GM-CSF) is produced by activated T lymphocytes and stimulates the formation of granulocytes and macrophages from multipotential hematopoietic stem cells. Granulocyte colony stimulating factor (G-CSF) is produced by activated monocytes and macrophages and stimulates the formation of granulocyte colonies from multipotential hematopoietic stem cells. Macrophage colony stimulating factor (M-CSF) is produced in fibroblast cells and stimulates the formation of macrophage colonies from multipotential hematopoietic stem cells. The fourth colony stimulating factor is interleukin 3 (multipotential CSF).

13.4 THE FUNCTIONS OF CYTOKINES IN THE IMMUNE AND HEMATOPOIETIC SYSTEMS

Cytokines are important in the activation of T cells, B cells and macrophages, the production and maturation of red blood cells, the killing of infectious agents and tumor cells and in producing the inflammatory response initiated by infection and trauma (Hamblin, 1988; Coleman *et al.*, 1989; Whitacre, 1990).

13.4.1 T CELL ACTIVATION

T cells are stimulated to multiply when they are activated by antigen presenting cells and interleukin 1. The mechanism involved in T cell

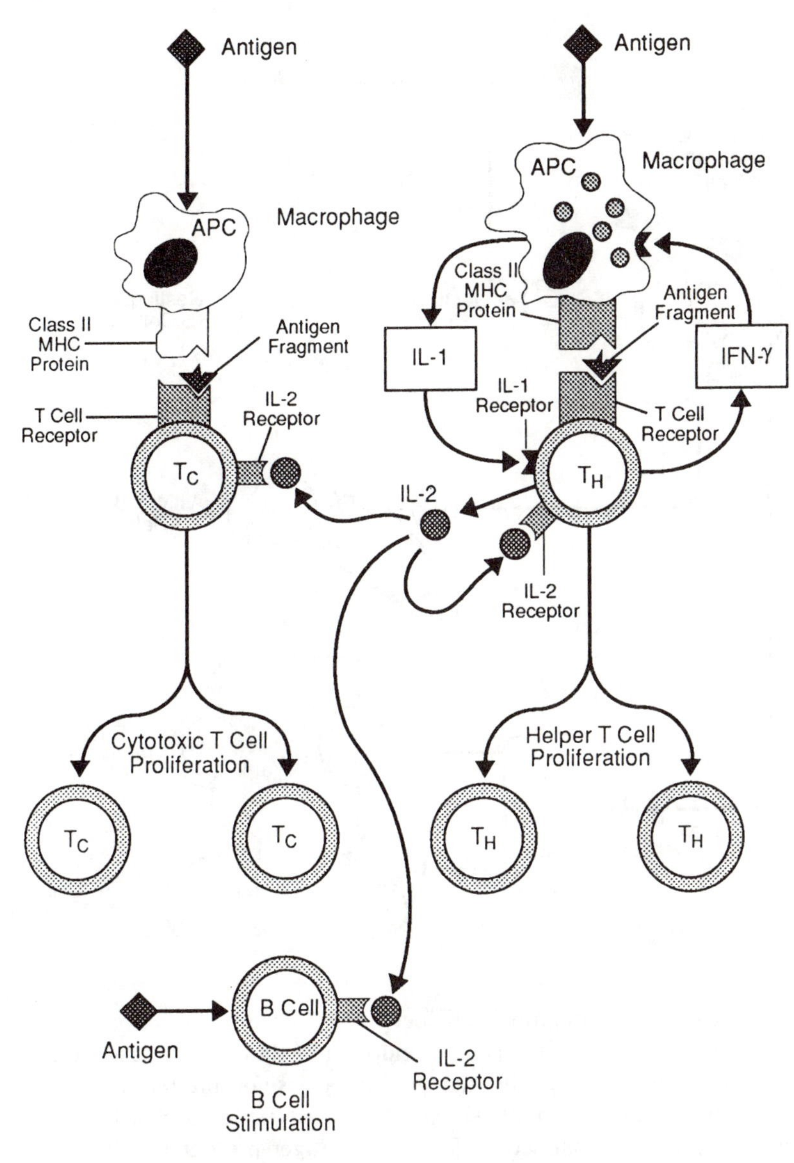

Figure 13.5. The release of IL-1, IL-2 and IFNγ by helper T cells. Helper T cells (T_H) are activated by macrophages (antigen-presenting cells) which present antigen fragments along with class II MHC proteins to the T cell receptors. The T_H cells then release IFNγ, which promotes MHC expression and stimulates IL-1 release from the macrophages. IL-1 stimulates the release of IL-2 by the T_H cell and up-regulates IL-2 receptors in the T_H cell. IL-2 acts in an autocrine fashion to stimulate T_H cell proliferation. IL-2 also stimulates the proliferation of cytotoxic T cells (T_C) and activates B cells to produce immunoglobulins. (Redrawn from Hamblin, 1988.)

activation is fairly complex and may best be understood with reference to Figure 13.5. Antigen presenting cells (e.g. monocytes and macrophages) detect antigens in the body and attach to them. They destroy these antigens by phagocytosis and digestion and then present antigen fragments, bound to major histocompatibility complex (MHC) proteins (which are self-recognition markers), to receptors on helper T cells and cytotoxic T cells. The antigen presenting macrophages also release interleukin-1, which stimulates the helper T cell.

The helper T cell has a complex antigen-MHC receptor as well as receptors for interleukin 1 and interleukin 2. Stimulation of the T cell receptors by the antigen-MHC complex, combined with the stimulation of the IL-1 receptor, activates the synthesis of interleukin 2 and the up-regulation of IL-2 receptors on the surface of the helper T cells. IL-2

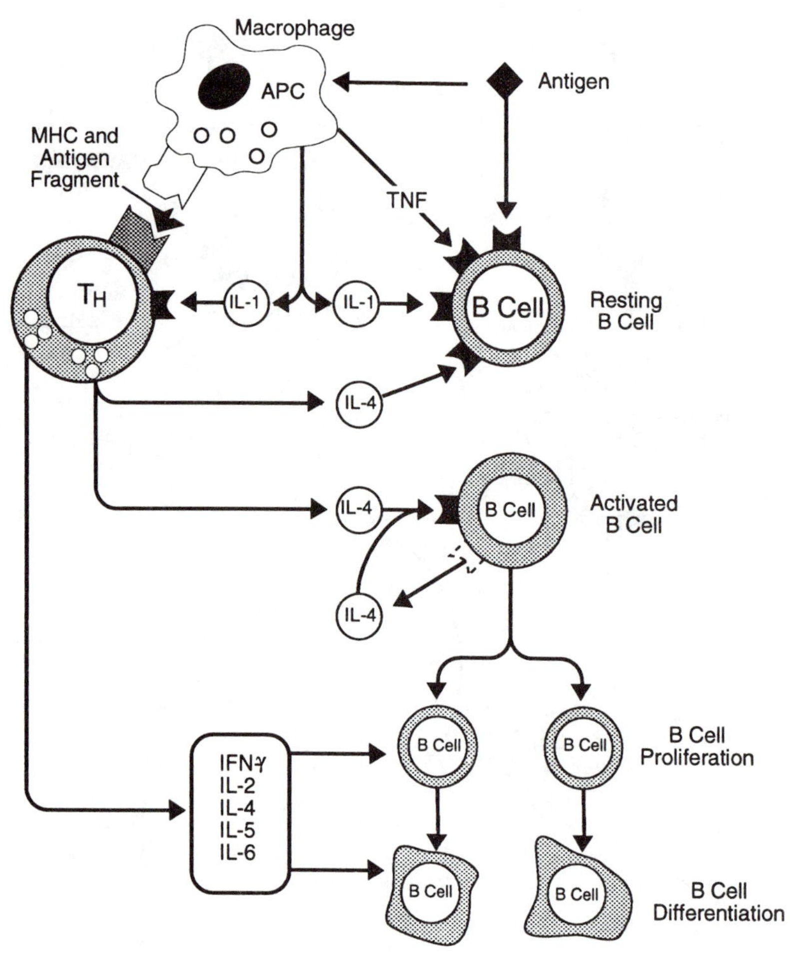

Figure 13.6. The role of cytokines in B cell activation, proliferation and differentiation. Resting B cells are activated by antigens plus IL-1 and tumour necrosis factor (TNF) from macrophages and IL-4 from helper T cells. Once activated, B cells produce their own IL-4, which has autocrine action. B cell division and differentiation are then stimulated by a number of cytokines from the helper T cell, including IFNγ, IL-2, IL-4, IL-5 and IL-6.

stimulates the proliferation of helper T cells by autocrine action and the proliferation of cytotoxic T cells by endocrine action. IL-2 released from T cells also stimulates B cells, whereas the release of interferon γ from T cells stimulates IL-1 synthesis and increased expression of the MHC proteins, thus increasing the efficiency of antigen presentation in macrophages. When the antigen is destroyed, the T cells are no longer stimulated, so they stop producing IL-2, the IL-2 receptors are down-regulated, and T cell proliferation stops (see Royer and Reinherz, 1987).

13.4.2 B CELL ACTIVATION

Cytokines secreted from macrophages and activated helper T cells stimulate the activation, proliferation and differentiation of B cells as depicted in Figure 13.6. IL-1, tumor necrosis factor and IL-4 'activate' the resting B cell, causing the synthesis and release of IL-4 and the up-regulation of cytokine receptors. IL-4 promotes B cell growth via autocrine action. The activated B cell is then stimulated to proliferate and differentiate by a host of cytokines as shown in Figure 13.6.

13.4.3 ACTIVATION OF MACROPHAGES

Resting macrophages are activated by antigens to release IL-1, which stimulates the T_H cell to release interferon γ. Interferon γ then further activates the macrophage (Figure 13.5). Once activated, the macrophage begins to function in phagocytosis and tumor cell killing and releases a number of cytokines, including IL-1, tumor necrosis factor and macrophage colony stimulating factor.

13.4.4 THE PRODUCTION AND MATURATION OF BLOOD CELLS (HAEMATOPOIESIS)

A number of cytokines are required for the normal production and maturation of blood cells in the bone marrow (Metcalf, 1991). Multipotential stem cells in the bone marrow are the precursors for all of the different types of blood cells (i.e. they are hematopoietic precursor cells). When stimulated by colony stimulating factors, these multipotential stem cells develop into colonies of specific blood cell types. Four cytokines are directly involved in haematopoiesis: IL-3 (multi-colony stimulating factor), granuloctyte-macrophage colony stimulating factor (GM-CSF), granulocyte colony stimulating factor (G-CSF) and macrophage colony stimulating factor (M-CSF). The production and maturation of blood cells occurs in a number of stages as summarized in Figure 13.7 (Hamblin, 1988; Metcalf, 1991). Tumor necrosis factor, IL-5, IL-9, lymphotoxin and interferon γ also regulate haematopoiesis.

One of the responses of the immune system to antigen is the stimulation of bone marrow cells to increase their production of blood cells. In response to antigen stimulation, activated macrophages and helper T cells produce CSFs, IL-1 and TNF. These last two cytokines then stimulate further production of CSFs from fibroblasts and other immune cells and these CSFs increase blood cell production.

13.4.5 CYTOXICITY

Some cytokines, such as TNF and LT, kill infectious agents and tumor cells directly. Others act indirectly, by activating the cytotoxicity of other cells: For example: IFNγ activates cytotoxic T cells; IFNγ and IL-2 activate tumor killing in macrophages and IFNγ, IL-1 and IL-2 facilitate the cytotoxic activity of NK cells.

13.4.6 INFLAMMATORY RESPONSE

Two cytokines, IL-1 and TNF, are important mediators of the inflammatory responses of tissue to infection or trauma. Both IL-1 and TNF are produced by macrophages in response to a variety of toxic stimuli (e.g. bacterial endotoxins) and are released into the circulation to activate target cells throughout the body and brain. IL-1 facilitates local inflammation and wound healing and acts on the brain to stimulate ACTH release, cause fever and induce sleep, all of which are adaptive responses

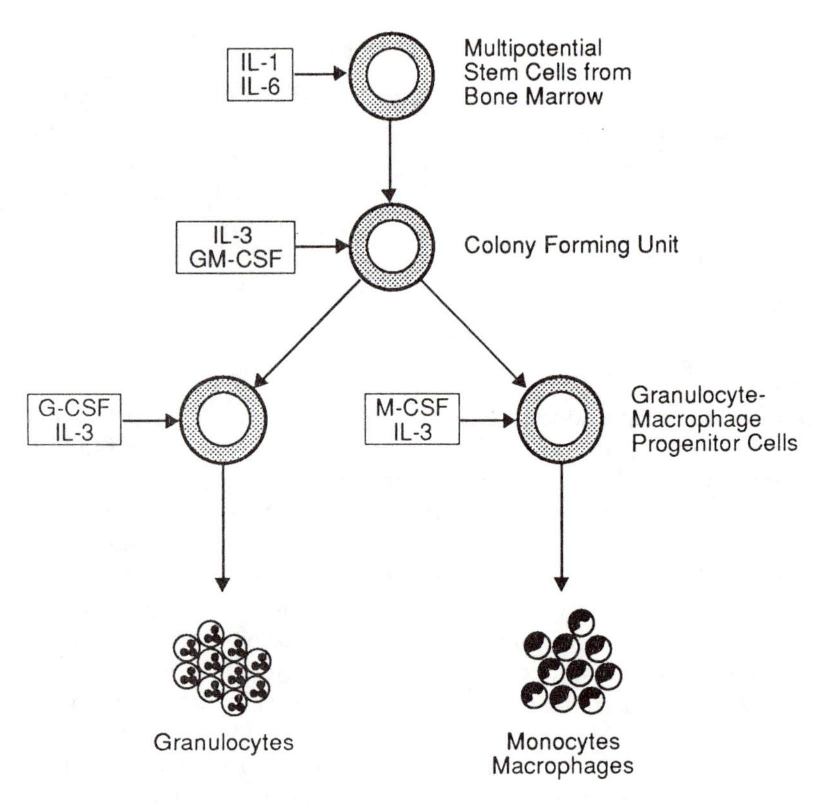

Figure 13.7. The role of cytokines in each stage of haematopoiesis. IL-1 and IL-6 act on multipotential stem cells in the bone marrow to produce colony forming units. IL-3 (multi-colony stimulating factor) and granulocyte-macrophage colony stimulating factor (GM-CSF) act on the colony forming units to produce granulocyte-macrophage progenitor cells. Macrophage colony stimulating factor (M-CSF) and IL-3 promote monocyte and macrophage formation from these progenitor cells, while granulocyte colony stimulating factor (G-CSF) and IL-3 promote the formation of granulocytic blood cells (neutrophils, eosinophils and basophils). (Redrawn from Metcalf, 1991.)

to infection. TNF also mediates inflammatory responses, but may do so indirectly, by stimulating the release of IL-1, prostaglandins, or corticosteroids, which then act on their receptors to modulate inflammatory responses (Beutler, 1990).

13.5 EFFECTS OF CYTOKINES AND OTHER 'IMMUNOMODULATORS' ON THE BRAIN AND NEUROENDOCRINE SYSTEM

As shown in Figure 13.8, the cells of the immune system can communicate with the brain and neuroendocrine system through the production of cytokines and through the synthesis and release of peptide hormones (Ballieux and Heijnen, 1987; Roszman and Brooks, 1988; Weigent *et al.*, 1990). Although Hall *et al.* (1985) suggested that the messengers of the immune system, including the cytokines, peptide hormones and thymic hormones, be termed 'immunotransmitters', it is more appropriate to call them 'immunomodulators' (Plata-Salaman, 1989) or 'immunoregulators' (Plata-Salaman, 1991) because they have modulator rather than transmitter actions on the brain and endocrine system. Before I examine the neuromodulatory effects of the immunoregulators, the localization of cytokines and their receptors in the brain is discussed.

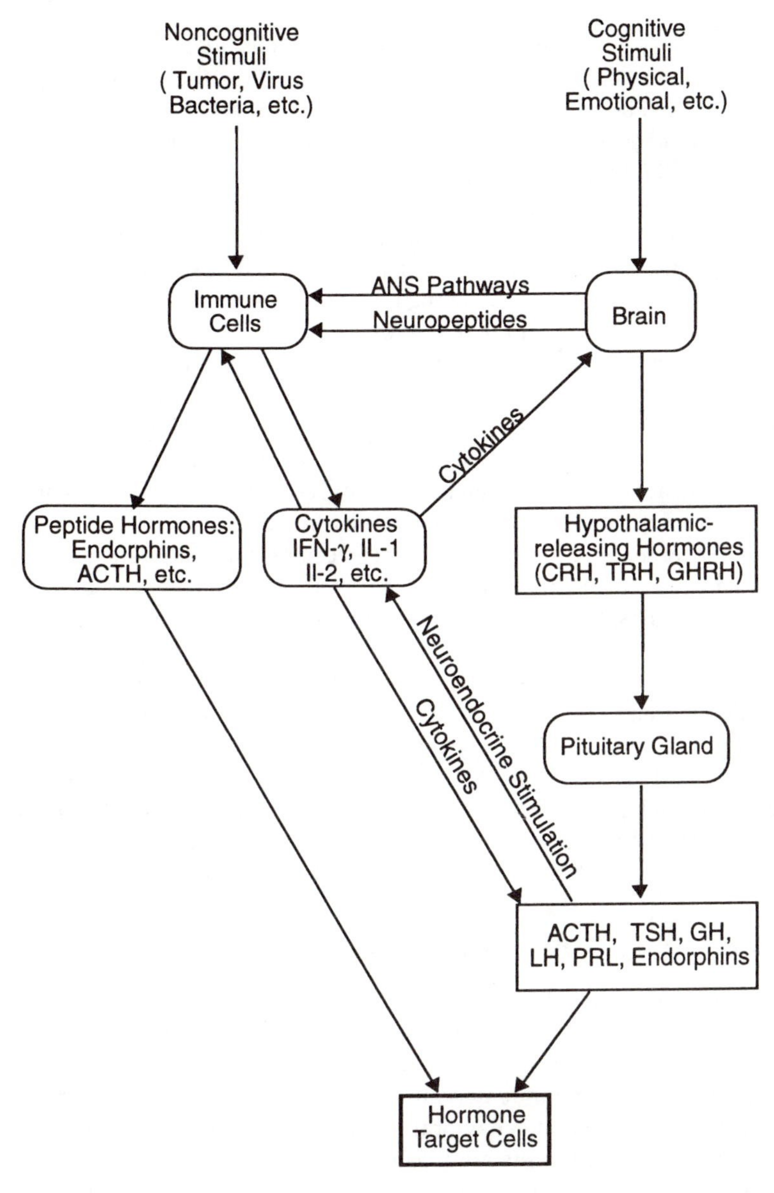

Figure 13.8. Inter-communication among the neural, endocrine and immune systems. The immune system is sensitive to non-cognitive stimuli, such as bacteria and viruses, which are not perceived by the central nervous system. Upon activation, cells of the immune system secrete cytokines and peptide hormones. The cytokines act on the brain and neuroendocrine system as well as on other immune cells. The peptide hormones released from the immune cells may stimulate immune cells or endocrine cells. Perception of cognitive stimuli by the brain can result in stimulation of the cells of the immune system by the peripheral nerves of the autonomic nervous system, the release of neuropeptides, or through the activation of the neuroendocrine system. In this way, the brain can indirectly perceive the presence of bacteria and viruses. One of the functions of the immune system, in fact, may be to make the brain aware of these stimuli through the release of cytokines. (Redrawn from Weigent, Carr and Blalock, 1990.)

13.5.1 LOCALIZATION OF CYTOKINES AND THEIR RECEPTORS IN THE BRAIN

Until recently, it was thought that cytokines produced in the body could not pass through the blood–brain barrier to enter the brain. However, like the neuropeptides, cytokines may enter the brain through the circumventricular organs (see Figure 11.14, p. 247), where there are gaps in the blood–brain barrier (Nathanson, 1989; Nistico and De Sarro, 1991). Banks *et al.* (1989) have shown that IL-1α is actively transported across the blood–brain barrier into and out of the brain.

Cytokines, including IL-1, IL-2, IL-6, TNF and the thymic hormones are synthesized in neural and non-neural cells in the brain (Plata-Salaman, 1991; Nistico and De Sarro, 1991). Like the neuropeptides, cytokine cell bodies and nerve fibers can be localized in the brain using immunohistochemical methods. IL-1, for example, has been localized in the infundibulum, median eminence, periventricular nucleus and other nuclei of the human brain, all of which are near the circumventrical organs (Breder, Dinarello and Saper, 1988). IL-6 is also distributed throughout the hypothalamus (Busbridge and Grossman, 1991) and bacterial endotoxins stimulate the release of IL-6 from the medial basal hypothalamus (Spangelo *et al.*, 1990).

Receptors for a number of cytokines are located throughout the brain. IL-1 receptors, for example, have their highest density in the olfactory bulbs, dentate gyrus, hippocampus, hypothalamus and choroid plexus of the rat brain (Farrar *et al.*, 1987; Lechan *et al.*, 1990; Rothwell, 1991b). There are at least three different forms of IL-2 receptors, some of which have been found in the hippocampus, striatum, and frontal cortex (Araujo *et al.*, 1989). Each of these receptors may activate different second messenger systems (Whitacre, 1990). IL-3 receptors occur in the olfactory bulbs, hippocampus and cortex (Farrar *et al.*, 1989). Like other peptide receptors, cytokine receptors are coupled to G-proteins and activate the cyclic AMP, inositol phospholipid and Ca^{2+} second messenger systems. For example, IL-1 activates the inositol phospholipid second messenger system, increasing protein kinase C and prostaglandin synthesis (Plata-Salaman, 1991). IL-3 receptors activate tyrosine kinase and protein kinase C second messenger systems (Mizel, 1989; Whitacre, 1990). IL-6 activates the cyclic AMP second messenger system (Nistico and DeSarro, 1991; see also Hadden, Hadden and Coffey, 1991).

13.5.2 NEUROMODULATOR EFFECTS OF CYTOKINES

There is growing evidence that the cytokines, like the neuropeptides, have a variety of neuromodulator functions. Cytokines modulate the firing rate of neurons, thus altering the release of neurotransmitters and regulating the autonomic nervous system, the neuroendocrine system, and the cognitive and behavioral functions of the brain (Roszman and Brooks, 1988). After rats are injected with sheep red blood cells (antigen), for example, there is an increase in the firing rate of the cells of the ventromedial hypothalamus. Infections increase the firing rate of cells of the preoptic area and paraventricular nucleus of the hypothalamus (Korneva, 1987). Following the injection of antigens, noradrenaline release is elevated in the paraventricular nucleus of the hypothalamus and serotonin release is elevated in the paraventricular and supraoptic nuclei. These changes are similar to those observed following other stressors (e.g. shock) and indicate that infections activate the same type of neural response as physical stressors (Dunn, 1989).

Neuromodulator actions of IL-1. IL-1 alters the firing rate of its hypothalamic target cells and stimulates the release of noradrenaline and serotonin in the hypothalamus (Breder *et al.*, 1988). IL-1 also modulates the release of adrenaline and VIP from the adrenal medulla. The effects of

Table 13.2. *Effects of cytokines and thymic hormones on the release of pituitary hormones*

	Pituitary hormones							
Cytokine	GH	PRL	ACTH	TSH	LH	FSH	β-END	Reference
Interferon γ	–	0↑	↑	↓	–	–	–	1, 9
IL-1	↑↓0?	↑↓0?	↑	↑↓?	↑↓?	0	↑	1, 3, 6, 8
IL-2	–	–	↑	–	–	–	↑	1, 2, 9
IL-6	↑	↑	↑	↑	↑	–	↑	8, 9
TNF	↑↓0?	↑	↑	↓	–	–	↑	1, 5
Thymic hormones	↑0?	↑0?	↑	0	↑	0	↑	2, 7, 8, 9

Notes:
↑ = stimulates release; ↓ = inhibits release; 0 = no effect; ? = conflicting results; – = not tested or unknown.
Source references: 1 = Scarborough, 1990; 2 = Blalock, 1989; 3 = Dunn, 1990; 4 = Merrill, 1990; 5 = Ricciardi-Castagnoli, *et al.*, 1990; 6 = Bazer and Johnson, 1989; 7 = Hall *et al.*, 1985; 8 = Spangelo and MacLeod, 1990; 9 = Plata-Salaman, 1991.

IL-1 modulation of hypothalamic nuclei are the induction of fever and slow wave sleep, the inhibition of feeding and the production of 'sickness' behavior, with reduced exploration (Plata-Salaman, 1991; Rothwell, 1991b).

Neuromodulator actions of IL-2. IL-2 inhibits the firing of neurons in the VMH and increases the firing of neurons in the PVN. IL-2 also inhibits acetylcholine release in a number of brain regions (Araujo *et al.*, 1989; Plata-Salaman, 1991). Acting at receptors in the locus ceruleus, IL-2 induces sleep, depression and mental confusion, which are often reported as side-effects when IL-2 is used to treat autoimmune disorders (Nistico and DeSarro, 1991).

Other neuromodulator effects of cytokines. As well as IL-1, TNF and interferon α stimulate sleep (Krueger *et al.*, 1990) and IL-6 and TNF inhibit food intake (Dunn, 1989; Plata-Salaman, 1989).

13.5.3 EFFECTS OF 'IMMUNOMODULATORS' ON THE NEUROENDOCRINE SYSTEM

The endocrine significance of the cytokines is now widely acknowledged (Rothwell, 1991a). Cytokines modulate the neuroendocrine functions of the hypothalamus as well as acting on the pituitary, thyroid, pancreas, adrenal glands and gonads to modulate hormone release (Scarborough, 1990; Kennedy and Jones, 1991). This section will focus on the effects of the cytokines and thymic hormones on the release of the hypothalamic and pituitary hormones (see Table 13.2).

Interferon

Interferon γ given by intracerebroventricular injection elevates CRH, ACTH and glucocorticoid release, suppresses TSH release and has either no effect or increases prolactin release in the rat (Blalock, 1989; Plata-Salaman, 1991).

Interleukin 1

Interleukin 1 elevates the release of ACTH, β-endorphin, and the glucocorticoids and has been shown to both elevate and inhibit TSH and LH release, but has no significant effect on FSH release. IL-1 has been reported to increase, decrease and have no effect on GH and prolactin secretion (Blalock, 1989; Dunn, 1990; Scarborough, 1990). The neuroendocrine effects of IL-1 have been studied most intensively in the hypothalamic-pituitary-adrenal system. As shown in Figure 13.9, IL-1β stimulates the release of CRH from the paraventricular nucleus of the hypothalamus (Navarra *et al.*, 1991) and may do this by elevating prostaglandin levels, which stimulate serotonin release in the hypothalamus (Katsuura *et al.*, 1990; Rothwell, 1991b). IL-1 is also released from the median eminence into the hypophyseal portal vessels to regulate ACTH release from the pituitary directly (Busbridge and Grossman, 1991) and may also act directly on the adrenal cortex, gonads and thyroid gland to regulate the release of their hormones (Adashi, 1990). Whatever its mechanism of action, IL-1 acts to signal to the brain that the immune system is being activated by an infection and triggers a neuroendocrine 'stress' response. The elevation of glucocorticoid levels by IL-1 during this stress response may provide a negative feedback mechanism to limit the responsiveness of the immune reaction and to down-regulate IL-1 production, since glucocorticoids inhibit immune reactions (Besedovsky, Del Ray and Sorkin, 1985).

Interleukin 2

IL-2 acts on receptors in the brain to activate the POMC system, elevating ACTH and β-endorphin release from the anterior pituitary gland (Merrill, 1990). IL-2 also stimulates vasopressin release from the PVN and thus increases water retention (Plata-Salaman, 1991).

Interleukin 6

IL-6 elevates ACTH, GH and prolactin release and may also elevate LH and TRH release (Plata-Salaman, 1991). Like IL-1, IL-6 appears to act in the hypothalamus to modulate CRH release, to stimulate ACTH release directly from the pituitary, and to stimulate the release of glucocorticoids directly from the adrenal cortex (Busbridge and Grossman, 1991; Navarra *et al.*, 1991).

Tumor necrosis factor

TNF elevates CRH, ACTH, β-endorphin and prolactin secretion and inhibits TSH release while its effects on GH release are controversial. TNF may also act on the ovary to regulate progesterone secretion (Adashi, 1990; Ricciardi-Castagnoli *et al.*, 1990; Scarborough, 1990).

Thymic hormones

Thymic hormones elevate plasma levels of ACTH, β-endorphin and LH, and may also elevate the release of GH and prolactin, but have no effect on

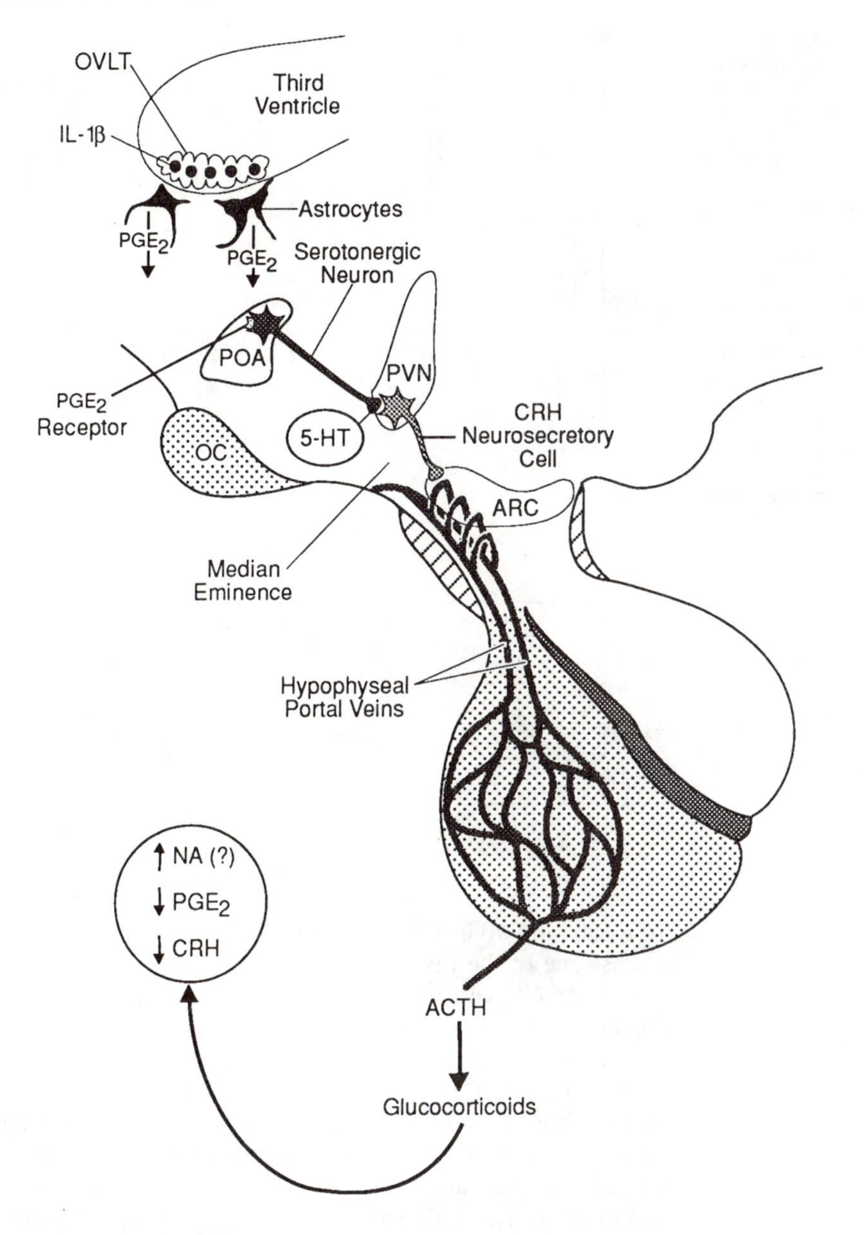

Figure 13.9. The release of corticotropin releasing hormone (CRH) by IL-1β. IL-1β enters the hypothalamus from the bloodstream through the vascular organ of the lamina terminalis (OVLT) in the third ventricle and stimulates prostaglandin (PGE_2) synthesis from astrocytes. Prostaglandin causes fever and also binds to PGE_2 receptors on serotonin neurons in the preoptic area of the hypothalamus (POA), which sends axons to the CRH neurosecretory cells in the paraventricular nucleus (PVN). Serotonin (5-HT) increases CRH release into the hypophyseal portal veins and causes increased ACTH and glucocorticoid release. The glucocorticoids then feed back to the brain to inhibit prostaglandin and CRH release, possibly by stimulating noradrenaline (NA) release. ARC = arcuate nucleus; OC = optic chiasm. (Redrawn from Katsuura *et al.*, 1990 and Rothwell, 1991b.)

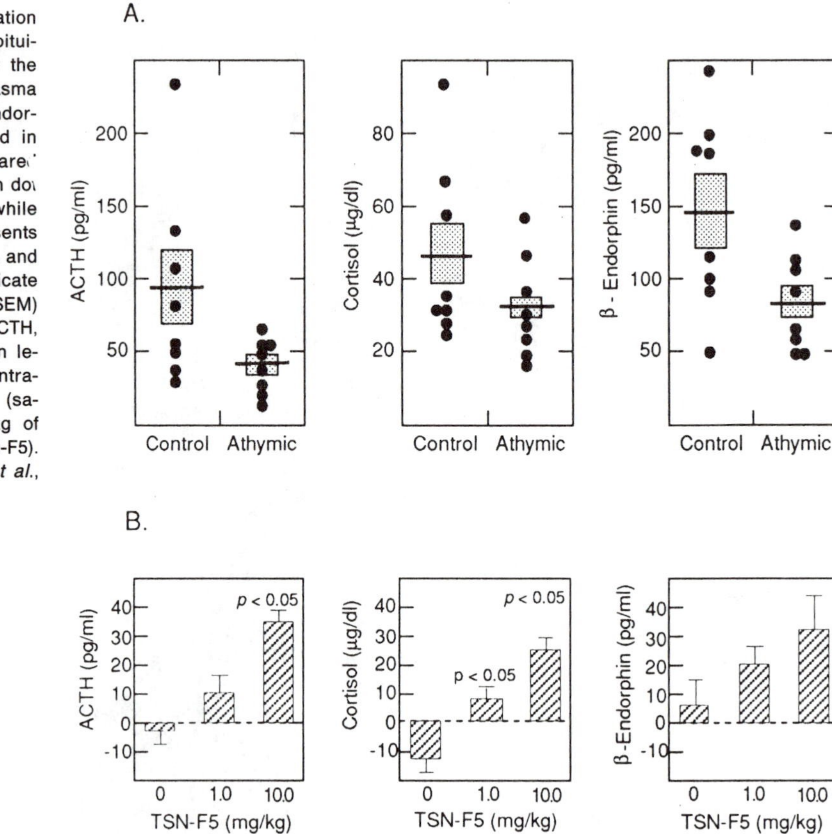

Figure 13.10. The regulation of the hypothalamic–pituitary–adrenal system by the thymus gland. (A) Plasma ACTH, cortisol and β-endorphin levels are reduced in athymic monkeys compared to control subjects. Each dot represents one subject, while the horizontal line represents the mean of all subjects and the shaded areas indicate ± 1 SEM. (B) Mean (± 1 SEM) increases in plasma ACTH, cortisol and β-endorphin levels in monkeys given intravenous injections of 0 (saline), 1.0 or 10.0 mg/kg of thymosin fraction 5 (TSN-F5). (Redrawn from Healy *et al.*, 1983.)

the release of FSH or TSH (Spangelo and MacLeod, 1990). For example, at least one of the thymic hormones (thymosin fraction 5) is able to elevate ACTH, corticosteroid, and β-endorphin release in monkeys (Figure 13.10). Removal of the thymus gland results in a decrease in plasma ACTH, corticosteroid, and β-endorphin levels in these monkeys (Healy *et al.*, 1983). The exact mechanism of the thymus–neuroendocrine interaction is unknown. There are receptors for the thymic hormones in the arcuate nucleus and median eminence of the hypothalamus, leading to the theory that the thymic hormones regulate the release of pituitary hormones via their action on the hypothalamic neurosecretory cells (Hall *et al.*, 1985; Rebar *et al.*, 1981). On the other hand, the thymic hormones may act primarily on non-neural brain cells (such as glial cells) involved in defense against neurological disease and the neuroendocrine changes might be side-effects of the activation of these cells. The release of cytokines and prostaglandins may also provide the mechanism for the thymus-induced stimulation of pituitary hormone release (Hall and Goldstein, 1984).

The thymic hormones are important regulators of steroid hormone release. Removal of the thymus gland reduces the secretion of both gonadal and adrenal steroids and the release of these hormones is elevated by injections of thymic hormones (Healy *et al.*, 1983; Hall *et al.*,

Table 13.3 *Hormones synthesized and released from cells of the immune system*

Cells	Hormones released	References
Thymus	Oxytocin, vasopressin	1, 2, 4
T cells	Enkephalins, TSH, hCG	1, 4
Lymphocytes	CRH, ACTH, β-endorphin, GH, prolactin	1, 3, 4, 6
Monocytes	ACTH, β-endorphin, substance P, somatostatin	6
Macrophages	ACTH, β-endorphin	1, 4
Leukocytes	VIP, somatostatin, LH, FSH	1, 4, 5
Spleen cells	ACTH, LH	4, 5

Source references: 1 = Blalock, 1989; 2 = Geenen, Legros and Franchimont, 1987; 3 = Galin, Le Boeuf and Blalock, 1990; 4 = Weigent and Blalock, 1987; 5 = Weigent *et al.*, 1990; 6 = Goetzl, Turch and Sreedharan, 1991.

1985). Different thymic hormones appear to influence the release of adrenal and gonadal steroids. For example, thymosin α1 stimulates the release of ACTH but not LH, while thymosin β4 stimulates LH, but not ACTH release. Thus, there is a complex interaction between the thymic hormones and the secretion of the steroid hormones (Hall and Goldstein, 1984; Marchetti, 1989; Hall, O'Grady and Farah, 1991).

13.5.4 THE PRODUCTION OF PEPTIDE HORMONES BY CELLS OF THE IMMUNE SYSTEM

As shown in Table 13.3, many cells of the immune system are able to synthesize and release peptide hormones which are identical with those produced in the hypothalamus and pituitary gland (Carr and Blalock, 1991; Goetzl *et al.*, 1991). For example, the thymus gland produces oxytocin and vasopressin, while leukocytes and mast cells produce somatostatin and vasoactive intestinal peptide (VIP). T cells can synthesize endogenous opioids, TSH and human chorionic gonadotropin, while lymphocytes produce GH and prolactin, as well as ACTH and β-endorphin. Both leukocytes and spleen cells produce LH, and other peptides (Weigent and Blalock, 1987; Blalock, 1989; Geenen *et al.*, 1989; Weigent *et al.*, 1990).

The peptide hormones are released from the cells of the immune system in response to stimulation from both antigens and hypothalamic hormones. For example, viral infections stimulate ACTH and β-endorphin release from lymphocytes, macrophages, and spleen cells. Lymphocytes, monocytes and macrophages all have CRH receptors and produce ACTH and β-endorphin in response to CRH release from the hypothalamus. The synthesis of these hormones by cells of the immune system is inhibited by glucocorticoids, indicating that hormone release from immune cells is regulated by negative feedback as it is in pituitary corticotroph cells (Blalock, 1989). Thus, the 'stress-induced' release of ACTH and β-endorphin can be stimulated from the pituitary gland by cognitive stimuli *and* from the cells of the immune system by non-cognitive stimuli (see Figure 13.8).

13.6 NEURAL AND ENDOCRINE REGULATION OF THE IMMUNE SYSTEM

The first indication that the brain could regulate immune function came from studies which found that damage to particular brain areas altered immune system responses (see Jankovíc, 1989, for a history of research on the neural control of the immune system). Lesions of the anterior hypothalamus, for example, decrease the proliferation of T cells and NK cells, while lesions of the hippocampus and amygdala enhance immune cell proliferation (Brooks *et al.*, 1982; Cross *et al.*, 1984; Roszman and Brooks, 1985). Lesions of the brain stem autonomic nuclei alter immune system function, with the effects depending on the particular nuclei lesioned (Felten *et al.*, 1991). Lesions of the cerebral cortex also alter immune function and, to complicate matters, lesions of the left cortex produce different effects on the immune system than lesions of the right cortex (Barnéoud *et al.*, 1988; Neveu, 1988).

Changes in neurotransmitter levels in the hypothalamus and brain stem regulate the responses of the immune system through changes in autonomic nervous system activity, thymus gland stimulation, and by altering neuroendocrine activity (Kordon and Bihoreau, 1989). Drugs which alter neurotransmitter levels also alter immune responses. For example, serotonin agonist drugs suppress antibody responses, while serotonin antagonists facilitate antibody responses (Rozman and Brooks, 1985). Serotonin may influence both T cells and B cells through its modulation of the neuroendocrine system or by acting directly on these cells. Serotonin elevates ACTH levels (and ACTH inhibits many immune system activities) and may mediate the release of TSH, LH and GH (Table 6.1, p. 93), so it may mediate immune responses by altering the release of these hormones (Hall and Goldstein, 1981).

The catecholamines have wide-ranging effects on the immune system. Dopamine stimulates T cell activity, but inhibits B cell activity. Reducing dopamine levels results in a decrease in T lymphocytes, but an increase in antibody levels. Since dopamine inhibits prolactin release, many of the actions of dopamine on the immune system may be mediated by changes in prolactin levels (Hall and Goldstein, 1981). Noradrenaline and adrenaline influence immune responses, both by their action as hormones from the adrenal medulla, and by their action as neurotransmitters in the hypothalamus, brain stem and sympathetic branches of the autonomic nervous system. Intercerebroventricular injection of 6-hydroxydopamine (6-OHDA), which decreases noradrenaline levels, impairs the production of antibodies in response to antigens (Roszman and Brooks, 1985). Acetylcholine may influence immune responses by regulating neurohormone release or through its action as a neurotransmitter in the parasympathetic branch of the autonomic nervous sytem (Hall and Goldstein, 1981).

As outlined in Figure 13.8, the neural and endocrine systems can modulate the immune system in three different ways (MacLean and Reichlin, 1981): (1) through the autonomic nervous system pathways; (2) through the release of hypothalamic and pituitary hormones; and (3) through the release of neuropeptides. Behavioral changes linked to the autonomic nervous system (e.g. behaviors that alter body temperature or blood volume) or neuroendocrine system (e.g. changes in feeding behavior or sleep–wake cycles) may also alter immune responses.

13.6.1 AUTONOMIC NERVOUS SYSTEM INFLUENCES ON THE IMMUNE SYSTEM

The autonomic nervous system innervates all of the tissues of the immune system shown in Figure 13.1: the bone marrow, the thymus gland, spleen and lymph nodes (Bulloch and Pomerantz, 1984; Bulloch, 1987; Felten *et al.*, 1985). Both the sympathetic (adrenergic) and the parasympathetic (cholinergic) branches of the ANS innervate immune tissues.

Adrenergic stimulation

The evidence for adrenergic sympathetic stimulation of the immune system is quite substantial. Noradrenergic nerve fibers are present in immune tissues, cells of the immune system have adrenergic receptors, and manipulation of noradrenaline release results in altered immune responses (Felten *et al.*, 1985, 1987; Nance, Hopkins and Bieger, 1987; Felten and Felten, 1988). In the bone marrow, catecholamines facilitate the production of multipotential stem cells, the precursors of both T cells and B cells (Hall and Goldstein, 1981). T cells, B cells and macrophages have β-adrenergic receptors (Dunn, 1989; Roszman and Carlson, 1991). Adrenergic stimulation promotes the differentiation and maturation of T cells, the growth of cytotoxic T cells and the production of antibodies by B cells (Marchetti *et al.*, 1990b). Adrenergic stimulation may also regulate the release of the thymic hormones, which stimulate the developing T lymphocytes by paracrine action (Bulloch, 1987; Hadden, 1987). This suggests that the thymus gland acts as a neuroendocrine transducer, responding to neural stimulation from the adrenergic nerves of the SNS by releasing thymic hormones.

Along with the catecholamines, the sympathetic branch of the SNS co-releases a number of neuropeptides, including VIP, CCK, neuropeptide Y, somatostatin, and others (MacLean and Reichlin, 1981; Dunn, 1989). These neuropeptides also regulate the action of the immune system (see Section 13.6.3).

Cholinergic stimulation

Cholinergic stimulation of the immune system is less pronounced than adrenergic stimulation, but cholinergic receptors have been found in the epithelial cells of the thymus and in the bone marrow (Dunn, 1989). Whether or not the thymus gland receives cholinergic stimulation is controversial, as Nance *et al.* (1987) found that only the sympathetic branch of the ANS innervated the thymus gland. T lymphocytes have cholinergic as well as adrenergic receptors. The most common effect of acetylcholine in the immune system is the activation of T cell proliferation, but acetylcholine may also accelerate the synthesis of antibodies by

B cells (Hadden, 1987; Maslinski, 1989). Table 13.4 provides an overview of the effects of the autonomic nervous system and the neuroendocrine system on the immune system.

13.6.2 EFFECTS OF HYPOTHALAMIC AND PITUITARY HORMONES ON THE IMMUNE SYSTEM

The first indication that hypothalamic and pituitary hormones could modulate immune responses was the finding that lesions of the anterior hypothalamus interfered with the immune response (Cross *et al.*, 1984). Since this demonstration, all of the hypothalamic and pituitary hormones have been shown to modulate the immune system, as have the hormones of the adrenal cortex, thyroid gland and the gonads (Blalock, 1985; Weigent and Blalock, 1987; Dunn, 1989; Hall and O'Grady, 1989). Cells of the immune system have receptors for more than 20 hormones and neuropeptides (Bost, 1988).

Neurohypophyseal hormones

Both oxytocin and vasopressin act on the thymus gland where they modulate the synthesis and secretion of thymic hormones and may thus function as T cell growth factors. Since oxytocin and vasopressin are both synthesized in the thymus, it is possible that they act in a paracrine manner in this gland (Gennen *et al.*, 1989). Oxytocin and vasopressin may also stimulate the release of interferon γ from cytotoxic T cells and act to regulate hormone production by the cells of the immune system (see Section 13.7).

GH and prolactin

Growth hormone and prolactin have wide-ranging effects on the immnune system (Berczi and Nagy, 1987; Johnson and Torres, 1988). Growth hormone stimulates the growth of the thymus gland and the proliferation and differentiation of T cells and modulates the actions of macrophages, B cells, and NK cells (Weigent and Blalock, 1987, 1990; Kelley, 1991). Dwarf mice, which lack GH, are also immunodeficient, and their immune system activity can be normalized by GH treatment (Weigent *et al.*, 1990). Prolactin enhances the responsiveness of the immune system by stimulating cytokine release from helper T cells (Cross and Roszman, 1989; Bernton, Bryant and Holaday, 1991). Prolactin and GH are also able to restore thymic function and T cell activity in aged rats whose thymus gland has deteriorated (Kelley *et al.*, 1987).

The hypothalamic-pituitary-thyroid system

TRH and TSH enhance the activity of the immune system by stimulating spleen cells and T cells. TSH increases antibody production and may do this by stimulating cytokine release from T cells (Weigent and Blalock, 1987; Blalock, 1989). Thyroid hormones (T3 and T4) also stimulate the release of thymic hormones and the maturation of T lymphocytes (Berczi, 1986c).

Table 13.4. *Summary of neural and hormonal effects on the immune system*

Neuroendocrine stimulus	Effects on immune system	Reference
Autonomic nervous system activity		
Adrenergic stimulation	Inhibits macrophage activity Stimulates multipotential stem cell development Stimulates T cell proliferation Stimulates antibody production from B cells	1, 2
Cholinergic stimulation	Stimulates multipotential stem cell development Enhances T cell proliferation Stimulates antibody production from B cells	1, 2, 3
Hypothalamic–pituitary (H-P) hormones		
Oxytocin and vasopressin	Enhance T cell growth	4
GH and PRL	Enhance T cell proliferation and differentiation	5
H-P-thyroid system (TRH and TSH)	Stimulate T cell maturation Stimulate antibody production from B cells	6
H-P-gonadal system		
Estrogen	Stimulates macrophage activity Suppresses T cell proliferation Stimulates antibody production from B cells	1, 7
Testosterone	Inhibits T cell proliferation Inhibits antibody production from B cells	7
Progesterone	Inhibits development of multipotential stem cells Inhibits macrophage activity Inhibits T cell proliferation Inhibits antibody production from B cells	8
HCG	Inhibits cytotoxic T cell activity Stimulates suppressor T cell activity Inhibits natural killer cells	6
H-P-adrenal system		
ACTH	Inhibits macrophage activity Inhibits T cell proliferation Stimulates B cell growth, but inhibits antibody production	9
Glucocorticoids	Inhibit multipotential stem cell colony formation Inhibit phagocytosis by macrophages Inhibit T cell proliferation and cytotoxic T cell activity Inhibit antibody production from B cells Inhibit natural killer cell activity	9
Neuropeptides		
Substance P	Stimulates phagocytosis by macrophages Stimulates T cell proliferation Stimulates antibody production from B cells	10
Somatostatin	Inhibits multipotential stem cell colony formation Inhibits T cell proliferation Inhibits antibody production from B cells	10
Endogenous opioid peptides (β-endorphin, enkephalins)	Inhibit phagocytosis by macrophages. Stimulate T cell proliferation and cytotoxic T cell activity Stimulate (β-endorphin) or inhibit (enkephalins) antibody production from B cells (depending on the state of the cell) Stimulate natural killer cell activity	6, 10
VIP	Inhibits phagocytosis by macrophages Inhibits T cell proliferation Inhibits antibody production from B cells Inhibits natural killer cell activity	10, 11
NGF	Stimulates cytokine release from T cells	

Source references: 1 = Hall and Goldstein, 1981; 2 = Hadden, 1987; 3 = Maslinski, 1989; 4 = Geenen *et al.*, 1989; 5 = Berczi and Nagy, 1987; 6 = Blalock, 1989; 7 = Grossman, 1990; 8 = Grossman, 1985; 9 = Bateman *et al.*, 1989; 10 = Weigent *et al.*, 1990; 11 = Nordlind, Mutt and Sundström, 1988; 12 = Levi-Montalcini, Aloe and Alleva, 1990; 13 = Teschemacher *et al.*, 1990.

The hypothalamic-pituitary-gonadal system

The hormones of the hypothalamic-pituitary-gonadal system modulate the immune system in a number of ways (Marchetti *et al.*, 1990a; McCruden and Stimson, 1991). The cells of the immune system have receptors for the gonadal steroids; there are sex differences in immune responses; gonadectomy and sex-hormone replacement alter the immune response, as do changes in estrogen and progesterone over the menstrual cycle; and the immune response is altered during pregnancy (Grossman, 1984, 1985, 1990; Hall and Goldstein, 1984; Berczi, 1986b). Because steroid hormone levels rise at puberty, the sex differences in immune responsiveness may be more obvious after puberty.

There are sex differences in the structure of bone marrow, the level of antibodies in the circulation, and in susceptibility to autoimmune diseases, infectious diseases and cancer. Female spleen cells produce more interleukin-2 (T cell growth factor) than male spleen cells and T cells from females are more responsive to interleukin-2 than those from males. Females have higher serum immunoglobulin (antibody) levels than males and develop stronger and longer lasting immune responses to antigens. Women are more prone to autoimmune diseases such as rheumatoid arthritis, in which 70% of the patients are female, and systemic lupus erythematosus (an autoimmune disease, which resembles rheumatoid arthritis and can affect many different organs and tissues), in which 90% of the patients are female. Some of these sex differences may be due to genetic differences in the sex chromosomes, but many others are due to immune system responses to sex differences in gonadal hormone levels, target cell sensitivity to hormones (i.e. receptor numbers) and in hormone metabolism (Berczi, 1986b).

In general, the evidence suggests that the gonadal steroids inhibit cell-mediated immunity and that removal of the gonadal steroids by gonadectomy stimulates cell-mediated immune responses. Like the corticosteroids, androgens and estrogens suppress thymus activity and may be responsible for the involution of the thymus after puberty. High levels of estrogens also suppress T cell and NK cell activity. Post-pubertal gonadectomy results in the enlargement of the thymus, spleen and lymph nodes in both sexes. The sex hormones may inhibit cell-mediated immunity by inhibiting the release of thymic hormones, thus inhibiting T cell development (Grossman, 1984, 1985; Hall and Goldstein, 1984). On the other hand, ovarian steroids, particularly estrogen, facilitate humoral immunity. Estrogens stimulate the production of antibodies while the effects of androgens on antibody production are controversial (Grossman, 1984, 1985). Estrogens also enhance phagocytosis by macrophages, but androgens have little effect (Hall and Goldstein, 1984; Pung *et al.*, 1985). Grossman (1990) has presented a model to explain the effects of sex differences in gonadal hormones on the immune system.

In human females, the activity of the immune system changes over the menstrual cycle. The levels of lymphocytes and white blood cells in the circulation are negatively correlated with estrogen levels, while the levels of monocytes and granulocytes are positively correlated with progesterone levels (Mathur *et al.*, 1979). As in other situations, estrogen and progesterone may act synergistically to influence the immune system (Hall and Goldstein, 1984). Finally, estrogens appear to activate autoimmune diseases while androgens may inhibit the development of autoimmune diseases (Grossman, 1984, 1985, 1990).

During pregnancy, the increased levels of gonadal steroids (as well as pituitary and placental gonadotropic hormones) suppress cell-mediated immune responses, thus help to prevent the rejection of the fetus (which is 'foreign tissue'). HCG, for example, is able to inhibit cytotoxic T cell and NK cell activity and may increase the activity of suppressor T cells, either by direct immunosuppressive action on these cells or by stimulating progesterone secretion (Grossman, 1984, 1985; Blalock, 1989). The size of the thymus gland and the number of T cells are both reduced during pregnancy. If this depression of cell-mediated immunity does not occur during pregnancy, the fetus may be aborted (Grossman, 1984, 1985). The interaction between the neuroendocrine and immune systems during pregnancy is very complex and the ability of the fetoplacental unit to stimulate cytokine release from the mother may facilitate implantation and normal pregnancy (see Harbour and Blalock, 1989).

The hypothalamic-pituitary-adrenal system

All three hormones of the hypothalamic-pituitary-adrenal system (CRH, ACTH, and the glucocorticoids) modulate immune system activity. CRH stimulates the production of IL-1 by monocytes, increases the proliferation of T cells, stimulates the release of IL-2, and promotes the up-regulation of IL-2 receptors on helper T cells (Singh *et al.*, 1990). ACTH inhibits the production of IFNγ by T cells and inhibits the production of IFNγ receptors on macrophages, thus blocking IFNγ activation of macrophages. ACTH also promotes B cell proliferation, but inhibits antibody production (Berczi, 1986a; Bateman *et al.*, 1989; Blalock, 1989).

Glucocorticoids have wide-ranging effects on the immune system (Hall and Goldstein, 1984; Berczi, 1986a; Bateman *et al.*, 1989; Munck and Guyre, 1991). In general, the glucocorticoids inhibit immune system function, but they also have some stimulatory effects. Corticosteriods inhibit the ability of macrophages to phagocytose foreign tissue and inhibit the release of cytokines from macrophages. Corticosteriods also suppress cellular immunity by inhibiting thymus gland development, inhibiting the release of thymic hormones, and inhibiting the development and differentiation of T cells in the thymus gland. Prolonged corticosteroid release may cause T cell death. Corticosteroids are also able to inhibit the release of cytokines (Hall and Goldstein, 1984). For example, glucocorticoids inhibit the release of IL-1 from macrophages, inhibit the release of IFNγ and IL-2 (T cell growth factor) from T cells, and inhibit the release of colony stimulating factors, thus inhibiting the development of monocyte (pre-macrophage) colonies in bone marrow. By inhibiting IL-2 release, glucocorticoids inhibit the development of cytotoxic T cells, but once these cells mature, they are resistant to the inhibitory effects of corticosteroids. Corticosteroids also inhibit the activity of NK cells (Bateman *et al.*, 1989).

Many of the effects of corticosteroids on the immune system are biphasic. For example, there is an initial inhibition of macrophage and T cell development when corticosteroids are released following a stressor stimulus, but the prolonged release of corticosteroids may stimulate the immune response. These biphasic responses to corticosteroids may also depend on the level of corticosteroids in the blood (Hall and Goldstein, 1984). High concentrations of corticosteroids inhibit the thymus, T cells and leucocytes and inhibit antibody production, whereas lower concentrations stimulate T cell differentiation, stimulate antibody production and modulate the immune response. These actions may be mediated by different corticosteroid receptors (see Chapter 9).

Corticosteroids as negative feedback signals for the immune system

Chapter 8 described the importance of negative feedback systems for the homeostatic regulation of hormone levels in the body. Since cells of the immune system produce cytokines (such as IL-1) and ACTH, both of which stimulate corticosteroid release, and since corticosteroids inhibit the activity of the immune system, it has been hypothesized that the immunosuppressive action of corticosteroids acts to provide negative feedback in the immune system, as shown in Figure 13.11 (Besedovsky *et al.*, 1986; Munck *et al.*, 1986; Dunn, 1989). The negative feedback effect of the corticosteroids could thus act to prevent immune (and other) stress-induced physiological mechanisms from overstimulating the body. For example, the immunosuppressive action of corticosteroid negative feedback may prevent the development of autoimmune diseases such as rheumatoid arthritis, in which the immune system attacks and destroys self cells as well as foreign cells (Dunn, 1989).

Summary

Figure 13.12 summarizes the effects of the anterior pituitary hormones on the immune system. In general, growth hormone, prolactin and the hormones of the pituitary-thyroid axis stimulate immune system function. The hormones of the pituitary-gonadal axis are generally inhibitory, but may also stimulate some immune responses, while the hormones of the pituitary-adrenal axis are inhibitory, especially at high doses. GH and prolactin act directly on T lymphocytes to stimulate their maturation and differentiation. Thyroxine stimulates the maturation of T cells in the thymus gland, while gonadal steroids and corticosteroids inhibit T cell development, possibly by inhibiting the production of the thymic hormones (Hadden, 1987).

13.6.3 EFFECTS OF NEUROPEPTIDES ON THE IMMUNE SYSTEM

As well as the hypothalamic and pituitary hormones, a number of neuropeptides modulate the activity of the immune system and the release of cytokines (Table 13.4). These include substance P, somatostatin, the endogenous opioid peptides, vasoactive intestinal peptide (VIP) and nerve growth factor (Payan *et al.*, 1986; Hartung, 1988; Nordlind *et

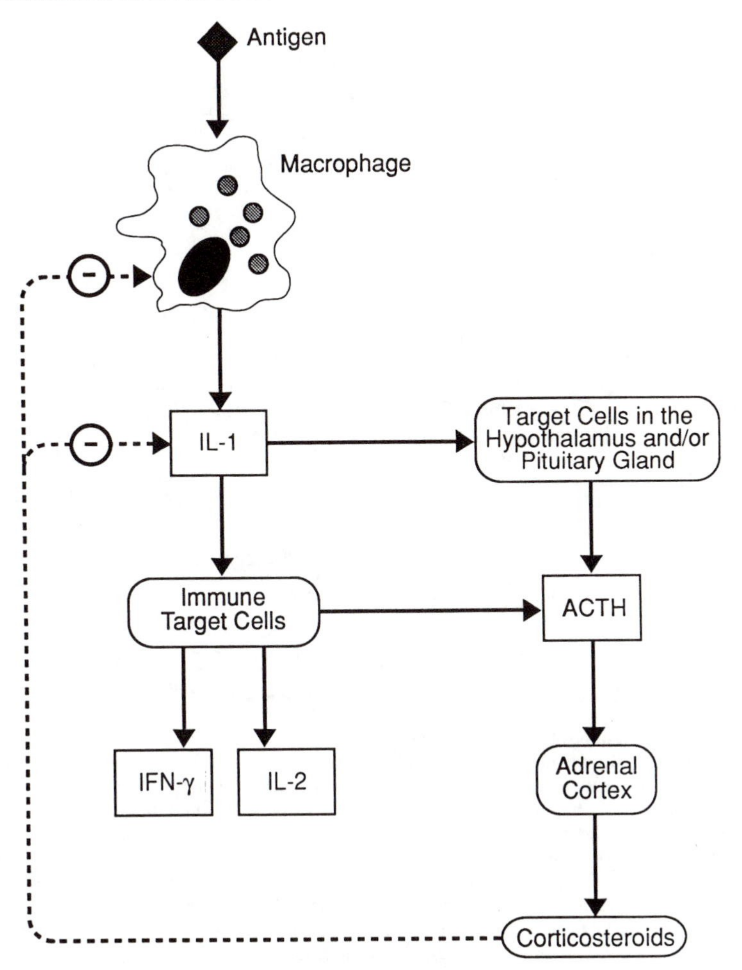

Figure 13.11. An interpretation of the negative feedback effects of corticosteroids on the immune system. IL-1 release from macrophages following antigen presentation stimulates both immune and neuroendocrine target cells. The immune target cells release cytokines such as IFNγ and IL-2, as well as ACTH and other peptides. ACTH from the pituitary and the immune cells stimulates corticosteroid release from the adrenal cortex. These corticosteroids provide negative feedback to inhibit immune activity and the release of cytokines and peptides from the macrophages and other cells of the immune system.

al., 1988; Bellinger *et al.*, 1990). These neuropeptides may be secreted from neurosecretory cells in the brain or co-released with neurotransmitters from neurons in the brain and peripheral nerves of the autonomic nervous system as discussed in Chapter 11. The thymus gland has receptors for β-endorphin, the enkephalins and substance P (Piantelli *et al.*, 1990) and T lymphocytes have receptors for somatostatin, VIP and substance P (Payan *et al.*, 1986; see also Blalock, Bost and Smith, 1985; Goetzl *et al.*, 1991).

The endogenous opioids have wide-ranging effects on the immune system, but these effects are complex and the results of *in vitro* studies can not always be replicated *in vivo* (Johnson and Torres, 1988; Sibinga and Goldstein, 1988; Blalock, 1989; Dunn, 1989; Carr and Blalock, 1991). β-endorphin and the enkephalins (met- and leu-) modulate the synthesis of IFNγ, enhance NK cell activity, promote B cell proliferation and stimulate the production of thymic hormones. β-Endorphin enhances antibody production by B cells, while the enkephalins suppress antibody production.

Substance P promotes phagocytosis by macrophages, stimulates T cell

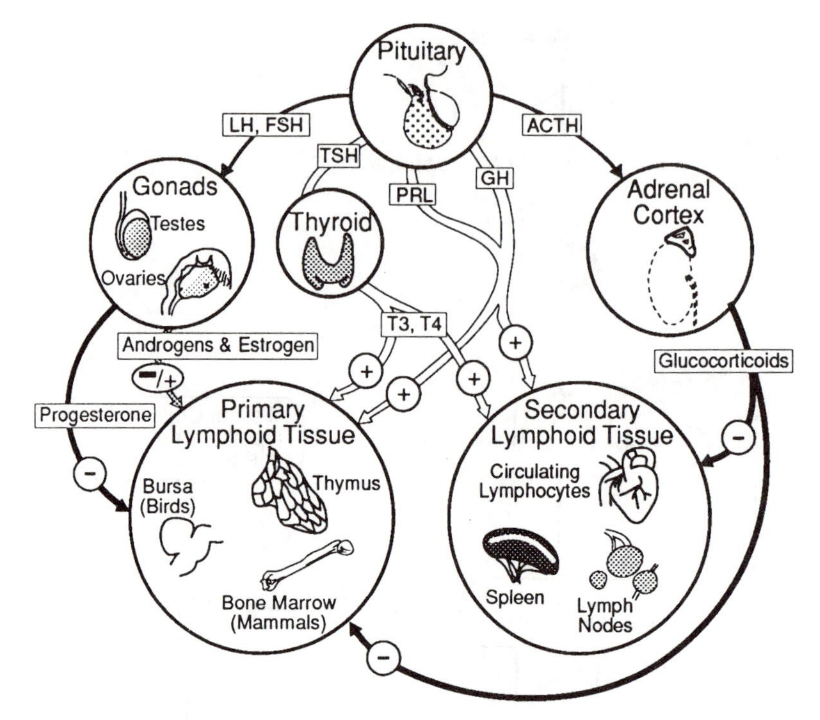

Figure 13.12. The control of hematopoietic tissue and immune function by the hormones of the pituitary gland. Growth hormone (GH) and prolactin (PRL) have a stimulatory effect on both the primary and secondary lymphoid tissues and are able to maintain hematopoietic and immune function in hypophysectomized animals. Placental lactogen has a similar effect. TSH, T_3 and T_4 also stimulate the pituitary and secondary lymphoid tissue. Adrenocorticotropic hormone (ACTH) stimulates the release of glucocorticoids which have inhibitory actions on the primary and secondary lymphoid tissues. The gonadotropic hormones (LH and FSH) stimulate the release of the gonadal steroids. Androgens and estrogen inhibit many immune functions, but can also stimulate some immune activities. Progesterone suppresses immune function, particularly during pregnancy. The sex hormones coordinate the responsiveness of the immune system during reproduction (e.g. transfer of immunoglobulins into eggs, uterine secretions, milk, the fetus, and the protection of the conceptus from maternal immunological insult). Prolactin, and the placental hormones may also contribute to the regulation of reproductive immune function. (Redrawn from Berczi and Nagy, 1987.)

proliferation and stimulates antibody production by B cells (McGillis *et al.*, 1991). Somatostatin, on the other hand, inhibits T lymphocyte proliferation, inhibits colony stimulation in bone marrow, and inhibits antibody production by B cells (Johnson and Torres, 1988; Payan and Goetzl, 1985; Payan, 1989). VIP inhibits T cell proliferation and the production of IL-2, inhibits NK cell activity and has a number of other regulatory effects on the immune system (Johnson and Torres, 1988; Nordlind *et al.*, 1988; Ottaway, 1991). Nerve growth factor (NGF) is released by the adrenergic neurons of the SNS which innervate the bone marrow, thymus gland, spleen and lymph nodes, and can influence the development of lymphocytes in these organs. NGF stimulates the proliferation and differentiation of lymphocytes and stimulates the release of IL-2 from T cells (Levi-Montalcini *et al.*, 1990; Thorpe, Jerrels and Perez-Polo, 1990). Other neuropeptides have also been located in the immune system, but their function is not yet known (Hadden, 1987; Bellinger *et al.*, 1990).

13.7 HYPOTHALAMIC INTEGRATION OF THE NEUROENDOCRINE–IMMUNE SYSTEM

If the immune response is considered as a mechanism for maintaining homeostasis among the cells of the body in response to antigen stimulation, then the 'immunomodulators' provide the communication signals for maintaining immune system homeostasis. As described previously in this chapter, these immunomodulators serve two functions: they inform the brain about the type of immune response being activated (an affector or sensory function), and they regulate the immune response (an effector function).

As shown in Figure 13.8, the immune system serves as a sensory system, responding to the presence of non-cognitive stimuli (such as bacteria, viruses, tumors and other antigens), which are not detected by the central or peripheral nervous systems. When these non-cognitive stimuli are detected, cell-mediated or humoral immune responses occur, depending on the type of antigen, and this information is sent to the brain via cytokine and peptide hormone release from the cells of the immune system. These immunomodulators enter the brain through the circumventrical organs and stimulate the hypothalamic neurons in the medial preoptic area, anterior and ventromedial hypothalamus, paraventricular nucleus and median eminence/arcuate nucleus. These hypothalamic nuclei then modulate the immune response by activating the autonomic nervous system, the neuroendocrine system, and the cognitive and behavioral correlates of these systems, as described in Section 4.1 (MacLean and Reichlin, 1981).

Antigens are stressful stimuli and the neuroendocrine response activated when the immune cells detect an antigen is a 'stress response' with the same characteristics as the stress response to a cognitive stimulus. In order to co-ordinate adaptive responses to stressful stimuli, whether cognitive or non-cognitive, the brain and immune system communicate via the neuroendocrine–cytokine messenger systems and the information from these messenger systems is integrated in the hypothalamus. The neuroimmune stress response involves the activation of the defense mechanisms of the immune system (macrophages, T cells, B cells, NK cells, etc.) which are modulated by the neuroendocrine system to ensure the rapid production and proliferation of these cells in the thymus, spleen and lymph nodes and the development of their precursor cells in the bone marrow (Weigent *et al.*, 1990).

Activation of the immune system alters neurotransmitter release and electrical activity in hypothalamic nuclei and damage to the hypothalamus results in abnormal neuroimmune responses as well as abnormal neuroendocrine responses (Besedovsky *et al.*, 1985; Smith and Blalock, 1986). The area of the hypothalamus which appears to have the greatest involvement in the neuroendocrine–immune response is the paraventricular nucleus (PVN). At least six lines of evidence support the central role of the PVN in the integration of the neuroendocrine–immune response (Daruna and Morgan, 1990).

1. The PVN has receptors for cytokines and thymic hormones, so these immunomodulators can directly stimulate PVN activity.
2. The PVN is involved in both the afferent and efferent sympathetic and parasympathetic pathways of the autonomic nervous system (ANS). Activation of the ANS results in rapid stimulation of the bone marrow, thymus gland, spleen and lymph nodes, as well as the adrenal medulla and other neuroendocrine transducer cells (e.g. the pineal gland). This causes the activation of both immune and endocrine cells by peripheral neurotransmitter, neuropeptide and hormone release.
3. The PVN contains magnocellular neurons which release oxytocin and vasopressin, both of which act on the thymus gland to influence T cell development.
4. The PVN contains parvicellular neurons which release TRH, CRH and other hypothalamic hormones. Activation of the neurosecretory cells of the hypothalamus by immunomodulators results in the release of hypothalamic and pituitary hormones (such as GH, PRL and TSH), which stimulate both endocrine and immune cells. As a result of neuroendocrine stimulation, there is an elevation of gonadal and adrenal steroid hormones which modulate the immune system. The glucocorticoids in particular, may provide negative feedback to inhibit immune responses.
5. The PVN receives input from the subfornical organ, a structure which has no blood–brain barrier and thus allows certain large molecules to pass between the circulation and the brain.
6. The PVN receives neural input from the neocortex, amygdala and hippocampus, areas of the brain which mediate cognitive functions including emotional arousal and learning. Cognitive stressors can influence immune system activity through their action on the hypothalamic neuroendocrine–immune integrating mechanisms. Isolation rearing, crowding, low dominance status, and social stress all influence the ability of animals to defend themselves against disease and infection (Salvin, Rabin and Neta, 1990). In humans, stressors such as depression, bereavement and exam stress all suppress immune responses (Calabrese, Kling and Gold, 1987). Both humoral and cell-mediated immune responses can also be conditioned to neutral stimuli (Ader and Cohen, 1985). Thus, the PVN may be able to integrate the cognitive responses to stressful psychosocial stimuli with the activation of the neuroendocrine–immune system, resulting in a psycho-neuroendocrine–immune system response (Daruna and Morgan, 1990).

Other functions of the hypothalamus are also influenced by neuroimmune activity. When the immune system is activated by certain antigens, such as bacterial pyrogens, body temperature is elevated, causing fever. Some cytokines (e.g. IL-1) also cause fever through their receptors in the hypothalamus. The hypothalamus then initiates responses from the autonomic nervous system (sweating or shivering), the neuroendocrine system (e.g. TSH and ACTH release) and behavioral responses to changes in body temperature. The hypothalamus also regulates the cardiovascular reflexes and control of the microvascular circulation may be an important mechanism for the neural modulation of immunity (MacLean and Reichlin, 1981).

Cytokines, such as IL-1 and IL-6, also alter hunger and feeding behavior mediated by the ventromedial and lateral hypothalamic nuclei, and changes in food intake alter neuroendocrine functions. For example, fasting reduces autonomic nervous system activity while feeding activates the release of insulin and the gastrointestinal peptides (Gross and Newberne, 1980; MacLean and Reichlin, 1981). Cytokines also produce cognitive, psychiatric and behavioral side-effects, such as fatigue, anorexia, sleep disruption and disturbances of emotional behavior (Denicoff *et al.*, 1987).

Virtually all of the cells of the hypothalamus show day–night or sleep–wake rhythms of activity. For example, GH and PRL secretion peaks during sleep and ACTH secretion is highest just before waking. Many of these hypothalamic cycles are regulated by a clock mechanism in the suprachiasmatic nucleus of the hypothalamus. Immune responses also show day–night cycles (Cove-Smith *et al.*, 1978; Knapp *et al.*, 1980) and these may be controlled by hypothalamic neuroendocrine rhythms (MacLean and Reichlin, 1981).

Neurotransmitters, peptide hormones and cytokines act via membrane receptors and G-protein transducers to activate second messenger systems in their target cells. These second messengers regulate ion channels in the target cell membrane and the transcription of genomic information via mRNA in the nucleus (see Chapter 10). The interactions between the neural, endocrine and immune systems may be mediated through common receptor and second messenger mechanisms (Su, London and Jaffe, 1988). In T lymphocytes, for example, β-adrenergic agonists, endotoxins, and thymic hormones all stimulate cyclic AMP production (Hadden, 1987; Carr and Blalock, 1989; Kordon and Bihoreau, 1989; Scarborough, 1990). Hormones, neurotransmitters and peptides may, therefore, regulate the responses of the cells of the immune system to antigens by regulating cell membrane permeability and protein synthesis through second messenger cascades. Cytokines act through G-protein coupled receptors and second messenger systems to control membrane permeability and synthesis of neurotransmitters and neuropeptides in the brain (Roszman and Brooks, 1988). Since the cells of the immune system can produce neuropeptides, it should not come as a surprise to find that neural cells can produce cytokines. IL-1, IL-3, interferons and thymosins may all be produced in the brain and neuroendocrine system and provide a mechanism for the direct control of immune cells by the brain (Blalock, 1989).

13.8 SUMMARY

This chapter has examined the interaction of the neuroendocrine system and the immune system. The immune system consists of a number of specific cell types, including the macrophages, T and B lymphocytes, and NK cells which control cell-mediated and humoral responses to antigens.

T cells mature in the thymus gland, the so-called 'master-gland of the immune system'. The thymus gland secretes a number of thymic hormones which are important regulators of immune system functions. T lymphocytes and other cells of the immune system produce the cytokines, which include interferon γ, the interleukins, tumor necrosis factor, lymphotoxin, and colony stimulating factors. These cytokines regulate T cell activation, B cell activation, the production of blood cells (haematopoiesis), cytotoxicity, and inflammatory responses. Cells of the immune system can regulate the brain and endocrine system through the actions of the cytokines on receptors in the brain and through the release of peptide hormones. Cytokines and thymic hormones stimulate neurotransmitter and neuropeptide release, thus regulating autonomic, neuroendocrine and behavioral actions. Neural regulation of the immune system occurs through autonomic nervous system activation and the release of hypothalamic and pituitary hormones. Complex neuroendocrine–immune system interactions occur between the thymus gland and the gonadal hormones and between cytokines and corticosteroid hormones. The corticosteroids appear to provide negative feedback to inhibit the release of IL-1 and suppress immune system activity, thus preventing some autoimmune diseases. Opioid peptides, substance P, somatostatin, and other neuropeptides also regulate the activity of the immune system. Neuroendocrine–immune system interactions are mediated through the hypothalamus, and the paraventricular nucleus (PVN) appears to play a central role in the integration of these three systems. Immune responses to antigens can be compared with neural responses to stressful cognitive stimuli as both initiate neuroendocrine stress responses. The activation of neuroendocrine and autonomic nervous activity in response to cognitive and non-cognitive stimuli is an adaptive response which maintains the homeostasis of the body in response to these stressors. Through the integrative activity of the hypothalamus, cognitive stressors such as crowding, exam stress and depression can alter the activity of the immune system, while interleukins and other cytokines can have cognitive and behavioral side-effects. Through the action of the brain, immune responses may also be conditioned to neutral (non-antigenic) stimuli. The integration of neural, endocrine and immune system activity may occur through common receptors and second messenger systems.

FURTHER READING

Ader, R., Felten, D. L. and Cohen, N. (eds.) (1991). *Psychoneuroimmunology*, 2nd edn. San Diego, CA: Academic Press.

Blalock, J. E. (1989). A molecular basis for bidirectional communication between the immune and neuroendocrine systems. *Physiological Reviews*, **69**, 1–32.

Coleman, R. M., Lombard, M. F., Sicard, R. E., and Rencricca, N. J. (1989). *Fundamental immunology*. Dubuque, IO: Wm. C. Brown.

Dunn, A. J. (1989). Psychoneuroimmunology for the psychoneuroendocrinologist: a review of animal studies of nervous system–immune system interactions. *Psychoneuroendocrinology*, **14**, 251–274.

Golub, E. S. and Green, D. R. (1991). *Immunology: A Synthesis*, 2nd edn. Sunderland, MA: Sinauer.

Hamblin, A. S. (1988). *Cytokines*. Oxford: IRL Press.

Whitacre, C. C. (1990). Immunology: A state of the art lecture. *Annals of the New York Academy of Sciences*, **594**, 1–16.

REVIEW QUESTIONS

13.1 What is the difference between T cells and B cells in the immune system?
13.2 What two immune functions does the thymus gland perform?
13.3 Which two cells of the immune system are most prominant in cytokine production?
13.4 Which cytokine stimulates fever, sleep, and prostaglandin synthesis and inhibits eating?
13.5 What is the main function of TNF and LT?
13.6 Which two neurohormones are produced in the thymus gland?
13.7 Name two of the hormones synthesized from POMC that are produced by lymphocytes.
13.8 Describe the steps involved in the immune-corticosteroid negative feedback system.
13.9 How do GH and prolactin influence the immune system activity?
13.10 How do progesterone and corticosteroids influence immune system activity?
13.11 Which area of the hypothalamus appears to regulate the neuroendocrine–immune system interaction?
13.12 Why is the immune response considered a stress response?

ESSAY QUESTIONS

13.1 Discuss the interactions between the immune system and the hormones of the hypothalamic-pituitary-adrenal axis.
13.2 Discuss the interactions between the immune system and the hormones of the hypothalamic-pituitary-gonadal axis.
13.3 Discuss the role of the ANS in regulating immune system function.
13.4 Discuss the mechanisms by which infections can activate the neuroendocrine system through the release of cytokines.
13.5 Discuss the functions of the thymic hormones in the neural, endocrine and immune systems.
13.6 Discuss how the neuroendocrine system suppresses the immune system during pregnancy.
13.7 Discuss the function of the peptide hormones produced by the cells of the immune system.
13.8 Discuss the neural, endocrine and immune functions of IL-1.
13.9 Discuss the effects of psychological stress on the immune system.

REFERENCES

Aarden, L. A., Brunner, T. K., Cerottini, J.-C. *et al.* (1979). Revised nomenclature for antigen-nonspecific T cell proliferation and helper

factors. *Journal of Immunology*, **123**, 2928–2929.
Adashi, E. Y. (1990). Do cytokines play a role in the regulation of ovarian function? *Progress in NeuroEndocrinImmunology*, **3**, 11–17.
Ader, R. and Cohen, N. (1985). CNS-immune system interactions: conditioning phenomena. *Behavioral and Brain Sciences*, **8**, 379–394.
Alberts, B., Bray, D., Lewis, J., Raff, M., Roberts, K. and Watson, J. D. (1989). *Molecular Biology of the Cell*, 2nd edn. New York: Garland.
Araujo, D. M., Lapchak, P. A., Collier, B. and Quirion, R. (1989). Localization of interleukin-2 immunoreactivity and interleukin-2 receptors in the rat brain: interaction with the cholinergic system. *Brain Research*, **498**, 257–266.
Ballieux, R. E. and Heijnen, C. J. (1987). Brain and immune system: a one-way conversation or a genuine dialogue? *Progress in Brain Research*, **72**, 71–77.
Barnéoud, P., Neveu, P. J., Vitiello, S., Mormede, P. and Le Moal, M. (1988). Brain neocortex immunomodulation in rats. *Brain Research*, **474**, 394–398.
Bateman, A., Singh, A., Kral, T. and Solomon, S. (1989). The immune-hypothalamic-pituitary-adrenal axis. *Endocrine Reviews*, **10**, 92–112.
Banks, W. A., Kastin, A. J. and Durham, D. A. (1989). Bidirectional transport of interleukin-1 alpha across the blood-brain barrier. *Brain Research Bulletin*, **23**, 433–437.
Bazer, F. W. and Johnson, H. M. (1989). Actions of lymphokines and cytokines on reproductive tissue. *Progress in NeuroEndocrinImmunology*, **2**, 50–54.
Bellinger, D. L., Lorton, D., Romano, T. D., Olschowka, J. A., Felten, S. Y. and Felten, D. L. (1990). Neuropeptide innervation of lymphoid organs. *Annals of the New York Academy of Sciences*, **594**, 17–33.
Berczi, I. (1986a). The influence of pituitary-adrenal axis on the immune system. In I. Berczi (ed.), *Pituitary Function and Immunity*, pp. 49–132. Boca Raton, FL: CRC Press.
Berczi, I. (1986b). Gonadotropins and sex hormones. In I. Berczi (ed.) *Pituitary Function and Immunity*, pp. 185–211. Boca Raton, FL: CRC Press.
Berczi, I. (1986c). The pituitary–thyroid axis. In I. Berczi (ed.), *Pituitary Function and Immunity*, pp. 213–219. Boca Raton, FL: CRC Press.
Berczi, I. and Nagy, E. (1987). The effect of prolactin and growth hormone on hemolymphopoietic tissue and immune function. In I. Berczi and K. Kovacs (eds.) *Hormones and Immunity*, pp. 145–171. Lancaster, PA: MTP Press.
Bernton, E. W., Bryant, H. U. and Holaday, J. W. (1991). Prolactin and immune function. In R. Ader, D. L. Felten, and N. Cohen (eds.) *Psychoneuroimmunology*, 2nd edn, pp. 403–428. New York: Academic Press.
Besedovsky, H. O., Del Rey, A. E. and Sorkin, E. (1985). Immune-neuroendocrine interactions. *Journal of Immunology*, **135**, 750s–754s.
Besedovsky, H., Del Rey, A., Sorkin, E. and Dinarello, C. A. (1986). Immunoregulatory feedback between interleukin-1 and glucocorticoid hormones. *Science*, **233**, 652–654.
Beutler, B. (1990). The tumor necrosis factors: cachectin and lymphotoxin. *Hospital Practice*, **25** (2), 45–56.
Blalock, J. E. (1985). Proopiomelanocortin-derived peptides in the immune system. *Clinical Endocrinology*, **22**, 823–827.
Blalock, J. E. (1989). A molecular basis for bidirectional communication

between the immune and neuroendocrine systems. *Physiological Reviews*, **69**, 1–32.

Blalock, J. E., Bost, K. L. and Smith, E. M. (1985). Neuroendocrine peptide hormones and their receptors in the immune system. *Journal of Neuroimmunology*, **10**, 31–40.

Bost, K. L. (1988). Hormone and neuropeptide receptors on mononuclear leukocytes. *Progress in Allergy*, **43**, 68–83.

Breder, C. D., Dinarello, C. A. and Saper, C. B. (1988). Interleukin-1 immunoreactive innervation of the human hypothalamus. *Science*, **240**, 321–324.

Brooks, W. H., Cross, R. J., Roszman, T. L. and Markesbery, W. R. (1982). Neuroimmunomodulation: neural anatomical basis for impairment and facilitation. *Annals of Neurology*, **12**, 56–61.

Bulloch, K. (1987). The innervation of immune system tissues and organs. In C. W. Cotman *et al.* (eds.) *The Neuro-Immune-Endocrine Connection*, pp. 33–47. New York: Raven Press.

Bulloch, K. and Pomerantz, W. (1984). Autonomic nervous system innervation of thymic-related lymphoid tissue in wildtype and nude mice. *Journal of Comparative Neurology*, **228**, 57–68.

Busbridge, N. J. and Grossman, A. B. (1991). Stress and the single cytokine: interleukin modulation of the pituitary-adrenal axis. *Molecular and Cellular Endocrinology*, **82**, C209–C214.

Calabrese, J. R., Kling, M. A. and Gold, P. W. (1987). Alterations in immunocompetence during stress, bereavement, and depression: focus on neuroendocrine regulation. *American Journal of Psychiatry*, **144**, 1123–1134.

Carr, D. J. J. and Blalock, J. E. (1989). From the neuroendocrinology of lymphocytes toward a molecular basis of the network theory. *Hormone Research*, **31**, 76–80.

Carr, D. J. J. and Blalock, J. E. (1991). Neuropeptide hormones and receptors common to the immune and neuroendocrine systems: bidirectional pathway of intersystem communication. In R. Ader, D. L. Felten, and N. Cohen (eds.) *Psychoneuroimmunology*, 2nd edn, pp. 573–588. New York: Academic Press.

Coleman, R. M., Lombard, M. F., Sicard, R. E. and Rencricca, N. J. (1989). *Fundamental Immunology*. Dubuque, IO: Wm. C. Brown.

Cove-Smith, J. R., Kabler, P., Pownall, R. and Knapp, M. S. (1978). Circadian variation in an immune response in man. *British Medical Journal*, **2**, 253–254.

Cross, R. J., Markesbery, W. R., Brooks, W. H. and Roszman, T. L. (1984). Hypothalamic-immune interactions: neuromodulation of natural killer activity by lesioning of the anterior hypothalamus. *Immunology*, **51**, 399–405.

Cross, R. J. and Roszman, T. L. (1989). Neuroendocrine modulation of immune function: the role of prolactin. *Progress in NeuroEndocrinImmunology*, **2**, 17–21.

Daruna, J. H. and Morgan, J. E. (1990). Psychosocial effects on immune function: Neuroendocrine pathways. *Psychosomatics*, **31**, 4–12.

Denicoff, K. D., Rubinow, D. R., Papa, M. Z., Simpson, C., Seipp, C. A.,Lotze, M. T., Chang, A. E., Rosenstein, D. and Rosenberg, S. A. (1987). The neuropsychiatric effects of treatment with interleukin-2 and lymphokine-activated killer cells. *Annals of Internal Medicine*, **107**, 293–300.

Deschaux, P. and Rouabhia, M. (1987). The thymus: key organ between endocrinologic and immunologic systems. *Annals of the New York Academy of Sciences*, **496**, 49–55.

Dinarello, C. A. and Mier, J. W. (1987). Lymphokines. *New England Journal of Medicine*, **317**, 940–945.

Donahue, R. E., Yang, Y.-C. and Clark, S. C. (1990). Human P40 T-cell growth factor (interleukin-9) supports erythroid colony formation. *Blood*, **75**, 2271–2275.

Dunn, A. J. (1989). Psychoneuroimmunology for the psychoneuroendocrinologist: A review of animal studies of nervous system–immune system interactions. *Psychoneuroendocrinology*, **14**, 251–274.

Dunn, A. J. (1990). Interleukin-1 as a stimulator of hormone secretion. *Progress in NeuroEndocrinImmunology*, **3**, 26–34.

Durum, S. K., Schmidt, J. A. and Oppenheim, J. J. (1985). Interleukin 1: an immunological perspective. *Annual Review of Immunology*, **3**, 263–287.

Farrar, W. L., Kilian, P. L., Ruff, M. R., Hill, J. M. and Pert, C. B. (1987). Visualization and characterization of interleukin 1 receptors in brain. *Journal of Immunology*, **139**, 459–463.

Farrar, W. L., Vinocour, M. and Hill, J. M. (1989). In situ hybridization histochemistry localization of interleukin-3 mRNA in mouse brain. *Blood*, **73**, 137–140.

Felten, D. L., Cohen, N., Ader, R., Felten, S. Y., Carlson, S. L. and Roszman, T. L. (1991). Central neural circuits involved in neural-immune interactions. In R. Ader, D. L. Felten and N. Cohen (eds.) *Psychoneuroimmunology*, 2nd edn, pp. 3–25. New York: Academic Press.

Felten, D. L. and Felten, S. Y. (1988). Sympathetic noradrenergic innervation of immune organs. *Brain, Behavior, and Immunity*, **2**, 293–300.

Felten, D. L., Felten, S. Y., Bellinger, D. L., Carlson, S. L., Ackerman, K. D., Madden, K. S., Olschowki, J. A. and Livnat, S. (1987). Noradrenergic sympathetic neural interactions with the immune system: structure and function. *Immunological Reviews*, **100**, 225–260.

Felten, D. L., Felten, S. Y., Carlson, S. L., Olschowka, J. A. and Livnat, S. (1985). Noradrenergic and peptidergic innervation of lymphoid tissue. *Journal of Immunology*, **135**, 755s–765s.

Galin, F. S., LeBoeuf, R. D. and Blalock, J. E. (1990). Characteristics of lymphocyte-derived proopiomelanocortin-related mRNA. *Annals of the New York Academy of Sciences*, **594**, 382–384.

Geenen, V., Legros, J.-J. and Franchimont, P. (1987). The thymus as a neuroendocrine organ. *Annals of the New York Academy of Sciences*, **496**, 56–66.

Geenen, V., Robert, F., Defresne, M.-P., Boniver, J., Legros, J.-J. and Franchimont, P. (1989). Neuroendocrinology of the thymus. *Hormone Research*, **31**, 81–84.

Goetzl, E. J., Turch, C. W. and Sreedharan, S. P. (1991). Production and recognition of neuropeptides by cells of the immune system. In R. Ader, D. L. Felten and N. Cohen (eds.) *Psychoneuroimmunology*, 2nd edn, pp. 263–282. New York: Academic Press.

Golub, E. S. and Green, D. R. (1991). *Immunology: A Synthesis*, 2nd edn. Sunderland, MA, Sinauer.

Gross, R. L. and Newberne, P. M. (1980). Role of nutrition in immunologic function. *Physiological Reviews*, **60**, 188–302.

Grossman, C. J. (1984). Regulation of the immune system by sex steroids. *Endocrine Reviews*, **5**, 435–455.

Grossman, C. J. (1985). Interactions between the gonadal steroids and the immune system. *Science*, **227**, 257–261.

Grossman, C. J. (1990). Are there underlying immune-neuroendocrine interactions responsible for immunological sexual dimorphism? *Progress in NeuroEndocrinImmunology*, **3**, 75–82.

Hadden, J. W. (1987). Neuroendocrine modulation of the thymus-dependent immune system: Agonists and mechanisms. *Annals of the New York Academy of Sciences*, **496**, 39–48.

Hadden, J. W., Hadden, E. M. and Coffey, R. G. (1991). First and 2nd messengers in the development and function of thymus-dependent lymphocytes. In R. Ader, D. L. Felten and N. Cohen (eds.) *Psychoneuroimmunology*, 2nd edn, pp. 529–560. New York: Academic Press.

Hall, N. R. and Goldstein, A. L. (1981). Neurotransmitters and the immune system. In R. Ader (ed.) *Psychoneuroimmunology*, pp. 521–543. New York: Academic Press.

Hall, N. R. and Goldstein, A. L. (1984). Endocrine regulation of host immunity. In R. L. Fenichel and M. A. Chirigos (eds.) *Immune Modulation Agents and their Mechanisms*, pp. 533–563. New York: Marcel Dekker.

Hall, N. R., McGillis, J. P., Spangelo, B. L. and Goldstein, A. L. (1985). Evidence that thymosins and other biologic response modifiers can function as neuroactive immunotransmitters. *Journal of Immunology*, **135**, 806s–811s.

Hall, N. R. S. and O'Grady, M. P. (1989). Regulation of pituitary peptides by the immune system: historical and current perspectives. *Progress in NeuroEndocrineImmunology*, **2**, 4–10.

Hall, N. R. S., O'Grady, M. P. and Farah, J. M, Jr (1991). Thymic hormones and immune function: Mediation via neuroendocrine circuits. In R. Ader, D. L. Felten, and N. Cohen (eds.) *Psychoneuroimunology*, 2nd edn, pp. 515–528. New York: Academic Press.

Hamblin, A. S. (1988). *Lymphokines*. Oxford: IRL Press.

Harbour, D. V. and Blalock, J. E. (1989). Lymphocytes and lymphocytic hormones in pregnancy. *Progress in NeuroEndocrinImmunology*, **2**, 55–63.

Hartung, H.-P. (1988). Activation of macrophages by neuropeptides. *Brain, Behavior, and Immunity*, **2**, 275–281.

Healy, D. L., Hodgen, G. D., Schulte, H. M., Chrousos, G. P., Loriaux, D. L., Hall, N. R. and Goldstein, A. L. (1983). The thymus-adrenal connection: thymosin has corticiotropin-releasing activity in primates. *Science*, **222**, 1353–1355.

Hsu, D.-H., Malefyt, R., de W., Fiorentino, D. F., Dang, M.-N., Vierira, P., De Vries, J., Spits, H., Mosmann, T. R. and Moore, K. W. (1990). Expression of interleukin-10 activity by Epstein–Barr virus protein BCRFI. *Science*, **250**, 830–831.

Hültner, L., Druez, C., Moeller, J., Uyttenhove, C., Schmitt, E., Rüde, E., Dormer, P. and van Snick, J. (1990). Mast cell growth-enhancing activity (MEA) is structurally related and functionally identical to the novel mouse T cell growth factor P40/TCGFIII (Interleukin 9). *European Journal of Immunology*, **20**, 1413–1416.

Jankovíc, B. D. (1989). The relationship between the immune system and the nervous system: old and and new strategies. In P. M. H. Mazumdar (ed.) *Immunology 1930–1980*, pp. 203–220. Toronto: Wall and Thompson.

Johnson, H. M. and Torres, B. A. (1988). Immunoregulatory properties of neuroendocrine peptide hormones. *Progress in Allergy*, **43**, 37–67.

Katsuura, G., Arimura, A., Koves, K. and Gottschall, P. E. (1990). Involvement of organum vasculosum of lamina terminalis and preoptic area in interleukin 1β-induced ACTH release. *American Journal of Physiology*, **258**, E163–E171.

Kelley, K. W. (1991). Growth hormone in immunobiology. In R. Ader, D. L. Felten, and N. Cohen (eds.) *Psychoneuroimmunology*, 2nd edn, pp. 377–402. New York: Academic Press.

Kelley, K. W., Brief, S., Westly, H. J., Novakofski, J., Bechtel, P. J., Simon, J. and Walker, E. R. (1987). Hormonal regulation of the age-associated decline in immune function. *Annals of the New York Academy of Sciences*, **496**, 91–97.

Kennedy, R. L. and Jones, T. H. (1991). Cytokines in endocrinology: their roles in health and in disease. *Journal of Endocrinology*, **129**, 167–178.

Kishimoto, T. (1989). The biology of interleukin-6. *Blood*, **74**, 1–10.

Knapp, M. S., Byrom, N. P., Pownall, R. and Mayor, R. (1980). Time of day of taking immunosuppressive agents after renal transplantation: a possible influence on graft survival. *British Medical Journal*, **281**, 1382–1385.

Kordon, C. and Bihoreau, C. (1989). Integrated communication between the nervous, endocrine and immune systems. *Hormone Research*, **31**, 100–104.

Korneva, E. A. (1987). Electrophysiological analysis of brain reactions to antigen. *Annals of the New York Academy of Sciences*, **496**, 318–337.

Krueger, J. M., Toth, L. A., Johannsen, L. and Opp, M. R. (1990). Infectious disease and sleep: Involvement of neuroendocrine-neuroimmune mechanisms. *International Journal of Neuroscience*, **51**, 359–362.

Lechan, R. M., Toni, R., Clark, B. D., Cannon, J. G., Shaw, A. R., Dinarello, C. A. and Reichlin, S. (1990). Immunoreactive interleukin-1β localization in the rat forebrain. *Brain Research*, **514**, 135–140.

Levi-Montalcini, R., Aloe, L. and Alleva, E. (1990). A role for nerve growth factor in nervous, endocrine and immune systems. *Progress in NeuroEndocrinImmunology*, **3**, 1–10.

Low, T. L. K. and Goldstein, A. L. (1984). Thymosin, peptidic moieties, and related agents. In R. L. Fenichel and M. A. Chirigos (eds.) *Immune Modulation Agents and their Mechanisms*, pp. 135–162. New York: Marcel Dekker.

MacLean, D. and Reichlin, S. (1981). Neuroendocrinology and the immune process. In R. Ader (ed.) *Psychoneuroimmunology*, pp. 475–520. New York: Academic Press.

Makinodan, T. and Kay, M. M. B. (1980). Age influence on the immune system. *Advances in Immunology*, **29**, 287–330.

Marchetti, B. (1989). Involvement of the thymus in reproduction. *Progress in NeuroEndocrinImmunology*, **2**, 64–69.

Marchetti, B., Morale, M. C., Guarcello, V., Cutuli, N., Raiti, F., Batticane,

N., Plaumbo, Jr, G., Farinella, J. and Scapagnini, U. (1990a). Cross-talk communication in the neuroendocrine-reproductive-immune axis. *Annals of the New York Academy of Sciences*, **594**, 309–325.

Marchetti, B., Morale, M. C. and Pelletier, G. (1990b). Sympathetic nervous system control of rat thymus gland maturation: autoradiographic localization of the β_2-adrenergic receptor in the thymus and presence of sexual dimorphism during ontogeny. *Progress in NeuroEndocrinImmunology*, **3**, 103–115.

Maslinski, W. (1989). Cholinergic receptors of lymphocytes. *Brain, Behavior and Immunity*, **3**, 1–14.

Mathur, S., Mathur, R. S., Goust, J. M., Williamson, H. O. and Fudenberg, H. H. (1979). Cyclic variations in white cell subpopulations in the human menstrual cycle: correlations with progesterone and estradiol. *Clinical Immunology and Immunopathology*, **13**, 246–253.

McCruden, A. B. and Stimson,W. H. (1991). Sex hormones and immune function. In R. Ader, D. L. Felten, and N. Cohen (eds.) *Psychoneuroimmunology*, 2nd edn, pp. 475–493. New York: Academic Press.

McGillis, J. P., Mitsuhashi, M. and Payan, D. G. (1991). Immunologic properties of substance P. In R. Ader, D. L. Felten, and N. Cohen (eds.) *Psychoneuroimmunology*, 2nd edn, pp. 209–223. New York: Academic Press.

Merrill, J. E. (1990). Interleukin-2 effects in the central nervous system. *Annals of the New York Academy of Sciences*, **594**, 188–199.

Metcalf, D. (1991). Control of granulocytes and macrophages: Molecular, cellular, and clinical aspects. *Science*, **254**, 529–533.

Mizel, S. B. (1989). The interleukins. *FASEB Journal*, **3**, 2379–2388.

Munck, A. and Guyre, P. M. (1991). Glucocorticoids and immune function. In R. Ader, D. L. Felten and N. Cohen, (eds.) *Psychoneuroimmunology*, 2nd edn, pp. 447–474. New York: Academic Press.

Munck, A., Naray-Fejes-Toth, A. and Guyre, P. M. (1986). Mechanisms of glucocorticoid actions on the immune system. In I. Berczi and K. Kovacs (eds.) *Hormones and Immunity*, pp. 20–37. Lancaster, PA: MTP Press.

Nance, D. M., Hopkins, D. A. and Bieger, D. (1987). Re-investigation of the innervation of the thymus gland in mice and rats. *Brain, Behavior and Immunity*, **1**, 134–147.

Nathanson, J. A. (1989). The blood-cerebrospinal fluid barrier as an immune surveillance system: Functions of the choroid plexus. *Progress in NeuroEndocrine Immunology*, **2**, 96–101.

Navarra, P., Tsagarakis, S., Faria, M. S., Rees, L. H., Besser, G. M. and Grossman, A. B. (1991). Interleukin-1 and -6 stimulate the release of corticotropin-releasing hormone-41 from the rat hypothalamus *in vitro* via the eicosanoid cyclooxygenase pathway. *Endocrinology*, **128**, 37–44.

Neveu, P. J. (1988). Cerebral neocortex modulation of immune functions. *Life Sciences*, **42**, 1917–1923.

Nistico, G. and De Sarro, G. (1991). Is interleukin 2 a neuromodulator in the brain? *Trends in Neurosciences*, **14**, 146–150.

Nordlind, K., Mutt, V. and Sundström, E. (1988). Effect of neuropeptides and monoamines on lymphocyte activation. *Brain, Behavior and Immunity*, **2**, 282–292.

Norman, A. W. and Litwack, G. (1987). *Hormones*. New York: Academic Press.

Oates, K. K. and Goldstein, A. L. (1984). Thymosins: hormones of the thymus gland. *Trends in Pharmacological Sciences*, **5**, 347–352.

Ottaway, C. A. (1991). Vasoactive intestinal peptide and immune function. In R. Ader, D. L. Felten, and N. Cohen (eds.) *Psychoneuroimmunology*, 2nd edn, pp. 225–267. New York: Academic Press.

Payan, D. G. (1989). Substance P: A modulator of neuroendocrine-immune function. *Hospital Practice*, **24** (2), 67–80.

Payan, D. G. and Goetzl, E. J. (1985). Modulation of lymphocyte function by sensory neuropeptides. *Journal of Immunology*, **135**, 783s–786s.

Payan, D. G., McGillis, J. P., Renold, F. K., Mitsuhashi, M. and Goetzl, E. J. (1986). The immunomodulating properties of neuropeptides. In I. Berczi and K. Kovacs (eds.) *Hormones and Immunity*, pp. 203–213. Lancaster, PA: MTP Press.

Piantelli, M., Maggiano, N., Larocca, L. J., Ricci, R., Ranelletti, F. O., Lauriola, L. and Capelli, A. (1990). Neuropeptide-immunoreactive cells in human thymus. *Brain, Behavior and Immunity*, **4**, 189–197.

Pierpaoli, W. and Sorkin, E. (1972). Hormones, thymus and lymphocyte functions. *Experientia*, **28**, 1385–1389.

Plata-Salaman, C. R. (1989). Immunomodulators and feeding regulation: A humoral link between the immune and nervous systems. *Brain, Behavior and Immunity*, **3**, 193–213.

Plata-Salaman, C. R. (1991). Immunoregulators in the nervous system. *Neuroscience and Biobehavioral Reviews*, **15**, 185–215.

Pung, O. J., Tucker, A. N., Vore, S. J. and Luster, M. I. (1985). Influence of estrogen on host resistance: increased susceptibility of mice to *Listeria monocytogenes* correlates with depressed production of interleukin 2. *Infection and Immunology*, **50**, 91–96.

Rebar, R. W., Miyake, A., Low, T. L. K. and Goldstein, A. L. (1981). Thymosin stimulates secretion of luteinizing hormone-releasing factor. *Science*, **214**, 671–690.

Ricciardi-Castagnoli, P., Pirami, L., Righi, M. *et al.* (1990). Cellular sources and effects of tumor necrosis factor-α on pituitary cells and in the central nervous system. *Annals of the New York Academy of Sciences*, **594**, 156–168.

Roszman, T. L. and Brooks, W. H. (1985). Neural modulation of immune function. *Journal of Neuroimmunology*, **10**, 59–69.

Roszman, T. L. and Brooks, W. H. (1988). Signaling pathways of the neuroendocrine-immune network. *Progress in Allergy*, **43**, 140–159.

Roszman, T. L. and Carlson, S. C. (1991). Neurotransmitters and molecular signalling in the immune response. In R. Ader, D. L. Felten, and N. Cohen (eds.) *Psychoneuroimmunology*, 2nd edn, pp. 311–335. New York: Academic Press.

Rothwell, N. J. (1991a). The endocrine significance of cytokines. *Journal of Endocrinology*, **128**, 171–173.

Rothwell, N. J. (1991b). Functions and mechanisms of interleukin 1 in the brain. *Trends in Pharmacological Sciences*, **12**, 430–436.

Royer, H. D. and Reinherz, E. L. (1987). T lymphocytes: ontogeny, function, and relevance to clinical disorders. *New England Journal of Medicine*, **317**, 1136–1142.

Salvin, S. B., Rabin, B. S. and Neta, R. (1990). Evaluation of immunologic assays to determine the effects of differential housing on immune reactivity. *Brain, Behavior and Immunity*, **4**, 180–188.

Scarborough, D. E. (1990). Cytokine modulation of pituitary hormone secretion. *Annals of the New York Academy of Sciences*, **594**, 169–187.

Schneider, D., Cohn, M. and Bulloch, K. (1987). Overview of the immune system. In C. W. Cotman *et al.* (eds.) *The Neuro-Immune-Endocrine Connection*, pp. 1–14. New York: Raven Press.

Sibinga, N. E. S. and Goldstein, A. (1988). Opioid peptides and opioid receptors in cells of the immune system. *Annual Review of Immunology*, **6**, 219–249.

Singh, V. K., Warren, R. P., White, E. D. and Leu, S. J. C. (1990). Corticotropin-releasing factor-induced stimulation of immune functions. *Annals of the New York Academy of Sciences*, **594**, 416–419.

Smith, E. M. and Blalock, J. E. (1986). A complete regulatory loop between the immune and neuroendocrine systems operates through common signal molecules (hormones) and receptors. In N. P. Plotnikoff, R. E. Faith, A. J. Murgo, and R. A. Good (eds.) *Enkephalins and endorphins. Stress and the Immune System*, pp. 119–127. New York: Plenum Press.

Spangelo, B. L., Judd, A. M., MacLeod, R. M., Goodman, D. W. and Isakson, P. C. (1990). Endotoxin-induced release of interleukin-6 from rat medial basal hypothalami. *Endocrinology*, **127**, 1779–1785.

Spangelo, B. L. and MacLeod, R. M. (1990). The role of immunpeptides in the regulation of anterior pituitary hormone release. *Trends in Endocrinology and Metabolism*, **1**, 408–412.

Su, T.-P., London, E. D. and Jaffe, J. H. (1988). Steroid binding at σ receptors suggests a link between endocrine, nervous, and immune system. *Science*, **240**, 219–221.

Teschemacher, H., Koch, G., Scheffler, H., Hildebrand, A. and Brantl, V. (1990). Opioid peptides: immunological significance? *Annals of the New York Academy of Sciences*, **594**, 66–77.

Thorpe, L. W., Jerrells, T. R. and Perez-Polo, J. R. (1990). Mechanisms of lymphocyte activation by nerve growth factor. *Annals of the New York Academy of Sciences*, **594**, 78–84.

Tonegawa, S. (1985). The molecules of the immune system. *Scientific American*, **253**(4), 104–112.

Trainin, N., Pecht, M. and Handzel, Z. T. (1983). Thymic hormones: inducers and regulators of the T-cell system. *Immunology Today*, **4**, 16–21.

Weigent, D. A. and Blalock, E. (1987). Interactions between the neuroendocrine and immune systems: Common hormones and receptors. *Immunological Reviews*, **100**, 79–108.

Weigent, D. A. and Blalock, J. E. (1990). Growth hormone and the immune system. *Progress in NeuroEndocrinImmunology*, **3**, 231–241.

Weigent, D. A., Carr, D. J. J. and Blalock, J. E. (1990). Bidirectional communication between the neuroendocrine and immune systems. Common hormones and hormone receptors. *Annals of the New York Academy of Sciences*, **579**, 17–27.

Weigle, W. O. (1989). Effects of aging on the immune system. *Hospital Practice*, **24**, 112–119.

Whitacre, C. C. (1990). Immunology: a state of the art lecture. *Annals of the New York Academy of Sciences*, **594**, 1–16.

14

Methods for the study of behavioral neuroendocrinology

Behavioral neuroendocrinology involves the study of the interactive effects of the steroid and peptide hormones, neuropeptides, cytokines and neurotransmitters on behavior. Previous chapters have mentioned the role of the hypothalamic nucleii in behavior (Section 4.1), the behavioral effects of neurotransmitter agonists and antagonists (Section 5.8), the neuroendocrine correlates of psychiatric disorders (Section 6.8), the behavioral functions of the steroid hormones (Section 9.9), the cognitive and behavioral effects of neuropeptides (Section 12.5), and the effects of the cytokines on the brain and behavior (Section 13.5). This chapter discusses the behavioral methods used for the study of neuroendocrinology, the neural and genetic mechanisms mediating the effects of hormones on behavior, and some of the special problems involved in conducting behavioral neuroendocrinology research.

Neuroendocrine research involves a number of specific methods such as radioimmunoassays (Chapter 8), autoradiography (Chapter 9), receptor binding assays (Chapter 10) and immunohistochemical techniques (Chapters 11 and 13). The study of behavioral neuroendocrinology relies on specific behavioral methodologies or behavioral bioassays. As discussed in Section 8.1, a bioassay measures physiological changes in an animal or cell culture to determine the concentration or potency of a hormone in the circulation. Thus, the size of a cock's comb is a bioassay for testosterone level and the size and weight of the adrenal glands are bioassays for the level of ACTH. A behavioral bioassay measures behavioral changes to determine the concentration or potency of a hormone.

14.1 BEHAVIORAL BIOASSAYS

A behavioral bioassay requires precise qualitative (verbal) descriptions of the behaviors of interest and accurate quantitative (mathematical) measures of the latency, frequency and duration of these behaviors. Thus, the measurement of behavior involves two stages: the observation and description of units of behavior and the quantitative measurement of these behavior units. Before these procedures can begin, however, one must determine which behaviors to record.

14.1.1 WHAT BEHAVIOR IS TO BE RECORDED?

Behavior can be classified into two broad groups: ethological (unconditioned) responses and conditioned responses, as shown in Table 14.1. Ethological studies involve the observation and recording of 'natural' behaviors, such as grooming, vocalizations, sexual, parental and aggressive behavior in a laboratory or field setting (Kelley, 1989; Krsiak, 1991). Studies involving conditioned responses usually involve training and testing procedures in the laboratory (Adams, 1986). Special categories of behavioral tests have been developed to measure the development of behavior (Zbinden, 1981) and the display of abnormal behavior (Baumeister and Sevin, 1990).

14.1.2 OBSERVATION AND DESCRIPTION

Before any neuroendocrine manipulation is carried out, the normal sequence of behavior must be observed and described in terms of the behavior units involved. These units include identifiable postures, motor acts, vocalizations, etc. which are stereotyped in form, duration or orientation and can be easily recognized by the observer. The sequence of sexual behavior of the male rat, for example, can be divided into a number of behavioral units such as approach female, investigate (sniff female), mount female without intromission, mount with intromission, mount with ejaculation, and ultrasonic vocalization (Brown and McFarland, 1979). Using such behavior units, the sequence of male rat copulatory behavior can be described, as shown in Figure 14.1. Other unconditioned behaviors such as grooming, aggression and parental behavior can be described using similar qualitative methods.

14.1.3 QUANTITATIVE MEASUREMENT OF BEHAVIOR UNITS

To measure changes in behavior associated with neuroendocrine activity, each behavior unit must be quantified by measuring its latency, fre-

Table 14.1. *Some examples of behavioral measures used in neuroendocrine research*

Ethological (unconditioned) responses
Locomotion (walking, swimming)
Grooming
Exploratory behavior (open field test)
Feeding, food preferences
Drinking, taste preferences
'Emotional behavior' (defaecation, freezing, escape, avoidance)
Scent-marking
Vocalizations (audible and ultrasonic)
Social behavior (huddling, play, sexual, parental, aggressive, defensive)
Conditioned responses
Habituation/dishabituation
Conditioned suppression (conditioned emotional response)
Passive and active avoidance conditioning
Taste/odor aversion tests
Maze learning for food or water reward
Operant conditioning for food or water reward
Brain stimulation reward learning
Developmental parameters
Developmental reflexes (self-righting, auditory startle, negative geotaxis)
Spontaneous movements (crawling, standing, climbing, swimming)
Homing to nest
Vocalizations
Exploration
Learning
Abnormal behaviors
Hyperactivity
'Illness' related behavior (vomiting, lethargy)
Feeding disorders (anorexia, obesity)
Affective disorders (anxiety, depression)
Neuromuscular disorders (dyskinesias)
Social disorders (extreme aggression or passivity)
Cognitive disorders
Developmental delays

quency, or duration within a specific test session (see Figure 14.1). The accurate quantitative measurement of behavior units is a skill which requires the knowledge of experimental design, sampling procedures and recording methods, and usually requires the use of some apparatus such as check sheets, event recorders, videotapes, audiotapes, sound analyzers, or automatic recording devices such as photocells or microswitches. A number of computer programs are now available for recording behavioral observations. If behavioral bioassays are to be reliable and valid measures of hormonal concentration or potency, considerable effort must be made to ensure the accuracy of the behavioral measures used (see Lehner, 1979; Martin and Bateson,1986; Donat, 1991).

14.2 CORRELATIONAL STUDIES OF HORMONAL AND BEHAVIORAL CHANGES

Behavioral changes can be correlated with naturally occurring hormonal fluctuations or the hormonal changes associated with endocrine tumors

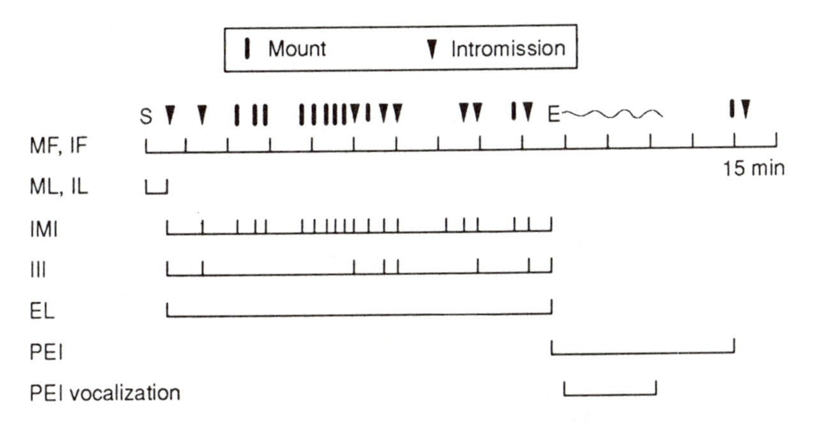

Figure 14.1. An example of the units of sexual behavior of a male rat and their temporal patterning. Behavioral units recorded: mount = mount female without intromission; intromission = mount female with intromission; ejaculation (E) = mount female with intromission and ejaculation; post-ejaculatory vocalization = 22-kHz ultrasonic vocalizations of the male. Quantitative measures recorded: mount latency (ML) = the time from the introduction of the female (S) until the first mount; intromission latency (IL) = the time from the introduction of the female until the first intromission; ejaculation latency (EL) = the time from the first mount to an ejaculation (this is called an ejaculatory series); mount frequency (MF) = the total number of mounts which occur in an ejaculatory series; intromission frequency (IF) = the number of mounts with intromission in each ejaculatory series, including the ejaculatory thrust; inter-mount-interval (IMI) = the time in seconds between each mount in an ejaculatory series; inter-intromission-interval (III) = the time between each intromission in an ejaculatory series; post-ejaculatory interval (PEI) = the time from the ejaculation to the first mount of the next ejaculatory series. There is a 22 kHz post-ejaculatory vocalization during the PEI which divides it into an absolute refractory period (time from ejaculation to the end of the song) and a relative refractory period (time from the end of the post-ejaculatory song to the first mount of the next ejaculatory series). (Redrawn from Brown and McFarland, 1979.)

or diseases (Beach, 1974, 1975). This type of study measures hormone levels and behavioral variables simultaneously and looks for a relationship between them, without any experimental manipulation of the hormones or the behavior.

14.2.1 NATURAL FLUCTUATIONS IN HORMONE LEVELS AND BEHAVIOR

Many behavioral changes occur in conjunction with normal fluctuations in the rate of hormone secretion. In female mammals, for example, changes in the circulating levels of estrogen and progesterone occur at puberty, at menopause, during the reproductive cycle and during the mating season in seasonally breeding animals. Behavioral changes can be correlated with these changes in gonadal hormone levels. Figure 14.2 shows the fluctuations in aggressive behavior in female mice which can be correlated with changes in estrogen and progesterone levels over their estrous cycles.

Hormone levels can also be correlated with daily or seasonal rhythms of daylight or temperature. For example, the timing of gonadal hormone release and reproductive behavior in seasonally breeding animals, such as ducks, can be correlated with seasonal changes in the light–dark cycle as shown in Figure 14.3.

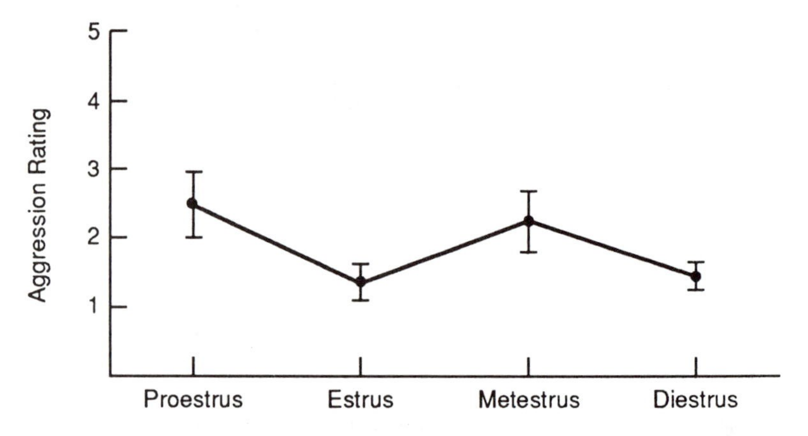

Figure 14.2. Mean aggression ratings (±SEM) of female house mice, showing the fluctuations which occur over the four phases of the estrous cycle. (Redrawn from Hyde and Sawyer, 1977.)

14.2.2 ENDOCRINE AND BEHAVIORAL ABNORMALITIES

Diseases or malfunctions of the endocrine system, occurring in adulthood or during development, can lead to correlated changes in behavior. Patients with Cushing's syndrome, for example, have excessive glucocorticoid secretion and a number of cognitive and behavioral symptoms, including depression, irritibility, and loss of recent memory. These symptoms improve when glucocorticoid secretion is normalized (Martin and Reichlin, 1987). Abnormal hormone secretion during prenatal development leads to a number of behavioral problems later in development. For example, disruption of gonadal hormone secretion during sexual differentiation may influence later social and sexual behavior (Gorski, 1991) and lack of thyroid hormones during early development leads to stunted bodily growth and mental retardation (Dussault and Ruel, 1987). Depression, anorexia and schizophrenia can also be correlated with alterations in the neuroendocrine system (Martin and Reichlin, 1987).

14.2.3 THE PROBLEM OF HORMONE SAMPLING DURING BEHAVIORAL STUDIES

In order to correlate hormonal and behavioral changes, it is necessary to measure hormone levels at the same time as behavior is observed. This can be done by direct measures of the hormones in the circulation or by indirect measures.

Direct (invasive) measures

Direct measures of hormone levels in the circulation depend on taking samples of blood or cerebral spinal fluid and then using bioassays or radioimmunoassays to measure the levels of the hormones of interest in these bodily fluids (see Section 8.1). While this procedure provides an accurate measure of the circulating level of the hormone, there is the disadvantage that such direct sampling methods disrupt the behavior being observed. For example, to correlate testosterone level with sexual behavior in male rats, one might take blood samples every 10 min. But

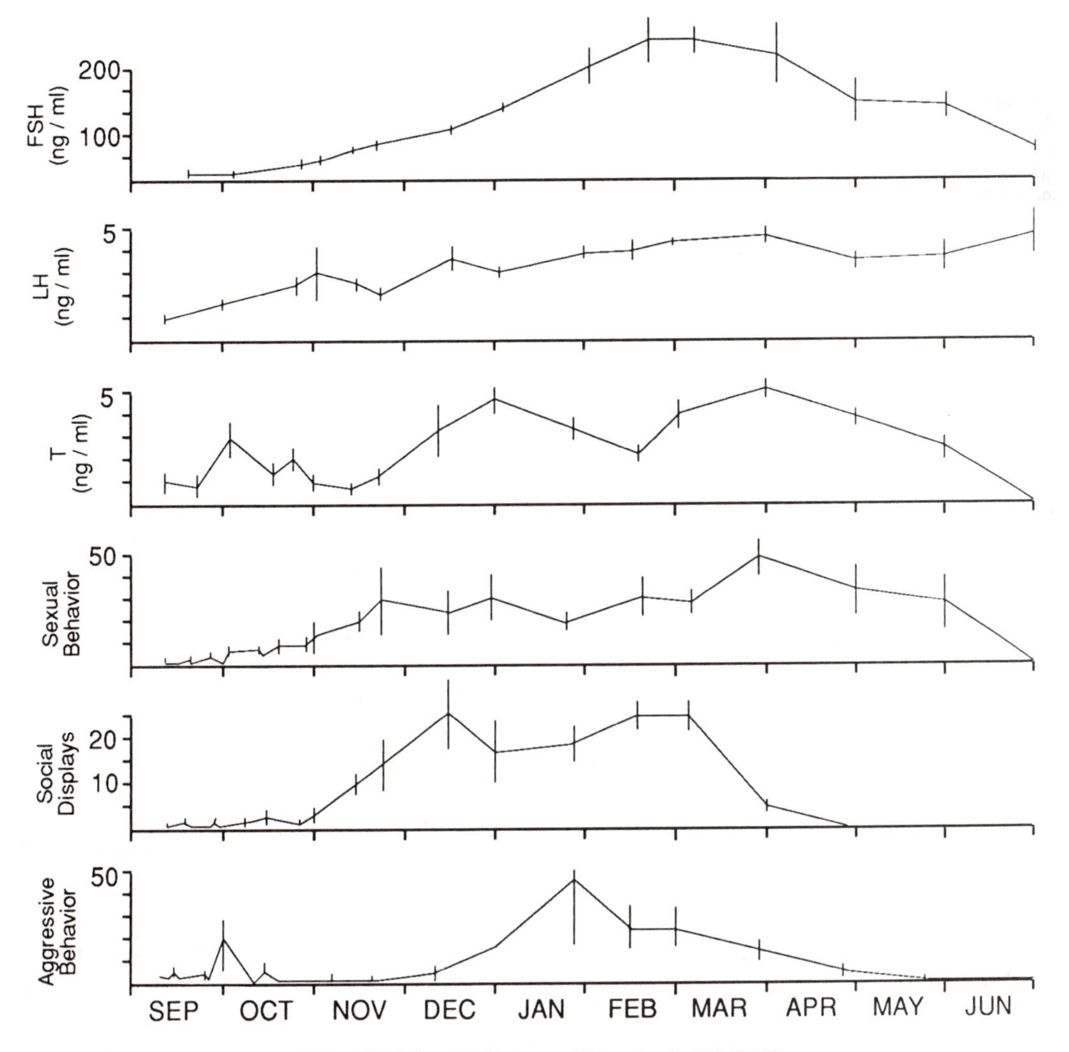

Figure 14.3. Annual variations in plasma FSH, LH and testosterone (T) levels correlated with annual changes in sexual behavior, social displays and aggressive behavior directed toward other males in a group of five male domestic ducks. Data shown are means ± standard errors. Sexual behavior is highly correlated with testosterone levels, while social displays are more highly correlated with FSH levels. Maximum testosterone levels correlate with the height of the breeding period during March and April. (Redrawn from Balthazart and Hendrick, 1976.)

taking these blood samples using a syringe or tail catheter involves restraining the animal and such restraint is stressful for the rat and disrupts the behavior being observed. The stress hormones released during the restraint period may confound the effects of testosterone (see Figure 14.11). To allow stress-free blood sampling during behavioral studies, many techniques have been developed for implanting chronic cannulae (Steffens, 1969; Singh and Avery, 1975). Figure 14.4 shows a simple intravenous cannula which can be used for taking blood samples from freely moving rats without disrupting them during behavioral studies (Terkel and Urbach, 1974).

Figure 14.4. A rat with a chronic intravenous cannula inserted. The drawing shows the path of the catheter to the right atrium, the plastic housing for the cannula, and a tube inserted into the cannula for blood withdrawal. (Redrawn from Terkel and Urbach, 1974.)

Indirect (non-invasive) measures

Because it is not always possible to collect blood samples by direct means during behavioral studies, a number of indirect methods of determining hormone levels are used. These include the analysis of hormones or their metabolites in the urine, saliva and feces (see Section 8.1) and the use of bioassays. Because many hormones or their metabolites are eliminated in the urine (Table 7.3), urine samples can be used as an assay for the levels of certain hormones. Methods for the analysis of urinary hormones or their metabolites are well established (Zarrow *et al.*, 1964) and provide an accurate estimate of the level of the hormone in the blood. Urinary hormone analysis is a practical, non-stressful method for monitoring hormone levels in children, adults, wild animals or animals in zoos (Andelman *et al.*, 1985). Steroid hormone levels can be analyzed from saliva samples, as discussed in Section 8.1, and the saliva can be collected using cotton swabs or chewing gum (Dabbs, 1991).

Biological changes associated with altered hormonal levels may be used as hormone bioassays for behavioral studies. For example, the vaginal secretions of female rodents are used as a bioassay for estrogen and progesterone levels (Everett, 1989). For behavioral studies, the phase of the estrous cycle of female rodents can be determined by taking vaginal smears, as shown in Figure 14.5, rather than blood samples (Zarrow *et al.*, 1964). Changes in behavior can then be correlated with vaginal cell type, as shown for aggressive behavior in Figure 14.2.

14.3 EXPERIMENTAL STUDIES. I. BEHAVIORAL RESPONSES TO NEUROENDOCRINE MANIPULATION

Correlational studies provide a description of the relationship between behavioral and neuroendocrine variables, and, if the correlation is high enough, allow prediction of the behavioral changes associated with changes in levels of particular hormones. It is not, however, possible to infer that the hormonal changes cause the correlated change in behavior. In order to show that a change in hormone level causes a change in behavior, or vice versa, systematic manipulation of the hormone or the behavior in a controlled experiment is necessary (Leshner, 1978). The

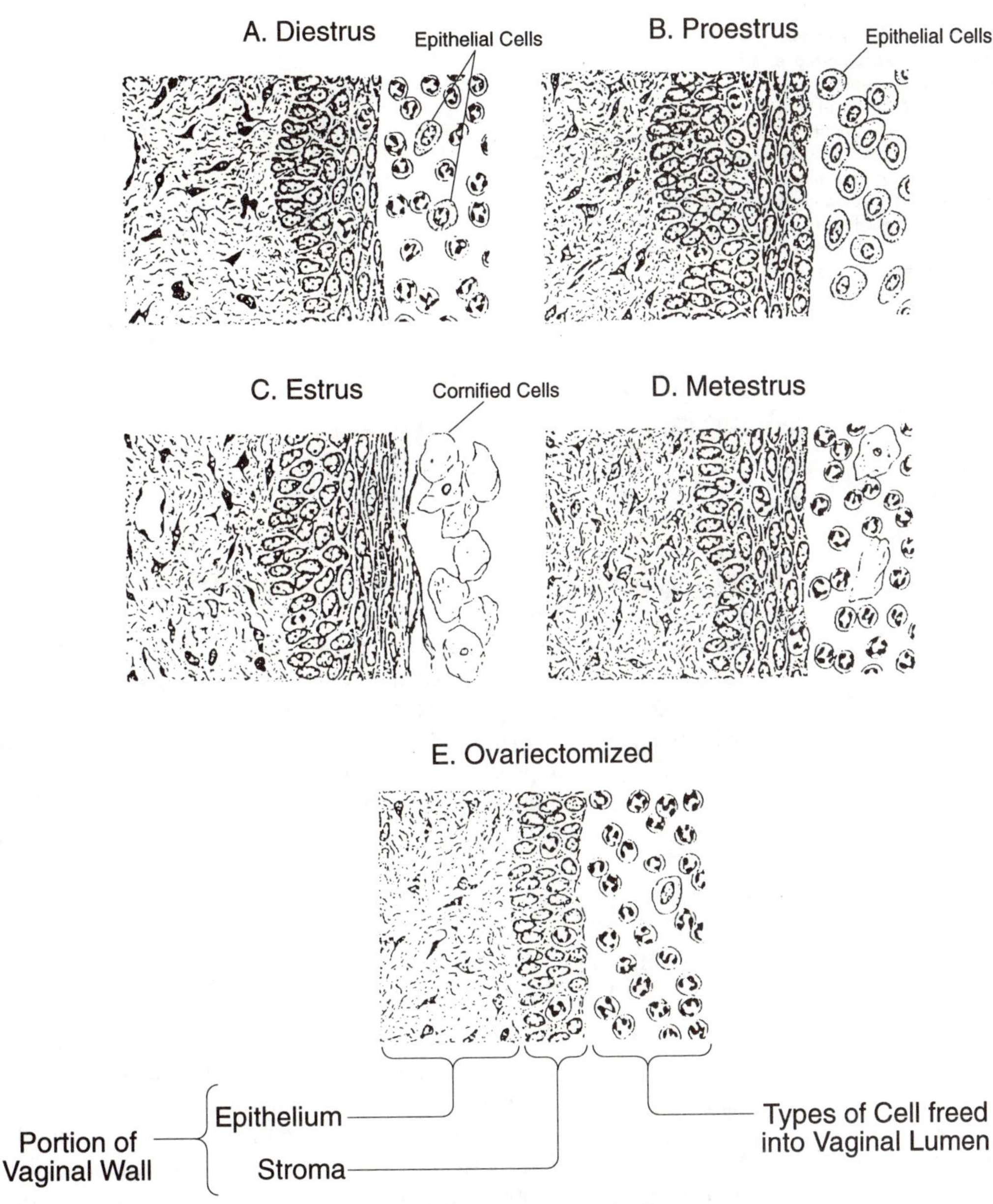

Figure 14.5. Sections through the vaginal wall of the rat during different stages of the estrous cycle, showing the corresponding types of cells which appear in smears obtained from the vaginal lumen. During diestrus (A) the vaginal mucosa is thin and there are only a few epithelial cells in the vaginal smear. In proestrus (B) the vaginal mucosa becomes dry and there are epithelial (early proestrus) and cornified cells (late proestrus) in the vaginal smear. At estrus (C) the vaginal smear contains masses of cornified cells. At metestrus (D) the vaginal mucosa becomes moist and the vaginal smear shows mixed cornified and epithelial cells.(E) An adult female which had been ovariectomized for six months. (From Turner and Bagnara, 1976.)

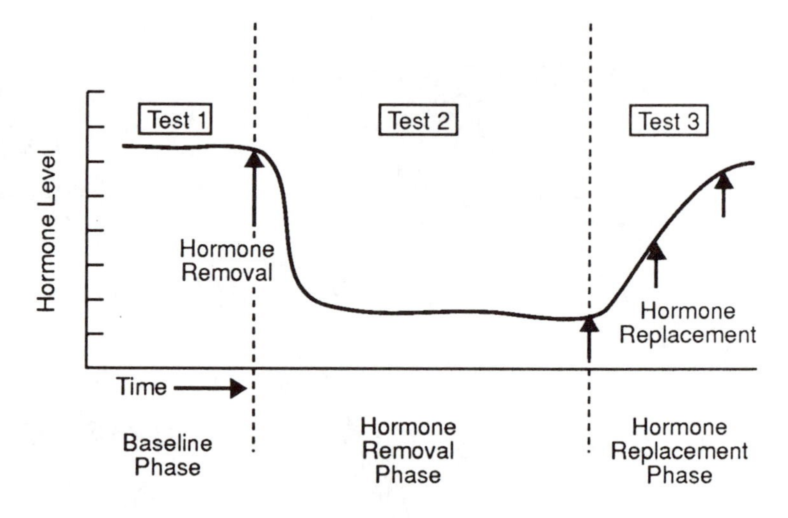

Figure 14.6. The three phases of the standard hormone removal and replacement experiment. Behavioral tests (bioassays) are conducted one or more times during the baseline, hormone removal and hormone replacement phases. A series of hormone replacement injections (arrows) is often given during the replacement phase.

level of a hormone can be manipulated as the independent variable in an experiment, and behavior measured as the dependent variable during each phase of the hormonal manipulation, or behavior can be manipulated as the independent variable, and hormonal responses measured as the dependent variable. While the majority of the experiments discussed in this chapter concern the effects of the steroid hormones on mammalian behavior, the same problems are faced in studying the effects of neuropeptides and cytokines on behavior (Kelley, 1989) and in studying behavioral neuroendocrinology in other species, such as birds (Balthazart, 1983).

14.3.1 THE STANDARD HORMONE REMOVAL AND REPLACEMENT EXPERIMENT

Three criteria must be met before the hormonal control of a specific behavior can be demonstrated. First, removal of the hormone or its source (e.g. the gland secreting that hormone) should modify the behavior. Second, replacement of the hormone by grafting a gland from another animal or by hormone replacement therapy should restore the behavior to its original level. Third, variations in hormone concentration should result in correlated changes in the frequency or intensity of the behavior. Also, the hormone replacement therapy must activate the behavior at dose levels which are physiologically meaningful, not at abnormally high levels (Balthazart, 1983). Experiments involving the systematic manipulation of hormone levels, therefore, have three phases: the baseline phase, the hormone removal phase, and the hormone replacement phase. There are behavioral tests in each phase, as shown in Figure 14.6.

An example of such an experiment is shown in Figure 14.7. Male gerbils use their ventral sebaceous gland to scent-mark their environment. To investigate the relationship between scent-marking frequency and testosterone levels, males were tested in the baseline (pre-castration)

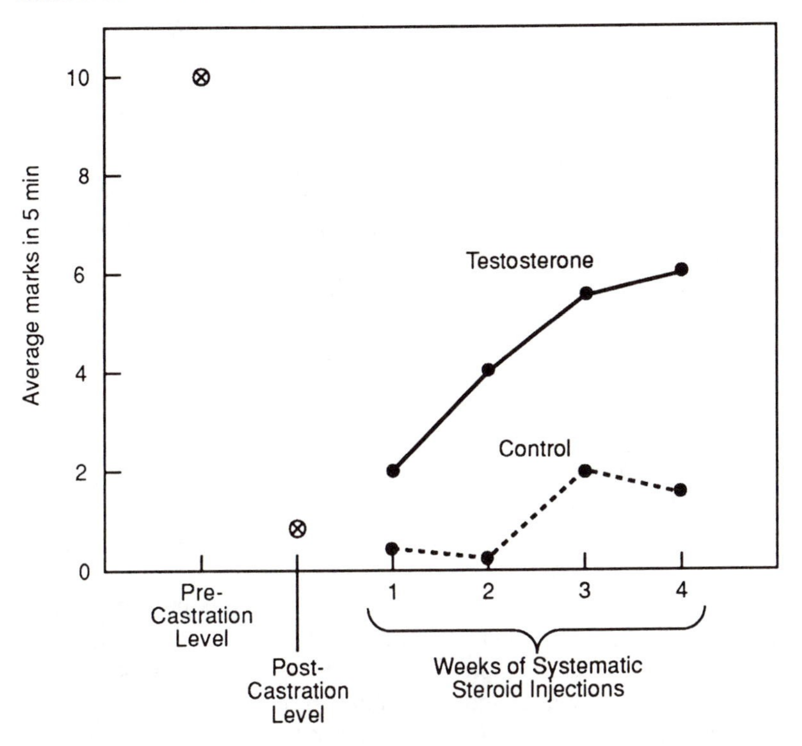

Figure 14.7. The effects of castration and injections of testosterone or the vehicle only (control) on the frequency of ventral gland scent-marking in a 5 min test period by adult male gerbils. (Redrawn from Yahr and Thiessen, 1972.)

period and were tested again after removal of the testes (post-castration). Testosterone replacement injections were then given for 4 weeks and scent-marking was recorded each week. Castration reduced the frequency of scent-marking in a 5 min period almost to zero and testosterone replacement increased scent-marking frequency. The control group of castrated males given only the vehicle and no testosterone injections did not show a significant increase in scent-marking frequency (Yahr and Thiessen, 1972). Conducting this type of experiment requires careful consideration of the methods to be used for hormone removal and replacement.

14.3.2 METHODS OF HORMONE REMOVAL

There are three general methods of hormone removal: surgical, pharmacological and immunological.

Surgical methods (gland removal or ablation)

The oldest type of experimental procedure in behavioral neuroendocrinology is to remove a gland (e.g. testes or thyroid) and observe physiological or behavioral changes in the subject. Thus, the testes of a rooster can be removed and the size of the comb measured (bioassay) or the amount of crowing, courtship, mating and aggressive behavior recorded (behavioral bioassay). If these are all found to decrease after castration, testes implants, testes extracts or purified hormone (testosterone) can be given in replacement therapy to see if this reverses the effect of the gland removal. Such an experiment was first conducted by Arnold A. Berthold in 1849 (Beach, 1981).

Surgical removal is possible for many endocrine glands (testes, ovaries, adrenal glands, thyroid, etc.), but surgical procedures are complex and, since many glands produce two or more hormones, elimination of the other hormones produced by the gland may confound the experiment. Ablation of the adrenal gland to eliminate glucocorticoids, for example, will also remove the mineralocorticoids, and, since it is difficult to ablate the adrenal cortex without also removing the adrenal medulla, the source of adrenaline and noradrenaline is also removed. Thus, the controlled study of the effects of individual hormones on behavior may require more selective mechanisms than gland ablation for the removal of hormones from the circulation.

Pharmacological methods

It is possible to use drugs to prevent hormone synthesis and release from endocrine glands or to act as hormone antagonists at target cell receptors (Landau, 1986). Cyproterone acetate, for example, is an androgen antagonist which blocks androgen receptors and thus prevents androgens from stimulating their target cells. Likewise, clomiphene and tamoxifen are estrogen receptor antagonists (Martin, 1985). Bromocriptine, a dopamine receptor agonist, inhibits prolactin secretion from the pituitary. Many drugs which alter neurotransmitter levels in the brain also alter the release of hormones because they alter the neural regulation of the hypothalamic-pituitary system as described in Chapter 6. Likewise, the neuropeptides and cytokines modulate the neuroendocrine system (Chapters 12 and 13), so drugs which alter these chemical messengers can also alter hormone release and behavioral responses.

Immunological methods

It is possible to use immunological methods to develop antibodies for hormones. Injection of these antibodies deactivates the hormone and prevents it from activating its receptors. Antibodies for estrogen, androgens, thyroid hormones, pituitary and hypothalamic hormones, and most other peptide hormones are available (see Martin, 1985 and Hadley, 1992).

14.3.3 METHODS OF HORMONE REPLACEMENT

Hormone replacement can be done by replacing an entire gland, by injecting purified hormones or by injecting synthetic hormones.

Gland replacement

In studies where an entire gland is surgically ablated, that gland or one from another animal can be replaced and, as long as it can be reattached to the circulatory system, hormone release can be reinstated. Thus, testes or ovaries can be removed and replaced in the body cavity and continue to secrete their hormones. One mechanism for elevating prolactin levels is to implant anterior pituitary glands from donor animals under the kidney capsule of the recipient. There, free from the inhibitory action of hypothalamic prolactin inhibiting factors, these transplanted pituitaries release high levels of prolactin into the bloodstream.

Injection of purified hormones

Rather than replacing an entire gland, it is possible to extract and purify the hormones from the ablated gland and inject these hormones in solution. Many of these purified hormones are taken from the endocrine glands of cattle (bovine), sheep (ovine), pigs (porcine) or horses (equine). Thus, purified ovine prolactin, porcine growth hormone or pregnant mare's serum gonadotropin (equine) are available for medical or scientific use. Some hormones, however, are species specific and in these cases, hormones can not be used for replacement therapy in other species. Human patients requiring growth hormone injections, for example, must have human growth hormone, which is purified from human pituitary glands taken from organ donors. Ovine and bovine growth hormones have only about 60% identity with human growth hormone in their amino acid structure and thus are inactive in humans. Monoclonal antibodies and genetic engineering methods can now be used to create human GH and other hormones (see Baxter and Gertz, 1991).

Injection of synthetic hormones

Knowledge of the chemical structure of hormones has enabled the production of synthetic hormones to serve as hormone replacement therapy in animals and humans. A wide variety of synthetic hormones are now available, some of which mimic exactly the effects of the natural hormones (full agonists) and some of which are only partial agonists, having some, but not all of the effects of the natural hormone, due to differences in their chemical structure (Gilman *et al.*, 1985). Some of the synthetic steroid hormones used in behavioral studies include testosterone propionate, ethinyl estradiol, norethindrone (a synthetic progestin) and dexamethasone (a synthetic glucocorticoid agonist).

14.3.4 FACTORS TO CONSIDER WHEN GIVING HORMONE REPLACEMENT THERAPY

Although the term 'hormone replacement' appears quite simple, there are a number of factors to be considered before giving hormone replacement therapy. These include the chemical preparation of the hormone, the dose to be used, the vehicle in which the hormone is given, the route of administration, and the timing of the injections.

Chemical preparation

There are many synthetic preparations of each hormone (Gilman *et al.*, 1985) and these may have different effects on behavior. Birth control pills, for example, utilize a number of different synthetic estrogens and progestins, some of which may have unwanted side-effects. Different hormone preparations may be substituted to avoid these side-effects. Likewise, some synthetic androgens are aromatizable to estrogen while

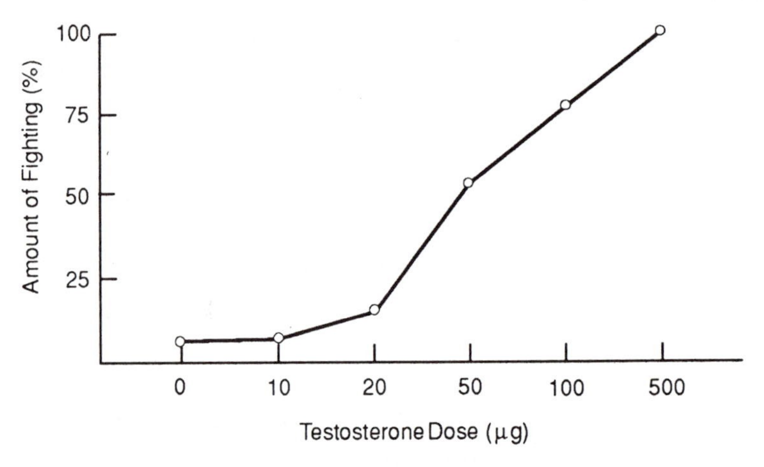

Figure 14.8. The percentage of castrated adult male mice fighting at each dose level of testosterone. (Redrawn from Edwards, 1969.)

others are not. These can be used to determine whether the behavioral effects of androgens are due to their action at androgen or estrogen receptors (see Chapter 9).

Dose

Once the type of hormone to be used is determined, the dose to be injected must be calculated. Since different doses of a hormone may have different effects on behavior, studies often give varying doses to different groups of subjects. Figure 14.8, for example, shows that fighting behavior is almost eliminated in castrated male mice and that an injection of 10 μg testosterone does not reinstate fighting. As the testosterone dose is increased, however, the percentage of males fighting increases. Many behaviors respond to hormone replacement in this dose-dependent fashion.

Vehicle

Hormones are administered in a vehicle such as distilled water, physiological saline, or oil. The selection of the vehicle depends on the hormone to be administered and the route of administration. Steroid hormones to be injected, for example, are prepared as suspensions in corn oil or peanut oil, while protein hormones and amino acids are dissolved in water or saline. The control group in the experiment is then injected with the vehicle alone without any hormone (see Figure 14.7).

Route of administration

Some hormones can be given orally, such as birth control pills, but others, such as insulin must be injected. Injections can be subcutaneous (s.c.), intramuscular (i.m.), or intraperitoneal (i.p.). Hormones can also be injected into the brain by intracerebral injections or into the cerebral spinal fluid (CSF) in the ventricles by intracerebroventricular injections. Direct injection of hormones into the brain or CSF requires special cannulation and microinjection techniques as shown in Figure 14.9. (see Myers, 1971; Singh and Avery, 1975; Baldwin *et al.*, 1979).

Hormones can also be implanted in pellets or controlled release

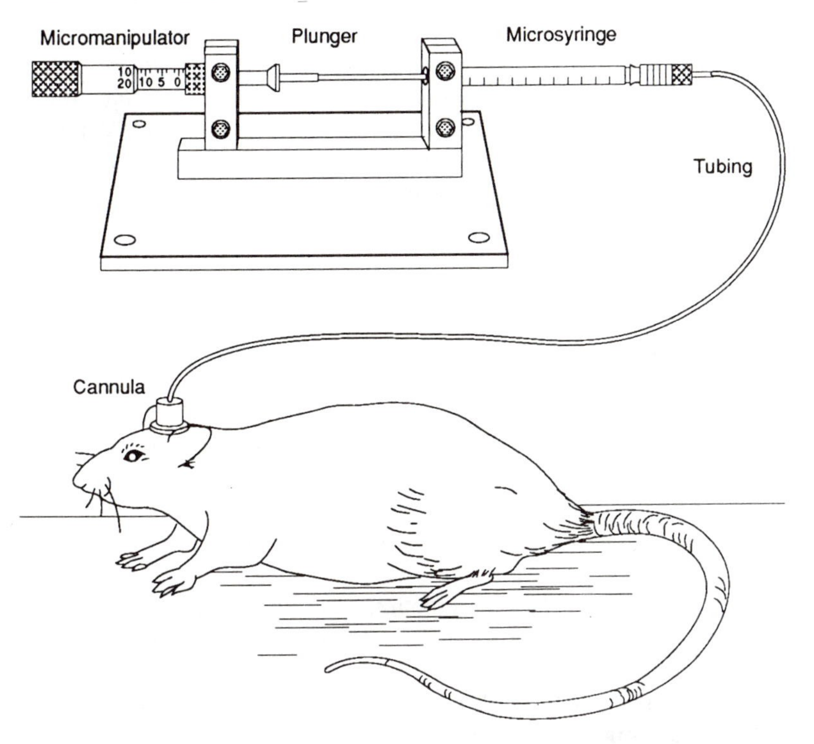

Figure 14.9. A microinjection unit being used to inject a hormone into the brain of a rat through a cannula implanted intracerebrally. (Redrawn from Singh and Avery, 1975.)

capsules made from silastic tubing. Because the silastic tubing is porous and allows the hormone to diffuse into the circulation at a constant rate, these capsules are often used for steroid hormone implants. Different lengths of tubing can be used for different hormone doses and these capsules may continue to secrete hormones at a constant rate for more than 2 months (Smith *et al.*, 1977). Pellets of hormone-impregnated beeswax can be implanted in specific brain areas to gradually release the hormone into the brain (Holman, 1980).

Timing of hormone replacement and behavioral testing

The timing of the hormone replacement therapy and the temporal patterning of the injections with respect to hormone removal and behavioral observations must also be determined. Hormone replacement can be given immediately after hormone removal, in which case there is little time for the circulating levels of hormones to decline before replacement is given. On the other hand, hormone replacement can be delayed for some days or weeks to allow baseline hormone levels to drop after gland removal.

When hormone levels are altered, there are compensatory changes in receptor numbers, with up- or down-regulation of receptors occurring, as discussed in Chapters 9 and 10. Once receptor numbers are down-regulated following gland removal, it may require one or more hormone injections before the receptors are up-regulated to their former baseline levels. Thus, some hormone pre-treatment may be necessary before behavioral changes occur in response to the hormone replacement therapy. Figure 14.7, for example, shows that the frequency of scent-marking by male gerbils in response to testosterone injection is low on the first week of replacement therapy and reaches an asymptote on the fourth week. How frequently the hormone injections are given depends on the active life of the hormone in the circulation. Some hormones can be injected once per day, others once every 2 days and others twice per day. Figure 14.10 shows the testosterone level in the blood of castrated male rats from half an hour to 24 hours after testosterone injection. The level of circulating testosterone reaches a peak 3 hours after the injection and then gradually declines.

Finally, the timing of the behavioral tests with respect to the hormone injections must be determined. Should tests be given immediately after the hormone injection or some hours later? While Figure 14.10 suggests that testing should be done 3 or more hours after testosterone injection, other hormones have very short half-lives and testing should be done sooner after the injection.

When hormone capsules are implanted, the level of hormone released may remain constant for up to 2 months (Smith *et al.*, 1977) and testing can be done for many days after the implant without repeated injections being necessary. While this has many advantages, there are some disadvantages in the use of sustained release capsules. Many hormones, such as LH-RH, are active only when released in pulses. Constant release of these hormones, as occurs with sustained release capsules, causes receptor down-regulation, thus the high constant levels of the hormone inhibit rather than stimulate their target cells. LH-RH pulses occur about once every 90 min (see Figure 7.4, p. 121) and injections this frequently are impractical. To solve this problem, osmotic mini-pumps can be implanted in an animal. These pumps can be 'programed' to release a specific dose of hormone every few minutes or hours, thus providing synthetic pulsatile hormone release.

14.4 EXPERIMENTAL STUDIES. II: NEUROENDOCRINE RESPONSES TO ENVIRONMENTAL, BEHAVIORAL AND COGNITIVE STIMULI

14.4.1 NEUROENDOCRINE RESPONSES TO ENVIRONMENTAL CHANGES

The neuroendocrine system of an individual can not be isolated from the environment in which that individual lives. Neuroendocrine rhythms are correlated with daily light–dark cycles and changes in day length over the seasons of the year, as shown in Figure 14.4, and the neuroendocrine system also responds to environmental stimuli. Figure 14.11, for example, shows the changes in ACTH and GH levels in rats following exposure to noise, restraint and immobilization. Immobilization stimulates greater ACTH release than restraint which, in turn, stimulates more ACTH release than noise. All three stimuli inhibit GH release, with immobilization producing the greatest inhibition and noise the least (Armario and Jolin, 1989). Likewise, food input stimulates the secretion

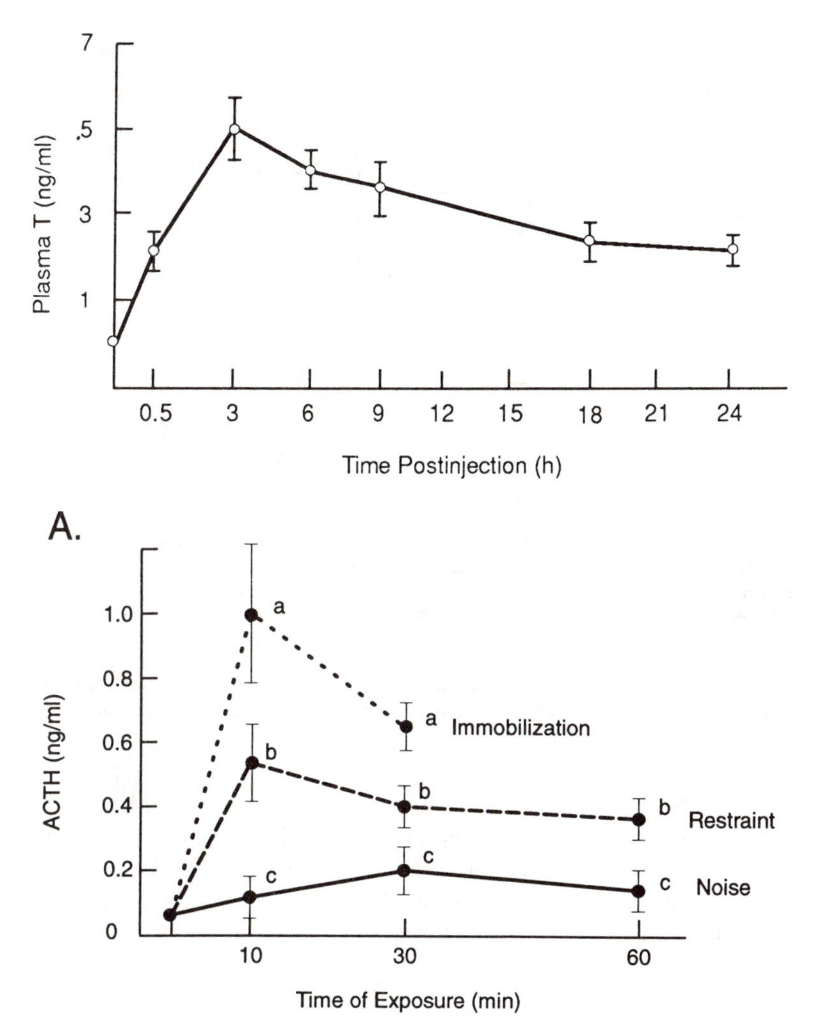

Figure 14.10. Plasma testosterone (T) levels in five adult castrated male rats immediately following a subcutaneous injection of testosterone propionate (100 mg/100 g body weight in 0.1 ml sesame oil). These data show the peak in plasma T following subcutaneous injection of the hormone and the change in T level for 24 h after the injection as means ± SEM. (Redrawn from Smith, Damassa and Davidson, 1977.)

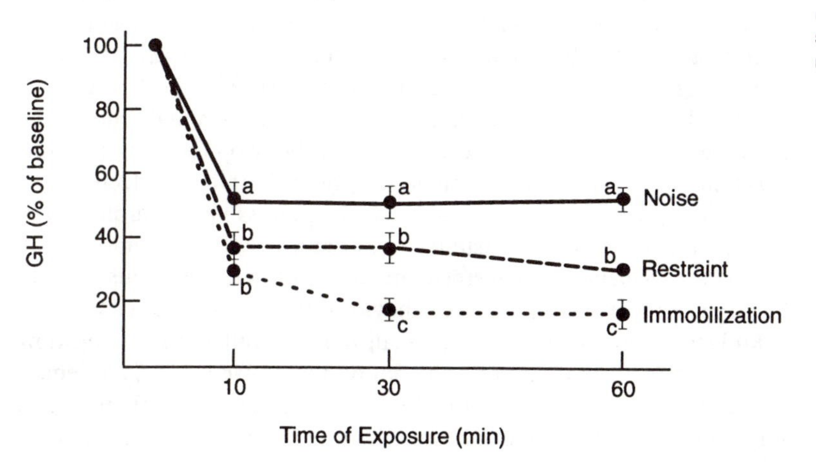

Figure 14.11. Neuroendocrine responses to environmental changes. (A) ACTH and (B) GH responses to noise, restraint in tubes, or immobilization in adult male rats. Groups which differed statistically are labeled with different letters. (A) Noise increased serum ACTH levels less than restraint and immobilization, which significantly increased serum ACTH at all three observation periods. The rise in ACTH was the highest in immobilized rats. The differences with respect to control values were always significant and are not indicated. (B) All three stressors reduced GH levels, with immobilization causing the greatest decrease. The GH response to stress was always significant. Results are means ± SEM for groups of 5–7 rats. (Redrawn from Armario and Jolin, 1989)

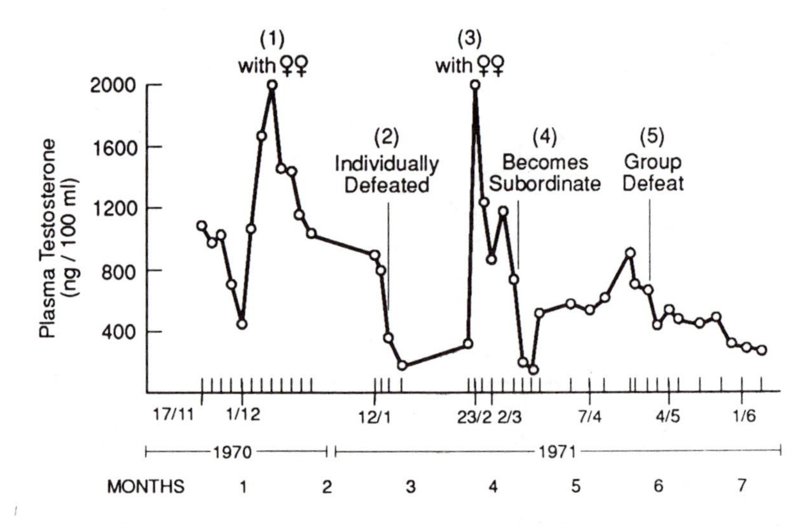

Figure 14.12. Neuroendocrine responses to social interactions. Plasma testosterone levels in a single male rhesus monkey over a 7-month period. He showed an increase in testosterone level on two occasions (1 and 3) following access to receptive females (♀♀) and a decline in testosterone level following defeat in three different circumstances. On the first occasion (2), he was defeated after a brief exposure to an all-male group of 34 animals. On the second occasion (4), he became the subordinate male in a newly formed group consisting of three other adult males and 13 females. On the third occasion (5), he was defeated, along with the three other males, by a large, well-established breeding group. (Redrawn from Rose, Bernstein and Gordon, 1975.)

of gastrointestinal hormones and some of these, such as bombesin and cholecystokinin act as neuropeptides to regulate eating (Gibbs and Smith, 1986). In these examples, the external stimuli alter the neuroendocrine system, which then modulates behavior. Social interactions with other animals can have similar effects.

14.4.2 NEUROENDOCRINE RESPONSES TO SOCIAL INTERACTIONS

Hormone levels can be altered in response to social interactions. Sexual interactions, for example, result in transient increases in LH and testosterone levels in males, while aggressive interactions result in a decrease in testosterone levels and an increase in corticosteroid levels (Harding, 1981). Figure 14.12 shows the change in testosterone level in a single adult male rhesus monkey over a 7 month period. This male showed an increased level of testosterone secretion on the two occasions when he was housed with a receptive female and showed prolonged decreases in testosterone secretion following defeat in aggressive encounters with dominant males (Rose, Bernstein and Gordon, 1975). Cohabitation with a female also elevates the testosterone levels of male rats, thus influencing their sexual and aggressive behavior (Flannelly and Lore, 1977).

Prolactin secretion in lactating female rats is stimulated by the suckling behavior of her pups. Figure 14.13 shows the prolactin levels of female rats from day 4 to 12 of lactation. Pups were separated from their mother for 4 hours on days 4, 8 and 12 and then replaced. Blood samples were

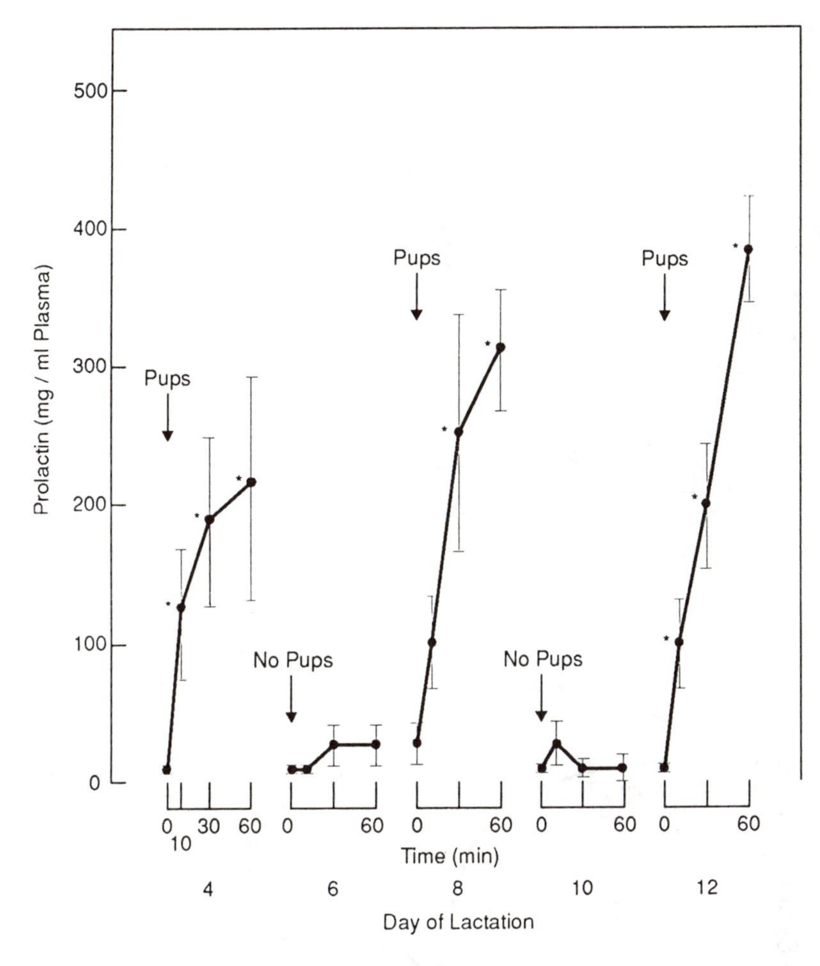

Figure 14.13. The effects of suckling on lactating female rats. Plasma prolactin levels increased in lactating female rats which had their pups removed for 4 h and then returned for 60 minutes on days, 4, 8, and 12. No pups were returned on days 6 and 10. Blood samples were taken before pups were returned (time 0) and 10, 30 and 60 minutes after pup return. Results are means ± standard error and * indicates a significant increase in prolactin level compared to that at time 0. (Redrawn from Samuels and Bridges, 1983.)

then taken from the mother by a cannula four times during the first hour after the pups were returned. The suckling behavior of the pups on their return stimulated the release of prolactin from the mother. On days 6 and 10, when no pups were returned, no increase in prolactin release occurred in the mother (Samuels and Bridges, 1983).

14.4.3 NEUROENDOCRINE REFLEXES

Many sensory stimuli 'trigger' the release of hormones in a reflex fashion. Chemical signals from conspecifics (pheromones) can 'prime' the reproductive system by triggering the release of gonadotropic hormones and can thus influence the timing of puberty, reproductive behavior, and pregnancy in many mammals. Odours of females also trigger the release of LH and testosterone in males (Brown, 1985b). Vaginal stimulation during mating triggers the release of prolactin and oxytocin in the female rat (Komisaruk and Steinman, 1986). Females of many species of mammals, such as cats and rabbits are induced ovulators and require tactile stimulation of the vagina during sexual behavior to trigger the LH release that stimulates ovulation. Without this external

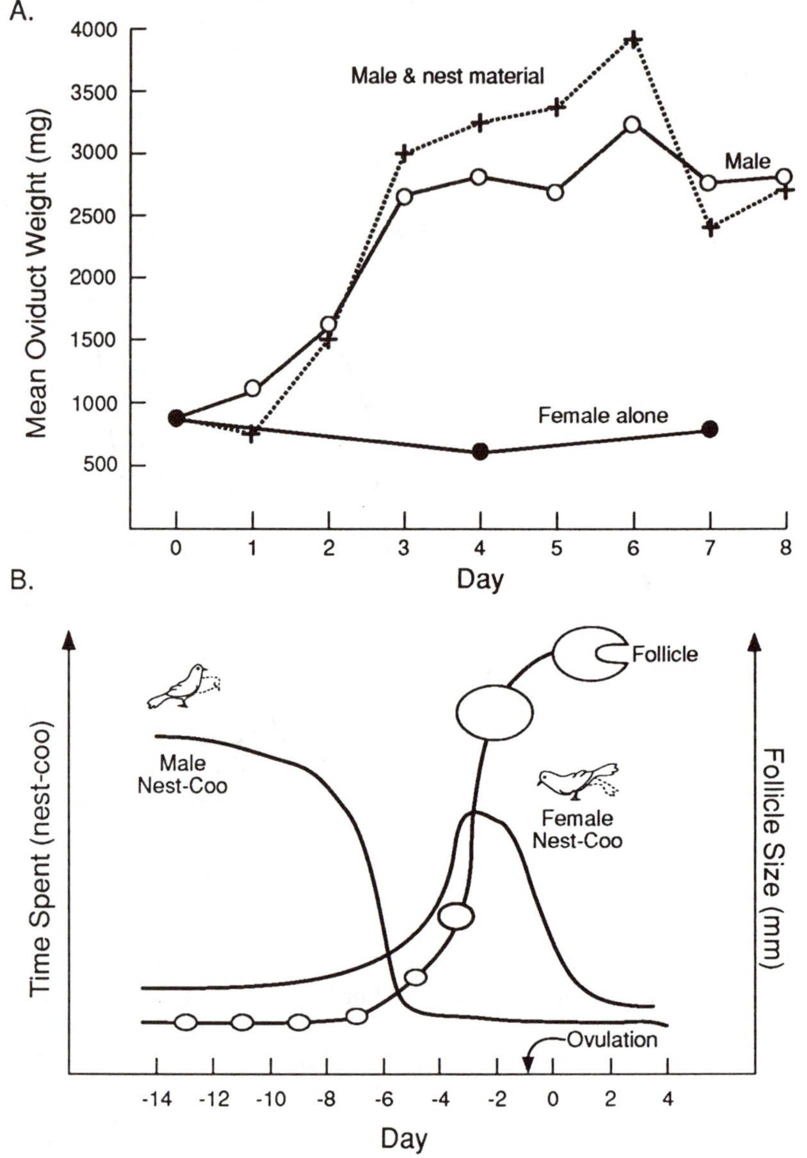

Figure 14.14. Neuroendocrine reflexes. (A) The effect of association with a mate or with a mate plus nesting material on oviduct growth in female ring doves. Each point is derived from tests of 20 birds. No individual bird is represented in more than one point. The abscissa represents the duration of the subject's association with a mate or with a mate plus nesting material, or (for the control group) the time spent alone in the test cage. 'Day 0' means that the bird was tested immediately upon being placed in the cage. (Redrawn from Lehrman *et al.*, 1961.) (B) A schematic representation of the sequence of the male and female ring dove's nest-cooing behavior in relation to the pattern of follicular development plotted with reference to ovulation. (Redrawn from Cheng, 1986.)

stimulation, they will not ovulate. The tactile stimulation of the nipple when pups are suckling from a lactating female rat triggers prolactin and oxytocin release (Figure 14.13). Likewise, the male ring dove acts as a stimulus to trigger the reflexive release of gonadotropic hormones in the female ring dove (Figure 14.14).

14.4.4 CHAINING OF NEUROENDOCRINE AND BEHAVIORAL RESPONSES

In some cases, a series of behavioral and neuroendocrine events may be linked in a chain reaction (Harding, 1981). Such hormone–behavior chains involve three factors – environmental stimulation of hormonal change, hormonal stimulation of behavioral change, and behavioral feedback on hormone levels – which lead to the next level of behavioral change (Beach, 1975; Leshner, 1978). The best example of such a hormone–behavior chain is the reproductive behavior of the female ring dove, in which social interactions with a male are essential for the timing of the transitions from one phase of the female's reproductive cycle to the next and the stimulation of ovulation. (Lehrman, Brody and Wortis, 1961; Lehrman, 1965). The availability of nest material in addition to the male's presence further accelerates ovulation, as shown in Figure 14.14A. When a female ring dove is paired with a male, the visual, auditory and tactile stimuli from the male stimulate the growth of her ovarian follicles and the corresponding secretion of estrogen and progesterone (Figure 14.14B). The nest-coo call of the female ring dove to the male during this courtship phase may also facilitate her follicular development through self feedback (Cheng, 1986). The increases in estrogen and progesterone levels stimulate female courtship and copulatory behavior and initiate nest building. Nest building and the continued presence of the male induce LH secretion which triggers ovulation and egg laying.

After ovulation, progesterone levels rise and estrogen levels fall, resulting in a decline in sexual behavior. In the presence of the nest and eggs, these hormonal changes stimulate the female to sit on the nest and incubate the eggs. The tactile stimulation received by sitting on the eggs stimulates prolactin release. Prolactin stimulates growth of the crop sac and secretion of 'crop milk' which is fed to the young doves when they hatch. Prolactin also induces parental behavior toward the hatchlings and their presence stimulates continued prolactin release. As the young birds grow and begin to feed on their own, they no longer stimulate prolactin release from the female. As her prolactin levels decline, there is a rise in FSH, which stimulates the growth of new follicles. In the presence of the male, more FSH and estrogen are secreted, stimulating follicular growth and another cycle of reproductive behavior begins. Similar neuroendocrine–behavior chains occur in the reproductive behavior of other birds, fish, amphibians and mammals (Brown, 1985a; Baggerman, 1968; Crews, 1980).

14.4.5 NEUROENDOCRINE RESPONSES TO COGNITIVE STIMULI

Because the neuroendocrine system can respond to changes in the central nervous system, the way that external stimuli are perceived can influence neuroendocrine responses. For example, cognitive factors in the emotional response to an external stimulus may determine the pattern of neuroendocrine responses to that stimulus. Emotional arousal involves a physiological response to a stimulus and the cognitive appraisal of that stimulus. The physiological response involves activation of the autonomic nervous system by the hypothalamus and limbic system, resulting in increased heart rate, blood pressure, temperature, perspiration, etc., as

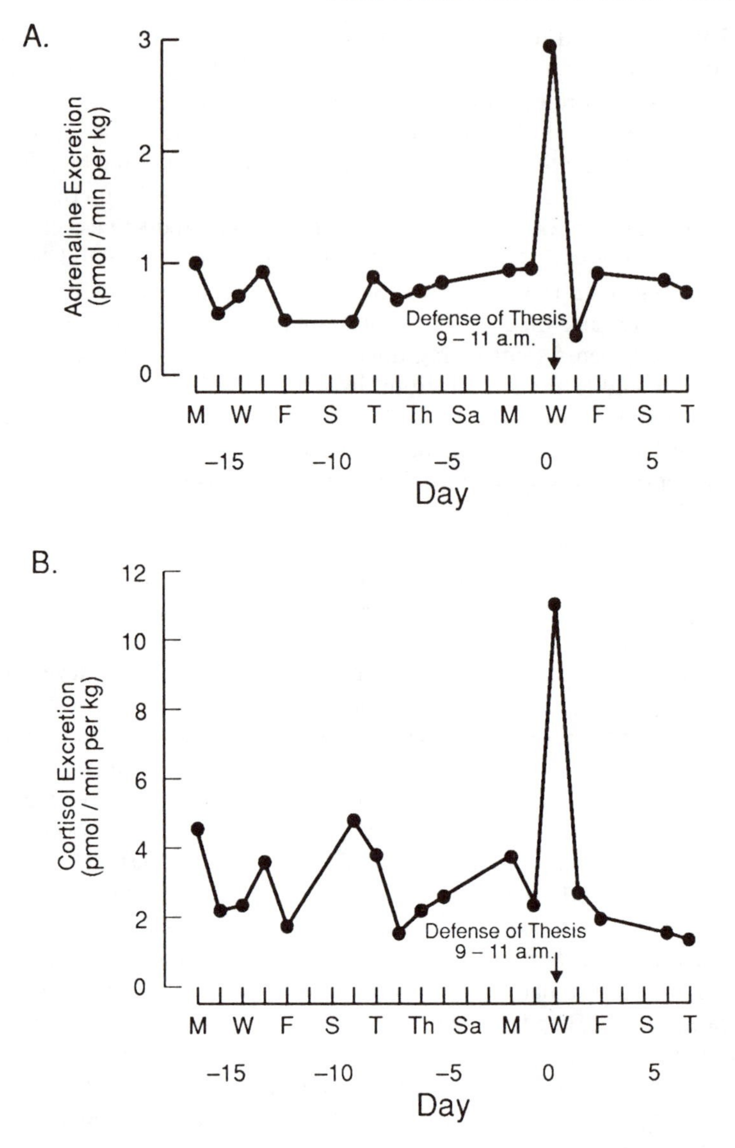

Figure 14.15. Neuroendocrine responses to emotional changes. (A) Adrenaline and (B) cortisol excretion of a male graduate student between 9:00 and 11:00 am on the days preceding, following, and coinciding with, the public defense of his Ph.D. thesis. (Redrawn from Johnsson, Collins and Collins, 1983.)

shown in Figure 12.11 (p. 285). Cognitive appraisal involves assessing the stimulus situation as positive, in which case the physiological arousal may be associated with joy or happiness; or negative, in which case the physiological arousal may be associated with fear or anger (Plutchik 1962). Different emotional states have been associated with different patterns of neuroendocrine responses (Henry, 1986).

Figure 14.15, for example, shows the levels of adrenaline and corticosteroid secretion in a 30-year-old male student on the days before, during, and after the oral examination of his Ph.D. thesis. Note the peak in these adrenal hormones on the morning of his examination and their return to baseline afterwards (Johansson, Collins and Collins, 1983). Extreme cases of emotional arousal, which result in pathological affective states, such as depression, anxiety disorders, or panic attacks, may be associated with a wide range of neuroendocrine responses (Carter, 1982; Prange *et al.*, 1987; Cameron and Nesse, 1988). Given that the adrenal hormones are released in stressful situations, one can infer that the changes in ACTH in rats in response to immobilization, restraint and noise shown in Figure 14.11 represent stress responses to these stimuli.

14.4.6 CONDITIONED HORMONE RELEASE

Since environmental stimuli and cognitive factors can both influence the release of neurotransmitters, hormones, neuropeptides and cytokines (Chapters 6, 12 and 13), it is not surprising to learn that hormone release can be conditioned to environmental stimuli. Ader (1976), for example, has shown that corticosteroid release can be conditioned in an illness-induced taste aversion test, in which saccharin taste is paired with drug-induced illness. The rats learned to avoid saccharin-flavored water and also produced an elevated corticosteroid secretion to the taste of saccharin. Decreases in corticosteroid release have been conditioned to stimuli associated with feeding and drinking in rats (Coover, Sutton, and Heybach, 1977); increased LH and testosterone secretion have been conditioned to stimuli associated with sexual behavior in male rats (Graham and Desjardins, 1980); and histamine release has been conditioned to odor stimuli in the guinea pig (Russell *et al.*, 1984). Immune responses can also be conditioned, indicating that the neuroimmune system can respond to conditioned stimuli as well as the neuroendocrine system (Ader and Cohen, 1991).

14.4.7 LONG- AND SHORT-TERM NEUROENDOCRINE CHANGES IN RESPONSE TO EXTERNAL STIMULI

The changes in hormone levels in response to environmental, behavioral, or cognitive stimuli may be long-term changes in the baseline secretion of a hormone, as occurred following defeat in aggressive encounters in the rhesus monkey (Figure 14.12), or short-term surges in hormone release, as occurs during exam stress in humans (Figure 14.15). The neuroendocrine responses to external stimuli may also trigger other physiological responses as occurs when pup-induced prolactin release stimulates milk production in the female rat (Figure 14.13).

Hormonal surges represent short-term neuroendocrine responses to external stimuli and are often associated with particular behavioral responses such as sexual behavior, maternal behavior or stress reactions. Chronic changes in the baseline secretion of hormones, on the other hand, are often associated with long-term changes in behavioral responsiveness. The lowered baseline levels of testosterone in the subordinate male rhesus monkey, for example, are associated with submissive behavior in a number of different situations, which results in the animal being treated differently by other monkeys in the group. There are two important distinctions to make between short-term stimulus-induced changes in hormone release and long-term changes in baseline hormonal levels with respect to their effects on behavior. First, an altered baseline level of hormone release means that the hormone level is changed before the animal or person enters a particular stimulus situation, whereas short-term changes in hormone release occur in response to a particular stimulus. Second, short-term hormonal changes occur within minutes of stimulus onset and quickly return to baseline, whereas changes in the baseline hormone levels develop slowly, taking hours or days to occur and result in new baseline hormone levels, which remain for days or weeks (Leshner, 1979). This difference can be seen by comparing the neuroendocrine stress responses of the male student in Figure 14.15 with those of the male rhesus monkey in Figure 14.12. The student's adrenaline and corticosteroid levels increased just before he defended his thesis, and when this stress was over, the hormone levels returned to baseline. When the monkey was placed with other monkeys, however, he remained with them and was under prolonged stress for over 3 months and his testosterone level remained depressed for this entire period.

14.4.8 SUMMARY

The timing of many neuroendocrine events is regulated by external stimuli. These can be environmental changes, social interactions, or cognitive processes. Some of these stimuli are cyclic, such as day–night and seasonal light cycle rhythms, some are short-term stimuli, such as noise, restraint or immobilization, and others are more chronic, such as long-term changes in social status. Often, the release of hormones does not occur without being triggered by external stimuli, as occurs with induced ovulation and lactation. The changes in neuroendocrine activity caused by these external stimuli alter the behavior of the animal, resulting in hormone-behavior chains, as occurs in the mating behavior of the ring dove. Because neuroendocrine activity can be conditioned to external stimuli, the animal or person may have conditioned hormone release in certain stimulus situations. These learned neuroendocrine responses may lead to chronic or acute changes in hormone secretion, depending on the conditioning stimulus. The result of these neuroendocrine responses to external stimuli is that the timing of certain neuroendocrine events, such as puberty, may be regulated by a host of environmental stimuli, such as nutrition, day–night cycles, stress, social interactions and pheromones from adult animals (see Brown, 1985b).

14.5 NEURAL AND GENOMIC MECHANISMS MEDIATING NEUROENDOCRINE–BEHAVIOR INTERACTIONS

Hormones, neuropeptides and cytokines do not stimulate behavior directly: they act as neuromodulators to alter the activity of their neural target cells. This neuromodulatory action regulates the membrane permeability and the genomic activity in the target cell, as was discussed in Sections 9.8, 10.6, 12.3 and 13.5. Changes in membrane permeability alter the electrophysiological activity of the cell and the release of neurotransmitters. Activation of the genetic material in the nucleus of the target cell leads to mRNA production and protein synthesis. Endocrine stimulation of neural target cells activates behavioral changes and hormone synthesis in adults, but during embryonic and neonatal development, hormonal stimulation modulates the growth and differentiation of neural and non-neural tissues, thus regulating the developmental process (Lauder, 1983; Arnold and Breedlove, 1985). This section discusses the neural and genomic mediators of hormonal modulation of behavior in adult and developing animals.

14.5.1 NEURAL MEDIATING MECHANISMS

Hormones can influence behavior by modulating three different aspects of neural activity. These involve the hormonal modulation of sensory receptors and sensory input to the brain, the motor pathways controlling behavior, and the central mechanisms responsible for the integration of sensory information with the organization of behavioral response patterns. These neural mechanisms have been studied using electrical stimulation of the brain, electrophysiological recording of neural activity, lesion techniques, and intracerebroventricular hormone implants (Beach, 1974, 1975; Leshner, 1978).

Sensory receptors and sensory input

Gonadal, thyroid and adrenocortical hormones modulate gustatory (taste), olfactory, tactile, and other sensory systems by acting directly on the peripheral sense organs and by affecting the processing of sensory stimuli in the brain (Gandelman, 1983; Henkin, 1975). Steroid hormones influence taste and smell sensitivity by altering the oral and nasal mucosa as well as modulating the neural pathways processing olfactory and taste stimuli (Gandelman, 1983). Disorders of thyroid and adrenocortical hormone secretion alter taste, smell and auditory perception (Henkin, 1975). Estrogen increases the size of the tactile sensory field of the pudendal nerve, increasing the sensitivity of the skin receptors around the genital area of the female rat during estrus (Komisaruk, Adler, and Hutchison, 1972). As discussed in Section 12.5.1, the endogenous opioids modulate the transmission of pain signals in the spinal cord, thus regulating pain reception.

Motor pathways

Hormones influence the motor pathways controlling a number of behavioral acts, including locomotion, sexual reflexes, and grooming behavior. Excessive grooming behavior in rats is stimulated by intracranial injections of ACTH and MSH (Spruijt and Gispen, 1983) and many neuropeptides modulate motor activity (see Table 12.4). Increased locomotor activity stimulated by estrogen secretion is characteristic of estrus in rats and many other mammals, including cows and can be used to predict sexual receptivity in these females (Burke and Broadhurst, 1966; Hurnik, King and Robertson, 1975). The spinal nucleus of the bulborcavernosis (SNB) in the male rat contains motor neurons which innervate the muscles around the penis. The size of this spinal nucleus and the number

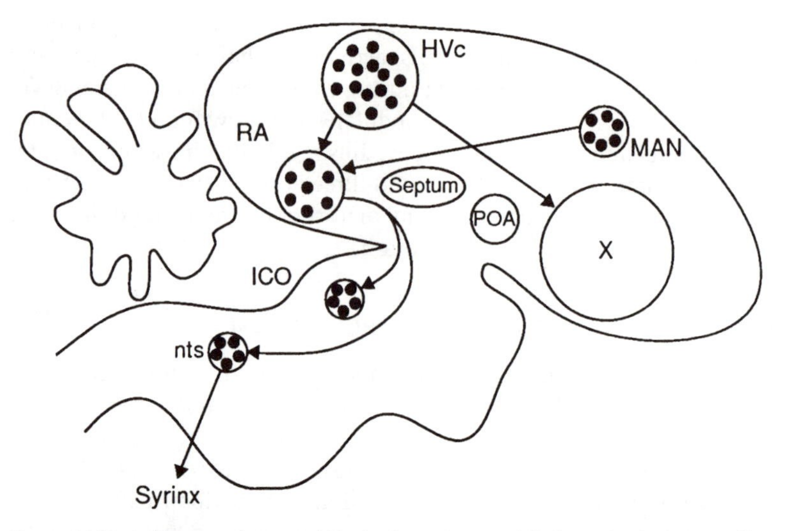

Figure 14.16. A diagram of some of the brain regions and their anatomical connections involved in vocal control in zebra finches and canaries. Black dots indicate the presence of cells labeled by androgens. These brain regions include the hyperstriatum ventrale pars caudale (HVc); the magnocellular nucleus of the anterior neostriatum (MAN); area X of the lobus parolfactorus (X); the nucleus robustus archistriatalis (RA); the nucleus intercollicularis (ICO); and the tracheosyringeal motor neurons of the hypoglossus (nts) which innervate the syrinx. The septum and preoptic area (POA) are also noted. (Redrawn from Arnold, 1981.)

of its dendritic connections, as well as the development of the penile muscles innervated by these nerves, are regulated by androgens, which are essential for penile erection and other penile reflexes in the male rat (Davidson *et al.*, 1978; Arnold and Jordan, 1988).

Central integrative functions

For many hormone-dependent behaviors, it is difficult to distinguish the effects of hormones on the sensory and motor pathways from those on the motivational or arousal mechanisms in the brain. The hormones seem to perform an integrative function, enabling each component of the behavioral system to interact with the others to produce functional sequences of behavior. The central integrative function of the neuroendocrine system in the control of bird vocalizations and in the arousal of sexual behavioiur in male and female rats are discussed as examples.

Figure 14.16 illustrates the androgen-sensitive neural areas of the brain of song birds which are involved in the arousal and motor control of singing (Arnold, 1981). Although it is difficult to separate the motivational and motor effects of androgens in stimulating singing in these birds, it is possible to do this with the vocalizations of ring doves (Cohen, 1983). Testosterone implants into the septum and anterior hypothalamic-medial preoptic area of the brain activate reproductive motivational systems and stimulate increased courtship and vocal behavior in ring doves, while testosterone implants into the nucleus intercollicularis (ICo) stimulate vocalizations, but not courtship behavior. Lesions of the vocal control system (area RA) and the motor neurons controlling the vocal cords (nts) alter vocal behavior in a number of behavioral contexts. Thus, testosterone activates singing as a component of courtship behavior when it stimulates receptors in the hypothalamus and limbic system and stimulates singing alone by acting at receptors in the motor control neurons.

The sexual behavior of the male rat has both arousal (motivational) and performance (motor) components and testosterone-sensitive areas of the anterior hypothalamic-medial preoptic area of the brain integrate these components into a functional sequence of behavior (Davidson and Trupin, 1975; Soulairac and Soulairac, 1978). The full integration of male sexual behavior depends on testosterone activating receptors in the AH-MPOA and limbic system of the brain as well as in the spinal neurons of the SNB and the penile striated muscles (Arnold and Jordan, 1988). A number of neuropeptides are also involved in the activation of male sexual behavior (see Table 12.4, p. 298).

The sexual behavior of the female rat involves the action of estrogen and progesterone on the olfactory pathways, the tactile receptors of the pudendal nerves of the genital area, and on the neurons in the AH-MPOA, ventromedial hypothalamus (VMH) and other limbic structures in the brain (Pfaff, 1989). The effects of estrogen on the neural components underlying female sexual behavior have been depicted in Figure 9.17. Neuropeptides are also involved in stimulating female sexual behavior and Figure 14.17 provides a model of the interactions among estrogen, progesterone, LH-RH, the endogenous opioid peptides, and the catecholamine neurotransmitters in the integration of the sensory, motivational and motor components of sexual behavior in the female rat (Crowley, 1986).

14.5.2 GENOMIC MEDIATING MECHANISMS

The hormonal activation of behavior is mediated by the stimulation of mRNA and protein synthesis in the target cell as well as by the release of neurotransmitters. When the steroid hormone–receptor complex binds to the acceptor site in the nucleus of the target cell, as shown in Figure 9.2 (p. 151), specific genetic information in the DNA is transcribed by mRNA synthesis (Blaustein, 1986). This genetic information is translated into the production of new proteins in the ribosomes of the endoplasmic reticulum in the cytoplasm of the cell (see Chapter 9). These new proteins can alter neural function and stimulate neural growth and differentiation. Neuropeptides regulate protein synthesis through the activation of second messenger cascades, as discussed in Chapter 10. Techniques such as *in situ* hybridization, filter hybridization (dot blots), complementary DNA and recombinant DNA analyses can be used to measure the genomic responses to neuroendocrine activation (Wilcox, 1986; Pfaff, 1989).

Using these methods, increased levels of mRNA can be detected in the estrogen target cells of the VMN of the female rat brain six to 24 hours after estrogen injection (Pfaff, 1989). The new proteins produced via this mRNA synthesis include progesterone receptors, muscarinic acetylcholine receptors, and α_1-noradrenergic receptors which increase the sensitivity of the target cell to neurotransmitter stimulation. In other target cells,

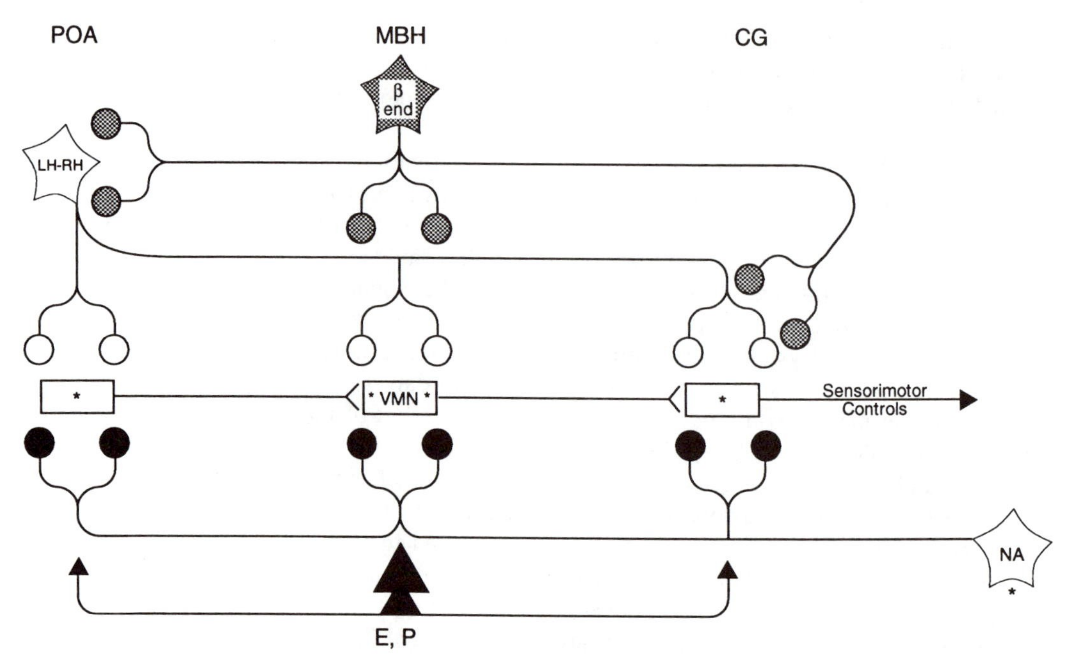

Figure 14.17. The sexual behavior of the female rat. A model of the functional interactions among catecholamine, LH-RH and opioid neurons that form part of the neural system controlling the lordosis reflex of female rats in response to ovarian hormones. The principal target neurons for estradiol (E) and progesterone (P) modulation of lordosis behavior is the ventromedial nucleus (VMN) in the medial basal hypothalamus (MBH). Additional sites for uptake and binding of ovarian hormones are also found in the preoptic area (POA) and mesencephalic central gray area (CG). Neurons which integrate and transmit information to lower brain stem and spinal sensorimotor centers (shown as open boxes) may be modulated directly by the steroid hormones. Some components of the POA, MBH and CG that influence lordosis are also shown. For example, LH-RH-containing neurosecretory cells may innervate the POA, VMN, and CG. LH-RH released from these nerve terminals may function as a neurotransmitter to stimulate lordosis behavior. Noradrenaline (NA) neurons, ascending from the medulla carrying sensory cues, also innervate the POA, VMN, and CG. The excitatory noradrenergic influence on lordosis may be mediated by enhanced LH-RH release in these areas. β-endorphin cells in the arcuate nucleus of the MBH, which project to the POA, VMN, and CG, may inhibit LH-RH release by inhibition of the release of NA in these three areas. (Redrawn from Crowley, 1986.)

estrogen stimulates the synthesis of LH-RH, oxytocin and enkephalin mRNA and the increased synthesis of these neuropeptides (Pfaff, 1989). The synthesis of these new peptides can result in changes in behavior and neuroendocrine activity. Figure 14.18 shows how the multiple effects of estrogen on the electrophysiological and genomic activity of its target neurons in the VMN may lead to changes in female sexual behavior.

14.5.3 NEUROENDOCRINE ORGANIZATION OF NEURAL DEVELOPMENT

The modulation of genomic activity in endocrine target cells during the prenatal and neonatal periods regulates the growth and development of these cells (Balazs, Patel and Hajos, 1975; Balazs, 1976; Lauder, 1983).

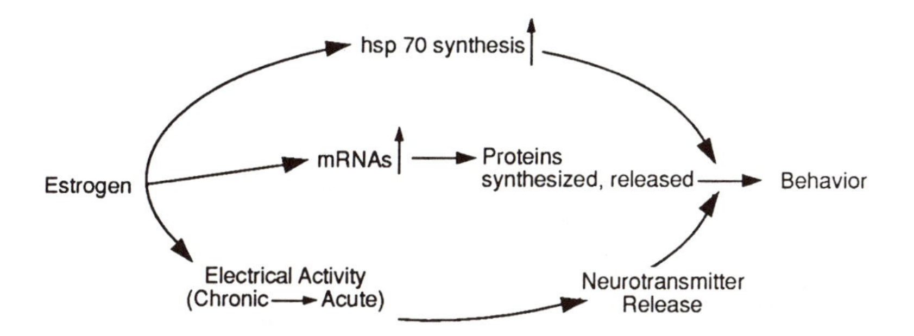

Figure 14.18. The effects of estrogen on the neural and genomic mechanisms of target cells in the ventromedial nucleus of the hypothalamus (VMN). Estrogen stimulates the synthesis of a specific protein (hsp 70) in the VMN which is transported to the midbrain central gray where it may facilitate other hormone-induced effects. The changes in electrical activity, neurotransmitter release and protein synthesis all influence behavior. (Redrawn from Pfaff, 1989.)

By stimulating mRNA and protein synthesis during 'critical periods' of brain development, hormones, neuropeptides and neurotransmitters can modulate nerve cell growth and differentiation, the synthesis of hormone and neurotransmitter receptors, and the number of synaptic connections formed by the cell, as shown in Figure 14.19A. Quantitative measures of nerve cell size, dendrite growth, synapse formation, myelination, and enzyme activity can be used to assay the effects of hormonal stimulation during cell development (Figure 14.19B). Measures of metabolic activity, such as the rate of conversion of glucose to amino acids and the rate of uptake of radio-labelled amino acids can also be used to determine the effects of neuroendocrine stimulation on target cell function during development (Balazs, 1976; Reboulleau, 1986).

As shown in Figure 14.19A, hormones, neuropeptides and neurotransmitters interact to modulate brain development during critical periods of neural growth and differentiation (Lauder and Krebs, 1986). These neuroregulators organize the development of the neural circuits mediating sensory input, motor activity, and the central integrative functions of the brain. As a result, the general metabolic state of the animal and the functioning of the neuroendocrine, autonomic and central nervous systems are all 'shaped' by the modulatory actions of hormone–neurotransmitter interactions during neural development (Dorner, 1983; Lauder, 1983).

Because these neuroregulators shape the development of the brain, any factors which disrupt their secretion during the embryonic period may alter neural development and thus have permanent effects on the brain and behavior. For example, a lack of thyroid hormones during embryonic development results in retarded neural growth, as shown in Figure 14.19C. If pregnant females suffer from nutritional deficiencies, take drugs which alter neurotransmitter or neuropeptide levels, are subjected to environmental toxins or stimuli which alter her steroid hormone release (such as severe stress) the neuroendocrine development of the fetus can be disrupted, resulting in perturbations in the modulation of brain development (Fein *et al.*, 1983; Boer and Swaab, 1985).

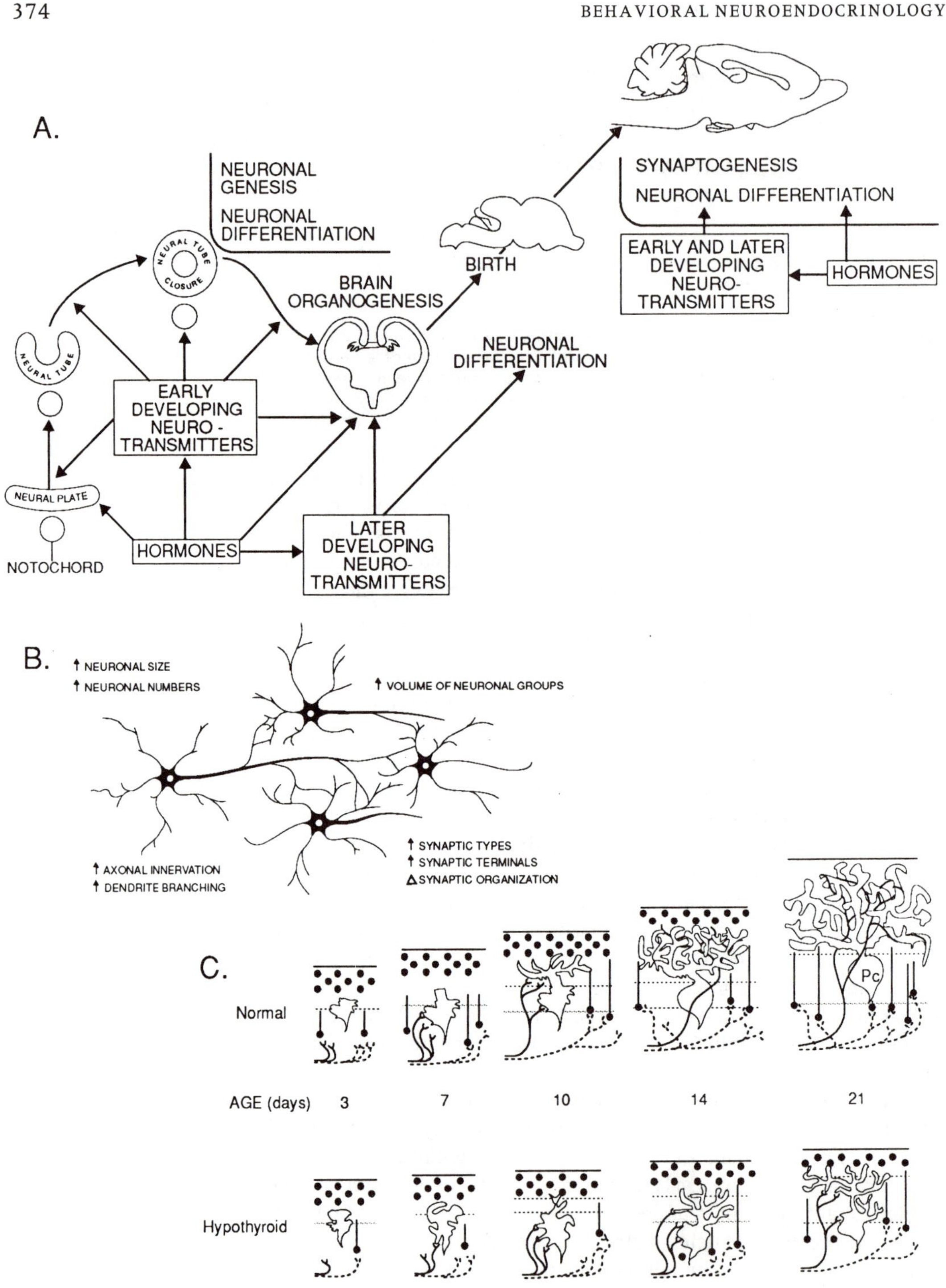

Figure 14.19. The role of the neuroendocrine system in neural development. (A) Hormones and neurotransmitters promote neural growth and differentiation at each stage of brain development and may have different effects, depending on when and where they appear in the developing nervous system. (Redrawn from Lauder and Krebs, 1986.) (B) The effects of

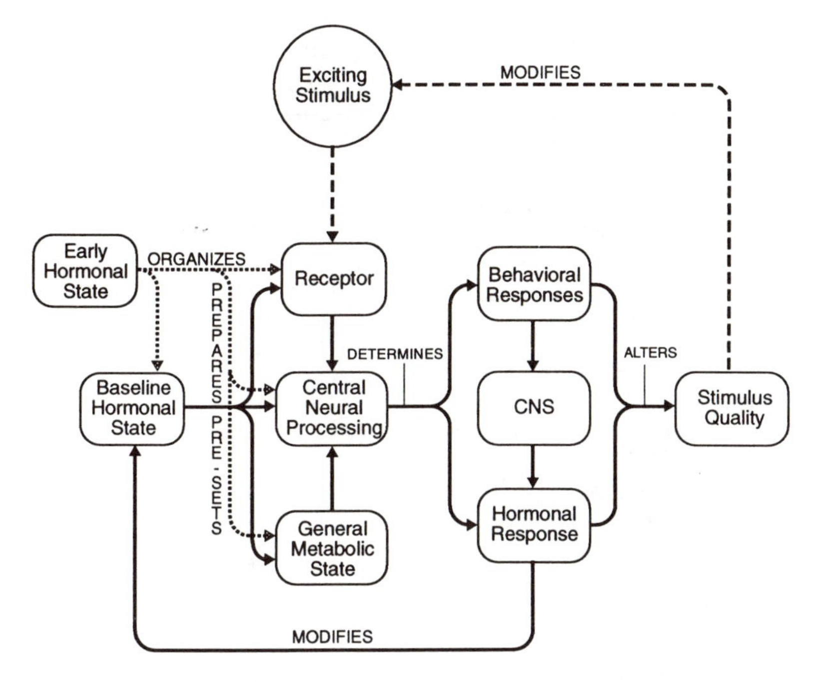

Figure 14.20. A model of neuroendocrine–behavior interactions showing the organizational and activational effects of hormones on behavior and the feedback effects of behavior on hormone responses and the baseline hormonal state. (----) denotes interactions between individuals (....) denotes the effects of hormones present early in development, and (—) denotes relationships within an adult individual. (Redrawn from Leshner, 1978.)

14.5.4 INTEGRATION OF NEUROENDOCRINE–BEHAVIOR–ENVIRONMENTAL INTERACTIONS

The interactions between the neuroendocrine system, the external environment and behavior involve a number of levels of integrative mechanisms as shown in Figure 14.20 (Leshner 1978). During embryonic development, hormones and other neuroregulators organize the development of the baseline hormonal state, the general metabolic state, the central nervous system pathways and the sensitivity of the organism's sensory receptors. In adulthood, the perception of an external stimulus by sensory receptors is modulated by baseline hormone levels as is the neural response to this stimulus. Through the neural mediating mechanism, the stimulus can elicit a behavioral and/or a neuroendocrine response and the hormones released can feed back to modulate both the ongoing behavioral response and the baseline hormonal state of the animal. The hormonal state and behavioral activity of the animal can, in turn, influence its quality as an stimulus in social interactions with other animals. Since the initial behavioral and hormonal responses of animals to external stimuli will depend on the organizational effects of the neuroendocrine system on baseline hormone levels, there are three levels of neuroendocrine–behavior interactions to consider: the neuroendocrine

neuroendocrine stimulation on particular neural circuits include modulation of the number and size of nerve cells, axonal growth, dendritic branching, synapse formation and synaptic organization. (Redrawn from Toran-Allerand, 1991.) (C) The development of Purkinje cells in the cerebellum of normal and hypothyroid rats from 3 to 21 days of age. Note the reduced size and dendritic branching in the cells of the hypothyroid rat. (Redrawn from Dussault and Ruel, 1987.)

Table 14.2. *Confounding variables which influence hormone–behavior interactions*

1. *Genetic differences between species*
2. *Individual differences*
   Genetic and other biological differences
   Organizational effects
   Uterine position
3. *Conditioning and experiential factors*
   Learning/conditioning
   Social experiences
   Dominance status
4. *Environmental factors*
   Day–night and annual cycles of hormone release
   Social stimuli and pheromones
5. *The stimulus situation*
   Test apparatus
   Time of day
   Subject–experimenter interaction
6. *The present state of the subject*
   Presence of other hormones or drugs
   Psychological expectancies
   Emotional and motivational state
   Nutritional variables – feeding cycles
   Sleep–wake cycles
7. *Interactions among hormones, neurotransmitters, neuropeptides, cytokines, environmental stimuli and behavior*

organization of behavior during development (organizational effects), the neuroendocrine modulation of behavior during adulthood (activational effects), and the behavioral modulation of the neuroendocrine system (behavioral feedback).

14.6 CONFOUNDING VARIABLES IN BEHAVIORAL NEUROENDOCRINOLOGY RESEARCH

Any research on the interactions between the neuroendocrine system and behavior must contend with a host of extraneous or confounding variables, some of which are listed in Table 14.2 (Beach, 1974, 1975).

14.6.1 SPECIES DIFFERENCES

Not all species show the same behavioral responses to hormonal changes. There are, for example, phylogenetic differences in the effects of gonadal hormones on sexual and parental behavior (Aronson, 1959; Brown, 1985a). Even closely related species may show different behavioral responses to hormonal manipulation. For example, genetically different inbred strains of mice differ in the level of sexual and aggressive behavior activated by gonadal hormones (Whalen, 1986). There are also phylogenetic differences in the perinatal organizational effects of hormones, particularly with reference to sexual differentiation. In mammals, for example, the 'neutral' sex is female and androgens are necessary for masculinization, while in birds, the neutral sex is male and estrogens are necessary for feminization (Adkins-Regan, 1981). Thus, the genotype of the animal may determine its sensitivity to hormonal stimulation and this makes it difficult to generalize from the effects of hormones on the behavior of one species to that of another. The discipline of comparative behavioral neuroendocrinology has developed to study evolutionary and ecological differences in the effects of hormones on behavior (Crews, 1986).

14.6.2 INDIVIDUAL DIFFERENCES

Just as there are genetic differences between species, there are biological and experiential differences between individuals of the same species and the effects of hormones on behavior may differ between individuals. There are, for example, large individual differences in the decline of sexual behavior after castration in male cats (Aronson, 1959). Genetic differences may account for some of these effects, but even animals from genetically identical inbred strains show differences in behavioral responses to hormones. One explanation for these differences is that the intrauterine environment may not be the same for every embryo, even if they are all members of the same litter. Perturbations in the prenatal levels of neuroregulators can alter the organization of neural development and lead to permanent changes in behavior. These prenatal effects may be very subtle. For example, individual embryos may be exposed to different levels of gonadal hormones during prenatal development, depending on their uterine position. From 6 to 16 mice may be born in the same litter, half of which develop in each uterine horn, as shown in Figure 14.21. In the uterus, mice can occupy a position between two males (2M), between two females (0M) or between a male and a female (1M). The level of gonadal hormones circulating in the fetus has been shown to differ, depending on the uterine position. This slight difference in steroid hormone level alters the sexual differentiation of the animal and its adult sexual and aggressive behavior as shown in Figure 14.22. Thus, differences in the behavior of genetically identical mice reared in the same environment may be due to their interuterine position or other subtle differences during embryonic development (Vom Saal, 1983).

14.6.3 LEARNING AND EXPERIENTIAL FACTORS

Since hormone release may be conditioned, specific experiences may be associated with neuroendocrine responses in individual subjects. Thus individual differences in social interactions or stressful experiences may result in different patterns of neuroendocrine reactivity. For example, animals reared in isolation have neuroendocrine and behavioral responses to novel stimuli different from those of animals reared in social groups (Brain and Benton, 1979), sexually experienced animals respond differently to opposite sex conspecifics than sexually naive animals, and dominant animals respond differently than subordinates.

Since hormone release can be conditioned to environmental cues

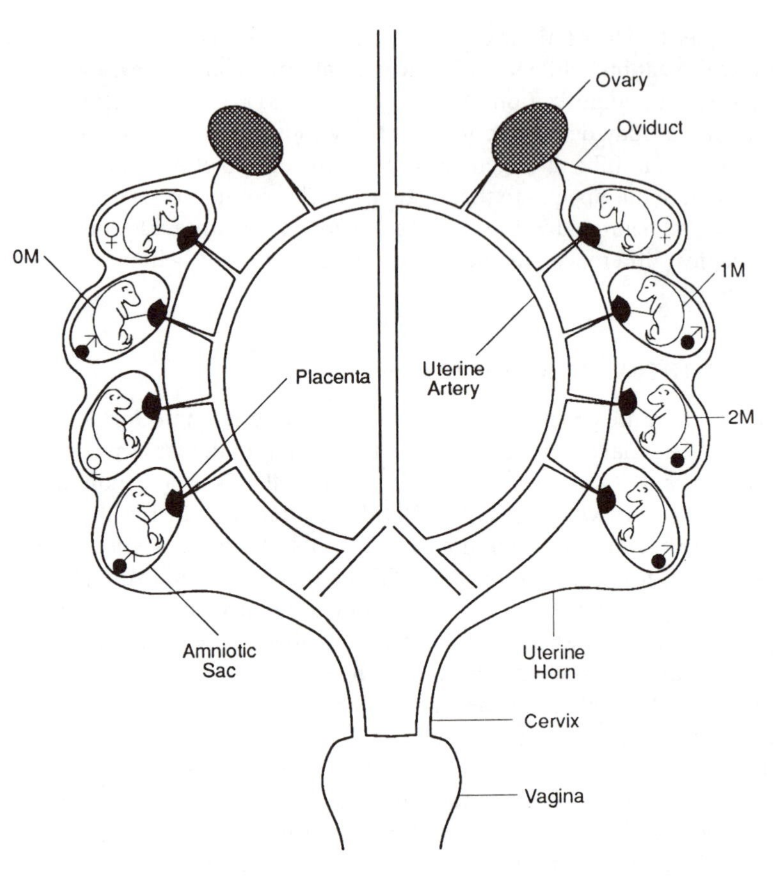

Figure 14.21. Intrauterine environment. A diagram of the uterine horns and uterine arteries of a pregnant mouse at term (21-day-old embryos). The intrauterine position of fetuses is determined by Caesarean delivery. 0M, 1M, and 2M refers to the number of male fetuses to which an individual is contiguous (2M between 2 males; 1M next to 1 male, 0M between 2 females). This scheme is used to identify both male and female fetuses, but only males are labeled in this diagram. (Redrawn from vom Saal, 1983.)

(Section 14.4.6), conditioned hormone release may occur independently of the stimuli being manipulated by the experimenter. If, for example, a male rat is conditioned to show an LH release to the chamber in which it has been tested with an estrous female, that male will show the LH release when placed in the test chamber, independently of the stimulus animal placed into the chamber by the experimenter. In this case, erroneous conclusions may be drawn about the effects of the stimulus animal on the hormone response of the test male. Likewise, sexually experienced male dogs may still show sexual responses to estrous females after castration while sexually naive males require hormonal stimulation before responding to the female. Thus, sexual experience may alter behavior independently of hormone levels.

14.6.4 ENVIRONMENTAL FACTORS

As discussed in Section 14.4, neuroendocrine activity is synchronized with environmental variables, such as day–night cycles and seasonal rhythms and can be influenced by subtle social stimuli, such as pheromones. The effects of light cycles and pheromones on neuroendocrine activity means that the housing conditions of the animals may influence their baseline hormonal levels and behavior independently of the vari-

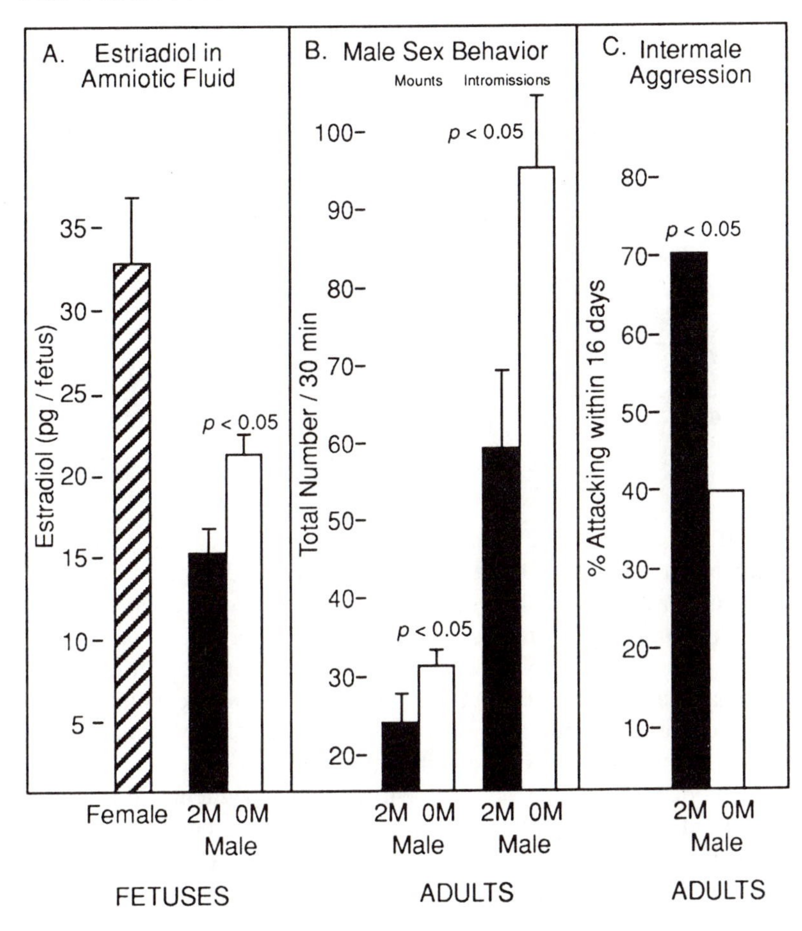

Figure 14.22. Effects of different uterine environments. See Figure 14.21 for 0M etc. (A) Amniotic fluid levels of estradiol (mean ± SEM) for female, 0M and 2M male mouse fetuses on day 17 of gestation. (B) Number of mounts and intromissions during a 30-min test with a sexually receptive 1M female for gonadally intact, 90-day-old 0M and 2M male mice (mean ± SEM20/group). (C) The percentage of neonatally castrated, 90-day-old 0M and 2M male mice (20/group) that exhibited a 5-sec biting attack toward a 1M male intruder within 16 days after the 0M and 2M males were implanted with a silastic capsule containing testosterone. 10-min tests for aggression were conducted every other day after implantation of the testosterone. Significance levels are for t-tests (A and B) and chi square tests (C). (Redrawn from Vom Saal, 1983.)

ables of interest to the experimenter. For example, seasonally breeding animals which mate in the long days of the spring will not mate during short days (8–12 hours of light per day) and may require 16 hours of light per day. Even when such seasonally breeding animals are held in constant laboratory conditions, they may show a breeding depression in the winter (Bermant and Davidson, 1974). Thus, experiments conducted at different times of the year may result in different behavioral responses because of the underlying seasonal change in hormone levels.

Animals also have a sleep–wake cycle, which is usually related to the environmental light–dark cycle. Changing the light–dark cycle results in correlated changes in neuroendocrine rhythms and behavioral responsiveness, such as occurs in 'jet-lag', which requires the traveller's physiological activity to shift to accommodate a change in the day–night cycle following a change in time zones. Housing animals under constant (24 hour) light alters neuroendocrine rhythms, and affects the timing of puberty, the estrous cycle of adult females and the sexual behavior of male rats (Schwartz, 1982; Fantie, Brown and Moger, 1984).

As a result of the daily activity and neuroendocrine rhythms, hormone treatment during the active, waking phase may have different effects than if given during the inactive sleep phase of the day–night cycle. Likewise,

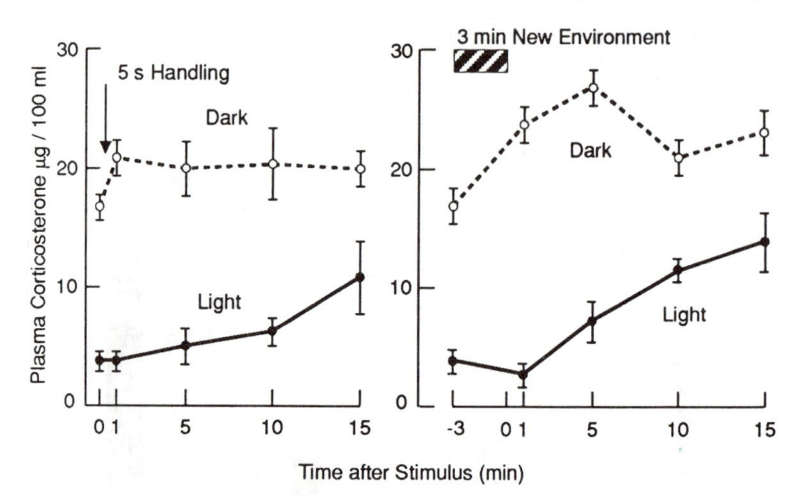

Figure 14.23. The effect of environmental factors on neuroendocrine activity. Levels of plasma corticosterone in male rats in response to (A) 5 s of handling and (B) 3 min in a novel environment (mean ± standard error). These responses differ during the light and dark phases of the day–night cycle. (Redrawn from Brown and Martin, 1974.)

stimuli presented at different times of the day may have different neuroendocrine effects. For example, corticosteroid levels in rats are much higher at the start of the dark phase of the light/dark cycle than they are at the start of the light phase. If rats are handled or placed into a novel environment during the light phase of the cycle, their corticosteroid response is much lower than if this same stimulation occurs during the dark phase of the cycle, as shown in Figure 14.23 (Brown and Martin, 1974). Thus, the effects of the light–dark cycle may have effects on the neuroendocrine control of behavior which are independent of the stimuli introduced in the experiment. Likewise, the social conditions under which the subjects live can influence neuroendocrine activity, as discussed in Section 14.4 (see Figure 14.11).

14.6.5 THE STIMULUS SITUATION

Behavioral responses to hormonal stimulation may vary depending on the specific stimuli present during the experiment. Behavior in a novel environment may differ from behavior in a familiar environment and the type of test cage, the method of stimulus presentation, the novelty of the stimuli and the design of the experiment can all influence the results (Johnston, 1981). Likewise, the time within the light–dark cycle that the animal is tested (Fantie *et al.*, 1984), the type of test used, and the behavior of the experimenter may all influence the animal's response. Finally, in tests of social behavior, the familiarity of the animals tested may influence the results. Male rats, for example, will mate with both familiar and unfamiliar females, but male gerbils are more likely to mate with a familiar estrous female (their mate) and will be aggressive toward an unfamiliar estrous female (Swanson, 1974).

14.6.6 THE PRESENT STATE OF THE SUBJECT

Many variables which could alter the state of the subject must be considered in the study of behavioral neuroendocrinology. When one hormone is of interest, the effects of other hormones or drugs which the subject may be taking must be considered. Women on birth control pills, for example, may show responses to certain stimuli that are different from those of women not taking birth control pills. Similarly, consumption of alcohol, caffeine, nicotine and antibiotic drugs may alter hormone–behavior interactions. For example, males given anti-psychotic drugs, such as the major tranquilizer thioridazine, may experience a number of sexual dysfunctions. The anticholinergic actions of these drugs interfere with both erection and ejaculation and their antidopaminergic properties may inhibit sexual motivation (Mitchell and Popkin, 1983).

Baseline levels of hormone secretion may also be determined by the social and cognitive situation of the subject, with subordinate animals or chronically depressed subjects showing different neuroendocrine activity than other subjects. Thus, the emotional state of the subject should be considered. Subjects under stress have high ACTH and corticosteroid levels and low LH and testosterone levels (Figures 14.11 and 14.12) and may respond to hormone treatment or external stimuli differently than non-stressed animals. Likewise, animals may respond differently when they are hungry or thirsty from when they are satiated after a meal.

14.6.7 NEUROENDOCRINE, ENVIRONMENTAL AND BEHAVIORAL INTERACTIONS

Unless hormones are given under controlled conditions, it may be difficult to distinguish between their effects and the effects of other confounding variables on behavior. It is, for example, difficult to demonstrate a one-to-one correspondence between neurotransmitters and neuropeptides on hormone release (Chapters 6 and 12) and behavioral changes because of the complex interactions among hormones, neurotransmitters, neuropeptides, environmental stimuli and behavior (Beach, 1974, 1975). Hormone release can also be synchronized with environmental or behavioral changes so that hormone–behavior interactions form a feedback chain. Hormones do not act alone. They interact with, and depend on, the presence of receptor proteins, enzymes, second messengers and essential precursors, such as cholesterol and amino acids. They interact with neurotransmitters, neuropeptides, cytokines and other hormones. They also form part of feedback loops; so manipulation of one hormone can stimulate or inhibit the release of other hormones.

The neural control of hormone release is a complex problem. Each of the pituitary hormones can be regulated by a multitude of neurotransmitters (Table 6.1, p. 93), neuropeptides (Table 12.2, p. 280) and cytokines (Table 13.2, p. 319). In addition, some neurotransmitters or neuropeptides can have opposite effects on the secretion of a hormone, depending on the mode of administration or on the levels of other hormones (see Section 6.6). Thus, to say that there is a neural mediation of hormone-behavior interactions does not imply that the nature of this neural mediation is easy to identify.

As with neurotransmitter–hormone interactions, there are many cases in which two or more hormones regulate the release of a third hormone.

As shown in Figure 4.3, both TRH and PRF can stimulate prolactin secretion. Likewise, LH, FSH, and prolactin can all regulate testosterone synthesis and secretion (Bartke *et al.*, 1978). Hormones may also act in pairs, with one hormone priming a target cell to respond to a second hormone. For example, estrogen 'primes' its target cells for progesterone stimulation by increasing the synthesis of progesterone receptors. Hormones can also have antagonistic actions on each other, so that an increase in one hormone may inhibit the secretion of others. High prolactin levels, for example, inhibit the release of gonadotropic hormones. These complex neuroendocrine interactions make it difficult to determine exactly what combination of neuroendocrine changes are associated with behavioral events. Many different neuroendocrine changes may cause the same behavioral change and a single environmental stimulus may cause different neuroendocrine reponses in different individuals or in the same individual at different times.

14.6.8 SPECIAL PROBLEMS WITH HUMAN SUBJECTS

All of the confounding variables listed in Table 14.4 apply to research on humans as well as animals, but there are other variables operating which further confound human behavioral neuroendocrinology research. If these confounding variables are not controlled, erroneous conclusions will be drawn about the hormone-behavior interactions. Research studies on animal subjects can control each of these variables to ensure that hormone levels are the only variables influencing behavioral changes. In research with human subjects, however, it is not possible to control all of these extraneous variables, so careful experimental designs are required to ensure that effects ascribed to hormones are not caused by some other factors. The special problems in doing research on human subjects include the choice of subjects, the design of experiments, the subject–experimenter interaction, and the social and political implications of neuroendocrine research. There are also special ethical principles to consider when conducting research on human subjects (Barber, 1976; American Psychological Association, 1981).

The choice of subjects

How are subjects selected for research in behavioral neuroendocrinology? Studies can be conducted on groups of volunteers, medical patients, or randomly selected groups. Each group of subjects has special considerations with respect to both experimental and ethical treatments. The ethics of psychological and medical experimentation now prevent many types of research which have been carried out in the past on populations of prisoners, racial minorities and unsuspecting medical patients (Edsall, 1969; Rutstein, 1969). No treatment can be given to human subjects which may cause unknown or detrimental effects.

Experimental design

The effects of hormones on behavior can be studied by three methods: direct observation and experimentation, indirect observation using questionnaires or subjective reports, and use of clinical medical reports. While it is relatively easy to conduct an experiment which presents different stimuli to subjects and involves taking blood samples to measure hormonal responses to these stimuli, it is difficult to design an experiment to alter hormonal levels in humans and look for behavioral changes. This is especially true when the effects of hormonal disruptions during prenatal development are of interest (Reinsch and Gandelman, 1978). For many studies on hormones and behavior in human subjects, therefore, medical patients with endocrine abnormalities are used as subjects. This poses problems with identifying, selecting and locating potential subjects from medical records and having them agree to take part in the study.

Once subjects have been selected, they must be matched in treatment and control conditions on factors such as age, sex, reproductive status, race, and medical condition. Patients on drug treatments, radiation therapy or undergoing psychiatric treatment may not be suitable for some research. If the patients being studied are involved in a treatment program, different physicians may be prescribing different treatment regimens for the same disorder, so standardizing the independent variables may not be possible. Randomization of patients into control and treatment conditions may not be possible as it may be unethical to withhold treatment from patients in the control group. As a result, techniques must be developed to find control groups matched to the patients with endocrine disorders. These may be sibling controls, in which the brother or sister of the patient serves as the untreated control. This allows for control of race, parental environment, sex, etc., and for partial control of heredity, but there may still be individual differences in biological, medical and psychological variables.

Statistical methods such as the analysis of covariance, multivariate procedures or correlatinal methods may be used to control covariates which may confound the hormonal effects. Factors attributable to age, sex, type of hormone given, dose levels, and treatment duration may be factored out using statistical methods rather than control groups (Reinsch and Gandelman, 1978). Problems of developing randomized clinical trials in medical research are discussed by Silverman (1985). It may be difficult to study behavior following hormone treatment, so subjects are often asked to complete questionnaires on their moods, thoughts, aggressive and sexual behaviors or feelings. How the results of these pencil and paper tests correlate with actual performance of sexual, aggressive and other behaviors is largely unknown.

Subject and experimenter bias

Subjects and experimenters both enter into experimental situations with expectancies which may bias the results. The experimenters have specific hypotheses to test and the subjects have expectancies of being cured of a disease or showing some altered behavior or psychological state. Many procedures exist to control for these expectancy effects. Experiments can be run in single- and double-blind designs and subjects in some groups may be given placebo treatments (Rosenthal, 1966). But the mood, temprament and personality variables of the subjects are not controllable nor are the social factors involved in the subject-experimenter interaction. The experimenter's sex, mannerisms, and general friendliness, as perceived by the subjects, can influence the behavior and the willingness of the subject to continue in the experiment. Also, the instructions given to the subject and what the subject thinks they are really 'supposed to do' can influence the results (Rosenthal, 1966).

Animal models

One way to study the effects of hormones on human behavior is to perform experimentally controlled studies with animals. Such animal models of human studies can follow the traditional experimental design discussed in Section 14.3, which looks at the effects of specific hormone parameters (chemistry, dose, timing, etc.) on behavior. On the other hand, one can attempt to mimic a human disorder in an animal and then test different methods of inducing and treating the disorder. This 'human analogue model' of hormonal research is designed to examine the etiology and treatment of specific biomedical disorders (Reinsch and Gandelman, 1978).

Social and political implications

Biomedical and sociobiological studies of human behavior have been the subject of criticism for fostering a biologically deterministic model of human behavior which is opposed by those who see behavior as determined by social learning and socio-cultural factors. Various critics have claimed that biological theories of human behavior support the capitalist system, are racist and sexist. These arguments against biological theories of human behavior are discussed by Ruse (1979). The interpretation of results from neuroendocrine research on human behavior have also been questioned. A biological interpretation is that behavioral changes are determined by neuroendocrine changes. A psychological interpretation is that emotional or cognitive responses to the neuroendocrine change alter the behavior, and a socio-cultural explanation is that socially acceptable responses to the neuroendocrine change are learned and, if this cultural attitude was changed, the response would be different. These arguments occur most often in sensitive areas such as the influence of hormones on sex differences in behavior and studies of premenstrual disorders. The fact that there may be sex and racial differences in the physiological responses to drugs and other neuroregulators does not seem to be a political issue.

Finally, the application of knowledge about the neuroendocrine control of behavior has produced a biotechnology with wide ranging social and political implications (Austin, 1972; Jaffe, 1973). How, for example, should hormonal birth control to be regulated? Should there be social, parental and religious control over who can use hormonal methods of birth control? Should birth control be forced upon people in overpopulated countries? Should mentally handicapped people be forced to use birth control? How should fertility drugs be used? Should hormonal treatments for inducing abortion be used? These are some of the questions raised by our ability to hormonally alter reproductive processes. Likewise, the social control of the male sex hormones is also controversial. How, for example, can the use of anabolic steroid hormones in sports be controlled? Should male sex offenders and extremely aggressive men be treated by castration or anti-androgen therapy? Many other issues in the use of neuroendocrine regulation for social and political purposes will develop as neuroendocrine technology becomes more widely applied to everyday life.

Summary

Research on the neuroendocrine control of human behavior is difficult because of the problems in designing carefully controlled experiments which eliminate confounding variables. Such considerations are extremely important because of the high level of neuroendocrine biotechnology in modern society. Taking drugs or hormonal treatment during pregnancy, birth control pills, hormone supplements in old age, medication for psychological or neurological disorders, cancer treatments and other medical treatments may all alter the neuroendocrine system, but how can the neuroendocrine effects of these treatments be determined? Much of the research on the neuroendocrine regulation of human behavior is from correlational studies, anecdotal case study reports, studies using medical patients, or from studies using questionnaires and self reports rather than behavioral data. As a consequence, much of the basic information in behavioral neuroendocrinology is from controlled experiments with animals or experiments with animal analogues of human disorders. It is usually assumed that the results of these animal studies can be generalized to humans, but this is often difficult because of species and other differences between animals and humans.

14.7 SUMMARY

This chapter has introduced some of the methods used to study behavioral neuroendocrinology. First, the behaviors to be recorded must be determined and careful observational and descriptive methods used to define the specific behavior units to be recorded. Accurate quantitative methods are then necessary for measuring the frequency, duration and patterning of these behavior units. Correlational studies can then be conducted to record the behavioral changes associated with natural fluctuations in hormone levels or disorders of the neuroendocrine system. Hormone levels for these correlational studies can be measured directly from blood samples or indirectly from urine, saliva or by using bioassays. Experimental studies can be of two types: hormones can be manipulated as the independent variable and behavioral responses recorded, or external stimuli can be manipulated and hormonal responses recorded. When neuroendocrine variables are manipulated, the experiments include a baseline, hormone removal and hormone replacement phase. These are required to demonstrate that it is the hormone which alters behavior and not some non-hormonal confounding factor. In these studies, careful consideration must be given to the methods of hormone removal and replacement. Consideration must also be given to the chemical composition of synthetic hormones used, the dose level, vehicle, route of administration and the timing of hormone replacement therapy with respect to behavioral testing. Neuroendocrine changes can also be measured in response to environmental stimuli, behavioral interactions, and cognitive functions. These stimuli can cause brief hormone surges, as occur in a neuroendocrine reflex, or long-term changes in baseline hormonal secretion. There can also be chains of hormone-behavior responses, in which neuroendocrine and behavioral stimuli interact to produce a sequence of coordinated behaviors, as occurs during reproduction in the ring dove. Neuroendocrine responses can also be conditioned to environmental stimuli.

The effects of hormones on behavior are mediated by neural and genomic mechanisms. Through their neuromodulatory actions on neurotransmitter release, hormones alter neural sensitivity to sensory input, regulate the motor pathways controlling behavior and modulate the central neural mechanisms which integrate complex behavioral responses. Hormones act through genomic mechanisms to stimulate mRNA and protein synthesis, resulting in cell growth and differentiation as well as long-term changes in behavior. The effects of hormones on neural and genomic mechanisms may occur in adulthood or during embryonic development, during which time they can alter the organization of the developing nervous system and have permanent effects on brain and behavior. Research into behavioral neuroendocrinology requires control of many confounding variables, such as species and individual differences, previous experience of the subjects, environmental factors, the stimulus situation, and the state of the subject. There are also complex interactions among hormones, neurotransmitters, neuropeptides, cytokines and environmental variables which may influence behavior. As a result, hormone secretion may be altered by changes in other hormones or neurotransmitters, synchronized with environmental changes, and modified by feedback effects from behavior.

Behavioral neuroendocrinological research with human subjects involves a number of special considerations of subject selection, experimental design, and experimenter and subject bias. Much of the research on hormones and behavior which is relevant to humans is done using animal models of human disorders. There are also a number of social and political questions which arise with respect to the use of the biotechnology resulting from knowledge about the neuroendocrine system and the ability to manipulate human hormones and behavior.

FURTHER READING

Adams, J. (1986). Methods in behavioral teratology. In E. P. Riley and C. V. Vorhees (eds.) *Handbook of Behavioral Teratology*, pp. 67–97. New York: Plenum Press.

Arnold, A. P. and Breedlove, S. M. (1985). Organizational and activational effects of sex steroids on brain and behavior: a reanalysis. *Hormones and Behavior*, **19**, 469–498.

Beach, F. A. (1974). Behavioral endocrinology and the study of reproduction. *Biology of Reproduction*, **10**, 2–18.
Henry, J. P. (1986). Neuroendocrine patterns of emotional response. In R. Plutchik and H. Kellerman (eds.) *Emotion: Theory, Research, and Experience*. Vol. 3: *Biological Foundations of Emotion*, pp. 37–60. Orlando, FL: Academic Press.
Komisaruk, B. R., Siegel, H. I., Cheng, M.-F. and Feder, H. H. (eds.) (1986). Reproduction: a behavioral and neuroendocrine perspective. *Annals of the New York Academy of Sciences*, **474**, pp. 1–465.
Lauder, J. M. (1983). Hormonal and humoral influences on brain development. *Psychoneuroendocrinology*, **8**, 121–155.

REVIEW QUESTIONS

14.1 What is a behavioral bioassay?
14.2 What is the difference between the qualitative and quantitative description of behavior?
14.3 What are two non-invasive or indirect measures of hormone levels which can be used in behavioral studies.
14.4 What are the three phases of the standard hormone and behavior experiment?
14.5 If one did not want to use surgical gland ablation to remove a hormone, what two other methods could be used?
14.6 What is the difference between subcutaneous and intracerebral hormone injection?
14.7 What are the advantages and disadvantages of using sustained release (silastic) capsules for hormone replacement?
14.8 What is the difference between long-term and short-term feedback effects of hormones in response to environmental stimuli?
14.9 What is the difference between neural and genomic mechanisms mediating the effects of hormones on behavior?
14.10 How can the intrauterine environment lead to individual differences in the behavior of genetically identical mice?
14.11 Why are hormone–hormone interactions important problems for the study of behavior?
14.12 What is a 'human analogue model'?

ESSAY QUESTIONS

14.1 Discuss the advantages, disadvantages and confounding variables in the use of sexual behavior in male rats as a behavioral bioassay for androgen levels.
14.2 Of what value is the study of behavioral changes during the menstrual cycle of adult human females for understanding the role of gonadal hormone changes in controlling behavior?
14.3 Discuss a way of adapting the standard hormone removal and replacement experiment for the study of how (a) anabolic steroids influence exercise performance in human males, or (b) how exercise level effects gonadal hormone secretion in human females.

14.4 Discuss the variables which must be controlled for a study on the hormonal control of the onset of parental (maternal) behavior in sheep (or rats).

14.5 Starting with Figure 13.11, discuss the effects of social stimuli on gonadal and adrenal steroid levels in monkeys.

14.6 Can pituitary hormone release be conditioned? Discuss, for example, conditioned release of LH in males or oxytocin in females.

14.7 Describe the neuroendocrine control of the lordosis reflex in the female rat.

14.8 Describe the hormonal control of the neuroanatomical changes underlying birdsong.

14.9 Discuss the role of androgens/estrogens as neural growth factors during fetal development.

14.10 Discuss the phenomenon of 'uterine position' as it relates to individual differences in sexual differentiation and subsequent sexually dimorphic behaviors.

14.11 Discuss the problems underlying the controlled study of hormones and human sexual behavior.

REFERENCES FOR CHAPTERS 14 AND 15

Adams, J. (1986). Methods in behavioral teratology. In E. P. Riley and C. V. Vorhees (eds.) *Handbook of Behavioral Teratology*, pp. 67–97. New York: Plenum Press.

Ader, R. (1976). Conditioned adrenocortical steroid elevations in the rat. *Journal of Comparative and Physiological Psychology*, **90**, 1156–1163.

Ader, R. and Cohen, N. (1991). The influence of conditioning on immune responses. In R. Ader, D. L. Felten and N. Cohen (eds.) *Psychoneuroimmunology*, 2nd edn. San Diego, CA: Academic Press.

Adkins-Regan, E. (1981). Early organizational effects of hormones: An evolutionary perspective. In N. T. Adler (ed.) *Neuroendocrinology of Reproduction: Physiology and Behavior*, pp. 159–228. New York: Plenum Press.

Andelman, S. J., Else, J. G., Hearn, J. P. and Hodges, J. K. (1985). The non-invasive monitoring of reproductive events in wild Vervet monkeys (*Cercopithecus aethiops*) using urinary pregnanediol-3α-glucuronide and its correlation with behavioural observations. *Journal of Zoology, London (A)*, **205**, 467–477.

American Psychological Association. (1981). Ethical principles of psychologists. *American Psychologist*, **36**, 633–638.

Armario, A. and Jolin, T. (1989). Influence of intensity and duration of exposure to various stressors on serum TSH and GH levels in adult male rats. *Life Sciences*, **44**, 215–221.

Arnold, A. P. (1981). Logical levels of steriod hormone action in the control of vertebrate behavior. *American Zoologist*, **21**, 233–242.

Arnold, A. P. and Breedlove, S. M. (1985). Organizational and activational effects of sex steroids on brain and behavior: a reanalysis. *Hormones and Behavior*, **19**, 469–498.

Arnold, A. P. and Jordan C. L. (1988). Hormonal organization of neural circuits. *Frontiers in Neuroendocrinology*, **10**, 185–214.

Aronson, L. A. (1959). Hormones and reproductive behavior: some phylogenetic considerations. In A. Gorbman (ed.) *Comparative Endocrinology*, pp. 98–120. New York: John Wiley.

Austin, C. R. (1972). The ethics of manipulating human reproduction. In C. R. Austin and R. V. Short (eds.) *Reproduction in Mammals. 5. Artificial Control of Reproduction*, pp. 141–152. Cambridge: Cambridge University Press.

Baggerman, B. (1968). Hormonal control of reproductive and parental behaviour in fishes. In E. J. W. Barrington and C. B. Jorgensen (eds.) *Perspectives in Endocrinology*, pp. 351–404. London: Academic Press.

Balazs, R. (1976). Hormones and brain development. *Progress in Brain Research*, **45**, 139–159.

Balazs, R., Patel, A. J. and Hajos, F. (1975). Factors affecting the biochemical maturation of the brain: effects of hormones during early life. *Psychoneuroendocrinology*, **1**, 25–36.

Baldwin, B. A., Hutchison, J. B., Parrott, R. F. and Steimer, Th. (1979). Hypothalamic uptake of tritiated testosterone in the sheep following intracerebroventricular infusion. *Journal of Neuroscience Methods*, **1**, 243–248.

Balthazart, J. (1983). Hormonal correlates of behavior. In D. S. Farner, J. R. King and K. C. Parkes (eds.) *Avian Biology*, vol VII, pp. 221–335. New York: Academic Press.

Balthazart, J. and Hendrick, J. (1976). Annual variation in reproductive behavior, testosterone and plasma FSH levels in the Rouen duck, *Anas platyrhynchos*. *General and Comparative Endocrinology*, **28**, 171–183.

Barber, B. (1976). The ethics of experimentation with human subjects. *Scientific American*, **234** (2), 25–31.

Bartke, A., Hafiez, A. A., Bex, F. J. and Dalterio, S. (1978). Hormonal interactions in regulation of androgen secretion. *Biology of Reproduction*, **18**, 44–54.

Baumeister, A. A. and Sevin, J. A. (1990). Pharmacologic control of aberrant behavior in the mentally retarded: toward a more rational approach. *Neuroscience and Biobehavioral Reviews*, **14**, 253–262.

Baxter, J. D. and Gertz, B. J. (1991). Gene expression and recombinant DNA in endocrinology and metabolism. In F. S. Greenspan (ed.) *Basic and Clinical Endocrinology*, 3rd edn, pp. 20–39. Norwalk, CN: Appleton and Lange.

Beach, F. A. (1974). Behavioral endocrinology and the study of reproduction. *Biology of Reproduction*, **10**, 2–18.

Beach, F. A. (1975). Behavioral endocrinology: an emerging discipline. *American Scientist*, **63**, 178–187.

Beach, F. A. (1981). Historical origins of modern research on hormones and behavior. *Hormones and Behavior*, **15**, 325–376.

Bermant, G. and Davidson, J. M. (1974). *Biological Bases of Sexual Behavior*. New York: Harper and Row.

Blaustein, J. D. (1986). Steroid receptors and hormone action in the brain. *Annals of the New York Academy of Sciences*, **474**, 400–414.

Boer, G. J. and Swaab, D. F. (1985). Neuropeptide effects on brain development to be expected from behavioral teratology. *Peptides*, **6**, Suppl. **2**, 21–28.

Brain, P. and Benton, D. (1979). The interpretation of physiological correlates of differential housing in laboratory rats. *Life Sciences*, **24**, 99–116.

Brown, G. M. and Martin, J. B. (1974). Corticosterone, prolactin, and growth hormone responses to handling and new environment in the rat. *Psychosomatic Medicine*, **36**, 241–247.

Brown, R. E. (1985a). Hormones and paternal behavior in vertebrates. *American Zoologist*, **25**, 895–910.

Brown, R. E. (1985b). The rodents. I. Effects of odours on reproductive physiology (primer effects). In R. E. Brown and D. W. Macdonald (eds.) *Social Odours in Mammals*, vol. 1, pp. 245–344. Oxford: Clarendon Press.

Brown, R. E. and McFarland, D. J. (1979). Interaction of hunger and sexual motivation in the male rat: a time-sharing approach. *Animal Behaviour*, **27**, 887–896.

Burke, A. W. and Broadhurst, P. L. (1966). Behavioural correlates of the oestrous cycle in the rat. *Nature*, **209**, 223–224.

Cameron, O. G. and Neese, R. M. (1988). Systematic hormonal and physiological abnormalities in anxiety disorders. *Psychoneuroendocrinology*, **13**, 287–307.

Carter, D. B. (1982). Affective disorders in endocrine disease. In A. A. Vernadakis and P. S. Timiras (eds.) *Hormones in Development and Aging*, pp. 637–643. New York: Spectrum Publications.

Cheng, M.-F. (1986). Individual behavioral response mediates endocrine changes induced by social interaction. *Annals of the New York Academy of Sciences*, **474**, 4–12.

Cohen, J. (1983). Hormones and brain mechanisms of vocal behavior in non-vocal learning birds. In J. Balthazart, E. Pröve and R. Gilles (eds.) *Hormones and Behaviour in Higher Vertebrates*, pp. 422–436. Berlin: Springer-Verlag.

Coover, G. D., Sutton, B. R. and Heybach, J. P. (1977). Conditioning decreases in plasma corticosterone level in rats by pairing stimuli with daily feedings. *Journal of Comparative and Physiological Psychology*, **91**, 716–726.

Crews, D. (1980). Interrelationships among ecological, behavioral and neuroendocrine processes in the reproductive cycle of *Anolis carolensis* and other reptiles. *Advances in the Study of Behaviour*, **11**, 1–74.

Crews, D. (1986). Comparative behavioral endocrinology. *Annals of the New York Academy of Sciences*, **474**, 187–198.

Crowley, W. R. (1986). Reproductive neuroendocrine regulation in the female rat by central catecholamine-neuropeptide interactions: a local control hypothesis. *Annals of the New York Academy of Sciences*, **474**, 423–436.

Dabbs, J. M. (1991). Salivary testosterone measurements: Collecting, storing, and mailing saliva samples. *Physiology and Behavior*, **49**, 815–817.

Davidson, J. M., Stefanick, M. L., Sachs, B. D. and Smith, E. R. (1978). Role of androgen in sexual reflexes of the male rat. *Physiology and Behavior*, **21**, 141–146.

Davidson, J. M. and Trupin, S. (1975). Neural mediation of steroid-induced sexual behavior in rats. In M. Sandler and G. L. Gessa (eds.) *Sexual Behavior: Pharmacology and Biochemistry*, pp. 13–20. New York: Raven Press.

Donat, P. (1991). Measuring behaviour: the tools and the strategies. *Neuroscience and Biobehavioral Reviews*, **15**, 447–454.

Dorner, G. (1983). Hormone-dependent brain development and behaviour. In J. Balthazart, E. Prove and R. Gilles (eds.) *Hormones and Behavior in Higher Vertebrates*, pp. 204–217. Berlin: Springer-Verlag.

Dussault, J. H. and Ruel, J. (1987). Thyroid hormones and brain development. *Annual Review of Physiology*, **49**, 321–334.
Edsall, G. (1969). A positive approach to the problem of human experimentation. *Daedalus*, **98**, 463–479.
Edwards, D. A. (1969). Early androgen stimulation and aggressive behavior in male and female mice. *Physiology and Behavior*, **4**, 333–338.
Erikson, C. J. (1986). Social induction of the ovarian response in the female ring dove. *Annals of the New York Academy of Sciences*, **474**, 13–20.
Everett, J. W. (1989). *Neurobiology of Reproduction in the Female Rat: A Fifty Year Perspective*. Berlin: Springer-Verlag.
Fantie, B. D., Brown, R. E. and Moger, W. H. (1984). Constant lighting conditions affect sexual behaviour and hormone levels in adult male rats. *Journal of Reproduction and Fertility*, **72**, 435–441.
Fein, G. G., Schwartz, P. M., Jacobson, S. W. and Jacobson, J. L. (1983). Environmental toxins and behavioral development. *American Psychologist*, **38**, 1188–1197.
Flannelly, K. and Lore, R. (1977). The influence of females upon aggression in domesticated male rats (*Rattus norvegicus*). *Animal Behaviour*, **25**, 654–659.
Gandelman, R. (1983). Gonadal hormones and sensory function. *Neuroscience and Biobehavioral Reviews*, **7**, 1–17.
Gandelman, R. (1984). Relative contributions of aggression and reproduction to behavioral endocrinology. *Aggressive Behavior*, **10**, 123–133.
Gibbs, J. and Smith, G. P. (1986). Satiety: the roles of peptides from the stomach and the intestine. *Federation Proceedings*, **45**, 1391–1395.
Gilman, A. G., Goodman, L. S., Rall, T. W. and Murad, F. (eds.) (1985). *Goodman and Gilman's the Pharmacological Basis of Therapeutics*. 7th edn. New York: Macmillan.
Gorski, R. A. (1991). Sexual differentiation of the endocrine brain and its control. In M. Motta (ed.) *Brain Endocrinology*, 2nd edn, pp. 71–104. New York: Raven Press.
Graham, J. M. and Desjardins, C. (1980). Classical conditioning: Induction of luteinizing hormone and testosterone secretion in anticipation of sexual activity. *Science*, **210**, 1039–1040.
Hadley, M. E. (1992). *Endocrinology*, 3rd edn. Englewood Cliffs, NJ: Prentice-Hall.
Harding, C. F. (1981). Social modulation of circulating hormone levels in the male. *American Zoologist*, **21**, 223–231.
Henkin, R. I. (1975). Effects of ACTH, adrenocorticosteriods and thyroid hormone on sensory function. In W. E. Stumpf and L. D. Grant (eds.) *Anatomical Neuroendocrinology*, pp. 298–316. Karger: Basel.
Henry, J. P. (1986). Neuroendocrine patterns of emotional response. In R. Plutchik and H. Kellerman (eds.) *Emotion: Theory, Research and Experience*. Vol. 3: *Biological Foundations of Emotion*, pp. 37–60. London: Academic Press.
Holman, S. D. (1980). A method for intracerebrally implanting crystalline hormones into neonatal rodents. *Laboratory Animals*, **14**, 263–266.
Hurnik, J. F., King, G. J. and Robertson, H. A. (1975). Estrous and related behaviour in postpartum Holstein cows. *Applied Animal Ethology*, **2**, 55–68.
Hyde, J. S. and Sawyer, T. F. (1977). Estrous cycle fluctuations in

aggressiveness of house mice. *Hormones and Behavior*, **9**, 290–295.

Jaffe, F. S. (1973). Public policy on fertility control. *Scientific American*, **229** (1), 17–23.

Johansson, G., Collins, A. and Collins, V. P. (1983). Male and female psychoneuroendocrine response to examination stress: a case report. *Motivation and Emotion*, **7**, 1–9.

Johnston, R. E. (1981). Attraction to odors in hamsters: an evaluation of methods. *Journal of Comparative and Physiological Psychology*, **95**, 951–960.

Kelley, A. E. (1989). Behavioural models of neuropeptide action. In G. Fink and A. J. Harmer (eds.) *Neuropeptides: A Methodology*, pp. 301–331. Chichester: Wiley.

Komisaruk, B. R., Adler, N. T. and Hutchison, J. (1972). Genital sensory field: enlargement by estrogen treatment in female rats. *Science*, **178**, 1295–1298.

Komisaruk, B. R. and Steinman, J. L. (1986). Genital stimulation as a trigger for neuroendocrine and behavioral control of reproduction. *Annals of the New York Academy of Sciences*, **474**, 64–75.

Krsiak, M. (1991). Ethopharmacology: a historical perspective. *Neuroscience and Biobehavioral Reviews*, **15**, 439–445.

Landau, I. T. (1986). Steroid hormone antagonists and behavior. *Annals of the New York Academy of Sciences*, **474**, 379–388.

Lauder, J. M. (1983). Hormonal and humoral influences on brain development. *Psychoneuroendocrinology*, **8**, 121–155.

Lauder, J. M. and Krebs, H. (1986). Do neurotransmitters, neurohumors, and hormones specify critical periods? In W. T. Greenough and J. M. Juraska (eds.) *Developmental Neuropsychobiology*, pp. 119–174. Orlando, FL: Academic Press.

Lehner, P. N. (1979). *Handbook of Ethological Methods*. New York: Garland Press.

Lehrman, D. S. (1965). Interaction between internal and external environments in the regulation of the reproductive cycle. In F. A. Beach (ed.) *Sex and Behavior*, pp. 355–380. New York: Wiley.

Lehrman, D. S., Brody, P. N. and Wortis, R. P. (1961). The presence of the mate and of nesting material as stimuli for the development of incubation behavior and for gonadotropin secretion in the ring dove (*Streptopelia risoria*). *Endocrinology*, **68**, 507–516.

Leshner, A. I. (1978). *An Introduction to Behavioral Endocrinology*. New York: Oxford University Press.

Leshner, A. I. (1979). Kinds of hormonal effects on behavior: a new view. *Neuroscience and Biobehavioral Reviews*, **3**, 69–73.

Martin, C. R. (1985). *Endocrine Physiology*. New York: Oxford University Press.

Martin, J. B. and Reichlin, S. (1987). *Clinical Neuroendocrinology*, 2nd edn. Philadelphia: F. A. Davis.

Martin, P. and Bateson, P. (1986). *Measuring Behaviour*. Cambridge: Cambridge University Press.

McCann, S. M. (ed.) (1988). *Endocrinology. People and Ideas*. Bethesda, MD: American Physiological Society.

Medvei, V. C. (1982). *A History of Endocrinology*. Lancaster, PA: MTP Press.

Meites, J., Donovan, B. T. and McCann, S. M. (1975) *Pioneers in Neuroendocrinology*. New York: Plenum Press.

Meites, J., Donovan, B. T. and McCann, S. M. (1978) *Pioneers in Neuroendocrinology II*. New York: Plenum Press.

Mitchell, J. and Popkin, M. (1983). The pathophysiology of sexual dysfunction associated with antipsychotic drug therapy in males: A review. *Archives of Sexual Behavior*, **12**, 173–183.

Myers, R. D. (1971). Methods for chemical stimulation of the brain. In R. D. Myers (ed.) *Methods in Psychobiology*, vol. 1. London: Academic Press.

Pfaff, D. W. (1989). Features of a hormone-driven defined neural circuit for a mammalian behavior. *Annals of the New York Academy of Sciences*, **563**, 131–147.

Plutchik, R. (1962). *The Emotions: Facts, Theories and a New Model*. New York: Random House.

Prange, Jr, A. J. Garbutt, J. C., Loosen, P. T., Bissett, G. and Nemeroff, C. B. (1987). The role of peptides in affective disorders: a review. *Progress in Brain Research*, **72**, 235–247.

Reboulleau, C. P. (1986). Hormonal aspects of the morphological differentiaton of neuronal clonal cell lines. *Annals of the New York Academy of Sciences*, **474**, 445–452.

Reinisch, J. M. and Gandelman, R. (1978). Human research in behavioral endocrinology: Methodological and theoretical considerations. In: G. Dorner and M. Kawakami (eds.) *Hormones and Brain Development*, pp. 77–86. Amsterdam: Elsevier.

Rose, R. M., Bernstein, I. S. and Gordon, T. P. (1975). Consequences of social conflict on plasma testosterone levels in rhesus monkeys. *Psychosomatic Medicine*, **37**, 50–61.

Rosenthal, R. (1966). *Experimenter Effects in Behavioral Research*. New York: Appleton-Century-Crofts.

Ruse, M. (1979). *Sociobiology: Sense or Nonsense?* Boston: D. Reidel.

Russell, M., Dark, K. A., Cummins, R. W., Ellman, G., Callaway, E. and Peeke, H. V. S. (1984). Learned histamine release. *Science*, **225**, 733–734.

Rutstein, D. D. (1969). The ethical design of human experiments. *Daedalus*, **98**, 523–541.

Samuels, M. H. and Bridges, R. S. (1983). Plasma prolactin concentration in parental male and female rats: effects of exposure to rat young. *Endocrinology*, **113**, 1647–1654.

Schwartz, S. M. (1982). Effects of constant bright illumination on reproductive processes in the female rat. *Neuroscience and Biobehavioral Reviews*, **6**, 391–406.

Silverman, W. A. (1985). *Human Experimentation: A Guided Step into the Unknown*. Oxford: Oxford University Press.

Singh, D. and Avery, D. A. (1975). *Physiological Techniques in Behavioral Research*. Monterey, CA: Brooks/Cole.

Smith, E. R., Damassa, D. A. and Davidson, J. M. (1977). Hormone adminstration: Peripheral and intracranial implants. In R. D. Myers (ed.) *Methods in Psychobiology*, vol. 3, pp. 259–279. New York: Academic Press.

Soulairac, A. and Soulairac, M. L. (1978). Relationships between the nervous and endocrine regulation of sexual behavior in male rats. *Psychoneuroendocrinology*, **3**, 17–29.

Spruijt, B. and Gispen, W. H. (1983). ACTH and grooming behaviour in the rat. In J. Balthazart, E. Prove and R. Gilles (eds.) *Hormones and Behaviour in Higher Vertebrates*, pp. 118–136. Berlin: Springer-Verlag.

Steffens, A. B. (1969). Method for frequent sampling of blood and continuous infusion of fluids in the rat without disturbing the animal.

Physiology and Behavior, **4**, 833–836.
Swanson, H. H. (1974). Sex differences in behaviour of the Mongolian gerbil (*Meriones unguiculatus*) in encounters between pairs of same or opposite sex. *Animal Behaviour*, **22**, 638–644.
Tausk, M. (1975). *Pharmacology of Hormones*, Chicago: Yearbook Medical Publishers.
Terkel, J. and Urbach, L. (1974). A chronic intravenous cannulation technique adapted for behavioral studies. *Hormones and Behavior*, **5**, 141–148.
Toran-Allerand, C. D. (1991). Organotypic culture of the developing cerebral cortex and hypothalamus: Relevance to sexual differentiation. *Psychoneuroendocrinology*, **16**, 7–24.
Turner, C. D. and Bagnara, J. T. (1976). *General Endocrinology*, 6th edn. Philadelphia: W. B. Saunders.
vom Saal, F. S. (1983). The interaction of circulating oestrogens and androgens in regulating mammalian sexual differentiation. In J. Balthazart, E. Prove and R. Gilles (eds.) *Hormones and Behaviour in Higher Vetebrates*. pp. 159–177. Berlin: Springer-Verlag.
Whalen, R. E. (1986). Hormonal control of behavior – a cautionary note. *Annals of the New York Academy of Sciences*, **474**, 354–361.
Wilcox, J. N. (1986). Analysis of steroid action on gene expression in the brain. *Annals of the New York Academy of Sciences*, **474**, 453–460.
Yahr, P. and Thiessen, D. D. (1972). Steroid regulation of territorial scent marking in the Mongolian gerbil (*Meriones unguiculatus*). *Hormones and Behavior*, **3**, 359–368.
Zarrow, M. X., Yochim, J. M. and McCarthy, J. L. (1964). *Experimental Endocrinology: A Sourcebook of Techniques*. New York: Academic Press.
Zbinden, G. (1981). Experimental methods in behavioral teratology. *Archives of Toxicology*, **48**, 69–88.

15

An overview of behavioral neuroendocrinology: present, future and past

15.1 THE AIM OF THIS BOOK

The aim of this book is to teach students the language and concepts of neuroendocrinology with an emphasis on how the neuroendocrine system influences behavior. It began with a classification of chemical messengers in the body as 'true' hormones, neurohormones, neurotransmitters, pheromones, parahormones, prohormones, growth factors and neuroregulators. As more became known about the neuroendocrine system, it was seen that these classifications are not clear cut and a single chemical could fit into two or more classes of messenger. While the classification of chemical messengers is useful to begin the study of neuroendocrinology, by the end it provides little help in understanding the different actions of peptides, steroids and neurotransmitters on different target cells.

The endocrine glands and the pituitary gland are generally thought to be the basis of the neuroendocrine system, but the traditional endocrine function of hormones acting on peripheral target cells provides only a small part of the neuroendocrine activity of a hormone such as testosterone, cholecystokinin or prolactin. These hormones also have significant effects on neural receptors and alter neural regulation of autonomic reflexes, behavior and emotional states. The hypothalamus provides the link between the traditional endocrine system and the brain and provides the mechanism for external factors to regulate the endocrine system. Thus, while the endocrine system appears to consist of a number of closed-loop feedback systems, with highly regulated physiological control over the synthesis, storage, release and deactivation of hormones, external stimuli can alter these systems. Environmental stimuli, social interactions and cognitive factors can greatly alter the functioning of the endocrine system by altering the 'classical' neurotransmitter pathways which regulate the release of hypothalamic hormones. Drugs, food, changes in light cycles, emotional arousal and other changes in the environment which alter the release of neurotransmitters in the hypothalamus can disrupt the hormonal control of feeding and drinking, arousal, sexual, aggressive and parental behavior, and responses to environmental stressors. Bacteria, viruses and other toxins activate the neuroendocrine system by stimulating the release of cytokines from the cells of the immune system.

While the 'classical' neurotransmitters are usually considered as the most important neurochemicals, the neuropeptides, steroid hormones and cytokines which modulate neural activity may prove to be equally, if not more important in determining neural function. As more is learned about the wide-ranging effects of neuropeptides on physiological and behavioral responses, it becomes clear that the functions of these hormones in the brain are significantly different from their actions in the body, yet they are often oriented toward the same goals. The gastrointestinal peptides, for example, help to digest food in the gut and regulate feeding and drinking behavior, body temperature and blood pressure in the brain. The sex hormones prepare the genitals for mating, develop the breasts for lactation, and act in the brain to regulate sexual arousal and sexual and parental behavior. The adrenal cortical and medullary hormones which increase heart rate and respiratory rate and facilitate muscular energy also prepare the brain to deal with stressful situations and to respond in the most adaptive way.

All of these actions result from the actions of the chemical messengers at receptors on their target cells. The biochemical changes in the target cells which result from the activation of receptors, and the resulting changes in ion channel permeability and second messenger cascades involve chains of enzyme reactions, protein synthesis, cell growth and differentiation, neuromuscular changes leading to behavior and neural changes leading to emotional arousal and memory. Hormones, neurotransmitters, neuropeptides and cytokines have their effects through a series of complex interactions, whereby a transmitter controls hormone release, but neuropeptides and other hormones control the release of neurotransmitters and the response of the postsynaptic cell to the transmitter. Cytokines are regulated by the neuroendocrine system and can act at receptors in these systems to influence both neural and hormonal responses.

The function of this book has been to introduce students to the complexity of the neuroendocrine system by presenting the necessary terminology and concepts in a gradual way. The peripheral endocrine glands, pituitary gland, hypothalamic hormones, neurotransmitters, neuropeptides and cytokines are the building blocks of the neuroendocrine system. Knowing about the synthesis, storage, transport, and release of each of these chemical messengers provides a mechanism for understanding the nature of each of the individual elements in the neuroendocrine system, while knowing about receptor mechanisms and second messenger cascades leads to an understanding of how these elements are interconnected. Knowing how the steroid hormones, neuropeptides and cytokines modulate the electrophysiological and genomic activity of their neural tareget cells leads to an understanding of how these neuroregulators can modulate the visceral, behavioral and cognitive functions of the brain.

This book has not gone into great detail on the functions of the hormones in medical, behavioral or psychiatric practice, but these have all been mentioned in various chapters. The function of this book is to provide the framework for more advanced study. The essay questions, for example, require extra reading from the specialized journals listed in Appendix 1, if they are to be answered with the most up-to-date information available. If, on finishing this book, you are able to read the references in scientific and medical journals without a dictionary, and are able to 'see' in your mind's eye the relationships among, the hypothalamic-pituitary-gonadal hormones, the inhibition of substance P in the pain pathways by the endogenous opioids, the action of a dopamine agonist drug on the release of prolactin, and the ability of interleukin 1 to alter the hypothalamic-pituitary-adrenal feedback system, then this book has served its function and you are ready for more advanced topics in neuroendocrinology.

15.2 THE HISTORY OF BEHAVIORAL NEUROENDOCRINOLOGY

Endocrinology has a long history (Medvei, 1982) and many of the discoveries about the neuroendocrine system have been important enough to be awarded the Nobel Prize for Medicine and Physiology. The biographies of many of the scientists who made these discoveries are available in a number of books (Meites, Donovan and McCann, 1975, 1978; McCann, 1988). Brief histories of the study of endocrinology are given by Turner and Bagnara (1976), and Hadley (1992). Gilman *et al.* (1985) and Tausk (1975) include a number of historical summaries of the discoveries of neuroendocrine phenomena and the drugs that influence them. The history of the study of hormonal influences on behavior, which is less well known, is described by Beach (1981) and Gandelman (1984). Table 15.1 gives some of the important milestones in the study of hormones and behavior from 1849 to the present. The study of hormones and behavior was unsystematic until 1948 when Frank Beach established the field with the publication of the book *Hormones and Behavior*. The history of research on hormones and behavior began in 1849 but was not developed into a scientific discipline until 1948. Since then the field has grown rapidly.

ESSAY QUESTIONS

15.1 Discuss the contributions of one of the following scientists (all deceased) to the study of behavioral neuroendocrinology: (a) Hans Selye, (b) Frank Beach, (c) Curt Richter, (d) Geoffry W. Harris, or (e) Daniel Lehrman.

15.2 Discuss the discovery of one of the following neuroendocrine phenomena: (a) the discovery of the endogenous opioids, (b) the discovery of the thymus hormones, (c) the discovery of insulin, (d) the discovery of the neurohyophyseal hormone system, (e) the discovery of secretin, and (f) the development of 'Dale's Principle'.

Table 15.1. *Milestones in the history of behavioral neuroendocrinology*

1849 Arnold A. Berthold castrated roosters and replaced their testes; observing changes in crowing, copulation and aggression.

1889–1894 Charles Edourd Brown-Sequard developed 'organo-therapy' by producing extracts of testes, thyroid and adrenal glands for treatment of human disorders, and became infamous for his rejuvenation experiments, injecting testes solution into elderly men.

1894–1910 Eugene Steinach experimented on the effects of testes removal and replacement on the sexual behavior of amphibia, birds, and mammals, and examined the effects of the gonads on sexual differentiation and the timing of puberty.

1898–1904 Feurth and Abel isolate adrenaline.

1905 E. H. Starling used the term 'hormone' with reference to secretin and Pende introduced the term 'endocrinology' a few years later.

1910 Cushing *et al.* discover the pituitary-gonadal link.

1914 H. H. Dale discovers acetylcholine's function as a neurotransmitter in the parasympathetic nervous system. He received the Nobel Prize in 1936.

1914 E. C. Kendall isolated thyroxine from thyroid gland extracts.

1917 Frank R. Lillie describes 'free martins' as the female twin of a normal male calf who is masculinized by androgens from the male's testes. Between 1917 and 1922 Lillie outlined a number of experiments to determine the role of perinatal gonadal hormones in sexual differentiation.

1917 C. R. Stockard and G. N. Papanicolau describe the estrous cycle of the female guinea pig and correlate the stages of development of the ovarian follicle with changes in cell types of the vaginal mucosa. In 1922, J. A. Long and H. M. Evans described a similar estrous cycle in rats and used vaginal smears to correlate vaginal estrus with mating behavior. In 1923 G. H. Wang showed a correlation between spontaneous activity and the estrous cycle of the rat.

1921 F. G. Banting, and G. H. Best discover insulin. Banting was awarded the Nobel Prize in 1923.

1923 Edgar Allen and Edward A. Doisey purify estrogen from the ovaries of mice and rats and show that estrogen injections into spayed females induce estrous behavior.

1925 F. H. Mashall and J. Hammond show that ovulation in rabbits is not spontaneous, but is induced by vaginal stimulation.

1930 C. Pfeiffer shows that neonatally castrated male rats have a feminine hypothalamic-pituitary-gonadal feedback system in adulthood.

1930 Popa and Fieldery discover the hypophyseal portal veins.

1933 B. P. Weisner and N. M. Sheard demonstrate that maternal behavior in rats is dependent on hormones from the pituitary gland and stimuli from the young.

1933–1941 Curt Richter shows the effects of hormones on locomotor activity, dietary selection, and on motivated behavior.

1934 Walter Hohweg discovers the positive feedback of estrogen on LH release and in 1938, he develops the orally acting estrogen, ethinyl estradiol, an essential component of the birth control pill.

1935 Oscar Riddle shows that prolactin is important in the control of maternal behavior.

1935 K. G. David isolates testosterone from the testes.

1936 Hans Selye begins his studies on the neuroendocrine response to stress and develops the concept of the 'general adaptation syndrome' of response to stress.

1937–1952 E. C. Kendall, T. Reichstein and Hench isolate adrenal corticosteroids. They win the Nobel Prize in 1950.

1937 Geoffry W. Harris shows that electrical stimulation of the hypothalamus alters pituitary hormone secretion. In 1948 Harris demonstrates conclusively the hypothalamic control of the pituitary. He publishes his book *Neural Control of the Pituitary Gland* in 1955.

1939 Philip Bard discovers the hypothalamic control of ovulation, emotional and sexual behavior in cats.

1940 J. G. Wilson, W. C. Young and J. B. Hamilton show that neonatal androgen injections masculinize female rats.

1940–1945 C. H. Li isolates LH, ACTH and GH.

1948 F. A. Beach publishes *Hormones and Behavior*, the first book of behavioral endocrinology.

1949 Wolfgang Bargmann discovers the neural connections between the hypothalamus and posterior pituitary.

1951 Ernst and Berta Scharrer describe hypothalamic neurosecretory cells in the brains of vertebrates and invertebrates.

1952 The Hodgkin–Huxley model of the axonal conduction of nerve impulses is developed. Hodgkin and Huxley win the Nobel Prize in 1963.

1955 CRF, the first hypothalamic hormone discovered simultaneously by M. Saffron and A. Schally and by R. Guillemin and Rosenberg. They win the Nobel Prize in 1977.

1955 Gregory Pincus gives his report on the first successful clinical trials of the birth control pill.

1956 A. E. Fisher shows that intracranial injections of hormones could stimulate sexual and maternal behavior in rats.

1958 Geoffry W. Harris and colleagues show that estrogen implanted directly into the anterior hypothalamus elicited sexual behavior in ovariectomized cats.

1959 Daniel Lehrman publishes his finding on the effects of external stimuli on hormones in birds.

1959 C. Phoenix and colleagues describe the organizational effects of prenatal androgens on the sexual behavior of the guinea pig. This paper established the importance of perinatal hormones on adult behavior.

1960 R. Yalow and Berson develop the radioimmunoassay for measuring hormone levels in the blood. Yalow wins the 1977 Nobel Prize.

1963 Soulairac demonstrates the hormone–neurotransmitter interaction in the control of rat sexual behavior.

1964 David de Wied discovers the effects of pituitary hormones on learning in rats.

1971 Schalley *et al.* isolate LH-RH.

1972 J. Terkel and J. Rosenblatt show that blood transfusions from a lactating female rat will induce maternal behavior in a virgin female rat.

1971–1975 Hughes, Kosterlitz *et al.* discover and identify the internal opiates.

1972–1978 Discovery of neuropeptides, factor S, neuropeptide Y, etc. in brain.

1978 A. I. Leshner writes first textbook on hormones and behavior, *An Introduction to Behavioral Endocrinology*.

1981 Robert Ader edits *Psychoneuroimmunology*, the first book on the subject.

1986 Rita Levi-Montalcini and Stanley Cohen win Nobel Prize in recognition of their discovery of nerve growth factor.

Source: from Beach, 1981, Gandelman, 1984 and other sources mentioned in the text.

15.3 Discuss the importance of one of the following scientists in the history of neuroendocrinology: (a) Arnold A. Berthold, (b) Charles Edourd Brown-Sequard, (c) Eugene Steinach, (d) F. H. Mashall, (e) W. C. Young, (f) C. H. Li, (g) Ernst and Berta Scharrer, and (h) A. Schally and R. Guillemin.

REFERENCES

References for Chapter 15 are listed at the end of Chapter 14.

Appendix

Journals in endocrinology, neuroendocrinology, psychoneuroimmunology and behavioral endocrinology

ENDOCRINOLOGY JOURNALS

Acta Endocrinologica
Advances in Steroid Biochemistry and Pharmacology
Annual Review of Biochemistry
Annual Review of Physiology
Biology of Reproduction
Clinical Endocrinology
Endocrine Reviews
Endocrinology
Frontiers in Hormone Research
General and Comparative Endocrinology
Journal of Clinical Endocrinology and Metabolism
Journal of Endocrinology
Journal of Reproduction and Fertility
Journal of Steroid Biochemistry
Peptides
Physiological Reviews
Recent Progress in Hormone Research
Seminars in Reproductive Endocrinology
Steroids
Trends in Endocrinology and Metabolism

NEUROENDOCRINOLOGY JOURNALS

Annual Review of Neuroscience
Brain Research
Brain Research Review
Developmental Brain Research
Frontiers in Neuroendocrinology
Neuroendocrinology
Neuroendocrinology Letters
Trends in Neuroscience
Trends in Pharmacological Sciences

PSYCHONEUROIMMUNOLOGY JOURNALS

Brain, Behavior and Immunity
Journal of Neuroimmunology
Psychoneuroimmunology
Progress in NeuroEndocrinImmunology

BEHAVIORAL ENDOCRINOLOGY JOURNALS

Behavioral and Neural Biology
Behavioral Neuroscience
Developmental Psychobiology
Hormones and Behavior
Neuroscience and Biobehavioral Reviews
Physiology and Behavior
Psychobiology
Psychoneuroendocrinology

Index